ADVANCES IN THE BIOSCIENCES

Volume 61

DYNAMIC PROPERTIES OF GLIA CELLS II

Cellular and Molecular Aspects

ADVANCES IN THE BIOSCIENCES

Latest volumes in the series:

Volume 46: NEUROBEHAVIORAL METHODS IN OCCUPATIONAL HEALTH
Editors: R. Gilioli, M. G. Cassitto and V. Foà

Volume 47: ADVANCED CONCEPTS IN ALCOHOLISM
Editor: H.-G. Tittmar

Volume 48: NEUROMODULATION AND BRAIN FUNCTION
Editors: G. Biggio, P. F. Spano, G. Toffano and G. L. Gessa

Volume 49: PSYCHOLOGICAL ASPECTS OF CANCER
Editors: M. Watson and T. Morris

Volume 50: PATHOPHYSIOLOGICAL ASPECTS OF CANCER EPIDEMIOLOGY
Editors: G. Mathé and P. Reizenstein

Volume 51: FRONTIERS IN HISTAMINE RESEARCH
Editors: C. R. Ganellin and J. C. Schwartz

Volume 52: ORO-FACIAL PAIN AND NEUROMUSCULAR DYSFUNCTION
Editors: I. Klineberg and B. Sessle

Volume 53: THE PINEAL GLAND: ENDOCRINE ASPECTS
Editors: G. M. Brown and S. D. Wainwright

Volume 54: OXYGEN TRANSPORT IN RED BLOOD CELLS
Editor: C. Nicolau

Volume 55: GERM CELL TUMOURS II
Editors: W. G. Jones, A. Milford Ward and C. K. Anderson

Volume 56: RECENT TRENDS IN MEDICAL GENETICS
Editors: K. M. Marimuthu and P. M. Gopinath

Volume 57: ADAPTIVE PROCESSES IN VISUAL AND OCULOMOTOR SYSTEMS
Editors: E. L. Keller and D. S. Zee

Volume 58: BRAIN TUMORS: BIOPATHOLOGY AND THERAPY
Editors: M. A. Gerosa, M. L. Rosenblum and G. Tridente

Volume 59: LEARNING & MEMORY: MECHANISMS OF INFORMATION STORAGE IN THE NERVOUS SYSTEM
Editor: H. Matthies

Volume 60: DISORDERS OF EATING BEHAVIOUR: A PSYCHONEUROENDOCRINE APPROACH
Editors: E. Ferrari and F. Brambilla

Volume 61: DYNAMIC PROPERTIES OF GLIAL CELLS II
Editors: T. Grisar, G. Franck, L. Hertz, W. T. Norton, M. Sensenbrenner and D. M. Woodbury

DYNAMIC PROPERTIES OF GLIA CELLS II

Cellular and Molecular Aspects

An International Symposium held in Liege, Belgium,
12–15 May 1985
Sponsored by
The International Society of Neurochemistry (ISN)
The Belgian Society of Neurology
The Belgian National Fund for Scientific Research (FNRS)
The University of Liege

A Satellite Symposium to the Tenth International Meeting of the International Society for Neurochemistry

Editors:

T. GRISAR and G. FRANCK
University of Liege, Belgium

L. HERTZ
University of Saskatchewan, Canada

W. T. NORTON
Albert Einstein College, New York, USA

M. SENSENBRENNER
Strasbourg, France

D. M. WOODBURY
University of Utah, Salt Lake City, USA

PERGAMON PRESS OXFORD · NEW YORK
BEIJING · FRANKFURT · SÃO PAULO · SYDNEY · TOKYO · TORONTO

U.K.	Pergamon Press, Headington Hill Hall, Oxford OX3 0BW, England
U.S.A.	Pergamon Press, Maxwell House, Fairview Park, Elmsford, New York 10523, U.S.A.
PEOPLE'S REPUBLIC OF CHINA	Pergamon Press, Qianmen Hotel, Beijing, People's Republic of China
FEDERAL REPUBLIC OF GERMANY	Pergamon Press, Hammerweg 6, D-6242 Kronberg, Federal Republic of Germany
BRAZIL	Pergamon Editora, Rua Eça de Queiros, 346, CEP 04011, São Paulo, Brazil
AUSTRALIA	Pergamon Press Australia, P.O. Box 544, Potts Point, N.S.W. 2011, Australia
JAPAN	Pergamon Press, 8th Floor, Matsuoka Central Building, 1-7-1 Nishishinjuku, Shinjuku-ku, Tokyo 160, Japan
CANADA	Pergamon Press Canada, Suite 104, 150 Consumers Road, Willowdale, Ontario M2J 1P9, Canada

First edition 1986

ISBN 0-08-032023-6

ISSN 0065-3446

In order to make this volume available as economically and as rapidly as possible the authors' typescripts have been reproduced in their original forms. This method unfortunately has its typographical limitations but it is hoped that they in no way distract the reader.

Printed in Great Britain by A. Wheaton & Co. Ltd., Exeter

The Glia of Liège—Part II

Eight years have gone by,
    we are back in Liège, "déjà vu".
The glial dreams that we dreamt in Liège
    have come to function, are true.

Glia have now come of age;
They have purpose, biochemistry and function.
We are grateful to the folks of Liège
For providing the impetus, a forum and sacretion.

But, alas! all is not clear with our glia,
And we do not as yet understand the glial-neuronal chatter.
So we vow to return to Liège, as we scatter
To our laboratories around the world,
To the cultures and explants, in the mood.

Please invite us again, dear Thierry,
So then all glial mysteres will be understood.

Bernard Haber

Contents

B. Developmental aspects :

Co-chair : S. Varon (San Diego - U.S.A.)
W.T. Norton (New York - U.S.A.)

C. Metabolism and neurotransmission :

D. Myelination

Foreword

This volume deals with cellular and molecular aspects of a nerve cell that was until recently a neglected subject in neurobiology.

For a long period of research on the CNS, the excitable cells were the inevitable leader in the neurobiological literature.

But "Times are changing". Glia cells are almost also "excitable" cells.

In 1977, the Liège group also organized a Symposium on the Dynamic properties of glia cells, reviewing: anatomy, metabolism, amino acids and transmitters, specific macromolecular compounds, ion homeostasis and glia cells reactions and aberrations.

But clearly if both of the major classes of glia cells (oligodendrocytes and astrocytes) appeared important partners with neurons in the properties of the nervous system, there are still a number of gaps in our knowledge and many unanswered questions in the molecular basis underlying these neuron-glia relationships.

However during recent years, much progress in this field has been achieved, not least as a result of the development of sensitive methods for the determination of specific molecular compounds of glia cells. Probably one of the most fundamental topics that has accumulated very important data is the developmental aspect of the neuron-glia relationships. Also, totally new results have arisen in the field of medicine, demonstrating molecular abnormality of glial metabolism associated with pathological conditions.

As a result, from these recent findings, the properties and evolutionary significances of glia cells became more and more clear. The old thought of a glia, passive support of adjacent neuron ("nerve glue"), is back to the story books.

Today, we at least know that neurons do not work adequately without glia.

We are therefore honored and we thank all scientists who accepted our invitation and presented their contributions in such a competent way to make once again glia cells happy with neurons during this Symposium.

T. Grisar.

Organizing Committee and List of Speakers

Organizing committee:

T. GRISAR (Liège, Belgium)
G. FRANCK (Liège, Belgium)
L. HERTZ (Saskatoon, Canada)
W.T. NORTON (New York, USA)
D.M. WOODBURY (Salt Lake City, USA)

List of speakers:

J. ALBRECHT	Polish Academy of Sciences, Medical Research Centre, 3 Dworkowa Street, 00784 Warsaw, Poland.
N. BAUMANN	I.N.S.E.R.M. Hôpital de la Salpêtrière - U 134, 47 Boulevard de l'Hôpital, 75651 Paris Cedex 13, France.
A. BENDER	University of Saskatchewan, Department of Pharmacology, Saskatoon, Saskatchewan, S7N OWO, Canada.
J.E. BOTTENSTEIN	The Marine Biomedical Institute, The University of Texas Medical Branch at Galveston, 200 Univ. Boulevard, Galveston, Texas 77550, USA.
M.J. BRAMMER	Institute of Psychiatry, Dept. of Biochemistry, De Crespigny Park, Denmark Hill, London, SE5 8AF, U.K.
J.P. DELAUNOY	Centre de Neurochimie du CNRS, 5 rue Blaise Pascal, 67084 Strasbourg, France.
J. de VELLIS	University of California, Dept. of Neurochemistry, 900 Veteran Avenue, Los Angeles, California 90024, USA.
L.F. ENG	Veteran Administration Medical Center Pathology Research (151 B), Palo Alto, California 94304, USA.
G. FRANCK	Centre Hospitalier Universitaire, Service de Neurologie (B33), Sart Tilman, 4000 Liege, Belgium.
M.S. GHANDOUR	Centre de Neurochimie du C.N.R.S., 5 rue Blaise Pascal, 67084 Strasbourg, France.
J. GHEUNENS	Laboratorium voor Neurochemie Universitaire Instellung Antwerpen Universiteitsplein 1, 2610 Wilrijk, Belgium.
R.M. GOULD	Institute for Basic Research in Developmental Disabilities, 1050 Forest Hill Road, Staten Island, New York 10314, USA.
T. GRISAR	Laboratoire de Biochimie Generale et Comparée, Place Delcour 17, 4020 Liège, Belgium.

D. GUILLAUME	Laboratoire de Biochimie Generale et Comparée, Place Delcour 17, 4020 Liège, Belgium.
B. HABER	Journal of Neuroscience Research, Neurochemistry Section, Marine Biomedical Institute, Univ. of Texas Medical Branch, Galveston, Texas 77550, USA.
E. HANSSON	University of Goteborg, Faculty of Medicine, Institute of Neurobiology, Fack S-400, 33 Goteborg, Sweden.
L. HERTZ	University of Saskatchewan, Dept. of Pharmacology, College of Medicine, Saskatoon, S7N OWO, Canada.
H.K. KETTERMANN	Institute fur Neurobiologie der Universitat Im Neuenheimer Feld 504, D-6900 Heidelberg 1, West Germany.
H.K. KIMELBERG	The Albany Medical College of Union University, Albany, NY 12208, USA.
G. LEVI	Istituto di Biologia Cellulare, Via G. Romagnosi 18/A, 00196 Roma, Italy.
P. MANDEL	Centre de Neurochimie du C.N.R.S., 5 rue Blaise Pascal, 67084 Strasbourg, France.
K.D. McCARTHY	The University of North Carolina at Chapel Hill, Faculty Lab., Office Building 231H, Chapel Hill, NC 27514, USA.
G. MOONEN	Universite de Liège, Institut d'Histologie, rue de Pitteurs, 20 Bât. L3, 4020 Liège, Belgium.
M. NORENBERG	University of Miami, School of Medicine, Dept. of Neuropathology, P.O. Box 016960, Miami, Florida 33101, USA.
W.T. NORTON	Albert Einstein College of Medicine of Yeshiva University, 1300 Morris Park Avenue, Bronx, NY 10461, USA.
F.X. OMLIN	Institut d'Histologie, Ecole de Medecine, Université de Lausanne, CH-1005, Lausanne, Switzerland.
A.J. PATEL	Medical Research Council, Developmental Neurobiology Unit, Institute of Neurology, 33 John's Mews, London, WC1N 2NS, U.K.
C.K. PETITO	The New York Hospital, Cornell Medical Center, Dept. of Pathology, 525 East 68th Street, NY 10021, USA.
B. PETTMANN	Centre de Neurochimie du C.N.R.S., 5 rue Blaise Pascal, 67084 Strasbourg, France.
J.F. PODUSLO	Depts. of Neurology and Biochemistry, Mayo Clinic & Mayo Foundation, Rochester, MN 55905, USA.
R. REYNOLDS	Inselspital Bern, Medizinische Universitäts Kinderklinik und Poliklinic, 3010 Bern, Switzerland.
A. SCHOUSBOE	University of Copenhagen, Dept. of Biochemistry A. Blegdamsvej 3C, DK-2200 Copenhagen N, Denmark.
W.E. SHASHOUA	Harvard Medical School, McLean Hospital, Belmont, Massachusetts 02178, USA.
D.H. SILBERBERG	University of Pennsylvania, Dept. of Neurology, 3400 Spruce Street, Philadelphia, PA 19104, USA.
A. SUZUMURA	University of Pennsylvania, Dept. of Neurology, 3400 Spruce Street, Philadelphia, PA 19104, USA.

S. SZUCHET	The University of Chicago, Dept. of Neurology, B.H. Box 425, 5841 South Maryland Avenue, Chicago, Illinois 60637, USA.
S. VARON	University of California, Dept. of Biology M-001, 7502 Hillside Drive, La Jolla, California 92037, USA.
A. VERNADAKIS	University of Colorado, Depts. of Pharmacology & Psychiatry, 4200 East Ninth Avenue, Denver, Colorado 80262, USA.
W. WALZ	University of Saskatchewan, Dept. of Pharmacology, Saskatoon, Saskatchewan, S7N OWO, Canada.
S. WATANABE	Shionogi Research Laboratory, Shionogy & Co. Ltd., Fukushima-ku, Osaka 553, Japan.
S. WHITE	University of Utah, School of Medicine, 410 Chipeta Way, Room 167, Salt Lake City, Utah 84108, USA.
B. WITTER	Institut fur Physiologische Chemie der Universitäts Koln, Joseph-Stelzmann Str. 52, 5000 Koln 41 (Lindenthal), West Germany.
D.M. WOODBURY	University of Utah, Dept. of Physiology, 410 Chipeta Way, Salt Lake City, Utah 84108, USA.

Session I

MOLECULAR ASPECTS OF GLIAL DEVELOPMENT

Co-chair : M. SENSENBRENNER (*Strasbourg, France*)
G. MOONEN (*Liège, Belgium*)

Chemically Defined Media for Growth and Differentiation of CNS Glial Cells

J. E. Bottenstein

Marine Biomedical Institute and Department of Human Biological Chemistry and Genetics, University of Texas Medical Branch, Galveston, Texas, USA

ABSTRACT

We have developed two different chemically defined media optimized for the growth and differentiation of neonatal rat brain astrocytes or oligodendrocytes. Our strategy was to first define the growth requirements of a continuous cell line expressing the appropriate glial phenotype. U251 MGsp cells contain glial fibrillary acidic protein (GFA) and can proliferate and be subcultured in G2 medium containing transferrin, fibroblast growth factor, hydrocortisone, sodium selenite, biotin, and a substratum modified with polylysine and fibronectin. Neonatal rat cerebral astrocytes also require these medium supplements but do not require a fibronectin-modified substratum. Continued growth beyond 5 days requires insulin, and epidermal growth factor markedly increases the number of astrocytes. Cerebral cells subcultured in G5 medium yield purified astrocytes (98%), which contain GFA and are more morphologically differentiated than in conventional serum-supplemented medium. CO-13-7 hybrid cells express the major biochemical properties of oligodendrocytes and can proliferate and be repeatedly subcultured in O1 medium containing insulin, transferrin, sodium selenite, biotin, and a substratum modified with polylysine and fibronectin. Neonatal rat brain cells can be maintained for several months and subcultured in O1 medium with the same substratum modification as CO-13-7 cells. These cultures contain 15-20 times more galactocerebroside-positive cells, no fibroblasts, and fewer astrocytes than conventional serum-supplemented cultures. O1 medium is also suitable for culturing isolated oligodendrocytes and for plating primary cells directly into it instead of into serum-supplemented medium initially. These two different chemically defined media (G5 and O1) have numerous advantages compared to conventional culture media and promote both the growth and expression of differentiated properties of CNS glial cells in the absence of fibroblast overgrowth. Our data suggest that G5 medium selects for astrocytes and O1 medium for oligodendrocytes from a mixed population of initial cells.

KEYWORDS

Chemically defined media; CNS glia; U251 MGsp cell line; CO-13-7 cell line; astrocyte; oligodendrocyte; glial fibrillary acidic protein; galactocerebroside; insulin; transferrin; selenium; biotin; epidermal growth factor; fibroblast growth factor; hydrocortisone; polylysine; fibronectin.

INTRODUCTION

In vitro studies of the regulation of growth and differentiation of neural cells have been hindered by the presence of serum and other poorly defined biological fluids in conventional culture medium. The problems associated with the use of these complex biological fluids are manifold. For example, concentrations of hormones, growth factors, trace elements, and other components vary between different preparations of these fluids; they contain factors that may inhibit the growth or expression of differentiated properties of specific cell types; it is often difficult to deplete them of specific factors in order to study their dose-dependent effects; and identification of important growth and differentiation regulating molecules is much more difficult. Thus, we have developed chemically defined media (CDM) devoid of serum for neural cell lines and their primary cell counterparts (Bottenstein, 1985a). The CDM we described for neuronal cells is not optimal for glial cells when starting with a mixed population of cell types. This paper describes the development of CDM that promote preferential growth of cells expressing an astrocytic or oligodendrocytic phenotype.

METHODS

Cell Culture. Experimental procedures and culture methods for U251 MGsp cells are described in detail by Michler-Stuke and Bottenstein (1982) and for cerebral cortex astrocytes derived from neonatal rats by Michler-Stuke, Wolff, and Bottenstein (1984).

Procedures and methods for CO-13-7 and neonatal rat brain cells are described by Bottenstein (1986). Briefly, stock cultures of CO-13-7 cells are maintained in Dulbecco's modified Eagle's medium (DME; Gibco 430-2100) supplemented with 5% (v/v) fetal calf serum (FCS), 5% (v/v) calf serum (CS), 1.2 g/l $NaHCO_3$, and 15 mM HEPES in a humid atmosphere of 95% air/5% CO_2 at 37°C. Cells are removed from culture vessels with trypsin/EDTA (0.05%/0.5 mM in PBS without Ca^{2+} or Mg^{2+}), an equal volume of soybean trypsin inhibitor (0.05% in Hank's balanced salt solution) is added, cells are centrifuged 1 min at low speed, the cell pellet is resuspended in DME, and cells are inoculated into culture vessels after hemocytometer counting. For growth experiments, the cells are counted with a Coulter counter after detachment with trypsin/EDTA.

Dissociated neonatal rat brain cultures are obtained as follows. Brains from decapitated rat littermates (1-3 days old) are minced in Ham's F12 medium (Gibco 430-1700) with 10% FCS or DME with 10% CS as indicated and antibiotics (50 µg/ml penicillin, streptomycin, and gentamycin; 10 µg/ml neomycin). Tissue fragments are dissociated by sequential sieving through 380 and 140 µm stainless steel mesh, followed by 30 µm nylon mesh. Cells are pelleted by centrifugation at low speed and resuspended in the above medium without gentamycin. Five X 10^6 cells/25 cm^2 are plated into tissue culture flasks. After 36 hr, plating medium is exchanged for DME with 10% CS and no antibiotics. The next day, unattached and loosely adherent cells are transferred to new culture vessels after centrifugation and resuspension in DME. When plating directly into O1 medium (Table 1), the culture surfaces are first modified with poly-D-lysine (M_r 30,000-70,000) and human fibronectin as described by Bottenstein and Sato (1980). Four X 10^6 cells/8 cm^2 are plated into culture vessels.

Isolated oligodendrocytes are prepared from 13-21 day old cultures initially plated in DME with 10% CS as above. To detach the oligodendrocytes from the primarily astrocytic bed layer, cells are incubated with Hank's balanced salt solution (HBSS) without Ca^{2+} or Mg^{2+} for exactly 4 min, with vigorous horizontal shaking at the end of each min. Detached cells are pelleted by centrifugation at low speed, resuspended in DME, and plated directly into O1 medium.

Concentrated stock solutions, storage, and sources of bovine insulin, human transferrin, and sodium selenite are described by Bottenstein (1984). A 500X concentrated solution of biotin (Calbiochem-Behring) was prepared in HBSS and stored at -20° C.

Indirect Immunofluorescence. These methods are as described by Norton *et al.* (1983). Rabbit anti-galactocerebroside (GalC) was provided by K. Fields (Albert Einstein College of Medicine) and mouse monoclonal anti-GalC by B. Ranscht (Massachusetts Institute of Technology). Mouse monoclonal anti-GFA was purchased from Labsystems and fluorochrome-labeled goat anti-IgG from Cappel.

RESULTS AND DISCUSSION

U251 MGsp Cells. A human glioma cell line (U251 MGsp) that contains glial fibrillary acidic protein (GFA), an astrocyte-specific marker, is able to proliferate in G2 medium (Table 1, 2). This medium also supports the division of C6 (2B subclone) rat glioma cells, which have characteristics of both astrocytes and oligodendrocytes and may represent a precursor cell for both cell types (Michler-Stuke and Bottenstein, 1982). G2 medium is unable to sustain the proliferation of LAN-1 human neuroblastoma or WI38 human fibroblast cells. Thus, with respect to these cell lines G2 medium appears to be glial-specific. Both U251 MGsp and C6 cells are more morphologically differentiated in this chemically defined medium than in serum-containing medium. They exhibit multipolar processes that are more highly branched.

Table 1. Composition of Chemically Defined Glial Media

Supplement	G2	G5	O1
Insulin, μg/ml	-	5	5
Transferrin, μg/ml	50	50	50
Hydrocortisone, nM	10	10	-
Biotin, ng/ml	10	10	10
Sodium selenite, nM	30	30	30
Fibroblast growth factor, ng/ml	5	5	-
Epidermal growth factor, ng/ml	-	10	-

Supplements are added to Dulbecco's modified Eagle's medium with 4500 mg/l glucose. The culture surfaces are modified with 2 μg/cm^2 poly-D-lysine precoating in all cases and with 0.6-1.2 μg/cm^2 fibronectin when G2 and O1 medium are used.

Table 2. Growth of Glial Cell Lines in Chemically Defined Media

Cell Line	Phenotype	Medium	Increase in Cell # day 3	Increase in Cell # day 5
U251 MGsp	astrocytic	G2	1.9	3.5
CO-13-7	oligodendrocytic	O1	2.9	5.0

Data from triplicate cultures are averaged and expressed as the increase in cell number relative to the cell inoculum. SD varied <5%. Media were changed on day 5. G2 and O1 media are described in Table 1.

Neonatal astrocytes. When we tested G2 medium on enzymatically dissociated neonatal rat cerebral cortex cells, we found that the medium was selective for astrocytes but they could not survive beyond 5 days (Michler-Stuke, Wolff, and Bottenstein, 1984). The addition of insulin increases the longevity of these cultures and epidermal growth factor (EGF) markedly increases the number of cells (G5 medium, Table 1). In contrast to U251 MGsp cells, there is no requirement for a fibronectin-modified substratum. Figure 1A illustrates the growth of astrocytes in these various conditions. Figure 1B shows the striking morphological differentiation of GFA-positive astrocytes in G5 medium: multipolar cells with very long processes. If EGF is removed from G5 medium, after about 2 weeks the morphology reverts to the less differentiated state seen in serum-containing medium: epitheliod-shaped cells with few if any short processes. Long-term growth is also possible in G5 medium; cells have been subcultured 4 times and remained viable after 2 months. Greater than 98% of the cells are GFA-positive after the first subculture, and there are no fibroblasts remaining in the cultures.

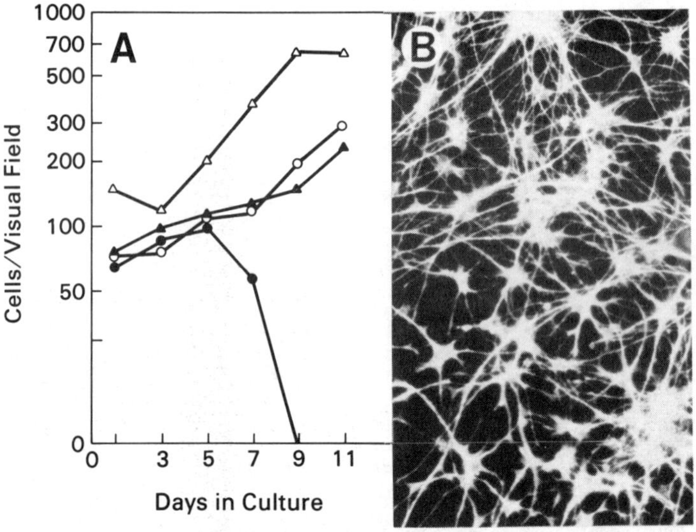

Fig. 1. Properties of neonatal rat astrocytes in chemically defined media. (A) Growth curve in various media. Cells were plated at 10^6 cells/8 cm^2 in Dulbecco's modified Eagles's medium and 10% fetal calf serum (○); G2 medium (●); G4 medium (▲); or G5 medium (△). G2 and G5 components are shown in Table 1. G4 medium is G2 medium plus insulin. SD of quadruplicate cultures was <10%. (B) Ten day old cultures of cells plated in G5 medium and immunostained for glial fibrillary acidic protein.

CO-13-7 Cells. These cells are a somatic cell hybrid of a calf oligodendrocyte fused with a C6 rat glioma cell and express the major biochemical properties characteristic of oligodendrocytes: expression of galactocerebroside (GalC), sulfatide, myelin basic protein, and 2',3'-cyclic nucleotide 3'-phosphohydrolase; glucocorticoid induction of glycerol phosphate dehydrogenase (McMorris et al., 1981). These cells proliferate in O1 medium (Table 1, 2) as do ROC-1 cells (Bottenstein, 1986), which are a hybrid of a rat oligodendrocyte and a C6 rat glioma cell (F. A. McMorris, personal communication). In contrast, the parent C6 rat glioma and 3T3 mouse fibroblast cells do not continue dividing in O1 medium, and B104 rat neuroblastoma cells are unable even to survive in it. Thus, with respect to these cell lines O1 medium appears to be oligodendrocyte-specific.

Neonatal Oligodendrocytes. Mechanically dissociated neonatal rat brain cells are initially plated in serum-containing medium, followed by two subcultures directly into O1 medium on polylysine and fibronectin-modified surfaces. Figure 2 shows these cells in chemically defined medium and serum-supplemented medium and corresponding fluorescence micrographs of GalC-stained cells. The percentage of GalC-positive oligodendrocytes is markedly increased, while astrocytes are reduced, and fibroblasts are not present (Bottenstein, 1985b, 1986). Random field counts of GalC-positive cells in identical cultures indicated a 15-20 fold increase in the number of GalC-positive cells in O1 medium compared to serum-containing medium. Figure 3 (A,B) shows cells that are plated directly into O1 medium after removal from the rat and subsequent dissociation. They have cell diameters (mean=10 μm) and morphology characteristic of oligodendrocytes. Fields of two different cellular densities are shown. These cells also express GalC and myelin basic protein as indicated by immunostaining (data not shown). The inset (Figure 3C) shows a process-bearing oligodendrocyte 29 days after being plated directly into O1 medium. Even after 30 days or longer in O1 medium, there is no overgrowth by non-oligodendrocytes.

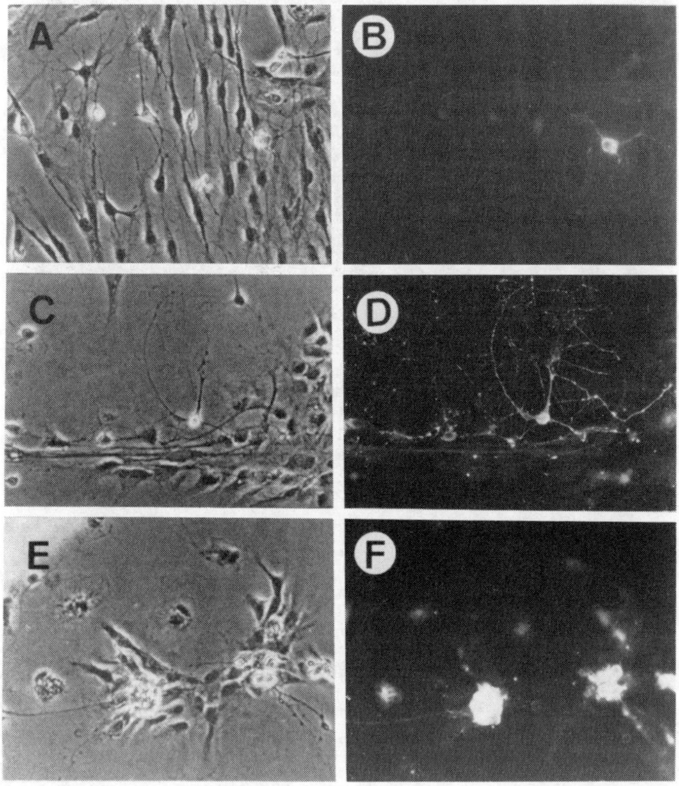

Fig. 2. Neonatal rat brain cells immunostained for galactocerebroside. Dissociated cells are shown after two subcultures and maintenance for 24 days in various media. (A,C,E) phase contrast micrographs; (B,D,F) corresponding fluorescence micrographs. (A,B) Dulbecco's modified Eagle's medium with 5% fetal calf serum and 5% calf serum; (C-F) O1 medium.

Oligodendrocytes can be isolated from a mixed culture using a differential adhesion method. Dissociated neonatal rat brain cells are plated into serum-supplemented medium and after 13 days, the overlaying oligodendrocytes are removed from the bed layer, consisting primarily of astrocytes (Figure 4A). The isolated oligodendrocytes are shown 2 days later in Figure 4B and the remaining astrocytes and other cells in Figure 4C. This procedure can be repeated several times on the initial culture, since it appears that precursor cells left behind give rise to new oligodendrocytes. Figure 5 shows cells isolated a second time from a 21 day old culture and plated sparsely directly into O1 medium; they are immunostained for GalC. Characteristic oligodendrocyte morphology is evident in the extensive branching of processes and their numerous membranous expansions.

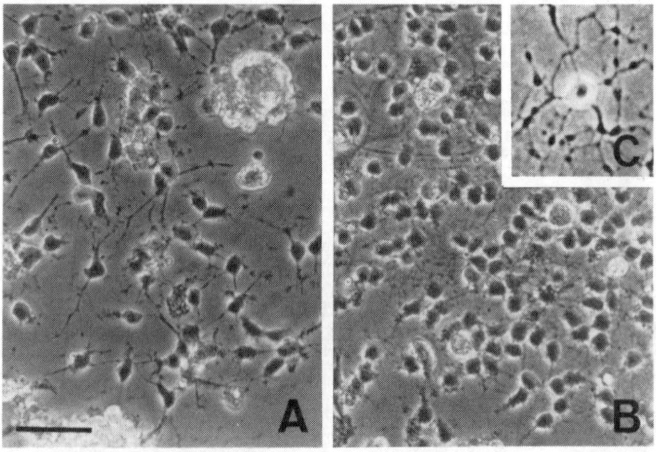

Fig. 3. Phase contrast micrographs of neonatal rat brain cells plated directly into O1 medium. (A,B) 18 days later and (C) 29 days later. Bar=50 µm (A,B); 39 µm (C).

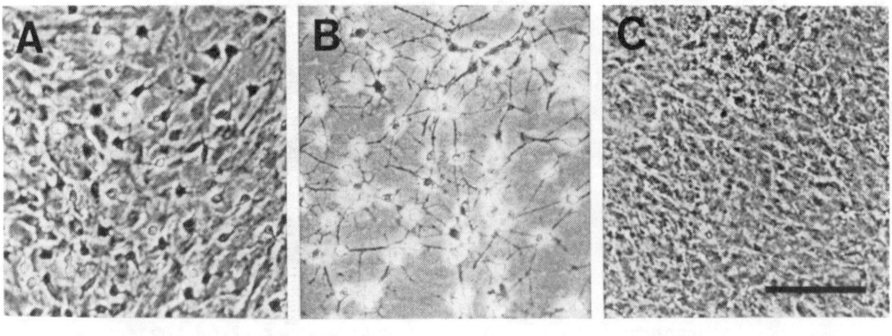

Fig. 4. Phase contrast micrographs of cells maintained in Dulbecco's modified Eagle's medium and 10% calf serum. (A) Mixed culture after 13 days in culture; (B) isolated oligodendrocytes obtained from (A) and shown 2 days later; (C) remaining bed layer of primarily astrocytes obtained from (A) and shown 2 days later. Bar=100 µm.

Summary. We have used continuous cell lines expressing glial phenotypes to develop chemically defined media for the preferential growth of astrocytes or oligodendrocytes. Proliferation and morphological differentiation of astrocytes are greatly enhanced in G5 medium compared to conventional serum-supplemented medium. In addition, the astrocytes continue to express high levels of GFA. Highly purified cultures of astrocytes are obtained without the use of cell separation techniques or initial plating into serum-containing medium. In O1 medium markedly increased numbers of oligodendrocytes are obtained with fewer astrocytes and no fibroblast overgrowth. Cells can be plated directly into O1 medium without initial culturing in serum-supplemented medium. Isolated oligodendrocytes can also be maintained in O1 medium and develop extensive morphological differentiation and expression of GalC. We are now using combined autoradiography and immunostaining for a variety of antigens, including A2B5 present on glial precursor cells that give rise to Type II astrocytes and oligodendrocytes, to assess the proliferation of cells in O1 medium.

The ability to culture CNS glial cells in chemically defined media (CDM) represents a major advance in neural cell culture methodology. The preferential growth of astrocytes or oligodendrocytes in G5 or O1 medium, respectively, is based on their different molecular growth requirements rather than on physical separation methods or selective toxicity. Furthermore, these cells continue to express differentiated properties characteristic of their cell type. Since the culture conditions are rigorously defined, variability between experiments is minimized and experimental results are easier to interpret. These media will be useful in further analyses of CNS glial cell proliferation and differentiation.

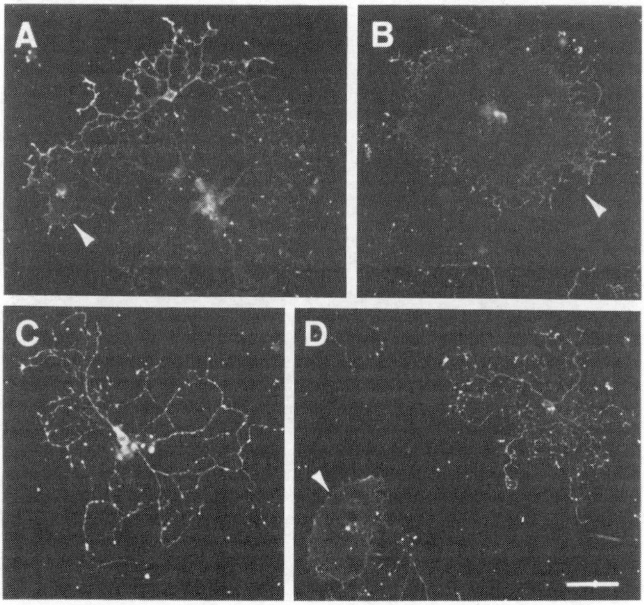

Fig. 5. Fluorescence micrographs of purified oligodendrocytes immunostained for galactocerebroside. Cells are shown 12 days after plating directly into O1 medium. Arrows indicate membranous expansions. Bar=100 μm.

ACKNOWLEDGEMENTS

The research described in this paper was supported in part by the National Institutes of Health (CA12800, CA22794, NS20375), the Concern Foundation, and the California Department of Health Services (GA80-00134).

REFERENCES

Bottenstein, J.E. (1984). In Methods in Molecular and Cell Biology, D. Barnes, D. Sirbasku, and G. Sato (eds.), Alan Liss, New York, vol. 4, pp. 3-13.

Bottenstein, J.E. (1985a). In Cell Culture in the Neurosciences, J.E. Bottenstein and G. Sato (eds.), Plenum, New York, pp. 3-43.

Bottenstein, J.E. (1985b). In Growth and Differentiation of Cells in Defined Environment (September 2-6, 1984), H. Murakami, I. Yamane, D. Barnes, J. Mather, I. Hayashi, and G. Sato (eds.), Springer-Verlag, Tokyo, pp. 25-30.

Bottenstein, J.E. (1986). Growth requirements in vitro of oligodendrocyte cell lines and neonatal rat brain oligodendrocytes, in press.

Bottenstein J.E. and G. Sato (1980). Exp. Cell Res., 129, 183-190.

McMorris, F.A., S. Miller, D. Pleasure and O. Abramsky (1981). Exp. Cell Res., 133, 395-404.

Michler-Stuke, A. and J.E Bottenstein (1982). J. Neuroscience Res., 7, 215-228.

Michler-Stuke, A., J. Wolff, and J.E. Bottenstein (1984). Intl. J. Develop. Neuroscience, 2, 575-584.

Norton, W., M. Farooq, K. Fields, and C. Raine (1983). Brain Res., 270, 295-310.

Glial Growth Factors

B. Pettmann*, G. Labourdette[+], M. Weibel* and M. Sensenbrenner*

*Centre de Neurochimie di CNRS, [+]INSERM U44, 5, rue Blaise Pascal, 67084 Strasbourg CEDEX, France

ABSTRACT

Growth factors active on glial cells from the central nervous system can be divided in two groups: those originating from the brain itself and those synthesized by other organs. Our interest was focused on the first category ; the glial growth factors synthesized (or found) in the brain. Such factors do exist since soluble brain extracts are able to elicit various typical effects on astroglial cells in primary culture : morphological changes, stimulation of the proliferation and of the synthesis of glial-specific proteins.
We isolated, from bovine brain, two factors called AGF1 and AGF2 (AGF : astroglial growth factor), of Mr of 17 500 and 18 500, respectively. The two factors elicit similar but not identical effects on the cells.
Polyclonal and monoclonal antisera were raised and characterized. All the antibodies obtained up to now react against both factors. Immunohistochemical localization on brain slices showed that AGFs are synthesized in neuronal cells exclusively. Comparison of the properties of AGF1 and AGF2 with other growth factors isolated from brain suggests that AGF1 is similar to the acidic FGF (fibroblast growth factor) isolated by Thomas, et al. (1984) and that AGF2 is similar to the basic brain FGF isolated by Gosporadowicz, et al. (1984).
Some growth factors isolated from other organs can, at least, stimulate the proliferation of the astroglial cells, but no extensive studies have been done on this subject.

INTRODUCTION

During brain ontogenesis, progenitor cells of the different neural cell types undergo differentiation, proliferation and maturation. These developmental processes can be regulated by internal programs and/or by environmental factors, such as cellular contacts, hormones, and growth factors. Research concerning growth factors has been facilitated by the availability of primary cell culture systems consisting of nearly pure populations of either neuronal (Pettmann, et al., 1979) or glial cells (Booher and Sensenbrenner, 1972). After the discovery of the nerve growth factor thirty years ago (Cohen, et al., 1954), studies were focused for a long time on this molecule and on other neurotrophic factors which were more or less characterized (Barde, et al., 1983). Only in the past ten years several groups have initiated studies to detect and purify factors, present

Advances in the Biosciences Vol. 61

in brain, active on glial cells (Sensenbrenner, et al., 1972 ; Lim, et al., 1973 ; Raff, et al., 1978). The work of our group started in 1972 with the observation that the addition of soluble brain extract to the culture medium induced a particular morphological change of the astroglial cells in culture (Sensenbrenner, et al., 1972).
We investigated this phenomenon further in two different ways : the first one was to characterize more extensively the effects of the brain extract on the astroblasts ; the second one was to try to purify the active molecules responsible for the effects.

CHARACTERIZATION OF THE EFFECTS OF BRAIN EXTRACTS ON ASTROBLASTS IN CULTURE

a) Morphological effect and species specificity

In control culture most astroglial cells are large flat cells with an irregular polygonal shape. In the presence of brain extract there is retraction of the cell body and appearance of long processes which give a fibrous aspect to the cells.
In the first study published by Sensenbrenner, et al. (1972) the brain extract and the cultures were from chick embryos. Later we observed that the morphology of astroglial cells from chick and rat brain in culture was modified as well by rat and bovine soluble brain extracts (Fig. 1) (Athias, et al., 1974 ; Pettmann, et al., 1980 ; Sensenbrenner, et al., 1980). Thus there is no evidence of species specificity for the activity present in the soluble brain extracts.

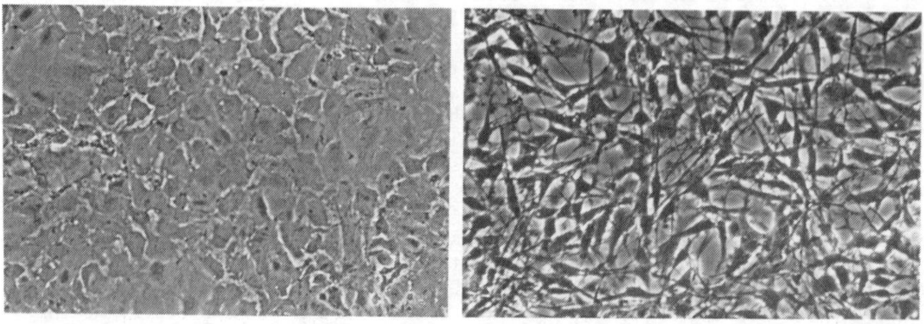

Fig. 1 - Effect of bovine brain extract on the morphology of rat astroblasts in primary culture.
The primary culture of astroblasts was prepared from brain hemispheres of newborn rats. Cells were dispersed with a syringe and grown in tissue culture petri dishes in a nutrient medium composed of Waymouth's MD 705/1 basal medium, 10 % fetal calf serum, 100 mg/l sodium pyruvate and antibiotics. To prepare bovine brain extract bovine brains were homogenized in 2 volumes of 35 mM citric acid, 0.5 mM PMSF. The suspension was centrifuged at 5000 g for 12 min. The supernatant was used as brain extract.
Left : control cultures after 20 days in vitro.
Right : similar culture treated from day 5 to day 20 in vitro.

b) Other effects of brain extracts

Brain extracts were found to stimulate the proliferation of rat glial cells as seen by the increase of the amount of DNA per dish and by the higher incorporation of radioactive iododeoxyuridine (Pettmann, et al., 1980, 1982 ; Kato, et al., 1981).
These extracts also stimulated the biochemical maturation of the cells. This maturation was defined by the levels of three astroglial-specific proteins, S100 (Cicero et al., 1970), glial fibrillary acidic protein (GFAP) (Eng, et al., 1970) and the glutamine synthetase (Norenberg, 1979). The levels or activity of these three "markers" were strongly increased (Pettmann, et al., 1981, 1982 ; Sensenbrenner, et al., 1980).
At the ultrastructural level, it was observed that the gliofilaments, which are dispersed in the cell bodies of the control cells, rearranged into large bundles in the cell body and the cell processes under the influence of brain extracts (Pettmann, et al., 1981).

ISOLATION OF THE ASTROGLIAL GROWTH FACTORS (AGFs)

To determine if all the different effects of the brain extracts were due to the same factor or to different molecules, we tried to isolate this or these factor(s). Preliminary studies showed that the activity is not species specific, is found mainly in the brain and in the pituitary gland and is not transient. So we chose adult bovine brain to start the purification.
The activity of the fractions at the different steps of the purification procedure was detected by observing the characteristic morphological effect 12 and 48 hours after their addition to a rat astroblast culture after 5 to 7 days in vitro.

a) Previous methods

We developed several methods of purification in the last years. Tissue extractions were performed in isotonic solution at neutral pH, or in various acidic buffers, or in a chloroform-methanol mixture. For further purification a variety of chromatographic steps in different buffers and pH conditions were used, like gel filtration (Ultrogel AcA 44 or AcA 54, Sephacryl S200, Biogel P200), ion-exchange (DEAE-Sephacel, DEAE-Sepharose, DEAE-Trisacryl, CM-Sepharose, CM-Trisacryl, SP-Trisacryl), adsorption chromatography (hydroxyapatite, HA-Ultrogel), chromatofocalisation, affinity chromatography (ConA-Sepharose and WGA-Sepharose, lysine-Sepharose, Procion-red agarose, Remazol-blue agarose, Cibacron-blue Sepharose, Ultrogel and Trisacryl, heparin-Sepharose) and hydrophobic interaction chromatography (octyl and phenyl-Sepharose). We published a first method of partial purification using chloroform-methanol extraction and then DEAE-Sephacel chromatography and two Ultrogel AcA 44 gel filtrations (Pettmann et al., 1980). A Mr of about 20 000 was found for the active fraction.
Later the FPLC (fast protein liquid chromatography) of Pharmacia was used with different columns Mono S, Mono P, Mono Q, and ProRPC. During this search for the best protocol we tried to avoid intermediate manipulations like dialysis, desalting or lyophilisation which resulted in major loss of activity, so we had to use compatible sequences of purification steps.
Other difficulties were encountered, particularly the change of properties after chromatographic steps. For instance when crude isotonic soluble extract is fractionated by gel filtration the active fraction is eluted at a Mr of 68 000, while after DEAE chromatography, the activity is eluted at a Mr of 18 000. This result suggests that in crude isotonic extract the factor is present as a complex which can be dissociated by ion-exchange chromatography.

The main problem was the low yields after chromatographic separations, probably due to non-specific interactions of the factor with the gels. The use of the FPLC system allowed a better recovery of the activity and a better fractionation. Our second best procedure included an acidic extraction in 35 mM citric acid containing 0.5 mM PMSF, a Blue Trisacryl M affinity chromatography in 20 mM Tris HCl, pH 7.5 (activity was eluted at 2M NaCl), an acidic precipitation at pH 2.5 and then three chromatographic steps : hydroxyapatite, Mono S (FPLC) and Pro RPC (FPLC). Two growth factors were isolated to homogeneity by this method, but with a low yield (Pettmann, et al., 1985).

b) Present method

This method is identical to the previous one for the acidic extraction, the Blue-Trisacryl affinity chromatography and the acidic precipitation. One more step of affinity chromatography on heparin-Sepharose allows the purification to homogeneity of one growth factor (AGF2). Another step of reverse phase chromatography on ProRPC gives the pure second growth factor (AGF1).

PROPERTIES OF THE ASTROGLIAL GROWTH FACTORS

a) Physicochemical properties

Analysis by SDS-PAGE showed that AGF1 and AGF2 have molecular weights of 17 500 and 18 500 respectively. AGF1 is acidic and has an isoelectric point around 5.5 ; AGF2 is a very basic protein (pI:9.5). Analysis of the amino acid composition was performed (Table 1), it shows clearly that the small molecule (AGF1) is not directly derived from the larger one (AGF2) since it contains a higher number of some amino acids like Thr, Ile, Leu and His.

Table 1. Amino acid composition of the AGF's and comparison with acidic FGF (Thomas, et al., 1984) and basic FGF (Gospodarowicz, et al., 1984).

Amino acid	AGF1	Acidic brain FGF	AGF2	Basic brain FGF
Asx	15	15	14	14
Thr	10	9	6	5
Ser	9	10	11	11
Glx	15	17	12	13
Pro	7	7	10	11
Gly	14	15	17	19
Ala	5	5	10	11
Cys	4	4	6	6
Val	4	5	7	6
Met	1	1	2	3
Ile	6	6	4	4
Leu	21	20	15	14
Tyr	8	8	8	7
Phe	7	7	9	9
His	6	6	3	2
Lys	14	14	16	17
Arg	8	6	14	12
Trp	1	1	1	1
Total	155	156	165	165
Mean deviations	5.3 %		10.8 %	

To allow better comparison the amino acid compositions of the acidic and of the basic FGFs were recalculated on the basis of molecular weights of 17 500 and 18 500, respectively. In these conditions the differences between AGF1 and the acidic FGF, and between AGF2 and the basic FGF, are not significant. In contrast, the mean deviation between AGF1 and AGF2 calculated on a basis of 165 amino acids was found to be 33 %, which is a very significant difference.

b) Biological properties

Both factors affect the morphology and stimulate the proliferation of the astroglial cells. But the effects on the morphology are not identical. After AGF2 treatment the cells are more straight and bipolar, while with AGF1 processes are more curvilinear. Effects of both factors on proliferation were difficult to compare since only AGF1 had to be purified with the ProRPC column which results in an important loss of activity.
The two factors were also able to stimulate the proliferation of meningeal cells and skin fibroblasts, and to modify their morphology. However they were inactive on lung fibroblasts.

COMPARISON OF AGFs WITH OTHER GROWTH FACTORS

a) Comparison with other factors from brain (Tables 2 and 3)

Other growth factors active on astroblasts and/or on other cell types have been purified from nervous tissues. Comparison of their aminoacid composition, when available, and of their physico-chemical properties suggests that there would be two classes of growth factors in the brain, or even only two growth factors, the acidic brain FGF and the basic brain FGF (Table 2). Only the glial growth factor (GGF) seems to be different from the others.

Table 2. Physico-chemical properties of the growth factors isolated from nervous tissues.

Group	Factor	Molecular weight	Iso-electric point	Elution from Heparin-Sepharose (molarity of NaCl)	References
1	AGF1	17 500	5.5	1.0	Pettmann, *et al.*, 1985
	FGF acidic	16 500	5.3	–	Thomas, *et al.*, 1984
	HGFα	16 000	acidic	1.0	Lobb and Fett, 1984
	HI	18 500	5.2	1.2	Conn and Hatcher, 1984
	aHDGF	18 000	5.	1.0	Klagsbrun and Shing, 1985
	RDGF	18 000	5.	1.0	D'Amore and Klagsbrun, 1984
	EDGF2	13 000	–	1.0	Courty, *et al.*, 1985
	ECGF	21 000	5.	–	Maciag, *et al.*, 1982
	GMF	13 000	5.4	–	Lim and Miller, 1984
2	AGF2	18 500	9.5	2.0	Pettmann, *et al.*, 1985
	FGF basic	16 500	9.5	1.5	Gospodarowicz, *et al.*, 1984
	HGFβ	18 000	basic	2.0	Lobb and Fett, 1984
	HII	19 300	5.2	1.4	Conn and Hatcher, 1984
	cHDGF	18 000	8.	1.4	Klagsbrun and Shing, 1985
	EDGF1	17 000	–	1.4	Courty, *et al.*, 1985
3	GGF	31 000	basic	–	Lemke and Brockes, 1984

Table 3. Target cells of the growth factors isolated from nervous tissues.

Factors	Origin	Target cells	Additional references
AGFs	Bovine brain	- **astroblasts** - skin fibroblasts - meningeal cells	
FGF acidic	Bovine brain	- 3T3 fibroblasts	
FGF basic	Bovine brain Bovine pituitary	- aortic vascular endothelial cells - Corneal endothelial cells - **astroblasts** - other cell types	Gospodarowicz, et al., 1977 Gospodarowicz, et al., 1978
HGFs	Bovine brain Bovine hypothalamus	- capillary endothelial cells - 3T3 fibroblasts	
Hs	Human brain	- umbilical endothelial cells	
HDGFs	Bovine hypothalamus	- capillary endothelial cells	
EDGFs	Bovine retina	- epithelial lens cells - 3T3 fibroblasts - human fibroblasts - **astroblasts**	Our observation
RDGF	Bovine retina	- capillary endothelial cells - umbilical vein endothelial cells - 3T3 fibroblasts	
GGF	Bovine brain Bovine pituitary	- Schwann cells - **astroblasts** - fibroblasts	Brockes, et al., 1980
GMF	Bovine brain	- **astroblasts** - skin fibroblasts - Schwann cells	Bosch, et al., 1984
ECGF	Bovine hypothalamus	- umbilical vein endothelial cells - 3T3 fibroblasts	

Abbreviations for the growth-factors

1) Derived from nervous tissues

AGF : Astroglial growth factor
ECGF : Endothelial cell growth factor
EDGF : Eye-derived growth factor
FGF : Fibroblast growth factor
GGF : Glial growth factor
GMF : Glia maturation factor
H : Human anionic endothelial cell growth factor
HDGF : Hypothalamus-derived growth factor (a: anionic; c: cationic)
HGF : Heparin-binding growth factor
RDGF : Retina-derived growth factor

2) Derived from other organs
EGF : Epidermal growth factor
GGPF : Glial growth-promoting factor
GSF : Glia cell stimulating factor
IL : Interleukin
PDGF : Platelet-derived growth factor

b) Comparison with other growth factors active on glial cells

Several growth factors isolated from various organs have been reported to affect astroglial cells in culture, they are putative growth factors for these cells. All these factors seem to be different from the AGFs by their properties (Table 4). Astroglial cells probably have specific receptors, for several different growth factors.

Table 4. Glial growth-promoting factors from other sources.

Factor	Origin	Molecular weight	Reference
EGF	Submaxillary gland	6 000	Westermark, 1976
GGPF	Activated T lymphocytes	6 000-10 000	Merrill , et al., 1984
GSF	Murine ⎡activated T	60 000-80 000	Fontana, et al., 1981a
	Human ⎣lymphocytes	10 000; 30 000	Fontana, at al., 1981b
IL-1	Monocytes	13 000	Giulian and Lachman, 1985
PDGF	Platelets	30 000	Heldin, et al., 1977

CELLULAR LOCALIZATION OF AGFs IN THE BRAIN

AGFs are extracted from brain but we do not know if they are synthesized in the brain, and if they are, in which cell type. To answer this important question we prepared specific antibodies against the AGFs and performed immunohistochemical staining on brain slices.

a) Monoclonal antibodies against the AGFs

We used the pure AGF1 and AGF2 for in vitro immunization of mice lymphocytes. About 15 clones of hybridomas were positive. All the monoclonal antibodies prepared from AGF1 or AGF2 were found to react with both factors by immunodotting and by immunoblotting. Some clones were grown as ascitic tumors in Balb/c mice. The ascitic fluids produced were much more efficient than the culture supernatants for the immunological reactions.

b) Localization of the AGFs in brain

Since the antibodies react with both factors we do not know which factor we localize on the immunohistochemical preparations, so we will say AGFs. On slices of different areas of rat brain and in cerebellum AGFs were found only in neurons, and in nearly all neurons (Fig. 2 a,c).
The cytoplasm and the processes of the neurons were labelled but not the nuclei (Fig. 2b). In the cerebellum too most if not all neurons were positive (Fig. 2c). On Fig. 2d we can distinguish heavily labelled Purkinje cells but also granules cells, Golgi cells and basket cells. Same result was observed with several monoclonal antibodies.

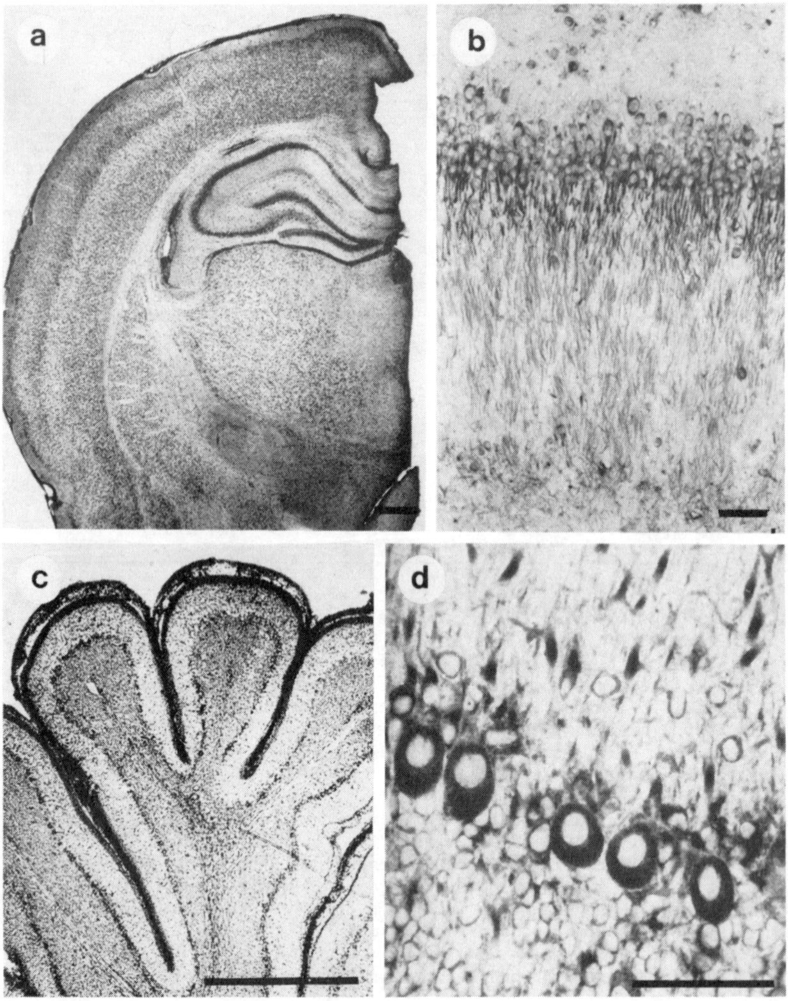

Fig. 2 - Immunohistochemical reaction of a monoclonal antibody against AGF on brain and cerebellum slices.
Twenty day-old rats were injected with Nembutal and then perfused with 4 % paraformaldehyde. Brains were left in the same solution overnight and then in PBS. Slices of 100 µm were cut and incubated in a 1:100 dilution of the ascitic fluid in PBS. After washing, the slices were incubated with goat-antimouse IgG coupled with horseradish-peroxidase. Activity was revealed with 4-chloro 1-naphtol.
a, brain; b, brain hippocampus ; c,d, cerebellum
Bar: a, 1 mm; b, 100 µm; c, 1 mm; d, 100 µm.

CONCLUSION

Astroglial cells in culture have receptors for a variety of growth factors, which are found outside the brain, like EGF and PDGF or inside the brain. We purified two different factors from bovine brain, AGF1 and AGF2, which are able to modify the morphology of the astroblasts, and to stimulate their proliferation and their biochemical maturation, as seen by an increase of the levels of the glial-specific proteins, S-100 protein and glutamine synthetase. Although other authors isolated various growth factors from brain, we believe that there are, in the brain, only or mainly, two growth factors, which are identical to the previously described acidic brain FGF and basic brain FGF. These two brain-derived factors, like the factors extracted from other organs are not specific for astroglial cells, they can react on many other cell types.

Monoclonal antibodies were raised against these factors and found to react with both of them, indicating that the two factors share, at least, one main antigenic site.

Immunohistochemical localization showed that the AGFs are localized exclusively in the neurons, in brain and cerebellum. The AGFs could play a role in the neuronal-glial cell interactions, during the development and/or in the adult animals. They could also be involved in pathological conditions like the reactive gliosis after brain injury.

ACKNOWLEDGEMENT

We thank Dr. Y. Boulanger for the determination of the amino acid composition of the factors, and Mrs. M.-F. Knoetgen for excellent technical assistance. We also thank Miss N. Volf for her efficient collaboration in the immunohistochemical studies. This work was supported by CNRS, INSERM and by grants from the CNRS (n° 951/511/01) and from the Association pour la Recherche sur le Cancer (N° 6451).

REFERENCES

Athias, P., M. Sensenbrenner and P. Mandel (1974). Differentiation, 2, 99- 106.

Barde, Y.-A., D. Edgar and M. Thoenen (1983). Ann. Rev. Physiol., 45, 601- 612.

Barritault, D., J. Plouet, J. Courty and Y. Courtois (1982). J. Neurosci. Res., 8, 477-490.

Benveniste, E.N., J.E. Merrill, S.E. Kaufman, D.W. Golde and J.C. Gasson (1985). Proc. Natl. Acad. Sci. USA, 82, 3930-3934.

Booher, J. and M. Sensenbrenner (1972). Neurobiology, 2, 97-105.

Bosch, E.P., J.C. Assouline, J.F. Miller and R. Lim (1984). Brain Res., 304, 311-319.

Brockes, J.P., G.E. Lemke and D.R. Balzer (1980). J. Biol. Chem., 255, 8374-8377.

Cicero, T.J., W. Cowan, B. Moore and V. Suntzeff (1970). Brain res., 18, 25- 34.

Cohen, S., R. Levi-Montalcini and V. Hamburger (1954). Proc. natn. Acad. Sci. U.S.A., 40, 1014-1018.

Conn, G. and V.B. Hatcher (1984). Biochem. Biophys. Res. Commun., 124, 262- 268.

Courty, J., C. Loret, M. Moenner, B. Chevallier, O. Lagente, Y. Courtois and D. Barritault (1985). Biochimie, 67, 265-269.

D'Amore, P.A. and M. Klagsbrun (1984). J. Cell Biol., 99, 1545-1549.

Eng, L.F., J.J. Vanderhaeghen and B. Gerstl (1971). Brain Res., 28, 351-354.

Fontana, A., R. Dubs, R. Merchant, S. Balsiger and P.J. Grob (1981a). J. Neuroimmunol., 2, 55-71.

Fontana, A., U. Otz, A.L. De Week and P.J. Grob (1981b). J. Neuroimmunol., 2, 73-81.

Giulian, D. and L.B. Lachman (1985) Science, 228, 497-499.

Gospodarowicz, D., J. Cheng, G.-M. Lui, A. Baird and P. Bohlent (1984). Proc. natl. Acad. Sci. U.S.A., 81, 6963-6967.

Gospodarowicz, D., G. Greenurg, H. Bialecki and B. Zetter (1978). In vitro, 14, 85-118.
Gospodarowicz, D., A.L. Mescher and C. Birdwell (1977). Exp. Eye Res., 25, 75-89.
Heldin, C.H., A. Wasteson and B. Westermark (1977). Exp. Cell Res., 109, 429-437.
Heldin, C.H., B. Westermark and A. Wasteson (1979). Proc. Natl. Acad. Sci. USA, 76, 3722-3726.
Kato, T., Y. Fukui, D.E. Turriff, S. Nakagewa, R. Lim, B.G.W. Arnason and R. Tanaka (1981). Brain Res., 212, 393-402.
Klagsbrun, M. and Y. Shing (1985). Proc. natl. Acad. Sci. U.S.A., 82, 805-809.
Lemke, G.E. and J.P. Brockes (1984). J. Neurosci., 4, 75-83.
Lim, R. and J.F. Miller (1984). J. Cell. Physiol., 119, 255-259.
Lim, R., K. Mitsunobu and W.K.P. Li (1973). Exptl. Cell Res., 79, 243-246.
Lobb, R.R. and J.W. Fett (1984). Biochemistry, 23, 6295-6299.
Maciag, T., G.A. Hoover and R. Weinstein (1982). J. Biol. Chem., 257, 5333-5336.
Merrill, J.E., S. Kutsunai, C. Mohlstrom, F. Hofman, J. Groopman and D.W. Golde (1984). Science, 224, 1428-1430.
Norenberg, M. (1979). J. Histochem. Cytochem., 27, 756-762.
Pettmann, B., G. Labourdette, G. Devilliers and M. Sensenbrenner (1981). Dev. Neurosci., 4, 37-45.
Pettmann, B., J.-C. Louis and M. Sensenbrenner (1979). Nature, 281, 378-380.
Pettmann, B., M. Sensenbrenner and G. Labourdette (1980). FEBS Letters, 118, 195-199.
Pettmann, B., M. Weibel, M. Sensenbrenner and G. Labourdette (1985). FEBS Letters, 189, 102-108.
Raff, M.C., E.R. Abney, A. Hornby-Smith and J.P. Brockes (1978). Cell, 15, 813-822.
Sensenbrenner, M., G. Devilliers, E. Bock and A. Porte (1980a). Differentiation, 17, 51-61.
Sensenbrenner, M., G. Labourdette, J.P. Delaunoy, B. Pettmann, G. Devilliers, G. Moonen and E. Bock (1980). Tissue Culture in Neurobiology, E. Giacobini, A. Vernadakis and A. Shahar, Eds., Raven Press, New York, pp. 385-395.
Sensenbrenner, M., N. Springer, J. Booher and P. Mandel (1972). Neurobiology, 2, 49-60.
Thomas, K.A., M. Rios-Candelore and S. Fitzpatrick (1984). Proc. natl. Acad. Sci. U.S.A., 81, 357-361.
Westermark, B. (1976). Biochem. Bioph. Res. Comm., 69, 304-309.

Astrocyte Subpopulations and Glial Precursors in Rat Cerebellar Cell Cultures

G. Levi*, V. Gallo*, G. P. Wilkin† and J. Cohen‡

*Istituto Superiore di Sanità, Laboratorio di Fisiopatologia di Organo e Sistema, Viale R. Elena 299, 00161 Roma, Italy
†Imperial College of Science and Technology, Biochemistry Department, Kensington, Lonson SW7, UK
‡Medical Research Council, Neuroimmunology Project, Zoology Department, University College London, London WC1E, UK

ABSTRACT

Two populations of astrocytes can be detected when postnatal rat cerebellar cells are cultured in a serum-containing medium. One has stellate shape, "neurone-like" GABA transport, surface antigens binding the monoclonal antibodies A2B5 and LB1 and scarce proliferative activity. It may correspond to fibrous astrocytes *in vivo*. The other population has epithelioid shape, proliferates rapidly and lacks GABA transport and antigens binding A2B5 and LB1. Several types of experiments (culturing of cerebellar cells in a chemically defined, serum-free medium, immunocytolysis with A2B5 antibodies and complement, addition of serum to serum-free cultures) indicate that the stellate astrocytes derive from A2B5- and LB1-positive bipotential precursors capable of GABA uptake which, in serum-free media, proliferate and differentiate into oligodendrocytes.

KEYWORDS

Cell lineage; oligodendrocytes; glial fibrillary acidic protein; surface antigens; GABA transport; neural cell differentiation.

INTRODUCTION

The cerebellum contains several types of astrocytes, which include the fibrous astrocytes in the white matter, the Bergmann glia, the velate protoplasmic and the smooth protoplasmic astrocytes in the grey matter (Palay and Chan-Palay, 1974). It is not know whether these cell types have different functions in the adult animal, although a specific guidance role for migrating granule cells has been ascribed to the Bergmann radial fibres during development (Rakic, 1971). The existence of antigenic differences among various classes of cerebellar astrocytes during development as well as in the adult animal (Schachner, 1982) points to the possibility that functional differences are also present. While it is known that cerebellar astrocytes derive from the primitive germinal zone of the roof of the fourth ventricle (see Wilkin and Levi, 1986, for review), little is known about the cell lineage of cerebellar glial cells. The difficulty in this type of studies has always been the lack of specific astroglial markers at early developmental stages and the lack of a precise catalogue of astrocyte functions.

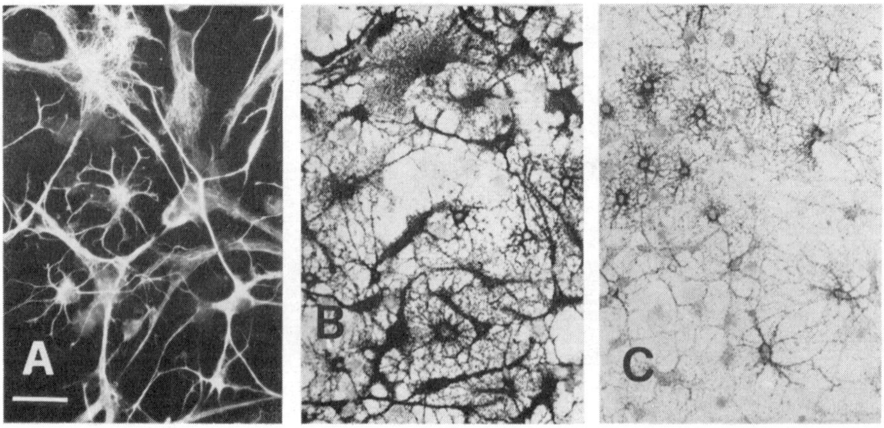

Fig. 1. Pictures of rat cerebellar astroglial cells cultured for 5 days in a serum-containing medium. Immunofluorescence staining with rabbit anti-GFAP antibodies and anti-rabbit rhodamine conjugate (A), and autoradiograms after incubation for 10 min in a Krebs-Ringer medium with 2 μCi/ml of ^{3}H-D-aspartate (B) or ^{3}H-GABA (C). Stellate astrocytes are visualized by all 3 procedures, astrocytes with other shapes are labelled by anti-GFAP and ^{3}H-D-aspartate only. Scale bar = 37 μm (A) and 100 μm (B and C).

In recent years, the availability of primary astroglial cell cultures has greatly increased the possibility of studying the functional and developmental features of these cells under controlled experimental conditions. At variance with most studies which utilized confluent cultures (for example, see Federoff and Hertz, 1977; Giacobini and Vernadakis, 1980; Pfeiffer, 1982), our initial strategy has been to analyze preconfluent stages, at which the cells still exhibit morphological heterogeneity and active proliferation when cultured in serum-containing media.

IDENTIFICATION OF TWO ASTROCYTE SUBPOPULATIONS IN RAT CEREBELLAR, SERUM-CONTAINING CELL CULTURES

Cells dissociated from 8-day postnatal rat cerebella (Levi et al., 1984) were plated at a density of 0.1-2.5 x 10^5 cells/cm^2 in Basal Medium Eagle or Dulbecco's Modified Eagle's Medium supplemented with 10% fetal calf serum (FCS) (Wilkin et al., 1983). After 4-6 days in vitro (DIV), astrocytes (identified immunocytochemically using anti GFAP (glial fibrillary acidic protein) antibodies)(Bignami et al., 1980) had two distinct morphological features. One cell type had a stellate shape, with many radially arranged fine processes, and was often present in small groups, while the other cells exhibited a polygonal, or epithelioid or a more irregular shape (Fig. 1A; Wilkin et al., 1983). At a previous culture stage (2 DIV) GFAP was poorly or not yet expressed in the stellate cells, which were also smaller in size; the other cells generally had an elongated morphology and were GFAP$^+$.

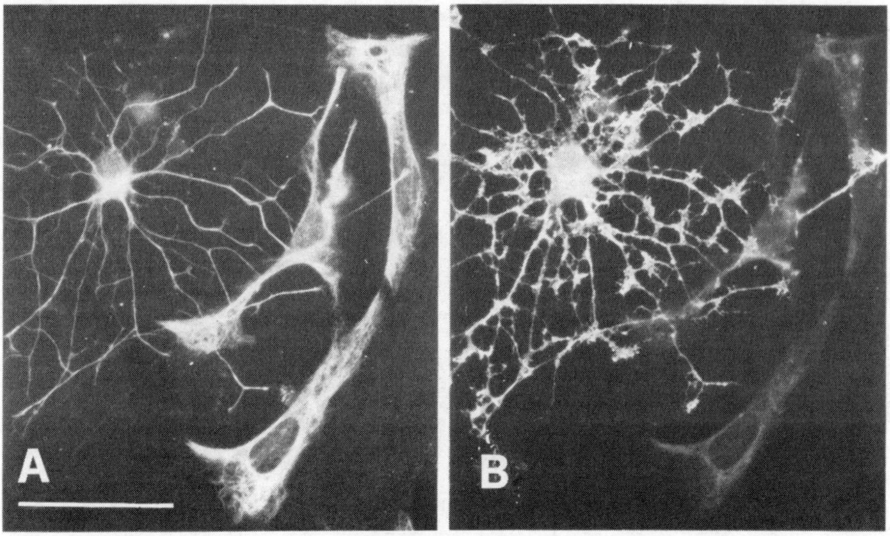

Fig. 2. Double staining of rat cerebellar astrocytes cultured in the presence of serum, with anti-GFAP polyclonal and LB1 monoclonal antibodies. Cultures at 4 DIV were exposed to LB1 antibodies followed by GAM-FITC, then fixed with 4% paraformaldehyde, permeabilized with acid alcohol and exposed to rabbit anti-GFAP antibodies followed by GAR-TRITC. The same microscopic field is shown under rhodamine (A) and fluorescein (B) optics. LB1 antibodies stain only astrocytes with a stellate morphology. Scale bar = 50 μm.

All the cells described avidly accumulated the non metabolized analogue of L-glutamate, ^{3}H-D-aspartate, but only the stellate cells accumulated also ^{3}H-GABA to a substantial degree (Fig. 1; Wilkin et al., 1983; Levi et al., 1983). At confluent stages, GFAP^{+} stellate cells were no longer detectable, and a very small percentage of cells capable of ^{3}H-GABA uptake was present, while ^{3}H-D-aspartate uptake was maintained (Wilkin et al., 1983). The ability to take up ^{3}H-GABA did not appear to be related to the morphology of the cell: in cultures exposed to dibutiryl cyclic AMP many of the polygonal cells acquired a stellate shape, but remained unable to accumulate ^{3}H-GABA (Wilkin et al., 1983). The progressive disappearance of stellate astrocytes as the cells approached confluency appeared to be related to at least two factors: cell to cell contact with the epithelioid cells and lower replication rate after the first few days in culture (Wilkin et al., 1983).

The GABA transport system present in cultured stellate cerebellar astrocytes could not be distinguished from a neuronal transport system on the basis of its sensitivity to a number of GABA analogues previously reported to be selective or predominant inhibitors of glial or neuronal GABA transport. In particular, the glial inhibitor beta-alanine was very scarcely effective, while the neuronal inhibitor cis-aminocyclohexane carboxylic acid was a potent blocker of ^{3}H-GABA uptake into stellate astrocytes (Levi et al., 1983; Johnstone et al., 1985).

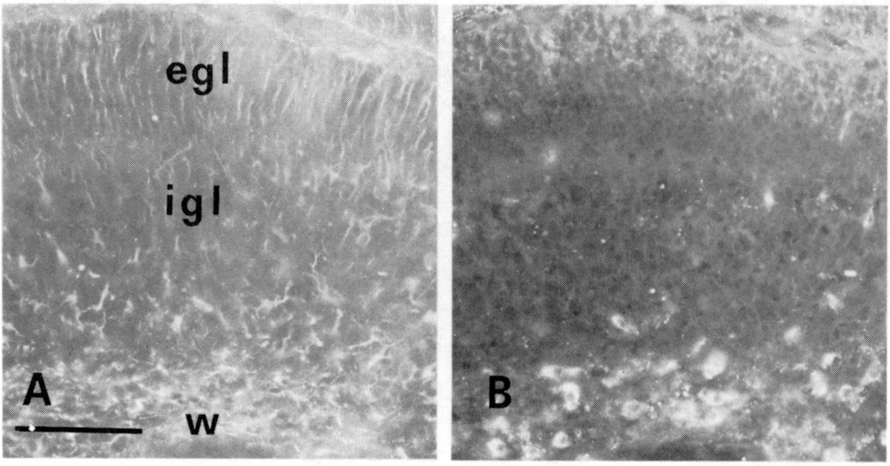

Fig. 3. Double staining of cryostat sections of cerebella from 8-day postnatal rats, with polyclonal anti-GFAP and monoclonal LB1 antibodies. Frozen 5-7 μm sections were obtained from cerebella fixed in situ by perfusion through the heart with 6% paraformaldehyde. The sections were kept in a humid box for 2 h, treated with sodium borohydride, and stained by exposing them in sequence to LB1 antibodies (16 h), GAM-FITC IgG (1 h), rabbit anti-GFAP serum (1 h) and SAR-TRITC IgG (1 h). The same microscopic field is shown under rhodamine (A) and fluorescein (B) optics. Staining of GFAP is present in the Bergmann fibres in the external granule cell layer (egl), in cells in the internal granule cell layer (igl) and in the putative white matter (w) in the central part of the folium (A). LB1 antibodies intensely stain cells in the central part of the folium (B). Light, punctate binding of LB1 is present also in the external part of the external granule cell layer. Scale bar = 100 μm.

Since the ability to transport GABA is a property of the cell membrane, we investigated whether stellate astrocytes had other membrane properties shared also by neurones. Neurones aro known to express surface antigens binding the monoclonal antibody A2B5 (Eisenbarth et al., 1979) and tetanus toxin (Raff et al., 1979). Moreover, Raff and his group have shown that, in rat optic nerve cultures, a subpopulation of astrocytes with a shape reminiscent of that of the cerebellar stellate astrocytes present in our cultures, also bound the antibody A2B5 and tetanus toxin (Raff et al., 1983a). The $A2B5^+$ cells, however, did not react with another monoclonal antibody, RAN2, which was instead bound by the non-stellate astrocyte population (Raff et al., 1984). Our experiments on cultured cerebellar astrocytes showed that the majority of the $GFAP^+$ stellate cells were $A2B5^+$ and $RAN2^-$ (Johnstone et al., 1985). Data obtained in cerebellar cell cultures from younger animals (2-day old) suggested that the antigen binding tetanus toxin is transiently expressed in the astrocytes with a stellate shape (Johnstone et al., 1985).

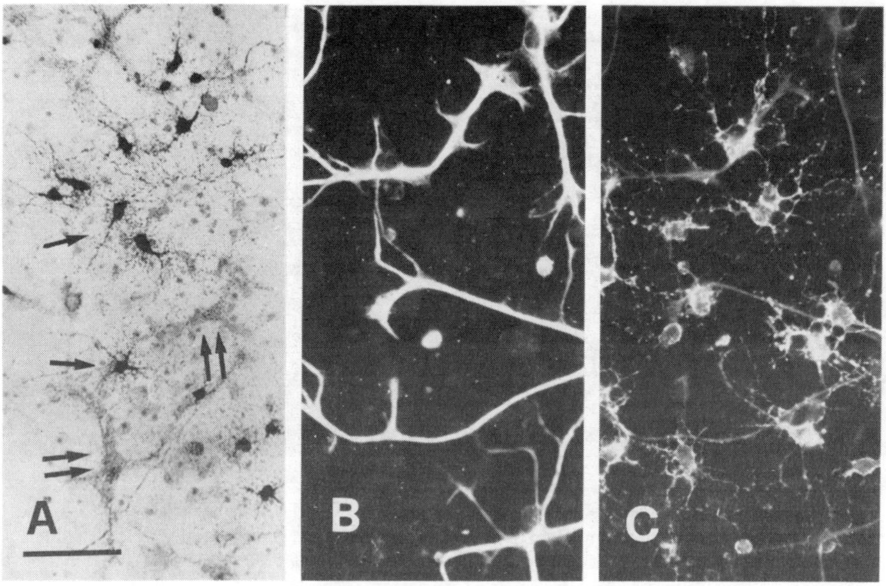

Fig. 4. Pictures of cerebellar cells cultured for 12 days in a serum-free, chemically defined medium. A: Autoradiographic uptake pattern of ^{3}H-GABA after incubation for 10 min in a Krebs-Ringer medium with 2 μCi/ml of ^{3}H-GABA. Heavy accumulation is present in a group of small stellate cells (single arrows), while only few silver grains are noticeable over the large, elongated astroglial cells (double arrows). B and C: Double staining with polyclonal anti-GFAP and monoclonal A2B5 antibodies. Staining procedure as in the legend for Fig. 2. The same microscopic field is shown under rhodamine optics, to visualize the GFAP staining (B), and under fluorescein optics, to visualize the cells labelled by A2B5 (C). The A2B5$^+$ cells are all GFAP$^-$ and viceversa. Scale bar = 100 μm (A) and 50 μm (B and C).

CEREBELLAR STELLATE ASTROCYTES MAY CORRESPOND TO FIBROUS ASTROCYTES

The studies of Raff's group have provided evidence in favour of the idea that the A2B5$^+$ astrocytes present in rat optic nerve cultures could be identified as fibrous astrocytes. This evidence was also based on immunocytochemical studies in tissue sections (Raff et al., 1984; Miller and Raff, 1984). The possibility that the cerebellar stellate astrocytes capable of GABA uptake present in our cultures, which also bind A2B5, correspond to fibrous astrocytes is supported by a set of data obtained with another monoclonal antibody, LB1, developed by one of us (J.C.). To raise LB1 monoclonals, Balb/c mice were immunized with whole cells from 10-day old cultures of embryonic rat spinal cord. Positive clones of hybridoma cells were identified by screening culture supernatants for cell surface binding antibodies,

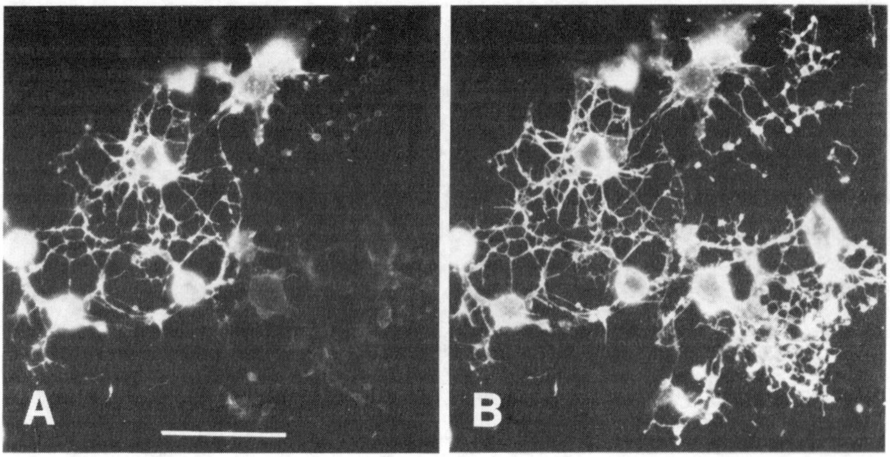

Fig. 5. Double staining with LB1 and anti-galactocerebroside (GC) monoclonal antibodies of cerebellar cells cultured for 12 days in a serum-free medium. Cells were exposed, in sequence, to the following: LB1 antibodies, goat IgGs raised against mouse IgM, anti-GC antibodies, rabbit IgG anti-mouse IgG subclass 3, and GAR-TRITC IgG. The same microscopic field is shown under fluorescein optics, to visualize LB1 antibodies (A), and under rhodamine optics, to visualize anti-GC antibodies (B). Several cells are stained by both antibodies, while 2 cells on the lower right side of the picture are labelled by anti-GC, but not by LB1 antibodies. Scale bar = 50 μm.

using indirect immunofluorescence techniques in embryonic rat spinal cord cultures. LB1 antibodies selectively stained the stellate, $GFAP^+$ cells found in cerebellar cultures (Fig. 2). Interestingly, $LB1^+$ cells could be detected in the developing white matter, in cryostat sections of 4-14 day postnatal rat cerebellar cortex, that is in the site where fibrous astrocytes are present in the adult animal (Fig. 3). The $LB1^+$ cells were much less numerous than the $GFAP^+$ cells (Fig. 3) and staining of the same cell with both antibodies was generally not seen (Wilkin, Cohen, Schneider and Hanley, in preparation).

EVIDENCE FOR THE PRESENCE OF AN $A2B5^+$, $LB1^+$ OLIGODENDROCYTE PRECURSOR CAPABLE OF GABA UPTAKE IN SERUM-FREE CULTURES

It is known that the phenotypic expression of cultured cells may vary depending on the culture conditions. In particular, serum factors can influence the survival, the differentiation and the proliferation of neural cells (Fischbach and Nelson, 1977). When dissociated cerebellar cells were cultured in a chemically defined, serum-free medium (CDM) (see Kingsbury et al., 1985; Levi et al., 1985 for details), the $GFAP^+$ cells showed almost negligible proliferative activity, while interneurones survived and accounted for about 90% of the cells present in the culture (Kingsbury et al., 1985). Moreover, the $GFAP^+$ cells expressed a morphology different from that observed in serum containing media. They were generally elongated, with several long, thick processes (Fig. 4B). These astrocytes were very modestly labelled by ^{3}H-GABA (Fig. 4A), but strongly labelled by ^{3}H-D-aspartate (Levi et al., 1985; Gallo et al., 1986). ^{3}H-GABA was instead accumulated by "colonies" formed by an increasing number of GFAP^{-} stellate cells (Fig. 4A; Levi et al., 1985; Gallo et al., 1986). Incorporation of ^{3}H-thymidine into the nuclei of these cells could be demonstrated autoradiografically (Levi and Gallo, unpublished data). This finding, taken together with the observation that the small stellate cells organized in "colonies" were not stained by anti-glutamate decarboxylase nor by anti-microtubule-associated protein 2 antibodies exclude that they are neurones, in particular GABAergic neurones (Gallo et al., 1986). The antibodies A2B5 and LB1 labelled groups of stellate cells that were strongly reminiscent of those cell groups labelled by ^{3}H-GABA (Fig. 4). Incidentally, A2B5 stained also neurones, but less intensely, whereas LB1 labelled neurones only at very early culture stages (data not presented).

In order to attempt the identification of the A2B5^{+}, LB1^{+}, GABA-uptake^{+} cells, the cultures were stained using a monoclonal antibody raised against galactocerebroside (GC), considered as a specific marker of oligodendrocytes (Raff et al., 1978). At plating, only a minimal proportion of cells bound this antibody. During the first week in culture the positive cells increased in number, but they maintained a scattered distribution (Levi et al., 1985). These cells were not stained by A2B5 nor by LB1 monoclonal antibodies. At later culture stages, anti-GC antibodies labelled not only scattered cells, but also cell groups of increasing size (Fig. 5B), which were morphologically very similar to the groups of A2B5^{+}, LB1^{+} cells (Levi et al., 1985). Preliminary double label experiments using LB1 and anti-GC monoclonal antibodies showed that at 10-12 DIV, within the groups of stellate cells, some cells could be labelled by both antibodies, others only by LB1 or by anti-GC antibodies (Fig. 5). At later culture stages the number of GC^{+} cells within each group appeared to increase further, while that of LB1^{+} cells tended to decrease. These observations suggest that the A2B5^{+}, LB1^{+}, GABA-uptake^{+} cell groups are oligodendrocyte precursors (Levi et al., 1985), and are in keeping with the data of Raff's group showing that the process-bearing cells present in rat optic nerve cultures differentiate into oligodendrocytes when maintained in serum-free media (Raff et al., 1983b).

The possibility that the A2B5^{+}, LB1^{+} cells are oligodendrocyte precursors is strongly supported by a set of immunological killing experiments. When cultures maintained in CDM were treated at 3 and 4 DIV with A2B5 antibodies and complement, no A2B5^{+}, LB1^{+} cells were detectable in the cultures at 6 DIV or at later stages, while the number of GFAP^{+} cells was similar to that of the corresponding control cultures, treated with complement alone (Levi et al., 1985). The disappearance of the LB1^{+} cells confirms that the antigens reacting with A2B5 and LB1 are present in the same cells. The immunocytolytic treatment with A2B5 prevented the appearance of the GC^{+} cell groups generally seen to develop after 6-7 days of cultivation in CDM. On the other hand, the number of scattered GC^{+} cells did not appear to be greatly affected (Levi et al., 1985).

Treating the cultures with anti-GC antibodies and complement at 3 and 4 DIV had virtually no effect on the development of the A2B5^{+}, LB1^{+} cell colonies nor on the appearance of GC^{+} cell groups after the 6th day. Also the scattered GC^{+} cells were scarcely affected. The apparent lack of effect of this treatment must be attributed to the fact that at 3-4 DIV most of the cells that will acquire an oligodendrocyte phenotype at later stages did not yet express GC on their membrane.

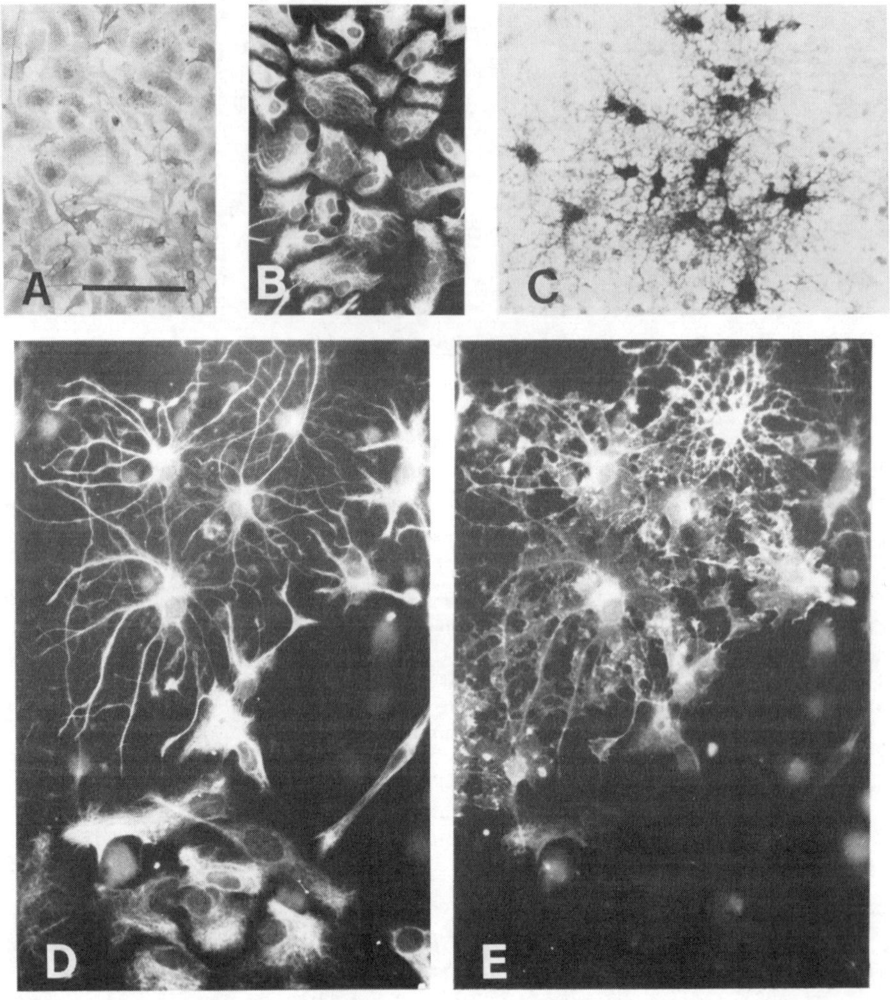

Fig. 6. Lack of ^{3}H-GABA accumulating stellate astrocytes in serum-free cultures treated with A2B5 and complement and then shifted to 10% FCS. A and B: cultures in CDM were treated at 3 and 4 DIV with A2B5 and complement (Levi et al., 1985). At 6 DIV, 10% FCS was added to the CDM and at 11 DIV the cells were processed for ^{3}H-GABA autoradiography (A) or GFAP staining (B) (details as in Fig. 1). No stellate astrocytes nor cells labelled by ^{3}H-GABA are present. C: ^{3}H-GABA is taken up by the stellate cells growing in control cultures, treated with complement and then shifted to FCS. D and E: double staining with anti-GFAP (D: rhodamine optics) and A2B5 (E: fluorescein optics) antibodies of control cultures (see C), same microscopic field. Many stellate cells bind both antibodies. Scale bar = 182 μm (A), 100 μm (B), 95 μm (C), 52 μm (D and E).

CEREBELLAR OLIGODENDROCYTE PRECURSORS CAN DIFFERENTIATE INTO $GFAP^+$ STELLATE ASTROCYTES WHEN CULTURED WITH SERUM

The experiments of Raff and Collaborators performed using rat optic nerve cultures have shown that the $A2B5^+$ oligodendrocyte precursors are bipotential: in fact they express an astroglial phenotype when cultured in a medium containing serum (Raff et al., 1983b; Temple and Raff, 1985). In agreement with this observation, when we added 10% FCS to cerebellar cells cultured in CDM for 6 days, we found that a large number of $A2B5^+$, $LB1^+$, GABA-uptake$^+$ cells became $GFAP^+$, while the expression of GC was prevented (Levi et al., 1985). Moreover, when cultures treated with A2B5 antibodies and complement were shifted to a serum-containing medium, no stellate $A2B5^+$ astrocytes capable of ^{3}H-GABA uptake developed, but only epithelioid astrocytes unable to transport ^{3}H-GABA (Fig. 6A and 6B). In contrast, in controls treated with complement alone, huge groups of stellate, ^{3}H-GABA-accumulating cells were present (Fig. 6C). These cells were positive for GFAP and bound the monoclonal antibodies A2B5 and LB1 (Fig. 6D and 6E; Levi et al., 1985).

CONCLUSIONS

Of the two types of astrocytes that we have characterized in serum-containing cultures of cells dissociated from postnatal rat cerebella, the one exhibiting a stellate morphology and a number of "neurone-like" features (type of GABA transport, binding of A2B5 antibodies and, transiently, of tetanus toxin) appear to derive from precursors which have similar characteristics, but lack the intermediate filament protein GFAP. Circumstantial evidence suggests that this type of astroglial cells may correspond to fibrous astrocytes. When cell were maintained in a serum-free medium, however, the above mentioned progenitor cells proliferated, forming large cell colonies, and differentiated into oligodendrocytes after about one week in culture. Thus, the original observation of Raff and his group on the existence in rat optic nerve cultures of a commun precursor for oligodendrocytes and a subpopulation of astrocytes (Raff et al., 1983b) can be extended to the rat cerebellum, and may be valid for other brain areas as well.

The scattered GC^+, $A2B5^-$, $LB1^-$ cells present in our cultures also at earlier stages may derive from postmitotic cells already committed to become oligodendrocytes at the time of cell dissociation and seeding. It can be suggested that the antigens binding A2B5 and LB1 antibodies are lost as the cells stop proliferating.

It is noteworthy that GABA transport ability is already present in a non-differentiated precursor cell. The transport of the amino acid in these cells had "neurone-like" features, as in the stellate astrocytes (Levi et al., 1983; Johnstone et al., 1985) deriving from them. Indirect evidence suggests that GABA is taken up also by the oligodendrocytes originating from the same precursor, and it is interesting that a similar type of GABA transport has been recently described in cultured oligodendrocytes from fetal mouse brain (Reynolds and Herschkowitz, 1984).

ACKNOWLEDGEMENTS

We thank Mrs. M.T. Ciotti (Institute of Cell Biology, CNR, Rome), Miss A. Schneider and Mr. S.R. Johnstone for collaboration. The anti-GC monoclonal antibody was a kind gift of Dr. B. Ranscht. This work was supported by a Grant of the Italian

National Research Council to G.L. (Progetto Finalizzato MPR, SP-3, Obbiettivo 26) and by a Grant of the Medical Research Council to G.P.W.

REFERENCES

Bignami, A., D. Dahl and D.C. Reuger (1980). In: Advances in Cellular Neurobiology, Vol. I, Academic Press, New York, pp. 285-310.

Eisenbarth, G.S., F.S. Walsh and M. Niremberg (1979). Proc. Natl. Acad. Sci. USA 76, 4913-4917.

Federoff, S. and L. Hertz, Eds. (1977) Tissue and Organ Cultures in Neurobiology Academic Press, New York.

Fischbach, G.D. and P.G. Nelson (1977). In: Handbook of Physiology, Section I. The Nervous System, Vol. I. Cellular Biology of Neurons. E.R. Kandel, Ed., Williams & Wilkins, Baltimore, pp. 719-735.

Gallo, V., M.T. Ciotti and G. Levi (1986). J. Neurosci. Res., in press.

Giacobini, E., A. Vernadakis and A. Shahar, Eds. (1980). Tissue Culture in Neurobiology, Raven Press, New York.

Johnstone, S.R., G. Levi, G.P. Wilkin, A. Schneider and M.T. Ciotti (1985). Dev. Brain Res., in press.

Kingsbury, A., V. Gallo, P.L. Woodhams and R. Balazs (1985). Dev. Brain Res. 17, 17-25.

Levi, G., G.P. Wilkin, M.T. Ciotti and S. Johnstone (1983). Dev. Brain Res., 10, 227-241.

Levi, G., F. Aloisi, M.T. Ciotti and V. Gallo (1984). Brain Res., 290, 77-86.

Levi, G., V. Gallo and M.T. Ciotti (1985). Proc. Natl. Acad. Sci. USA, in press.

Miller, R.H. and M.C. Raff (1984). J. Neurosci., 4, 585-592.

Palay, S. and V. Chan-Palay (1974). Cerebellar Cortex, Cytology and Organization, Springer, Berlin.

Pfeiffer, S.E., Ed. (1982). Neuroscience Approached through Cell Culture, Vol. I CRC Press, Boca Raton, FL.

Raff, M.C., R. Mirsky, K.L. Fields, R.P. Lisak, S.H. Dorfman, D.H. Silberberg, N.A. Gragson, S. Liebowitz and M. Kennedy (1978). Nature, 274, 813-816.

Raff, M.C., K.L. Fields, S. Hakomori, R. Mirsky, R.M. Pruss and J. Winter (1979). Brain Res., 174, 283-308.

Raff, M.C., E.R. Abney, J. Cohen, R. Lindsay and M. Noble (1983a). J. Neurosci. 3, 1289-1300.

Raff, M.C., R.H. Miller and M. Noble (1983b). Nature, 303, 390-396.

Raff, M.C., E.R. Abney and R.H. Miller (1984). Dev. Biol., 106, 53-60.

Rakic, P. (1971). J. Comp. Neurol., 141, 283-312.

Reynolds, R. and N. Herschkowitz (1984). Brain Res., 322, 17-31.

Schachner, M. (1982). Trends in Neuroscience, 5, 225-228.

Temple, S. and M.C. Raff (1985). Nature, 313, 223-225.

Wilkin, G.P. and G. Levi (1986). In: Astrocytes, S. Federoff and A. Vernadakis, Eds., Academic Press, New York, in press.

Wilkin, G.P., G. Levi, S.R. Johnstone and M.T. Ciotti (1983). Dev. Brain Res., 10, 265-277.

Monoclonal Antibodies to the Developing Cerebellar Glial Cells

M. S. Ghandour and G. Gombos

Centre de Neurochimie du CNRS, 5 rue Blaise Pascal, 67084
Strasbourg CEDEX, France

ABSTRACT

Monoclonal antibodies were raised by using a crude membrane preparation from 10-12 day old rat cerebella as immunogen; nine were selected for their specificity for cerebellar astrocytes. We report here on the characterization of the antigens corresponding to three of these monoclonal antibodies, each produced by a different hybridoma (called H8, H15 and H19 respectively). These antigens were localized in cerebellum at different stages of development and in primary cell cultures of astrocytes by immunohistochemistry. In Western blots MAb H8 appears to recognize two major polypeptide bands of M.W. 68 ± 2 KD and 58 ± 2 KD, exclusively present in the crude cerebellar membrane fractions while MAbs H15 and H19 recognize major bands of M.W. 67 ± 2 KD and 43 ± 2 KD respectively present in electrophoregrams of both soluble and particulate fraction proteins. The MAbs H8 and H15 are specific for protoplasmic astrocytes while MAb H19 labels both protoplasmic and fibrous astrocytes.

KEYWORDS

Glial cells; cerebellar development; monoclonal antibodies; astrocyte membranes, immunocytochemistry.

INTRODUCTION

The hybridoma technology and the production of monoclonal antibodies (MAbs) has been used extensively in neurobiological studies. In particular MAbs have been used as probes for revealing new cell-type specific molecules, developmentally regulated antigens, and in general, molecules which could present a particular interest for the study of cellular interaction because of their localization at the surface of cells. Monoclonal antibodies revealing specific molecules for different cell types in the nervous tissue and/or various CNS areas have been obtained in different laboratories (Eisenbarth et al, 1979; Barnstable, 1980; Matthew et al, 1981; Schachner et al, 1981; Zipser and McKay, 1981; Sternberger et al, 1982, Hirn et al, 1983) (for review see McKay et al, 1981). In recent years, increasing attention is being paid to the possible role of glial cells in histogenesis and to their developmental relationship with other nervous tissue elements.

Splenic lymphocytes derived from Balb/c mice immunized with a 10-12 day old cerebellar crude particulate fraction (Ghandour et al, 1984) were fused with myeloma cells. Antibody-secreting hybridomas were obtained and cloned. Nine different hybridoma clones, each producing MAbs specific for glial cells were selected according to their binding capacity to the particulate fraction of young cerebella and according to the immunocytochemical localization of the corresponding antigen in cerebellar tissue sections (Table I). In this paper we discuss three of these MAbs (called MAbs H8, H15 and H19) with the morphological and biochemical characterization of the corresponding antigens during cerebellar development. All the three MAbs bind to cerebellar astrocytes.

MATERIAL AND METHODS

Immunogen preparation :

The crude membrane fraction used in the present experiments were obtained after homogenization of 10-12 day old rat cerebella in 10 mM Tris-HCl, pH 7.4 buffer (1:10 w:v). One volume of 0.64 M sucrose was added to the homogenate. After centrifugation at 12,000 g for 20 min, the supernatant was centrifuged at 100,000 g for 2 h. The pellet was washed twice with the same buffer. This washed pellet is defined as "crude membrane fraction" while the 100,000 g supernatant is defined as "soluble fraction".

Cell fusion and screening test :

Three Balb/c mice were immunized with the particulate fraction. Each mouse received two subcutaneous injections at 3 week intervals using 0.5 mg protein in 0.5 ml in phosphate buffered saline (PBS) and 0.5 ml of complete Freund Adjuvant (Bio-Mérieux). Then two intraperitoneal injections with the same immunogen at the same intervals were performed. Finally 3-4 days before the cell fusion, each mouse received 250 µg protein by intravenous injection (Ghandour, 1984). Splenocytes were fused with cells of the P3X63-Ag8-653 myeloma line using polyethylene glycol (PEG) according to the procedure described by Kohler and Milstein, 1976; Galfré et al, 1977.

The tissue specificity of each monoclonal antibody (MAb) was determined by a dot -immunobinding assay (Hawkes et al, 1982) and by a differential binding test of each MAb to particulate fractions derived from cerebellum and liver of 10-12 day old rats (Ghandour et al, 1984). The cellular specificity was tested using immunocytochemical techniques.

Preparation of cell cultures :

Astrocytes in primary cell cultures were prepared from newborn rat brain as described by Booher and Sensenbrenner (1972).

Immunohistochemistry :

Preparation of tissue sections : Cerebellar sections were prepared as previously described (Ghandour et al, 1979, 1980). Briefly, cerebella were fixed by transcardiac perfusion with 4 % paraformaldehyde and 0.25 % glutaraldehyde. The isolated cerebella were post-fixed for 1 h at 4°C in the same fixative. 100 µm thick vibratome sections were obtained and then preincubated overnight in phosphate buffer saline (PBS) containing 3 % BSA.

Preparation of cell cultures : The cells were washed twice with PBS and then fixed for 1 h at 4°C. After washes, the cells were treated with 5 % acetic acid in 96 % ethanol for 10 min at -20°C.

Immunostaining : Tissue sections and cell cultures were incubated first 1-2 h at room temperature with a 1/20 dilution, in PBS, of hybridoma supernatants concentrated about 20-25 fold by ammonium sulphate precipitation (50 percent saturation) and then with fluoresceine isothiocyanate (FITC) or horseradish peroxidase (PO) - conjugated with sheep anti-mouse IgG (Pasteur Production, Paris). The peroxidase activity was revealed with 0.01% 3,3'-diaminobenzidine 4-HCl (DAB) and 0.001% H_2O_2 for 10 min.

Double immunostaining of cell cultures : For the double labelling technique in cell culture, after fixation as described above, the cultures were incubated for 1 h at room temperature in PBS containing one of the MAbs H8, H15 and H19 at 1/20 dilution together with a 1/100 dilution of rabbit immune serum directed against GFAP. After washes the cells were then incubated for 1 h in PBS containing both FITC-conjugated sheep anti-mouse IgG and tetramethyl rhodamine isothiocyanate (TRITC)-conjugated sheep anti-rabbit IgG. Tissue sections and cells in culture were examined with a Leitz Orthoplan microscope equipped with epifluorescence attachements.

Preparation of samples for electrophoresis : Fresh membranes and soluble fractions were boiled for 10 min in the buffer system of O'Farrel (1975), containing 2 % SDS and 5 % 2-mercaptoethanol.

10 % polyacrylamide gel electrophoresis (PAGE) in the presence of sodium dodecylsulfate (SDS) : PAGE-SDS was performed on 140 x 140 x 3 mm vertical gel slabs using the buffer systems described by O'Farrell (1975). Electrophoresis was carried out overnight at 80 V. Transfer of the electrophoretic pattern to nitrocellulose sheets (Millipore HA 0.45 μM) was performed according to the Western blot technique as described by Towbin et al (1979). After transfer the nitrocellulose sheets were incubated first with 3 % bovine serum albumine (BSA) for 1 h, then with hybridoma supernatant for 1 h and finally with 1:1000 dilution in PBS of horseradish peroxidase-conjugated sheep antibodies directed against mouse IgG. Peroxidase activity was revealed with 4-chloro-1-naphtol at 0.018 % in PBS and 0.01 % H_2O_2.

RESULTS AND DISCUSSION

Biochemical identification of the antigens recognized by the MAbs H8, H15 and H19 :

As described above, proteins from rat cerebellar soluble and particulate fractions obtained after PAGE-SDS in the presence of mercaptoethanol were transferred to nitrocellulose sheets and immunostained with different MAbs. The following results were obtained :

MAb H8 : This antibody labels two polypeptide bands of molecular weight 68 ± 2 KD and 58 ± 2 KD respectively present exclusively in the blot of the particulate fraction (Fig. 1b) while no labelling is observed in the soluble fraction (Fig. 1a). During cerebellar development, a polypeptide band of molecular weight 62 ± 2 KD is transiently present in the soluble fraction. This band is weak at birth, increases in the blot of cerebellar soluble fraction of 7 day old rats and disappears after the end of the second postnatal week. The immunostain of the polypeptide band of 58 ± 2 KD increases during development while no significant changes are observed for the 68 ± 2 KD band. Preliminary experiments show that the 68 and 58 KD antigens can be solubilized by treatment of the crude membrane fraction with 1 % Triton-X100. Cerebellar particulate fractions prepared

TABLE I

SUMMARY OF THE IMMUNOCYTOCHEMICAL LOCALIZATION OF THE GLIAL-SPECIFIC ANTIGENS RECOGNIZED BY MAbs PRODUCED BY NINE DIFFERENT HYBRIDOMAS

Hybridoma	Immunohistochemical localization in 12 day old rat cerebella		
	Protoplasmic astrocytes	fibrous astrocytes	oligodendrocytes
H8	+++	+	–
H14	+++	+	–
H15	++++	±	–
H19	+++	+++	–
H20	++	++	–
H21	++	++	–
H25	++	++	–
H36	++++	++	–
H89	++	++	–

++++ very strong staining; +++ strong staining; ++ moderate staining; + weak staining; ± very weak staining; – no staining.

and frozen two weeks or more before electrophoresis show several low molecular weight (mainly around 36 KD and 23 KD) bands in 10 % PAGE-SDS. This indicates extensive proteolysis of the 68 and 58 KD proteins which however could not be prevented by adding protease inhibitors such as phenylmethyl sulfonyl fluoride (PMSF) and aprotinin to the homogenization buffer.

MAb H15 : The antibody binds to the antigen very weakly on nitrocellulose sheets : in contrast cerebellar tissue sections strongly immunostained with the same antibody. After SDS-PAGE a major polypeptide band of molecular weight 68 ± 2 KD is stained in both "crude" membrane and soluble fractions (Fig. 1c and d). The weak binding of MAb H15 to its antigens on nitrocellulose sheets can be explained by two possibilities. The first is that the treatment with SDS denaturated or modified the antigenic site recognized by the antibody. This is unlikely, since the immuno-dot binding test with fresh cerebellar fractions without SDS treatment gives similar results. The second explanation is that the binding of the antigen to the nitrocellulose sheet masks totally or partially the antigenic site for the antibody.

MAb H19 : The antibody labels one major protein band of molecular weight 43 ± 2 KD in the particulate fraction and one band of the same molecular weight in the soluble cerebellar fractions (Fig. 1e and f). The membrane bound form is Triton-X100 soluble. A relatively high level of the antigen is already present at birth, increases until the end of the second week and then remains constant.

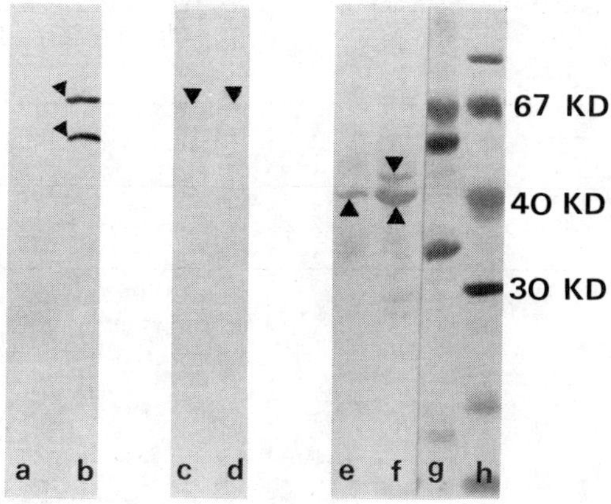

Figure 1 : Western blots of 10 % PAGE-SDS of soluble (a, c, and f) and particulate protein fractions (b, d and e) from adult rat cerebella, immunostained with MAbs H8 (a and b), H15 (c and d) and H19 (e and f). Lanes g and h represent molecular weight standards. Arrow heads show the immunoreactive polypeptides bands with the different MAbs.

Immunocytochemical localization in cerebellar tissue sections :

Using the MAb H8 in 12 day old rat cerebellar sections, the Bergmann fibres in the molecular layer, their end feet on the pia mater and their cell bodies in Purkinje cell layer are immunostained (Fig. 2). The astrocytes in the cerebellar granular layer are also stained (Fig. 3) in particular those astrocytes and their processes surrounding individual or groups of granule cells (Fig. 4). In contrast fibrous astrocytes in the white matter are very weakly stained. Some ependymal cells and the epithelium of choroid plexus are also immunoreactive (not shown). The most intense staining obtained in cerebellum is between the 7th and the 15th day after birth. Adult astrocytes are very weakly stained. Since a relative high level of the antigen is detectable by Western blot in adult cerebellar particulate fraction, the weak staining in adult cerebellar tissue sections could be explained by the fact that in the adult membranes, the antigen or at least the antigenic site is masked by other membrane components. The treatment of fixed cerebellar section with 0.1 - 0.2 % Triton X-100 has no effect on the immunostaining.

The MAb H15 intensely stains the Bergmann fibres in newborn rat cerebellum and other astrocytes in the medullary layer (Fig. 5). Astrocyte processes surrounding Purkinje cell perikarya and in the granular layer are strongly stained at postnatal day 12 (Fig. 6). The labelling becomes diffuse in the forming molecular layer and in the adult molecular layer (not shown). The

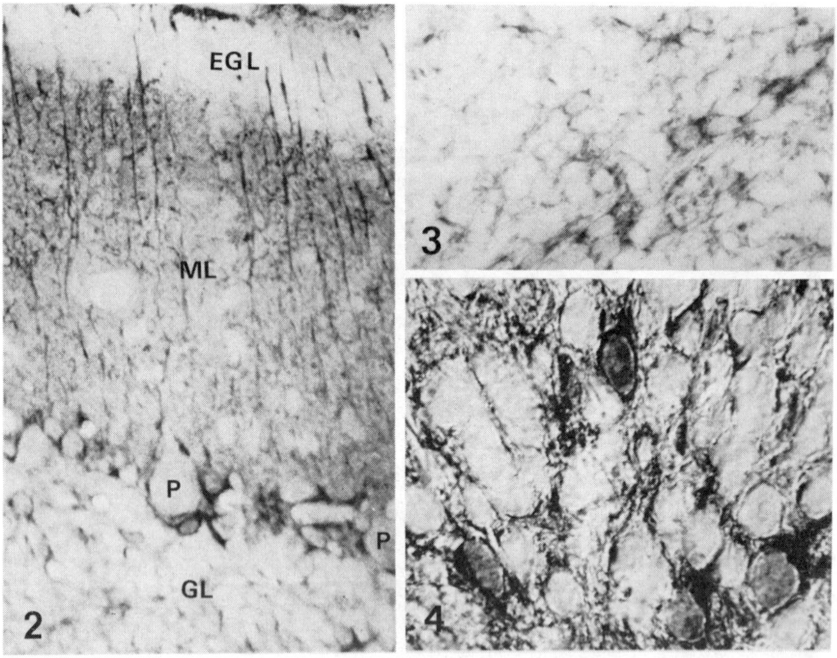

Figures 2, 3 and 4 : Immunohistochemical localization of the antigen(s) recognized by MAb H8 in 12 day rat cerebellar sections.
EGL, external granular layer; ML, molecular layer; P, Purkinje cell; GL, granular layer; WH, white matter.

Fig. 2 : Bergmann fibres in the molecular layer and other astrocytes in Purkinje cell and granular layers are stained (X260).

Fig. 3 : Astrocytes in the granular layer are stained (X260).

Fig. 4 : Interferential phase contrast micrograph showing a number of astrocytes and their processes in the granular layer heavily stained (X1000).

fibrous astrocytes in the white matter, the ependymal cells and the epithelium of choroid plexus are not stained. This antibody can be used as useful marker for protoplasmic cerebellar astrocytes including Bergmann glia. During development the staining of astrocytic perikarya decreases while the staining of astrocyte processes strongly increases.

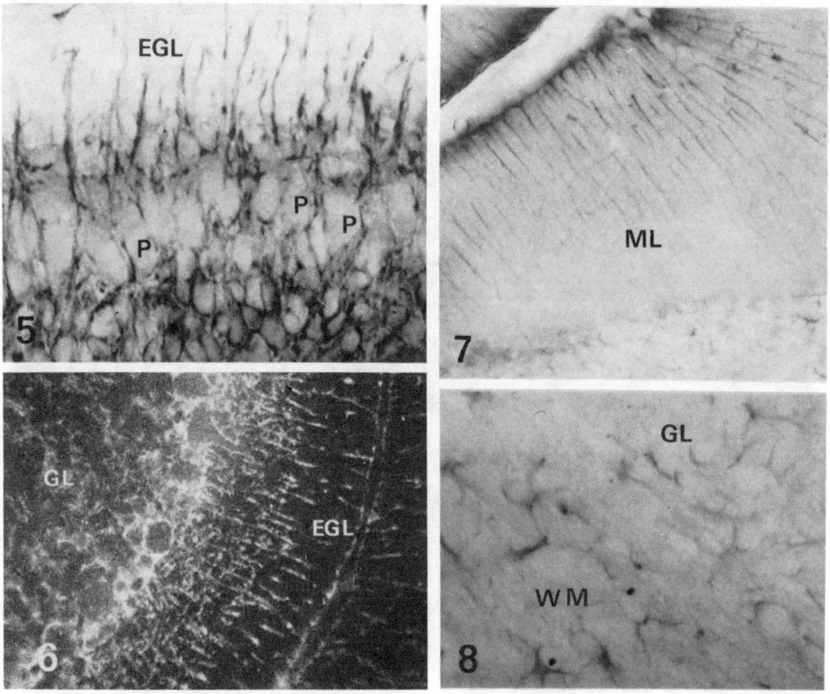

Figures 5, 6, 7 and 8 : Immunohistochemical localization of antigens recognized by MAb H15 (Figs. 5 and 6) and MAb H19 (Figs. 7 and 8).

Fig. 5 : In newborn rat cerebellum, Bergmann fibres and other astrocytes are heavily stained by MAb H15 (X260).

Fig. 6 : 12 day old rat cerebellum treated with MAb H15 : Bergmann glia and their cell perikarya surrounding the Purkinje cell bodies are immunofluorescent (X100).

Figs. 7 and 8 : Adult rat cerebellum treated with MAb H19 : Bergmann fibres, other protoplasmic astrocytes and the fibrous astrocytes in the white matter are labelled (Fig. 7 X100, Fig. 8 X260).

The MAb H19 labels both protoplasmic and fibrous astrocytes. Astrocyte development can be followed using this antibody from newborn cerebellum to the adult. In adult rat cerebellum (Figs. 7 and 8) the astrocyte distribution obtained with MAb H19 is identical to that obtained with GFAP. In addition, ependymal cells and the choroid plexus epithelium are stained (not shown).

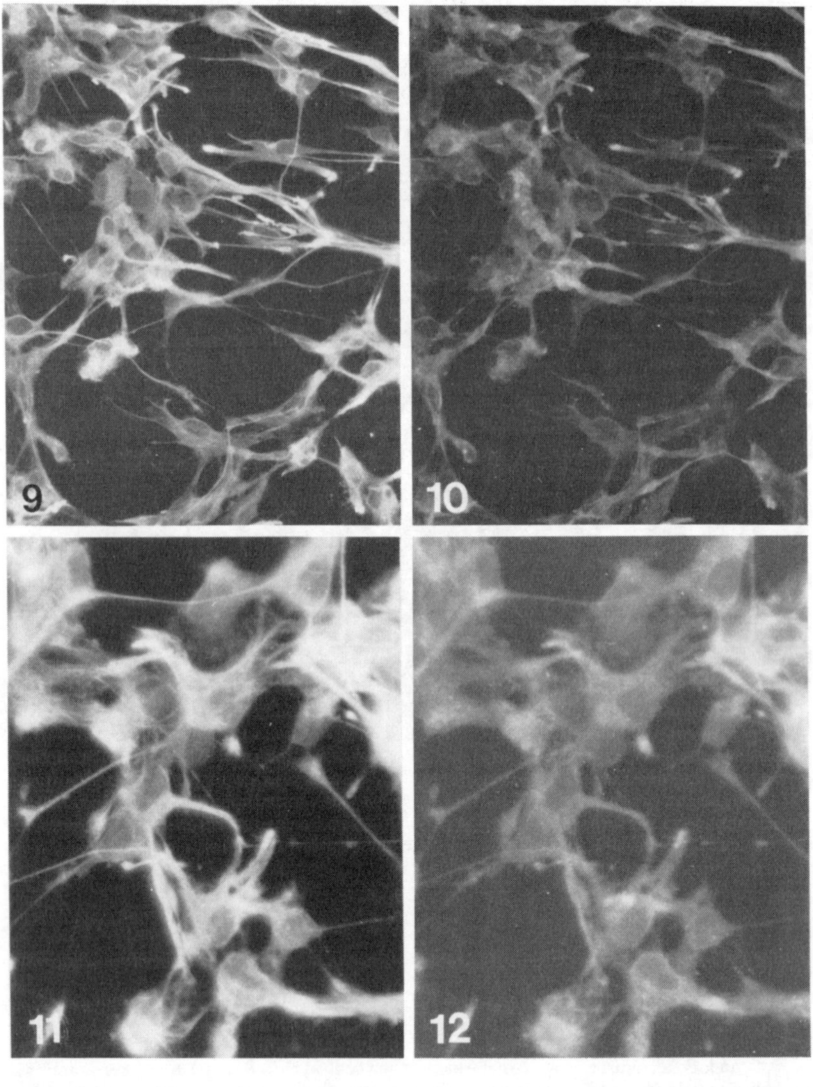

Figures 9, 10, 11 and 12 : Double labelling immunofluorescence staining of 12 day old astrocytes in primary cell cultures using an antiserum directed against GFAP together with either MAb H8 or MAb H19.

Figs. 9 and 10 : Astrocytes stained for the GFAP (Fig. 9) using rabbit antiserum and then TRITC-conjugated sheep anti-rabbit IgG. The same cells were treated with MAb H8 and then revealed with FITC-conjugated sheep anti-mouse IgG (Fig. 10). All the $GFAP^+$ cells are also MAb $H8^+$ (X260).

Figs. 11 and 12 : Astrocytes, fluorescent for GFAP revealed as shown above (Fig. 11) are also fluorescent when treated with MAb H19 (Fig. 12), hence all $GFAP^+$ astrocytes are also MAb $H19^+$ (X 500).

Immunocytochemical localization in astrocytes in primary cell cultures :

Astrocytes in primary cultures are stained with MAb H8 and MAb H19 (Figs 10 and 12). No significant changes in the immunostaining pattern are observed during development from the 7th to the 30th day in vitro except that the staining, particularly for MAb H8 becomes more granular in individual astrocytes. Cultured astrocytes are not stained with MAb H15. This lack of staining may be due to low levels or even to the absence of the corresponding antigen in astrocytes in primary cell cultures.

The double labelling immunocytochemical technique using a polyclonal antibody directed against GFAP shows that all $GFAP^+$ cells (Fig. 9) are also MAb $H8^+$ cells (Fig. 10). Similarly all $GFAP^+$ cells (Figs. 11) are also MAb $H19^+$ cells (Fig. 12).

Most glial cell markers and particularly those of astrocytes available in the literature are considered to be components of the soluble cytoplasm or of the intermediate filaments (for review see Lazarides, 1980, McKay et al, 1981, Ghandour et al, 1983). The advantage of the markers described here is the localization of the antigens recognized by them in the cell membranes. Our preliminary experiments with immunocytochemical staining of living astrocytes in primary cell cultures suggest that the corresponding antigens are partially present or even absent from the surface of astrocytes. These date together with the biochemical data and the immunocytochemistry in cerebellar section indicate that these antigens are accessible on the cytoplasmic face of the astrocytic plasma membrane.

This short description of the antigens recognized by MAbs, H8, H15 and H19 and of their cellular localization illustrate the potential importance of these MAbs for studies on normal astrocyte development. The expression of these astrocytic antigens during abnormal situations and in the CNS of mutant mice might also shed some light on their roles during development.

ACKNOWLEDGEMENT

The authors would like to thank Dr. E. Bock (Protein Laboratory, Copenhagen, Denmark) for the generous gift of immune serum to GFAP, Dr. O.K. Langley for reading the manuscript, Mrs Y. Schladenhaufen for technical assistance and Mrs. S. Boyle-Boiseaux for typing the manuscript. M.S. Ghandour is Chargé de Recherche 1ère Classe and G. Gombos, Directeur de Recherche 2ème Classe au CNRS. This work was partially supported by the Action Concertée DGRST-MIR, Décision n° 82-E-0127 and an INSERM contrat de recherche externe n° 846012.

REFERENCES

Barnstable, C.J. (1980). Nature (Lond.) 286, 231-235.

Booher, J. and M. Sensenbrenner (1972). Neurobiology, 7, 502-522.

Eisenbart, G.S., F.S. Walsh and M. Nirenberg (1979). Proc. nat. Acad. Sci. USA, 76, 4913-4917.

Galfré, G., S. Howe, C. Milstein, G. Butcher and J. Howard (1977). Nature (Lond.) 266, 550-552.

Ghandour, M.S., O.K. Langley, G. Vincendon and G. Gombos (1979). J. Histochem. Cytochem., 27, 1634-1637.

Ghandour, M.S., O.K. Langley, G. Vincendon, G. Gombos, D. Filippi, N. Limozin, C. Dalmasso and G. Laurent. (1980) Neuroscience, 5, 559-571.

Ghandour, M.S., Langley, O.K. and Clos, J. (1983). Int. J. Dev. Neuroscience, 1, 411-425

Ghandour, M.S., B. Foucaud and G. Gombos. (1984). Neuroscience Lett. 51, 119-125.

Hawkes, R., E. Niday and A. Matus (1982). Proc. nat. Acad. Sci. USA, 79, 2410-2414.

Hirn, M., M.S. Ghandour, H. Deagostini-Bazin and C. Goridis (1983). Brain Res. 265, 87-100.

Kohler, G. and G. Milstein (1976). Europ. J. Immunol. 6, 511-519.

Lazarides, E. (1980). Nature (Lond.) 283, 249-256.

Matthew, W.D., L. Tsavaler and L. Reichardt (1981). J. Cell Biol. 91, 257-269.

McKay, R., M. Raff and L.F. Reichardt (1981). Monoclonal antibodies to neural antigens. Cold Spring Harbor Reports in the Neurosciences, Vol. 2.

O'Farell, P.H. (1975). J. biol. Chem. 250, 4007-4021.

Schachner, M., S.K. Kim and R. Zehnle (1984). Develop. Biol. 83, 328-338.

Sternberger, L.W., L. Harwell and N. Sternberger (1982). Proc. nat. Acad. Sci. USA 79, 1326-1330.

Towbin H., T. Staehelin and J. Gordon (1979). Proc. Nat. Acad. Sci. USA 79, 4350-4354.

Zipser, B. and R. McKay (1981). Nature (Lond.) 289, 549-554.

Properties of Cultures Derived from Mature, Isolated Oligodendroglia

W. T. Norton*, M. Farooq, F.-C. Chiu*
and J. E. Bottenstein[+]

*Department of Neurology, Albert Einstein College of Medicine, Bronx, New York 10461, USA
[+]Marine Biomedical Institute and Department of Human Biological Chemistry and Genetics, University of Texas Medical Branch, Galveston, Texas 77550, USA

ABSTRACT

Conditions are described in which bovine oligodendroglia can be obtained in mixed culture with fibroblasts only, with a mixture of fibroblasts and astrocytes, or with astrocytes only. The latter cultures, grown in O1 serum-free medium, can be manipulated to obtain pure cultures of astrocytes.

KEYWORDS

Oligodendrocytes, astrocytes, glial fibrillary acidic protein (GFAP), vimentin, defined media, galactosylceramide, immunofluorescence.

INTRODUCTION

Most _in vitro_ studies of astrocytes and oligodendrocytes have been performed with primary cell cultures obtained from newborn rodent brain using methods based on the pioneering studies of Booher and Sensenbrenner (1972). Such cultures can consist of up to 95% astrocytes together with variable amounts of glial precursor cells and partially to fully differentiated oligodendrocytes. Many examples of the utility of such cultures can be found in the present volume. For reviews of these systems and earlier results see Hertz, _et al_, 1985; Norton, 1983; Pfeiffer, 1982; Pfeiffer, 1983; and Pfeiffer, 1984.

In the past several years we and others have shown that viable oligodendrocytes can be isolated in bulk from mature animals. A high percentage of these cells survive in culture for weeks to months (Szuchet, _et al_, 1980; Lisak, _et al_, 1981; Gebicke-Harter, _et al_, 1981; Norton, _et al_, 1983). Since they are obtained from mature animals they are isolated as fully differentiated cells and appear to retain their differentiated properties _in vitro_. Thus these cultures furnish complementary systems to neonatal primary cultures.

Attempts to culture astrocytes from mature animals have been less successful. Astrocyte-enriched fractions obtained by bulk isolation techniques are not viable and cells dissociated from adult tissue usually grow only reluctantly in culture (Lindsay, _et al_, 1982), although it is possible to obtain from mature animals mixed cultures that contain astrocytes (e.g. Kennedy and Lisak, 1972; Vernadakis, et al, 1984). If, however, prior to dissociation, gliosis is induced in the brain by a stab wound (Lindsay, et al, 1982) or by a kainic acid lesion (Singh and van Alstyne, 1978), then proliferating cultures of astroglia can be obtained.

We describe here properties of cultures derived from purified oligodendroglial preparations isolated from mature bovine brain. Under some conditions astrocytes will proliferate and highly enriched cultures of astrocytes can be obtained.

METHODS

Oligodendroglia were isolated from adult bovine brain white matter by the procedure of Farooq, et al (1981), or from rat forebrain by the method of Snyder, et al (1980). The cells were plated in substrate-coated plastic dishes in MEM plus 15% calf serum. Two substrates were evaluated: polylysine (PL) and an extracellular matrix prepared from a primary astrocyte culture (astrocyte extracellular matrix, AECM). The latter was prepared by allowing primary cultures from neonatal rat brain to grow to confluency and then treating them with Triton X-100 and dilute NH_4OH. Cells were maintained either in MEM plus 15% calf serum or switched to the 01 serum-free defined medium of Bottenstein after varying periods. The 01 medium is described in Dr. Bottenstein's chapter in this volume and contains only MEM, biotin, transferrin, sodium selenite and insulin. Cells were identified by immunofluorescent staining of the specific antigens: galactosylceramide (GC) and myelin basic protein (MBP) for oligodendrocytes, glial fibrillary acidic protein (GFAP) for astrocytes, and fibronectin for fibroblasts, using procedures we have described (Norton, et al, 1983).

RESULTS

Oligodendroglial fractions isolated from adult bovine brain consist initially of >90% oligodendrocytes (Farooq, et al, 1981). We have previously shown that these cells can be plated on PL-coated dishes with an efficiency of 55-80% (Norton, et al, 1983). When grown in serum-containing medium the dishes become filled after two weeks with a continuous mat of rapidly growing fibroblasts and some astrocytes, on top of which grow aggregates of oligodendroglia. When cytosine arabinoside is added at 5 days in vitro (DIV) astrocyte proliferation is completely prevented and fibroblast growth is inhibited, resulting in cultures that are approximately 90% oligodendroglia and 10% large fibroblasts. The oligodendroglia are well dispersed and extend a network of processes. With time the oligodendrocytes tend to aggregate. At 4 months in vitro cell death and detachment become evident. All of the oligodendrocytes are GC and MBP-positive, and GFAP and fibronectin-negative. The fibroblasts are fibronectin-positive and negative for the other antigens (Norton, et al, 1983).

If plated on AECM instead of PL initial cell growth is faster but eventually the cultures resemble those on PL. The AECM effect is much more pronounced with rat oligodendrocytes than with bovine oligodendrocytes.

When bovine cell preparations are plated in MEM plus serum and switched after 3-4 DIV to the serum-free 01 medium a completely different cellular composition is obtained. Under these conditions fibroblast proliferation is completely inhibited and cultures consisting only of astrocytes and oligodendrocytes are obtained (Fig. 1A). The astrocytes are fibrous in appearance, evenly spread with thick processes. The oligodendroglia are also evenly spread and have extensive fine processes. These cells eventually tend to form aggregates. The astrocytes are GFAP-positive and MBP, GC and fibronectin-negative. The oligodendrocytes are MBP and GC-positive and GFAP and fibronectin-negative (Fig. 1B-E). Only rarely is a fibronectin-positive fibroblast detected.

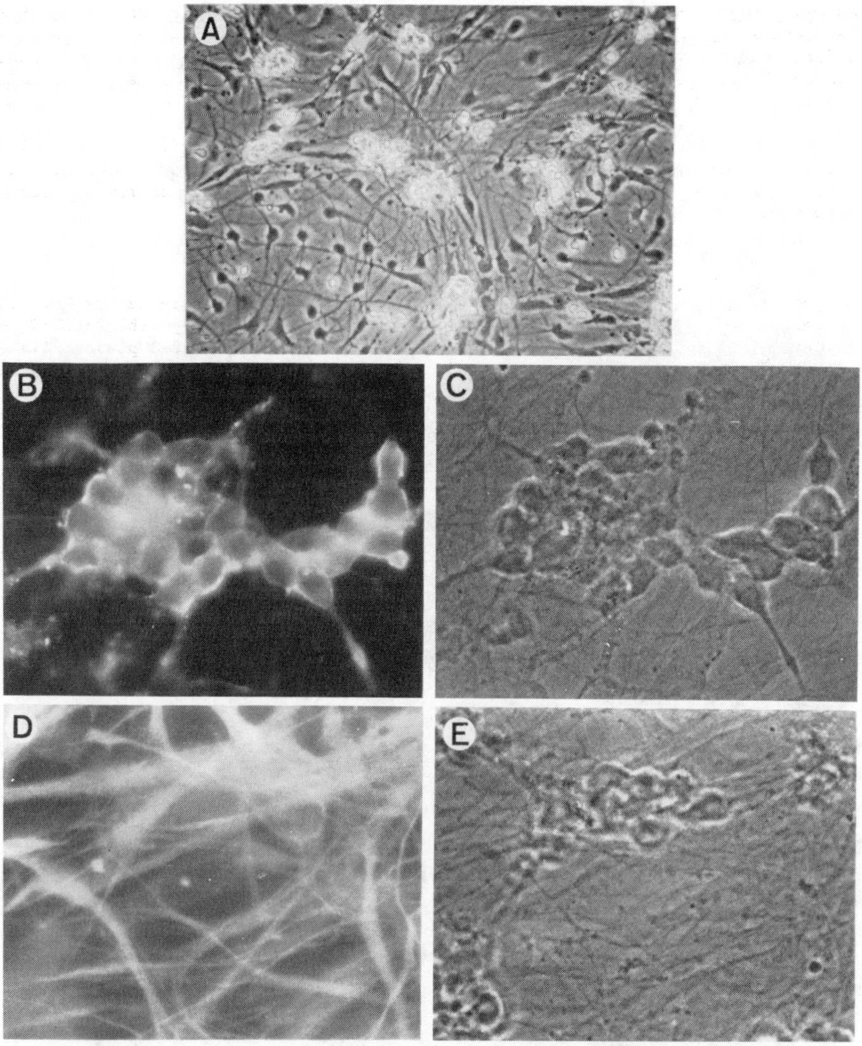

Fig. 1. Bovine oligodendrocyte fractions cultured in 01 defined medium.

A. Cultures grown on AECM for 15 days in 01 defined medium. Phase bright oligodendrocytes, many in aggregates, growing on large, phase-dark astrocytes. (X 210).

B. A similar culture, 23 days in 01 medium. Immunofluorescent staining with anti-MBP. A clump of oligodendrocytes is positive; astrocytes negative. (X 680).

C. Phase micrograph of the same field as B. (X 680).

D. The same culture as B. Immunofluorescent staining with anti-GFAP. The astrocyte fluorescence is partially obscured by clumps of oligodendrocytes growing on top. (X 680).

E. Phase micrograph of the same field as D. (X 680).

These cultures in 01 medium can be purified by selective adhesion to produce nearly pure cultures of adult bovine astrocytes. To accomplish this, cultures, after 2-3 weeks in vitro, are replated with the aid of trypsin-EDTA and the cells allowed to adhere for 75 min. The astrocytes attach to the substrate much more rapidly and firmly than the oligodendrocytes and the latter can be removed by shaking the flasks at 200 rpm for 20 min. The removed cells can be replated to yield an oligodendrocyte-enriched culture. The adherent cells continue to proliferate in 01 medium. These are more than 95% GFAP-positive cells, containing only a few MBP-positive cells and essentially no fibroblasts (Fig. 2). Astrocyte growth would probably be better in G5 medium which contains additional growth factors (Michler-Stuke, et al, 1984), and this is being evaluated.

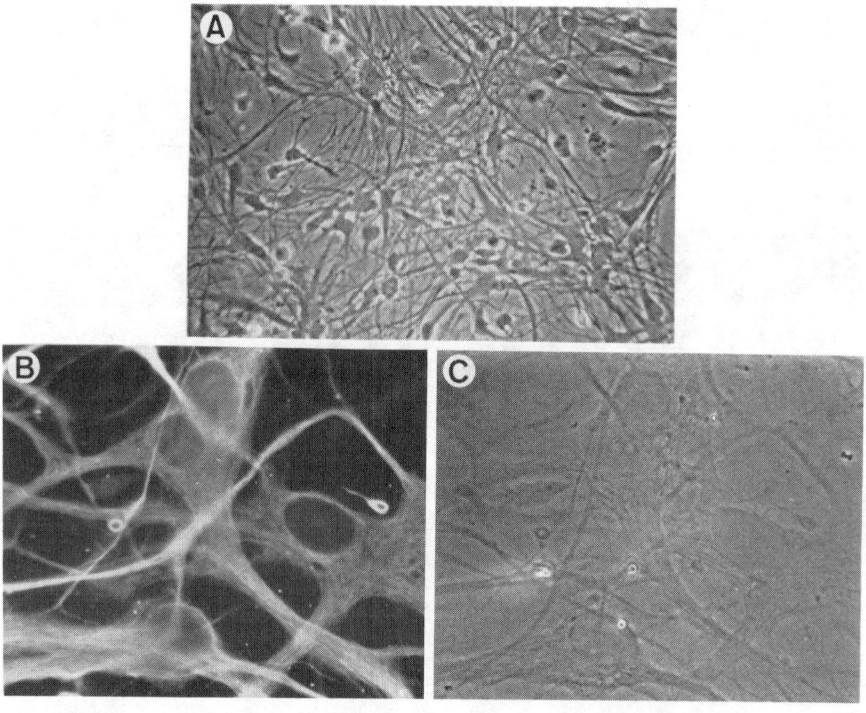

Fig. 2. Purified bovine astrocyte cultures in 01 defined medium.

A. After a total of 43 DIV, oligodendrocytes removed after 28 DIV. Note homogeneity of culture. (X 210).

B. GFAP immunofluorescence of a similar culture after a total of 56 DIV. (X 680).

C. Phase micrograph of the same field as B. (X 680). Note that all cells are GFAP-positive.

One of the characteristics of astrocytes cultured from neonatal rats is that they have much more of the vimentin intermediate filament protein than they do of GFAP (Chiu, _et al_, 1981; Chiu and Goldman, 1984; Goldman and Chiu, 1984). This has been considered an indicator of immaturity. Cultures of the purified bovine astrocytes were treated to obtain cytoskeletal fractions, which were subjected to SDS-PAGE (Chiu, _et al_, 1981). One prominent band at 58,000 daltons and a smaller band at 50,000 daltons were observed along with a much smaller amount of actin at 43,000 daltons (Fig. 3). Immunoblots with specific antisera showed that the 58,000 M.W. band was vimentin and the 50,000 M.W. band GFAP (not shown). Thus the composition of the intermediate filaments of these astrocytes grown from mature animals resembled that found in astrocytes from neonatal animals.

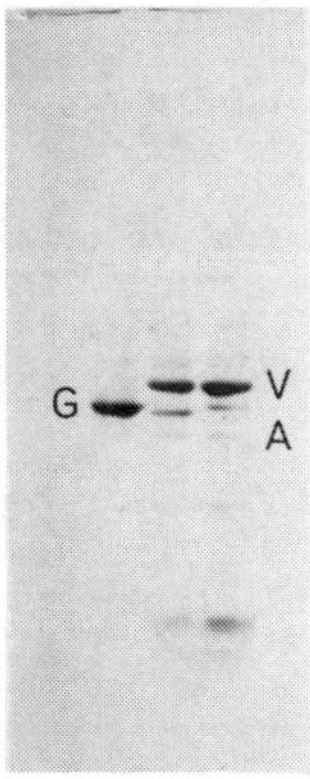

Fig. 3. _SDS-polyacrylamide gel electrophoresis of astrocyte cytoskeletal preparations, stained with Coomassie Blue_.

lane 1. Purified, reassembled GFAP from rat spinal cord, 3 μg (G).
lane 2. Cytoskeletal preparation from cultures of astrocytes purified from mixed cultures obtained from mature bovine brain, 12 μg.
lane 3. Cytoskeletal preparation from rat primary astrocyte cultures (Chiu and Goldman, 1984), 15 μg.

The major band in lanes 2 and 3 is vimentin (V) 58,000 daltons. Note that bovine GFAP (lane 2) has a slightly lower molecular weight than the rat GFAP (lanes 1 and 3) (Chiu, _et al_, 1981). The faint band below GFAP in lanes 2 and 3 is actin (A), 43,000 daltons.

Culture experiments like those described above have also been performed with purified oligodendrocytes from 30 day old rats. They produce similar but not identical cultures, which contain cell types other than astrocytes and oligodendrocytes.

DISCUSSION

Four distinct types of cultures can be obtained from enriched preparations of oligodendrocytes isolated from mature bovine brain. When plated in conventional serum-containing medium the oligodendrocytes send out processes within a week and the cultures remain relatively pure until sometime during the second week *in vitro*, when fibroblasts and astrocytes begin to proliferate rapidly and become the predominant cell types. If cytosine arabinoside is added to such cultures at 5 DIV, astrocytes do not appear and, although fibroblast proliferation is halted, the few fibroblasts present become hypertrophic and can account for a significant percentage of the bulk of the cultures (Norton, *et al*, 1983). If the cultures in serum-containing medium are switched to the serum-free medium at 3 DIV, then mixed cultures of astrocytes and oligodendroglia can be obtained. Thus depending on culture conditions bovine oligodendroglia can be obtained in mixed culture with fibroblasts only, with a mixture of fibroblasts and astrocytes or with astrocytes only. The fourth type of culture is purified astrocytes obtained by selective adhesion after subculture of cells in serum-free medium. The cells that are shaken off these cultures consist of a mixture of astrocytes and oligodendrocytes enriched in the latter cell type. When these are plated the astrocytes continue to proliferate. We have not yet obtained pure cultures of mature oligodendrocytes in long term culture. However, during the first week *in vitro* the bulk-isolated cells remain approximately 90% oligodendroglia. The cultures described here provide a source of oligodendroglia, all of which, unlike those in primary neonatal culture, express both GC and MBP.

These cultures also provide a means to obtain purified cultures of astrocytes in defined medium. The most common marker for the identification of astrocytes, both *in vitro* and *in situ*, is GFAP, an intermediate filament protein of 50,000 daltons. Immature astrocytes in the developing nervous system, however, express vimentin before they express GFAP (Dahl, *et al*, 1981; Dahl, 1981; Schnitzer, *et al*; Bignami, *et al*, 1982; Bovolenta, *et al*, 1984). Vimentin is a fairly ubiquitous intermediate filament protein of 58,000 daltons, found in cells of mesenchymal derivation and in a variety of cells in tissue culture. As astrocytes mature and become fully differentiated most lose vimentin and retain GFAP (Dahl, 1981; Bovolenta, *et al*, 1984) although some, especially in optic nerve, continue to express both proteins into adulthood (Yen and Fields, 1981; Shaw, *et al*, 1981). Therefore we were surprised to find that the cultured astrocytes obtained from mature brain contained more vimentin than GFAP, and thus by this criterion resembled astrocyte cultures obtained from neonatal animals rather than mature astrocytes *in vivo*. It is not yet clear whether the bovine astrocytes arise from a small population of glial precursor cells that do not differentiate fully in culture, or from mature astrocytes, present as contaminants in the oligodendroglial fraction, that partially dedifferentiate in culture.

ACKNOWLEDGEMENT

This work was supported by U.S.P.H.S. grants NS-02476, NS-03356 and NS-20375.

REFERENCES

Bignami, A, T. Raju, and D. Dahl (1982). *Dev. Biol., 91*, 286-295.
Booher, J., and M. Sensenbrenner (1972). *Neurobiology 2*, 97-105.
Bovolenta, P., R.K.H. Liem and C.A. Mason (1984). *Dev. Biol. 102*, 248-259.
Chiu, F.-C., and J.E. Goldman (1984). *J. Neurochem. 42*, 156-174.
Chiu, F.-C., W.T. Norton and K.L. Fields (1981). *J. Neurochem. 37*, 147-155.
Dahl, D. (1981). *J. Neurosci. Res. 6*, 741-748.
Dahl, D., D.C. Rueger, A. Bignami, K. Weber and M. Osborn (1981). *Eur. J. Cell Biol. 24*, 191-196.
Farooq, M., W. Cammer, D.S. Snyder, C.S. Raine and W.T. Norton (1981). *J. Neurochem. 36*, 431-440.
Gebicke-Harter, P.J., H.-H. Althaus, P. Schwartz and V. Neuhoff (1981). *Dev. Brain Res. 1*, 497-518.
Goldman, J.E., and F.-C. Chiu (1984). *J. Neurochem. 42*, 175-184.

Hertz, L., B.H.J. Juurlink and S. Szuchet (1985). In Handbook of Neurochemistry, 2nd ed., Vol. 8 (A. Lajtha, Ed.), Plenum Press, New York, pp. 603-661.

Kennedy, P.G.E., and R.P. Lisak (1980). Neurosci. Letts. 16, 229-233.

Lindsay, R.M., P.C. Barber, M.R.C. Sherwood, J. Zimmer and R. Raisman (1982). Brain Res. 243, 329-343.

Lisak, R.P., D.E. Pleasure, D.H. Silberberg, M.C. Manning and T. Saida (1981). Brain Res. 223, 107-122.

Michler-Stuke, A., J.R. Wolff and J.E. Bottenstein (1984). Int. J. Develop. Neurosci. 2, 575-584.

Norton, W.T. (1983). Adv. Cell. Neurobiol. 4, 3-55.

Norton, W.T., M. Farooq, K.L. Fields and C.S. Raine (1983). Brain Res. 270, 295-310.

Pfeiffer, S.E. (1984). Adv. Neurochem. 5, 233-298.

Pfeiffer, S.E. (Ed.) (1982). Neuroscience Approached Through Cell Culture, Vol. 1, CRC Press, Boca Raton.

Pfeiffer, S.E. (Ed.) (1983). Neuroscience Approached Through Cell Culture, Vol. 2, CRC Press, Boca Raton.

Schnitzer, J., W.W. Franke and M. Schachner (1981). J. Cell Biol. 90, 435-447.

Shaw, G., M. Osborn and K. Weber (1981). Eur. J. Cell Biol. 26, 68-82.

Singh, V.K., and D. Van Alstyne (1978). Brain Res. 155, 418-421.

Snyder, D.S., C.S. Raine, M. Farooq and W.T. Norton (1980). J. Neurochem. 34, 1614-1621.

Szuchet, S., K. Stefansson, R.L. Wollman, G. Dawson and B.G.W. Arnason (1980). Brain Res. 200, 151-164.

Vernadakis, A., D. Mangoura, N. Sakellaridis and S. Linderholm (1984). J. Neurosci. Res. 11, 253-262.

Yen, S.-H., and K.L. Fields (1981). J. Cell Biol. 88, 115-126.

Metabolism of Cultured Oligodendrocytes: Modulation by Cell-Substream Interaction

S. Szuchet

Department of Neurology and The Brain Research Institute, The University of Chicago, Chicago, Illinois 60637, USA

ABSTRACT

The overall metabolism of mature oligodendrocytes was examined as a function of both state of attachment and time in culture. It was found that adherence of oligodendrocytes to a polylysine substratum transduces an "on" signal(s) that initiates the brisk synthesis of myelin components. The latter accumulate over time in culture.

KEYWORDS

Myelin basic protein; proteolipid protein; 2',3' cyclic nucleotide phosphodiesterase; myelin associated glycoprotein; myelinogenesis; glycoproteins.

INTRODUCTION

Most of what is known about the *in vivo* and to a large extent *in vitro* metabolism of oligodendrocytes has been derived from studies on the incorporation of components into myelin (for reviews see Benjamins, 1984; Morell and Toews, 1984). Even though, as pointed out by Morell and Toews (1984), the assumption that is made is non trivial, this approach has proven informative on the myelinogenic properties of these cells. However, such an approach represents a too narrow port to view the biosynthetic repertoire of oligodendrocytes.

Long-term cultures of pure oligodendrocytes afford a good model to study the biology of these cells as they adapt and grow in an artificial environment, deprived of the influences that other central nervous system (CNS) cell types might exert on them. By examining what these cells in isolation *do* or *do not* do under a particular set of experimental conditions one may be able to infer what these influences might be. I have taken such a route (reviewed by Szuchet, 1986).

Adhesion phenomena, be it at the level of cell-cell or cell-substratum, play a pivotal role in regulating cell behavior. What role do cell-cell and cell-substratum interactions play in events preceding, during and following myelination in the CNS is not known. On the other hand, there is ample evidence that these types of interactions are of paramount importance for myelination in the peripheral nervous system. This is so both *in vivo* and *in vitro* (Salzer and Bunge, 1980; Salzer *et al*, 1980a&b).

In vitro, cell-substratum interaction influences cell shape, movement, metabolism, and proliferation (reviewed by Grinnell, 1978; Vasiliev and Gelfand, 1978). The sequence of events that lead to the attachment of cells to substrata has been explored extensively. While such studies have provided an understanding of the mechanics of a change in, say, cell shape, they have failed to give an insight into the molecular mechanisms whereby adhesion processes are transduced into events such as cell differentiation, cell division and/or regulation of cell phenotypes (Letourneau et al, 1980).

Here I present evidence that interaction with a competent substratum influences phenotype expression in cultured oligodendrocytes. Further, I will demonstrate that the signal arising from this interaction turns on the synthesis and post-translational modification of myelin specific components.

OLIGODENDROCYTE METABOLISM

It is the general consensus that oligodendrocytes assemble and maintain myelin. It is also known that the composition of myelin is altered with maturation. Given these two premises, it follows that oligodendrocytes must adapt their metabolism first, for making myelin (myelin-forming cells) and subsequently, for preserving it (myelin-maintaining cells).

Current understanding of the in vitro metabolism of oligodendrocytes has been adequately reviewed (Pleasure et al, 1984; Pfeiffer, 1984; Hertz et al, 1985, 1986; Szuchet and Dumas, 1983; Szuchet, 1986) and therefore will not be further described. Herein, I shall concentrate on those aspects of the metabolism of mature oligodendrocytes that have a bearing on my thesis that adherence of mature oligodendrocytes to a polylysine substratum generates an "on" signal(s) for the active synthesis of myelin components. This change in the metabolism is consistent with the concept that in vitro mature oligodendrocytes undergo a transition towards myelin-forming cells. Indeed cultured oligodendrocytes assemble multilamellar membranes that have the biochemical and ultrastructural characteristics of myelin. Because this takes place in the total absence of neurons, I refer to this process (Szuchet, 1986) as myelin palingenesis (reformation). The work to be described stems largely from my laboratory.

Source of Oligodendrocytes

All the work in my laboratory has been performed with ovine oligodendrocytes isolated from young but post-myelination brains. The procedures for cell isolation (Szuchet et al, 1980a) and culture (Szuchet et al, 1980b) as well as the characterization of these cells (for reviews see Szuchet and Stefansson, 1980; Szuchet and Dumas, 1983; Hertz et al, 1985) have been described in detail. Succinctly, oligodendrocytes are isolated as follows: white matter is trypsinized, disrupted by passage through nylon and metal screens from 350μm down to 30μm; myelin is removed by taking advantage of its low density and the crude cell suspension is resolved on a linear sucrose gradient into three bands (1 → 3 from top to bottom). Only cells from band 3 (Szuchet et al, 1980a) were used for the work described here.

We maintain oligodendrocytes under two sets of experimental conditions (Szuchet and Yim, 1984). I shall be referring to these cells as B3.f and B3.fA.

B3.f: these are oligodendrocytes from band 3 plated onto culture petri dishes and kept for 3-5 days in a floating state since they do not adhere to plastic plates. For control experiments B3.f have been maintained up to 14 days. Morphologically B3.f cells do not differ from freshly isolated cells (Fig.1a); and

B3.fA: these are B3.f oligodendrocytes transferred to polylysine coated plates to which they adhere. After attachment, these cells undergo drastic changes in their morphology (Fig.1b), ultrastructure, and metabolism (see below).

We have kept oligodendrocytes for 90 days and longer with better than 98% purity as judged immunocytochemically with a variety of oligodendrocyte specific markers (Szuchet and Yim, 1984).

Figure 1: Phase-contrast micrographs of live oligodendrocytes.

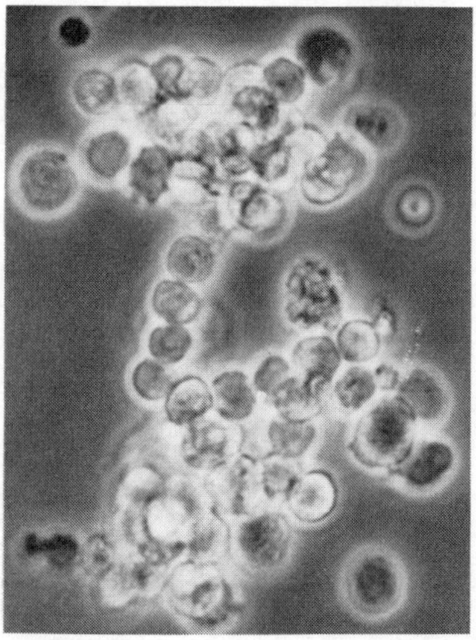

a) B3.f cells after 4 days *in vitro*.

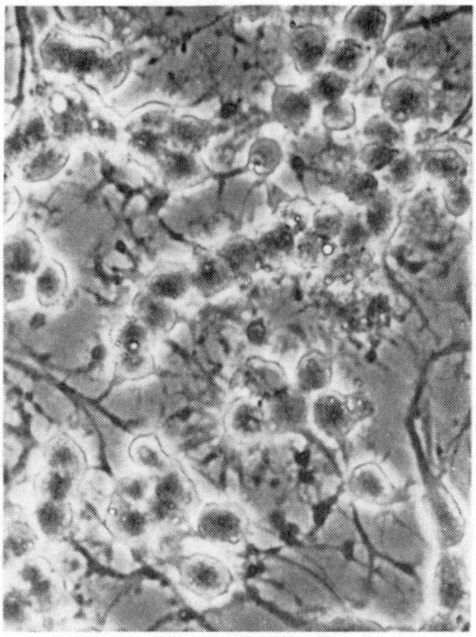

b) B3.fA cells after 25 days in culture. Note network of processes.

Lipids and Glycolipids

The lipid and glycolipid composition of oligodendrocytes differs from that of myelin. Myelin is enriched, relative to oligodendrocytes, in certain phospholipids (e.g., phosphatidylethanolamine; PtdEtn), cholesterol and glycolipids: galactocerebrosides (GC) and sulfatides (Pleasure *et al*, 1984; Morell and Toews, 1984; Norton, 1982). Moreover, as myelin matures, it enriches in these components. This process is measured by two indices: the ratio of GC to phosphatidylcholine (PtdCho) and the ratio PtdEtn:PtdCho, both of which increase with maturation. The period of active myelination is marked by an enhanced synthesis of myelin-specific components that peaks halfway through myelination and then subsides. Once myelination has been accomplished, the synthesis of these components is reduced to levels that are compatible with myelin maturation and maintenance.

Szuchet *et al* (1983) investigated the lipid and glycolipid metabolism of post-myelination oligodendrocytes by following the incorporation of a variety of precursors. These studies were performed on B3.f cells after 4-5 days *in vitro* and on B3.fA cells as a function of time in culture, at weekly intervals for 5 weeks. Their results are summarized in Table I. This is a comparison of the synthesis of phospholipids, neutral lipids, GC and sulfatides by B3.f and B3.fA cells as measured by the incorporation of [^{3}H]glycerol, [^{14}C]OAc, [^{3}H]galactose and $^{35}SO_4^{2-}$, respectively. It is apparent that with the sole exception of neutral lipids (essentially cholesterol), B3.f cells have a low incorporation of these precursors (relative to B3.fA cells) into lipids and glycolipids. Each precursor follows a different time course: during the first week following cell attachment there is a marked augmentation in the synthesis of phospholipids: synthesis of neutral lipids continues to be high while synthesis of glycolipids is only moderate at this time; but after 3 weeks in culture, B3.fA cells have a vigorous uptake of precursors into lipids and glycolipids. When the synthesis of

Table I Incorporation of [^{3}H]glycerol into phospholipids, [^{14}C]acetate into neutral lipids [^{3}H]galactose into galactocerebrosides and $^{35}SO_4{}^{2-}$ into sulfatides by B3.f and B3.fA cells as a function of time in culture.

		B3.fA (weeks in culture)				
Precursors[a]	B3.f	1	2	3	4	5
[^{3}H]glycerol	+	+++	+++	++++	+++	
[^{14}C]acetate	++	++	+++	++++	++++	
[^{3}H]galactose	+	+	+	+		++
$^{35}SO_4{}^{2-}$	+	++	+++	++++		+++++

[a] Cultures were double labeled with carrier-free $H_2{}^{35}SO_4$ at 15μCi/ml (782mCi/mmol); and [^{3}H]galactose at 1μCi/ml (11.5Ci/mmol) or with [2-^{3}H]glycerol at 1.5μCi/ml (500mCi/mmol) and [1-^{14}C]OAc at 2μCi/ml (56.7mCi/mmol) for a total of 72hr and processed as described by Szuchet et al (1983).

individual phospholipids was examined, it was found that the two ratios PtdEtn: PtdCho and GC:PtdCho increased over time in culture. Szuchet et al (1983) correlated the biochemical data with the morphological development of the cultures. They noted that high synthesis of phospholipids and cholesterol coincided temporally with active membrane assembly required for process extension. Differentiation of these newly formed membranes, as assessed by the incorporation of specific glycolipids, occurred at a time when a network of processes was already formed.

Recently, Vartanian, Szuchet, and Dawson (in preparation) extended these studies to include up to 9-week-old cultures; they also performed a more detailed characterization of the components synthesized than had been done by Szuchet et al (1983). Thus, the former authors were able to show that: a) significant changes in the lipid and glycolipid metabolism of B3.fA cells were already apparent 24h after oligodendrocyte attachment; b) ~40% of the PtdEtn synthesized was in the form of plasmalogen. The latter finding further stresses the fact that the in vitro system closely parellels the behavior of oligodendrocytes in situ (Norton, 1982).

In summary, it can be stated that there are changes in the lipid and glycolipid metabolism of cultured oligodendrocytes that comprise a shift toward the synthesis of myelin components. These changes take place following cell adhesion to a substratum as well as with time in culture.

Proteins

Yim et al (1986) followed the protein metabolism of oligodendrocytes as a function of the state of adherence (i.e.; B3.f and B3.fA) and over time in culture for 8 weeks. Cultured oligodendrocytes take up labeled amino acids; e.g., [^{3}H]leu, [^{35}S]met or [^{35}S]cys; into a number of polypeptide chains that cover a spectrum of apparent molecular weights from 240,000 down to 14,000. The highest uptake was seen in a component with M_r=44,000 followed in decreasing order by M_r=59,000, 56,000, 30,000, etc. (Fig.2). This was so at all times examined and held for both B3.f and B3.fA cells. Quantitative analysis of the fluorograms by laser densitometry revealed that synthesis of most proteins was essentially (within experimental errors) unaffected by the state of cell attachment or time in culture. Thus from the perspective of the metabolism of these proteins and with the proviso that this conclusion may only be valid within the limits of

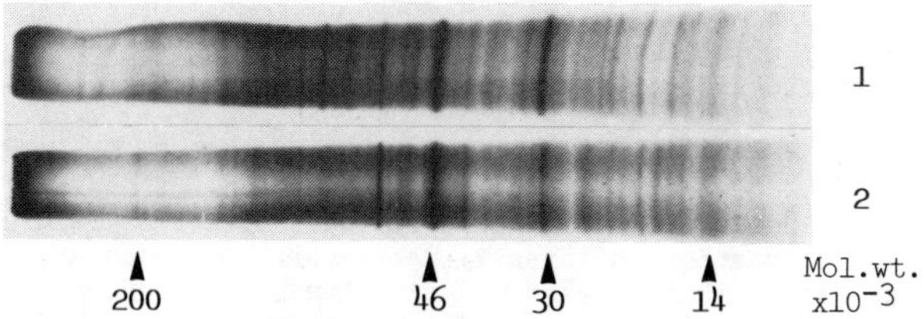

Figure 2: Incorporation of [^{3}H]leu by B3.f and B3.fA cells, 4d after isolation and at weekly intervals thereafter for 8 weeks, oligodendrocyte cultures were labeled with [^{3}H]leu for a total of 16-20h. Cell homogenates were resolved on discontinuous 7-25% gradient SDS-PAGE and polypeptide chains detected by fluorography. Lane 1, B3.f cells 4d; lane 2, B3.fA after 40d.

resolution of a one-dimensional gel electrophoresis, B3.f and B3.fA cells behave similarly. However, a restricted number of proteins was uncovered, whose expression by B3.f cells was significantly different from that of B3.fA cells. These components share two common characteristics: a) with two exceptions, they have M_r<30,000; b) major changes in their metabolisms occur in the first week following cell attachment (Fig.3).

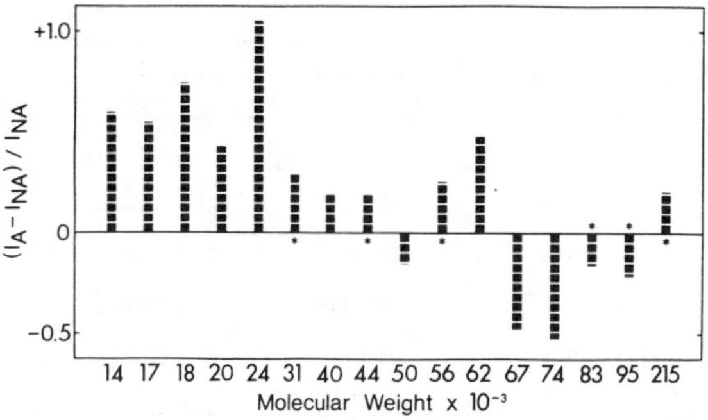

Figure 3: Normalized incorporation. Data from Fig.2 have been normalized as $(I_A-I_{NA})/I_{NA}$ where I_A is the average incorporation into each component by B3.fA cells and I_{NA} represents the the same average but for B3.f cells. Note that zero indicates "steady state." Unmarked bars denote >95% certainty; *, 90-55%. Components with M_r=18K, 21K, 24K, 43-45K, and 56K are MBP, DM-20, PLP, CNPase-actin and tubulin, respectively.

Yim et al (1986) identified a number of the synthesized polypeptide chains by combining the techniques of fluorography with immunoblotting (Towbin et al, 1979) and using a barrage of antisera against myelin and cytoskeletal proteins. With this approach they were able to localize MBP, CNPase, MAG, actin, α-actinin and tubulin (Table II). Because specific antisera against PLP and DM-20 were not available, these authors resorted to a different strategy in order to detect if these proteins were synthesized by cultured oligodendrocytes. PLP and DM-20 were identified as the [^{35}S]cys-labeled polypeptide chains that were also acylated by [^{14}C]palmitic acid, co-migrated with sheep myelin PLP and DM-20 on SDS-polyacrylamide gel electrophoresis, and had M_r=24,000 and 21,000 respectively. The results are outlined in Table II. It is of note that prominent among the components that undergo significant changes in their expression after cell adhesion and/or over time in culture figure myelin-specific proteins. These observations led Yim et al (1986) to conclude that the myelinogenic properties of B3.f and B3.fA cells are distinct.

B3.f and B3.fA cells are distinguished from each other by two factors: a) their physical state: i.e., floating vs. attched; b) time in culture. To ascertain whether one or both of these factors are responsible for the observed differences in the myelinogenic properties, we examined if keeping B3.f cells for an extended time in culture (i.e., beyond 5d) would influence their synthesis of MBP

(Vartanian et al, 1986), PLP and DM-20 (Yim et al, 1986). It emerged from these studies that time in culture alone does not change the behavior of B3.f cells but attachment to a substratum does. Thus, adherence surfaced as a major factor that signals the enhanced synthesis of myelin proteins. For example, within the first 24h following attachment of B3.f cells to polylysine coated plates there is a doubling in the rate of synthesis of MBP whereas prolonging the maintenance of B3.f cells for an equivalent time does not have any measurable effect (Vartanian et al, unpublished). However, there is a net accumulation of MBP in B3.fA cells over time in culture. Since myelin proteins undergo post-translational modifications; e.g.; MBP and CNPase are phosphorylated; PLP and DM-20 are acylated; the above studies were extended to investigate the effect of adhesion to substratum and time in culture on these events. Again it was adhesion of substratum rather than time in culture that triggered the post-translational modifications of myelin proteins. An interesting case is that of CNPase: B3.f and B3.fA cells synthesize CNPase to nearly the same extent, yet phosphorylation of CNPase is effected only by B3.fA cells. These results, taken in conjunction with the observations on the lipid and glycolipid metabolism of these cells, strengthen the notion that following their attachment to a substratum, oligodendrocytes turn on the biosynthetic machinery for the synthesis of myelin components.

Table II Identification of proteins synthesized by B3.f and B3.fA cells.[a]

Proteins[b]	B3.f[c]	B3.fA[d]	$M_r x10^{-3}$	Source
Myelin-specific				
MBP	+	+++	18;23	AT Campagnoni
CNPase	+++	+++	43;45	TJ Sprinkle
MAG	++	+++	100;110	SL Miller[e]
PLP	+	+++	24	see text
DM-20	+	+++	21	see text
Cytoskeletal				
Actin	++	++	44	Y Schoenfeld[c]
α-Actin	++	++	100	see text
Tubulin	+++	+++	56	F Luca[c]

[a] Adopted from the data of Yim et al (1986). Note that number of crosses represent relative quantities; i.e., B3.f vs. B3.fA for each protein. Two proteins with an equal number of crosses does not imply equal amounts.
[b] Abbreviations: MBP: myelin basic protein; CNPase: 2',3', cyclic nucleotide phosphodiesterase; MAG: myelin associated glycoprotein; PLP: proteolipid protein.
[c] In culture from 3 to 14d.
[d] Attached from 5 to 56d.
[e] Monoclonal antibodies, all others are polyclonal.

Glycoproteins

Administration of [^{3}H]fucose to oligodendrocyte cultures results in its uptake into a discrete number of glycoproteins (Fig.4); the major components span a range of molecular weights from ∿113,000 down to 78,000, although minor components outside this spectrum are clearly discernable. This pattern appears simple when compared to the pattern that is obtained when protein precursors such as leu or met are used (cf. Fig.4 to Fig.2). We followed the fate of glycoproteins in oligodendrocyte cultures for a period of 8 weeks (Szuchet and Yim, 1984). The total incorporation of [^{3}H]fucose per mg protein at each time point measured, normalized relative to the uptake at 5d (i.e., B3.fA:B3.f), increased with time up to about 35d _in vitro_ and then declined. Addition of tunicamycin abolished the incorporation of [^{3}H]fucose indicating that: a) the carbohydrate moiety is N-linked, and b) this is a case of _de novo_ synthesis.

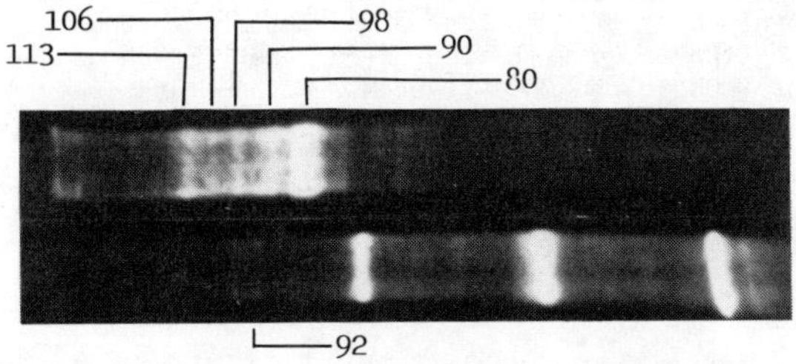

Figure 4: Autofluorogram showing fucose incorporation by cultured oligodendrocytes. Cell homogenate was resolved on a 7-25% gradient SDS-PAGE. Numbers give apparent mol.wts.X10^{-3}.

Figure 5 illustrates the sequential changes in the incorporation of [^{3}H]fucose that occur over a period of 8 weeks. Starting at 5d (B3.f cells) it is seen that a group of 6 major well-resolved components are fucosylated (numbered 4 to 11); they represent a consecutive series of polypeptide chains with no more than the 35,000 difference in the apparent mol.wts. between members at each extreme. Two of these bands are rather broad, with an indication that they are doublets (8-9; 10-11). The intensity of label appears to be comparable in all of the bands except bands 10-11 which are distinctly darker than the others (Fig.4). This could be taken to mean higher (relatively speaking) rates of synthesis or alternatively high fucose content. The expression of these components changes over time (Fig.5).

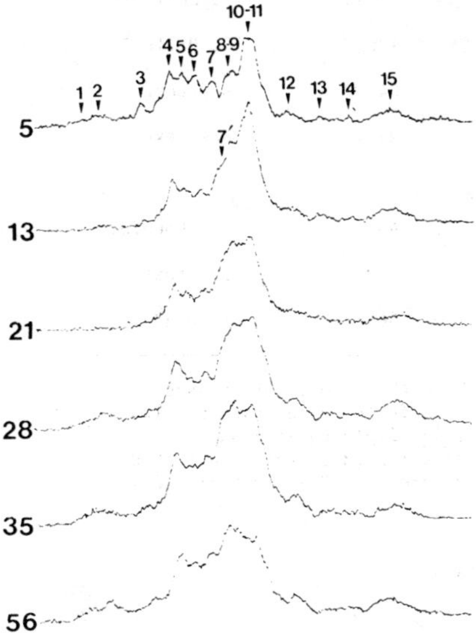

Figure 5: Densitometric scans of fluorograms from [^{3}H]fucose incorporation by oligodendrocytes as a function of time _in vitro_. Numbers on the left are days in culture. Numbers on top identify components that were resolved unambiguously.

The glycoprotein composition of CNS myelin has been studied extensively (Quarles, 1979). CNS myelin is notorious for its scanty content of glycoproteins: a major glycoprotein, MAG, and a number of minor glycoproteins (Poduslo _et al_, 1977; Zanetta _et al_, 1977). Two of these have been identified; a fucosylated component migrating with the same apparent mobility as proteolipid protein (M_r≈25K); the other localized in the Wolfgram protein zone, 42-46K. In accordance with this, when the <u>in vivo</u> incorporation of labeled fucose was followed by examining the fucosylated compounds in a myelin fraction, a profile dominated by a major peak with some minor ones was obtained; the major peak was identified as MAG. This pattern is different from those shown in Fig.5. When we used an anti-MAG monoclonal antibody (Miller <u>et al</u>, 1984) to identify MAG on immunoblots of [^{3}H]fucose labeled compounds, a single band (M_r=100,000) corresponding in position to component 5 (Fig.5) was stained on the fluorograms from B3.f cells; but two distinct bands (M_r=100,000 and 110,000) overlapping with bands 4 and 5 (Fig.5) were recognized by this antiserum on flurograms from B3.fA cells. The intensity of staining increased with time in culture. It should be noted that a shift (decrease) of ∿10K in the mol.wt. of MAG with maturation of rat myelin has been reported (Quarles <u>et al</u>, 1973). Two MAG-positive bands were also found in sheep myelin (unpublished observations). The fact that upon attachment of oligodendrocytes, both the "immature" and "mature" forms of MAG were expressed whereas only the "mature" form was found in floating cells is in line with the findings on the lipid and protein metabolism.

Secretory Glycoproteins

If, after labeling oligodendrocyte cultures with [^{3}H]fucose for 16h in a medium that is free of serum proteins, the medium is examined for the presence of [^{3}H]fucose-labeled proteins, 4 major components are found that have apparent mol. wts. from 125,000 down to 25,000 (Fig.6). The same polypeptide chains are detected if [^{3}H]leu is used as the precursor in lieu of [^{3}Hfucose, suggesting that all secreted proteins are glycoproteins (Fig.6). Moreover, the secreted glycoproteins appear to constitute a class of macromolecules that is different from the fucosylated proteins that are found in the cell homogenates (cf. lanes 3&4, Fig.6).

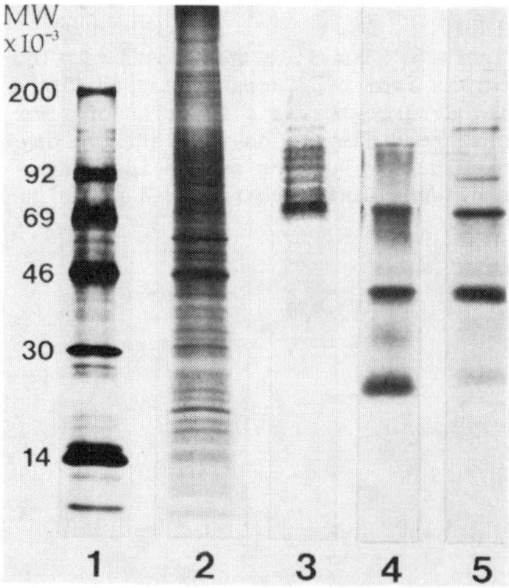

Figure 6: Autofluorograms showing [^{3}H]leu and [^{3}H]fucose labeled polypeptide chains secreted to the medium by cultured oligodendrocytes. Numbers give apparent mol.wts. Lanes: 1, mol.wt.std.; 2&3, [^{3}H]leu and [^{3}H]fucose labeled polypeptide chains in cell homogenate, respectively; 4&5, [^{3}H]fucose and [^{3}H]leu labeled proteins secreted to the medium, respectively.

Post-translational sulfation is emerging as a mechanism of routing proteins destined for secretion (Huttner, 1982; Hille <u>et al</u>, 1984). We have, therefore, investigated (Szuchet and Yim, unpublished) if the secreted proteins are also sulfated by labeling cultures with $^{35}SO_4{}^{2-}$. The results showed that all the glycoproteins found in the medium were also sulfated (Fig.7), an indication that they are bona fide secretory glycoproteins and not adventitious products of membrane shedding. A single sulfated band was detected in the cell homogenate, its M_r suggests that it is MAG.

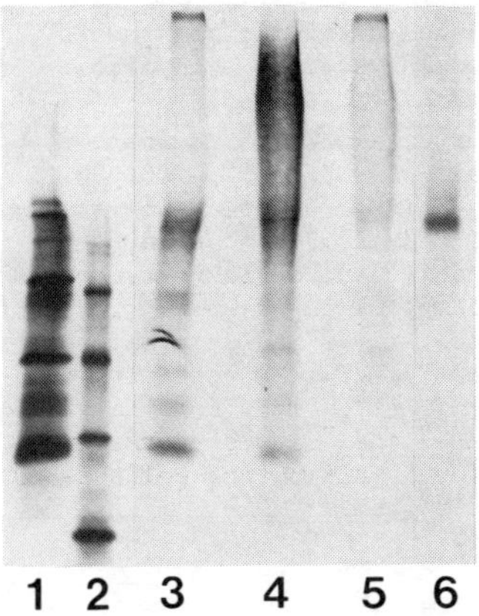

Figure 7: Incorporation of $^{35}SO_4^{2-}$ into proteins by cultured oligodendrocytes. Cells were labeled for 16h in a serum-free medium. Medium was collected, dialyzed exhaustively to remove excess label, freeze-dried and resolved on 7-25% gradient SDS-PAGE. Lanes: 1, [^{3}H]leu-secretory proteins; 2, mol.wt.std.; 3,4& 5, $^{35}SO_4^{2-}$-fluorogram of secretory proteins from 63d, 10d and 4d-old cultures, respectively; 6, $^{35}SO_4^{2-}$-fluorogram from cell homogenate.

It was of interest to examine if the pattern of secretion had a temporal component to it. For this, we followed secretion as a function of time in culture using both [^{3}H]fucose and [^{3}H]leu as precursors. A polypeptide chain with M_r=45,000 is of late appearance: it is barely seen in cultures that are less than 15d old but becomes prominent at later time points. This is so for both labels suggesting that it is a case of delayed synthesis (Fig.8). Preliminary data indicate that sulfation may be a better marker than either fucose or leu to follow the time and/or substratum modulation of secretion.

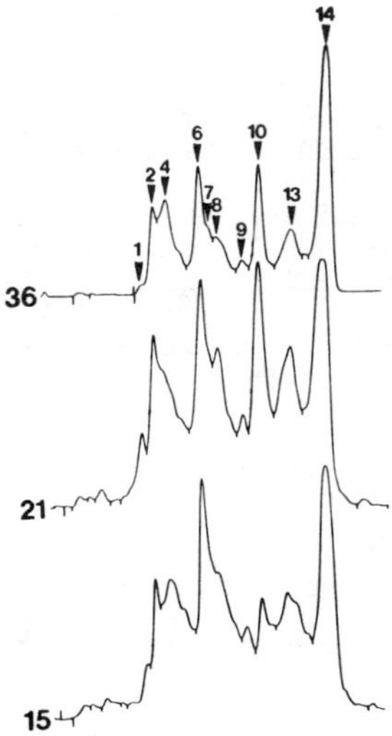

Figure 8: Densitometric traces of [^{3}H]fucose-fluorograms showing secretory proteins. Numbers on left give time in culture. Note the change in the expression of component 10 with time in culture.

Two pathways for secretion have been described: regulated and constitutive (Kelly, 1985). From the fact that the majority of secretory glycoproteins are not detected in the cell homogenate, it can be concluded that these proteins are not concentrated in the cell cytosol and therefore are released constitutively. The possible role of these proteins is now under investigation.

CONCLUSIONS

1. mature oligodendrocytes have the potential to regenerate;

2. if plated under favorable conditions they recapitulate myelinogenic events;
3. cell-substratum interaction appears to transduce the "on" signal(s) for the synthesis of myelin components;
4. over time in culture, there is an accumulation of myelin components;
5. cultured oligodendrocytes afford a good model system to study the biology of these cells and the manner in which they assemble myelin.

ACKNOWLEDGEMENT

I thank Dr. S.H. Yim, Messrs. P.E. Polak, T. Vartanian and M. Lang for their contributions. Supported by grants RG-1223-C4 from the Natl MS Soc and BNS 8304177 from NSF. Thanks are due to Ms. T.L. Smith for typing this manuscript.

REFERENCES

Benjamins JA (1984). Adv Neurochem 5, 87-124.
Grinnell F (1978). Int Rev Cytol 53, 67-129.
Hertz L, BHJ Juurlink, and S Szuchet (1985). In: Handbook of Neurochemistry (A Lajtha, ed), Plenum Publishing Corporation, New York, pp.603-661.
Hertz L, BHJ Juurlink, S Szuchet and W Walz (1986). Neuromethods: Neurochemistry I, General Techniques (AA Boton, CB Baker, eds), Humana Press, Clifton, New Jersey, pp.117-167.
Hille A, P Rosa and WB Huttner (1984). FEBS 177, 129-134.
Huttner WB (1982). Nature 299, 273-276.
Kelly RB (1985). Science 230, 25-32.
Letourneau PC, PN Ray and MR Bernfield (1980). In: Biological Regulation and Development (RF Goldberger, ed), Plenum Press, New York, pp.339-376.
Miller SL, D Pleasure, M Herlyn, B Atkinson, C Ernst, TG Tachovsky and L Baird (1984). J Neurochem 43, 394-400.
Morell P and AD Toews (1984). Adv Neurochem 5, 47-86.
Norton WT (1981). Adv Neurol 31, 93-121.
Pfeiffer SE (1984). Adv Neurochem 5, 233-298.
Pleasure D, SU Kim and DH Silberberg (1984). Adv Neurochem 5, 175-196.
Poduslo JF, JL Everly and RH Quarles (1977). J Neurochem 28, 977-986.
Quarles RH (1979). Complex Carbohydrates of Nervous Tissue (RU Margolis, RK Margolis, eds), Plenum Press, New York, pp.209-233.
Quarles RH, JL Everly and RO Brady (1973). J Neurochem 21, 1177-1191.
Salzer JL and RP Bunge (1980). J Cell Biol 84, 739-752.
Salzer JL, AK Williams, L Glaser and RP Bunge (1980a). J Cell Biol 84, 753-766.
Salzer JL, RP Bunge and L Glaser (1980b). J Cell Biol 84, 767-778.
Szuchet S (1986). Glial-Neuronal Communication in Development and Regeneration (A Althaus, W Seifert, eds), Plenum Press, New York, in press.
Szuchet S and M Dumas (1983). Multiple Sclerosis (JP Antel, ed), Neurology Clinics of North America, Saunders, New York, pp.729-755.
Szuchet S and K Stefansson (1980). Adv Cell Neurobiol 1, 314-346.
Szuchet S and SH Yim (1984). J Neurosci Res 11, 131-144.
Szuchet S, BGW Arnason and PE Polak (1980a). J Neurosci Methods 3, 7-19.
Szuchet S, K Stefansson, RL Wollmann, G Dawson and BGW Arnason (1980b). Brain Res 200, 151-164.
Szuchet S, SH Yim and S Monsma (1983). Proc Natl Acad Sci (USA) 80, 7019-7023.
Towbin H, T Staehelin and J Gordon (1979). Proc Natl Acad Sci (USA) 76, 4350-4354.
Vartanian T, SH Yim, S Szuchet and G Dawson (1986). Biochemistry of Glial Cells (T Grisar, ed), Pergamon Press, New York, this volume.
Vasiliev JM and IM Gelfand (1978). Int Rev Cytol 50, 159-274.
Yim SH, S Szuchet and PE Polak (1986). Submitted.
Zanetta JP, MS Ghandour, G Vincendon, R Eberhart, LL Sarlieve and G Gombos (1977). J Neurochem 29, 359-363.

Developmental Changes in the Glycoproteins of Nerve Cells in Culture as Probed with Lectins

J. Chapman and M. Rumsby

Department of Biology, University of York, York YO1 5DD, UK

ABSTRACT

Changes occurring in the plasma membrane glycoproteins of cultured nerve cells during growth and development are being probed with lectins. Glycoproteins, resolved by SDS-PAGE, are transferred on to nitrocellulose to increase the sensitivity of lectin probing. Further sensitivity is obtained by the use of biotinylated lectins and a Streptavidin-peroxidase conjugate. Lectin binding by nerve cell glycoproteins at 4 days post seeding is weak but more lectin-binding components appear as cells grow in culture for over four weeks.

KEYWORDS

Glycoproteins, nerve cells, culture, lectins, biotin-Streptavidin, growth, development.

INTRODUCTION

Different cell classes in the central nervous system interact closely with each other in contacts which are vital for function. Interactions between cells at their external surfaces are through carbohydrate-containing lipids and proteins on one cell surface interacting with, perhaps, complementary carbohydrate-binding proteins on an adjacent cell surface (Barondes, 1981). Glycoproteins and glycolipids at cell surfaces undergo changes during cell growth and differentiation; such cell surface groups are also unique to a particular cell class. Developmental changes in the gangliosides of the CNS have been characterised to some extent (e.g. Hilbig et. al., 1984). Less is known however about individual glycoproteins of nerve cells. As an approach to this problem we are looking at the surface glycoproteins of nerve cells in culture and how such components change during growth and development. Total particulate membrane preparations from cell cultures (Walker et. al., 1984) should contain all external surface membrane glycoproteins. Most of the carbohydrate-containing proteins in cells are located at the surface.

METHODS

Primary cultures of rat brain cells are grown as described by Walker et al.(1984) for periods up to 32 days. Total particulate preparations are prepared from cell homogenates in water at 4, 11, 18 and 31 days after culturing always from the same batch of cells. Proteins, resolved by SDS-PAGE on 6-30% gels, are transferred to nitrocellulose by electroblotting. Blots are then blocked with PBS/ 0.05% Tween 20/ 2% bovine serum albumin(IgG-free) and are then probed with biotinylated lectins (1μg/ml, Sigma). After repeated washing lectin binding is detected with peroxidase conjugated to Streptavidin(Amersham Int.,1:4000 dilution in PBS/Tween). Diluents and wash medium contain 1mM Ca^{2+}, Mg^{2+} and Mn^{2+} (as chlorides) to improve lectin binding (Leathem and Atkins, 1983).

RESULTS AND DISCUSSION

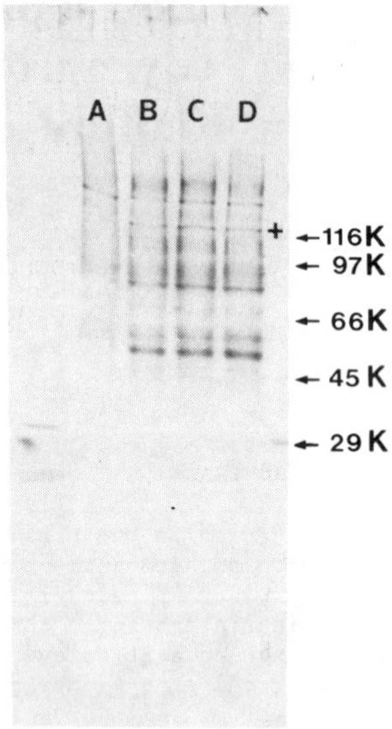

FIG.1. Con A-binding glycoproteins in total particulate preparations of brain cells in culture after resolution by SDS-PAGE and probing with biotinylated-Con A as described under Methods. Lanes A,B,C and D are brain cells harvested at 4, 11, 18 and 31 days after seeding. Equal amounts of membrane protein were applied to each lane. Arrows indicate positions of molecular weight markers.

The high affinity of the biotin-avidin reaction allows the use of very low concentrations of biotinylated lectins to probe blots. Further, the peroxidase-Streptavidin conjugate can also be used at very low dilutions giving a further saving on reagents - with no loss of sensitivity. We find that the presence of 2% albumin (IgG-free) in the blocking medium is important in reducing the artifactual binding of lectins to proteins and in minimising background staining. Some lectins (e.g. Con A) have a broad carbohydrate specificity. For accurate interpretation of the carbohydrate content of glycoproteins, therefore, a range of carefully selected lectins have to be used. In the initial results shown in Fig. 1 it is clear that several Con A-binding glycoproteins are present in the nerve cell total particulate preparations; several prominent bands, as well as some minor ones, are present. Though equal amounts of protein (3.5μg) have been applied to each lane it is clear that cells from 4 day cultures contain 1, fewer Con A-binding components and 2, reduced levels of those Con A components which are also present at later stages of growth. Binding of Con A indicates the presence of glycoproteins with mannose and glucose residues and perhaps GlcNAc but other lectins with narrower specificity are being used to define these bands more precisely. Use of the GalNAc-specific lectin Vicia villosa reveals that several GalNAc-containing glycoproteins are present. The Con A-binding component with a molecular weight of about 120,000(Fig.1,*) is also detected with Vicia villosa indicating the presence of GalNAc residues. With all lectins tested so far glycoproteins in 4 day old cells are markedly reduced in concentration compared with later cultures indicating an effect of growth and brain cell differentiation on glycoprotein expression.

REFERENCES

Barondes,S.H.(1981). Ann.Rev.Biochem. 50, 207-231

Hilbig,R.,G.Lauke and H.Rahmann (1984). Dev.Neurosci. 6, 260-270.

Leathem,A.J.C. and N.J.Atkins (1983). In Techniques in Immunocytochemistry,Vol.2 (Ed.G.R.Bullock and P.Petrusz) Academic Press, London, pp.39-70.

Walker,A.G.,J.Chapman,C.B.Bruce and M.G.Rumsby(1984/85).J.Neuroimmunol.7, 1-20.

Cyclic AMP Enhances the Stability of Microtubules in Cultured Astrocytes

J. Ciesielski-Treska, J. F. Goetschy, G. Ulrich and D. Aunis

Unité INSERM de Neurochimie Normale et Pathologique (U44), Centre de Neurochimie, 5 rue Blaise Pascal, 67000 Strasbourg, France

ABSTRACT

Present results demonstrate that the proportion of cold/Ca^{2+}-stable microtubules of astrocytes in primary culture increases after long exposure to dibutyryl-cyclic AMP.

KEY WORDS

Astrocytes ; short and long effect of dibutyryl-cyclic AMP ; cold/Ca^{2+}-labile microtubules ; stable microtubules.

Several observations indicate that microtubules (MT) are involved in the elaboration and maintenance of processes of astrocytes cultured in the presence of dibutyryl-cyclic AMP (dbcAMP). Astrocyte processes contain arrays of MT that are oriented parallel to their axis, and colchicine which promotes MT depolymerization inhibits the extension of processes (Ciesielski-Treska, et al, 1982). After combined treatment with dbcAMP and taxol, which promotes the polymerization of MT and stabilizes them, astrocyte processes are unusually broad and short. In early stages of the response to dbcAMP, astrocyte processes retract within several hours after dbcAMP withdrawal but they are maintained if the exposure to dbcAMP is longer than 4 days (Goetschy, et al, 1985). It has been shown that cytoplasmic MT are heterogeneous with regard to stability to depolymerizing conditions such as treatment with colchicine, low temperature or Ca^{2+} (Brinkley, et al, 1980). We have recently demonstrated that a fraction of MT in astrocytes growing in culture are cold stable (Hesketh, et al, 1984). In order to examine whether dbcAMP affects the stability of MT, we determined the solubility properties of tubulin under conditions that permitted the separation of unpolymerized and polymerized tubulin (Black, et al, 1984).

Dissociated cells from cerebral hemispheres of neonatal rats were cultured in serum supplemented Waymouth medium for 3 weeks and then exposed to 1 mM dbcAMP for 2 or 8 days. Parallel cultures were treated with 0.5 μM taxol for 4 days. Unpolymerized tubulin was extracted in MT stabilizing buffer containing Triton X-100. The detergent insoluble fraction which contained polymerized tubulin was further fractionated into cold/Ca^{2+}-soluble and -insoluble tubulin, corresponding to tubulin of labile and stable MT respectively. Each fraction was electrophoresed, immunoblotted and tubulin was revealed with monoclonal antibodies (Amersham, code N.356,357).

Fig. 1 shows that the detergent-soluble and cold/Ca^{2+}-soluble fractions represent the major pool of tubulin in control and 2 day dbcAMP treated cultures.

However in astrocyte cultures exposed to dbcAMP for 8 days and in taxol treated cultures the majority of tubulin is detergent insoluble (i.e. polymerized) and a substantial proportion of tubulin remains in cold/Ca^{2+}-insoluble fractions. These results idicate that after long exposure to dbcAMP the proportion of astrocyte cold/Ca^{2+}-stable MT increases. The relative stability of astrocyte processes observed after short and long exposure to dbcAMP is in part dependent on the polymerization / depolymerization properties of microtubules.

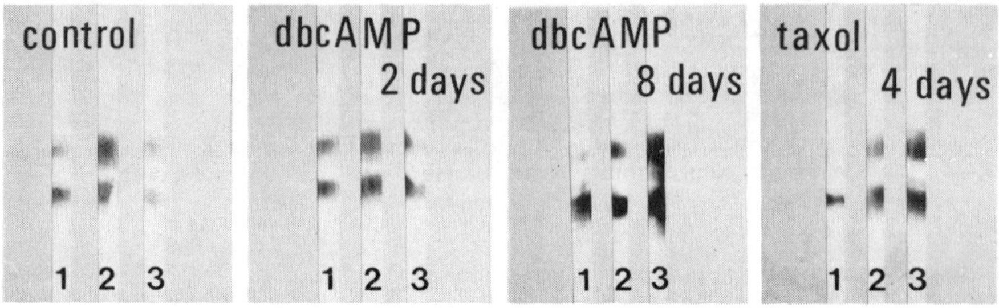

Fig. 1. Immunoblots of detergent soluble (1), cold/Ca^{2+}-soluble (2) and cold/ Ca^{2+}-insoluble (3) fractions stained with monoclonal antibodies to α and β tubulin.

REFERENCES

Black, M.M., J.M. Cochran and J.T. Kurdyla (1984). Brain Res., 295, 255-263.

Brinkley, B.R., S.M. Cox and D.A. Pepper (1980). Cytogenet. and Cell Genet., 26, 165-174.

Ciesielski-Treska, J., M.F. Bader and D. Aunis (1982). Neurochem. Res., 7, 275-286.

Goetschy, J.F., G. Ulrich, D. Aunis and J. Ciesielski-Treska (1985). J. Neurocytol., submitted.

Hesketh, J.E., J. Ciesielski-Treska and D. Aunis (1984). Neurosci. Lett., 51, 155-160.

Oligodendroglial and Astroglial Cell Clones Derived from Established Cultures of 2-day Postnatal Rat Cerebellum

P. L. Debbage

Physiologisch-Chemisches Institut der Universität, Koellikerstrasse 2, D-8700 Würzburg, FRG

ABSTRACT

Oligodendroglial precursor cells have been cloned; some of their properties are described here. Different types of astroglial clones were also obtained from the same original cell line. The properties of one of the astroglial clones are also described. Cellular interactions occuring between the oligodendroglial and the astroglial clonal lines are described.

KEYWORDS

Oligodendroglia; astroglia; precursors; clonal cell lines; cell interactions.

INTRODUCTION

Ontogenetic and physiological relationships between cells of early stages in the different glial lineages are being analysed using cell culture techniques (Raff et al., 1984; Schachner, 1982; Sensenbrenner et al., 1982). Such studies are facilitated if the cultures contain well-defined cell types (ideally, clones) and are maintained in serum-free chemically-defined media. Such a culture system is reported here: glial clones have been obtained, and they can be maintained either in serum-supplemented or serum-free medium.

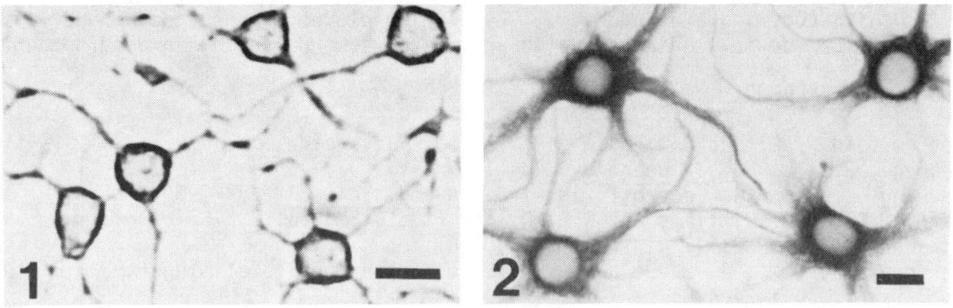

Fig. 1. Cells of the oligodendroglial clone 19, visualized in bright-field microscopy of a living culture. Scale bar, 10 μm.
Fig. 2. Cells of the astroglial clone 44, immunoperoxidase-labelled for glial acidic fibrillary protein (GFAP), visualized in bright-field microscopy. Scale bar, 10 μm.

RESULTS

Fig. 1 shows cells of the oligodendroglial clone 19, one of five similar clones established in long-term culture during a period of eight months. Fig. 2 shows cells of the astroglial clone 44, one of 40 clones established in long-term culture for longer than one year. These astroglial clones vary in their morphologies and in their content of GFAP. All the clones were established from a long-term parent cell line rich in cells identified as "oligodendroblasts" (Debbage, 1985), in which approximately 95% of cells were oligodendroglial and 5% astroglial. The astroglia could be cloned by limiting dilution, in serum-supplemented media, and with cloning efficiencies exceeding 40%. It was not possible to clone the oligodendroglia unless medium conditioned over either the parent cell line or the astroglial clonal lines was used, and cloning efficiencies were low, approximately 15%.

Both oligodendroglial and astroglial clones proliferate actively in serum-supplemented medium (5% fetal calf serum), and proliferation is very slow or ceases in Bottenstein and Sato's (1979) serum-free medium. However, in serum-free medium both clonal types develop more extensive and elaborate morphologies than in the serum-supplemented media. The oligodendroglial clones require astroglial-conditioned media for their survival, and if exposed to high concentrations of serum (10% - 20%) they generate small numbers of cells with astroglial morphologies. In contrast, the astroglial clones are stable, and it has not been possible to induce them to generate oligodendroglial cells.

All the clones are A2B5$^+$, S-100$^+$ and galactocerebroside$^-$. The astroglial clones label strongly for GFAP in a filamentous pattern (Fig. 2), and the oligodendroglial clones can be induced to label diffusely and weakly for GFAP if exposed to 10% serum in the culture medium. The ultrastructure of the oligodendroglial clones appears identical with that of cells typically described as "oligodendroblasts" (Debbage, 1985), and that of clone 44 resembles that of immature astroglia.

DISCUSSION

The oligodendroglial clones have a trophic requirement for astroglial factors; the astroglial clones have no such requirement. The phenotypic plasticity of the oligodendroglial clones resembles that of "bipotential glial progenitor" cells (Raff et al., 1984). Maturation of the oligodendroglia did not occur under the conditions reported here, including the absence of serum. It could not be induced by the addition of dibutyryl cyclic AMP: neuronal factors may be required. The cells of the parent line are not tumorigenic, and exhibit minimal chromosomal changes from normal rat chromosomes (Debbage et al., in preparation). Cells similar to the oligodendroglia reported here generate myelin when implanted into host brains (Doering and Federoff, 1984). These clones may provide a model system suitable for use in studies of cellular interactions during brain development.

REFERENCES

Bottenstein, J.E. and G.H. Sato (1979) Proc. Nat. Acad. Sci. U.S.A. 76, 514-517.
Debbage, P.L. (1985) J. Neurol. Sci. (In Press).
Doering, L.C. and S. Federoff (1984) J. Neurol. Sci. 63, 183-196.
Raff, M.C., B.P. Williams and R.H. Miller (1984) EMBO J. 3, 1857-1864.
Schachner, M. (1982) J. Neurochem. 39, 1-8.
Sensenbrenner, M., J.-P. Delaunoy, G. Labourdette and B. Pettmann (1982) Biochem. Soc. Trans. 10, 424-426.

Stimulation of Rat Primary Mixed Brain Cell Cultures with Lipopolysaccharide and Lectins

S. M. Roitt, M. G. Rumsby and A. J. Suckling

Department of Biology, University of York, York YO1 5DD, UK

ABSTRACT

Lipopolysaccharide (LPS) and two T lymphocyte mitogens, Concanavalin A (Con A) and phytohaemaglutinin (PHA) were added to the culture medium of 12/13-day-old rat primary mixed brain cell cultures and the stimulatory effect on the synthesis of RNA by glial cells was investigated. Glial cell numbers vary between culture wells and therefore incorporation of radiolabelled uridine has been related to the DNA content of cultures. DNA measurement was achieved by use of the fluorochrome dye 33258 Hoechst. The results show that whilst LPS stimulated RNA synthesis in the glial cell cultures the lectins Con A and PHA did not.

KEY WORDS

Concanavalin A, DNA, Glial Cells, 33258 Hoechst, Lipopolysaccharide, Mitogens, Peripheral Blood Lymphocytes, Phytohaemaglutinin, RNA, Tritiated Uridine.

INTRODUCTION

Immune mechanisms leading to the pathological changes occurring in autoimmune demyelination in experimental allergic encephalomyelitis (EAE) are being investigated. Experiments are underway to determine whether peripheral blood lymphocytes (PBLs) on stimulation liberate lymphokines which directly effect different classes of glial cells. As a first step, the supernatants of PBLs stimulated with either Con A or PHA are being added to glial cells and the effects investigated. Within such supernatants small amounts of free lections may persist. It has, therefore, been necessary to determine the effects of the lectins alone on the mixed brain cell cultures so that any potential stimulatory effect can be compensated for. An investigation was carried out to see if RNA synthesis in the cultures was increased in response to the lectins. This study was compared with a known stimulator of glial cell RNA synthesis, LPS (Fontana _et al._,1981). In order to compare the results between increased cultures more precisely the DNA content of each culture was determined with 33258 Hoechst which binds specifically to DNA. This data has been linked to the amount of radioactive label incorporated.

METHOD

The cerebra from 1/2-day-old rat pups were prepared for culture by the method of

Walker et al. (1984/5). Cells were grown in 24 well plates for 12-13 days. Following macrophage removal the culture medium was changed. Cultures were then supplemented with increasing concentrations of the mitogens and grown for a further 72 hours. During the last 2 hours of culture 0.5uCi of tritiated uridine were added. Triplicate samples were then removed for scintillation counting and DNA measurement. This was carried out by the method of Cesarone et al. (1979) with 33258 Hoechst. Results could then be expressed as counts per minutes of tritiated base per microgram of DNA sample.

RESULTS AND DISCUSSION

The two graphs below show the responses of the cultures to PHA and Con A respectively. Neither lectin causes an increase in RNA synthesis over the given range of concentrations.

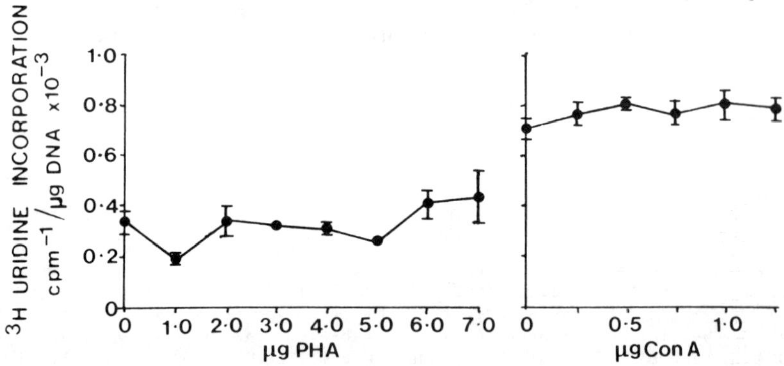

The responses of the glial cultures to PHA and Con A contrast with that of LPS which we found gave an 8 fold increase in RNA synthesis. This result confirms the findings of Fontana (1981) who found that LPS increases the amount of RNA synthesised by glial cells.

Our brain cells are cultured in this work in 24 well plates and cell numbers between wells vary. We thus feel it is important to express data on a DNA basis to eliminate errors due to variation in cells per well. For example, in one experiment the cellular DNA content per well of cells grown for 12-13 days varied from 10ug DNA at the lower limit to 27.5ug DNA at the upper limit. The use of 33258 Hoechst is a straightforward method for measuring DNA and is sensitive to less than 0.5ug DNA.

ACKNOWLEDGEMENT

We wish to thank the Multiple Sclerosis Society of Great Britain and Northern Ireland for their generous financial support.

REFERENCES

Cesarone, C.F., C. Bolognesi and L. Santi (1979). Analytical Biochem., 100, 188-197.

Fontana, A., R. Bosshard, C. Dahinden, P. Grob and A. Grieder (1981). J.Neuroimmunol., 1, 343-352.

Walker, A.G. J. Chapman, C.B. Bruce and M.G. Rumsby (1984/5). J.Neuroimmunol. 7, 1-20.

Microtubule-Associated Proteins (MAPs) and Astrocyte Differentiation

D. Couchie, C. Fages, A. M. Bridoux, B. Rolland, M. Tardy and J. Nunez

INSERM U 282, Hôpital H. Mondor, 94010 Créteil, France

Abstract

Immunoblotting and immunohistochemical techniques were used to identify the MAPs of the astrocyte during *in vitro* differentiation.

Key words

Microtubules, microtubule-associated proteins, brain development, astrocyte differentiation.

Introduction

Brain microtubules contain several high molecular weight (>200K) MAPs (Sloboda *et al.*1975; Murphy *et al.*1977) and a group of protein (TAU) of 58-65 K (Cleveland *et al.*1977). At an adult stage TAU is composed with several closely spaced bands whereas at early stages of development two peaks of 48 K and 62 K are observed (Mareck *et al.*1980). Immunoblotting analysis was used a) to characterize these entities at different stages of brain development b) to identify the MAPs present in cultured astrocytes. Two polyclonal antibodies prepared against adult brain MAP_2 (∿ 300K mol wt) and adult TAU were used.

Results

1. Changes in composition and immunochemical properties of the MAPs during brain development (Fig.1). At a fetal stage a 62 K band was detected by both antisera in the "TAU" region of the immublots whereas a 48 K band was detected only by the anti-TAU serum (Fig.1 lanes 1a, 1b). At day 3 and 6 after birth the 48 K and the 62 K bands were detected only by the anti-TAU and the anti-MAP_2 sera respectively (Fig.1 lanes 2a, 2b). Both the 48 K and 62 K bands were replaced at adulthood by the 4-5 closely spaced bands which reacted only with the anti-TAU serum (Fig.1 lanes 3a, 3b). Two transitions seem therefore to occur during brain development i.e. a) from a fetal to a newborn stage b) before adulthood (Couchie and Nunez 1985a).

2. Characterization of the MAPs of the astrocyte during *in vitro* differentiation (Fig.2). *In vitro* cultured astrocytes (Bardakdjan *et al.* 1979) contain several MAPs which were detected by immunoblotting analysis (Couchie *et al.*1985b). Anti-MAP_2 serum detected a 300 K band (i.e. similar to MAP_2) and three components of 380 K, 165 K and 83 K which seem to be specific of the astrocyte (Fig.2 lane 2). The 83 K entity was also stained by the anti-TAU serum which also revealed a 63 K band (Fig.2 lane 1). The dividing astroblasts contain two times less immunoreactive material than the mature cells.

Both antisera revealed in the differentiated cells a dense fibrous network. The reaction was very diffuse at early stages of the culture. Morphological differentiation, which began after serum deprivation and which was accelerated by forskolin (a drug which induces cAMP accumulation) led to high labeling both of the cell body and of the astroglial processes. In the presence of colchicine the

staining regressed, the processes shortened and the cell returned to a less differentiated morphology.
In conclusion a) several MAPs of the astrocyte differ from those found in both the mature and the immature brain b) microtubule assembly seems to be required for astroglial differentiation.

FIGURE 1 FIGURE 2

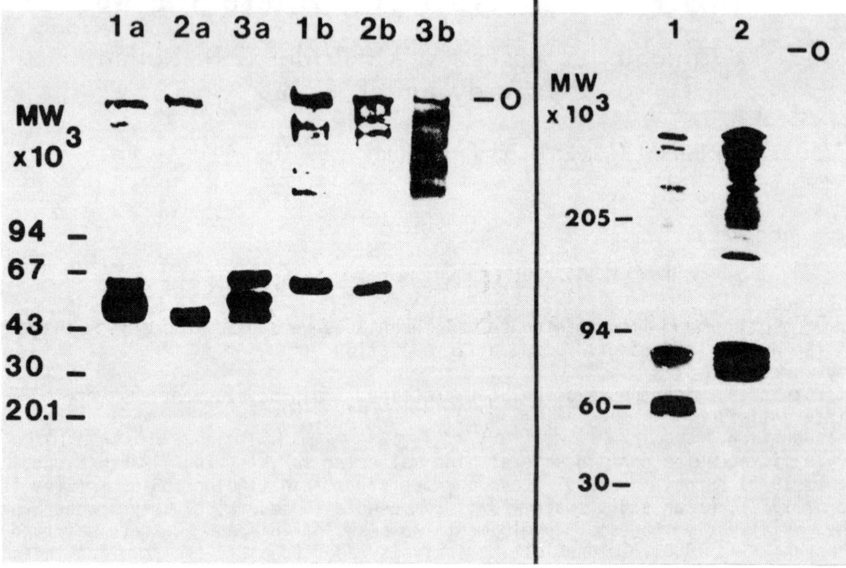

Fig.1 Slab gel electrophoresis of rat brain MAPs. 1: fetal (16 days); 2: 3 days; 3: 35 days. The MAPs were revealed by autoradiography after immunoblotting as described in Couchie and Nunez (1985a). Two antisera were used: one raised against adult TAU (lanes 1a, 2a, 3a) and the other against adult MAP_2 (lanes 1b, 2b, 3b).
Fig.2 Slab gel electrophoresis of differentiated astroglial extracts. The MAPs were revealed by autoradiography after immunoblotting as described in Couchie *et al.* (1985b) with anti-TAU (lane 1) or anti-MAP_2 (lane 2) sera.

References
Sloboda R.D., Rudolph S.A., Rosenbaum J.L. and Greengard P. (1975) Proc. Natl. Acad. Sci. USA 72, 177-181.
Murphy D.B., Vallee R.B. and Borisy G.G. (1977) Biochemistry 16, 2598-2605.
Cleveland D.W., Hwo S.Y. and Kirschner M.W. (1977) J. Mol. Biol. 116, 207-225.
Mareck A., Fellous A., Francon J. and Nunez J. (1980) Nature 284, 353-355.
Couchie D. and Nunez J. (1985a) FEBS Lett. 188, 331-335.
Bardakdjan J., Tardy M., Pimoule C. and Gonnard P. (1979) Neurochem. Res. 4, 519-529.
Couchie D., Fages C., Bridoux A.M., Rolland B., Tardy M. and Nunez J. (1985b) J. Cell Biol. in press.

Session II

DETECTION AND ROLE OF GLIAL CELLS SPECIFIC PROTEINS

Co-chair : L. ENG (*Palo Alto, USA*)
A. LOWENTHAL (*Antwerpen, Belgium*)

Glial Fibrillary Acidic Protein (GFAP) in Human Astrocytes

J. Gheuens, P. Cras, R. Crols, M. Noppe and A. Lowenthal

Laboratory of Neurochemistry, Born Bunge Foundation, University of Antwerp, Belgium

ABSTRACT

Alpha-albumin, or glial fibrillary acidic protein (GFAP) is a constituent of the astroglial fiber, and a marker of astrocytes.

Its clinical applications are described. Immunohistological demonstration of GFAP is of use in neuropathology, in particular in the study of brain tumors. Various conditions associated with increased GFAP levels in cerebrospinal fluid have been identified.

In addition, the existence of several molecular weight forms of GFAP in humans, and their possible respective role is discussed.

KEYWORDS

Alpha-albumin; astrocytes; astroglial tumors; cerebrospinal fluid; glial fibrillary acidic protein; monoclonal antibody.

INTRODUCTION

In 1962, a specific protein was described in aqueous extracts from normal human central nervous system (Van Sande, 1962). Alpha-albumin, as this protein was termed, had a typical topographical distribution in the brain (Chamoles et al., 1970), and was increased in different pathological conditions (Zeman et al., 1970). In 1971, GFAP was described in extracts from gliotic tissue (Eng et al., 1971). GFAP was soon identified as an intracellular protein of astrocytes (Ludwin et al., 1976). GFAP was shown to exist in a water soluble and a water-insoluble form (Eng, 1982). It has since been demonstrated that the earlier alpha-albumin preparations were heterogenous and that after further purification (Noppe et al, 1979a), alpha-albumin was immunologically indistinguishable from GFAP. Purified alpha-albumin is now considered to be water-soluble GFAP (Gheuens et al., 1979; Eng, 1980, 1982).

Since its discovery, GFAP has stayed of interest in neurology and neurobiology mainly for two reasons. First, GFAP is considered to be a good marker of cells of astroglial lineage. It can hence be used to determine the astrocytic nature of cells, and even to assess astrocytic involvement in certain pathological conditions. Second, GFAP is known to be a constituent of an intermediate filament specific to astrocytes, the astroglial fiber (Dahl and Big-

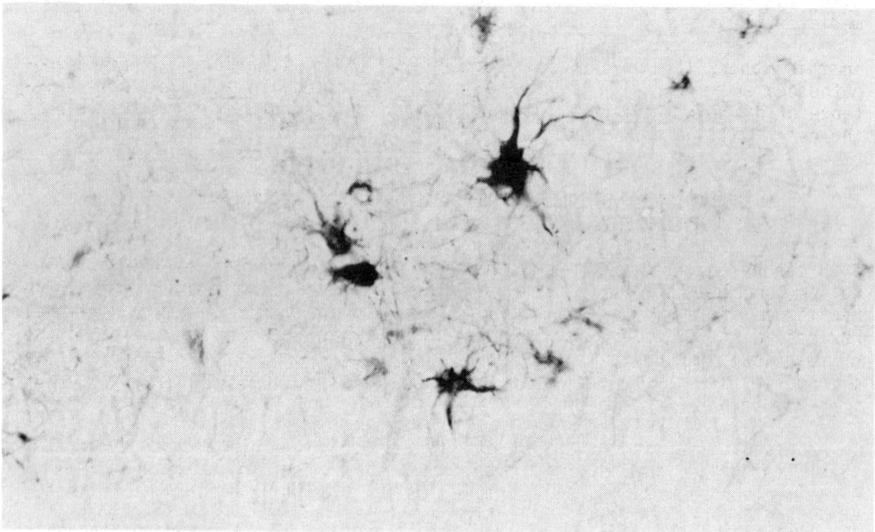

Fig. 1 : Large reactive astrocytes surrounding a senile plaque in Alzheimer's disease. Cytoplasm of cells is strongly labelled. (Polyclonal anti-serum directed to GFAP, unlabelled antibody bridge technique, x 420).

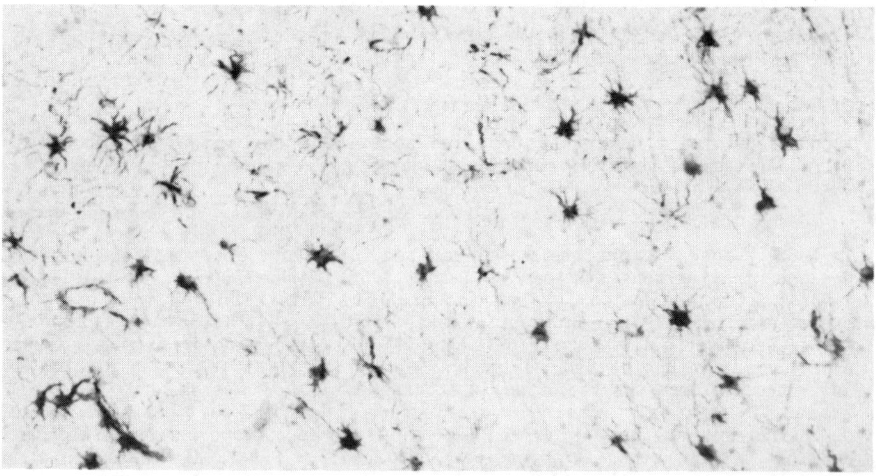

Fig. 2 : Numerous spider-like reactive astrocytes in the vicinity of a rapidly expansive brain tumor. These reactive cells show abundant cytoplasm and slender extensions. They are intensely GFAP-positive (G1 monoclonal anti-GFAP, avidin-biotin complex (ABC) method, haematoxylin counterstain,x 275).

nami, 1983). A better understanding of GFAP structure and expression in astrocytes may therefore contribute to our knowledge on the function of the intermediate filament system in general.

In this paper, the various applications of GFAP as a marker in human neuropathology and clinical neurology will be overviewed. In addition, the existence of different molecular weight forms of GFAP, and their possible role in the cell, will be discussed.

IMMUNOHISTOLOGICAL DEMONSTRATION OF GFAP IN GENERAL NEUROPATHOLOGY.

GFAP can be demonstrated readily in tissue sections, using specific polyclonal or monoclonal antibody and a variety of immunohistological methods (indirect immunofluorescence, indirect horse radish peroxidase, soluble peroxidase-anti peroxidase complex method, avidin-biotin complex method). It is of practical importance that GFAP is relatively fixation resistant, so that immunohistological studies of it can still be performed after routine fixation and paraffin embedding.

Immunohistological demonstration of GFAP has been used in the study of reactive gliosis in multiple sclerosis, in brain edema (Szymas et al., 1984), Alzheimer's disease (Fig. 1), brain stem gliosis of sudden infant death patients (Becker et al., 1985), etiopathogenesis of perinatal telencephalic leukoencephalopathy (Takashima et al., 1984), and Huntington's disease (Vacca et al., 1984). The extent of reactive gliosis, which denotes the proliferation of astrocytes in response to a number of stimuli, is correlated with the amount of damage afflicted to the central nervous system. In addition, it depends in some respects on the etiology of the noxious event. The extensive reactive gliosis surrounding a syrinx, senile plaques, vascular malformations and central nervous system lymphomas is well known. Glial outgrowth along spinal nerve roots has first been shown by immunohistochemical studies of GFAP in Werdnig-Hoffmann's disease (Brock et al., 1984), and has subsequently been demonstrated in amyotrophic lateral sclerosis (Kimura et al., 1984). The specificity and possible pathogenetic role of this phenomenon is now under discussion.

IMMUNOHISTOLOGICAL DEMONSTRATION OF GFAP IN BRAIN TUMORS.

GFAP has gained particular importance in the characterisation of human brain tumors, as illustrated by the substantial number of papers that have appeared on this subject (Eng et al., 1978; McKeever et al., 1982; Tascos et al., 1982; Trojanowki et al., 1983; Bonnin et al., 1984; DeArmond et al., 1980). GFAP has been demonstrated immunohistologically in all variants of astrocytoma (McKeever et al., 1982), astroblastoma (Rubinstein, 1982), some ependymomas (Kleinman et al., 1984), in the astrocytic component of oligoastrocytoma, ganglioglioma (Cras et al., in press), angioglioma (Bonnin et al., 1983), gliosarcoma (Schiffer et al., 1984), medulloblastoma (Palmer et al., 1981; Mannoji et al., 1981), oligodendroglioma (Herpers et al., 1984), choriod plexus papilloma (Rubinstein et al., 1981) , and hemangioblastoma (McComb et al., 1982). The demonstration of GFAP by some in non-astrocytic human tumors has implications for our understanding of the histopathogenesis of such tumors (Herpers et al., 1984).

In addition, GFAP has allowed interesting observations on astrocytic tumors. First, the number of GFAP-positive cells appears to be inversely correlated with the malignancy grade of an astrocytoma, which could possibly contribute to assessment of prognosis, or even therapeutic decisions. Second, cells in the vicinity of blood vessels in malignant astrocytic tumors are more GFAP-

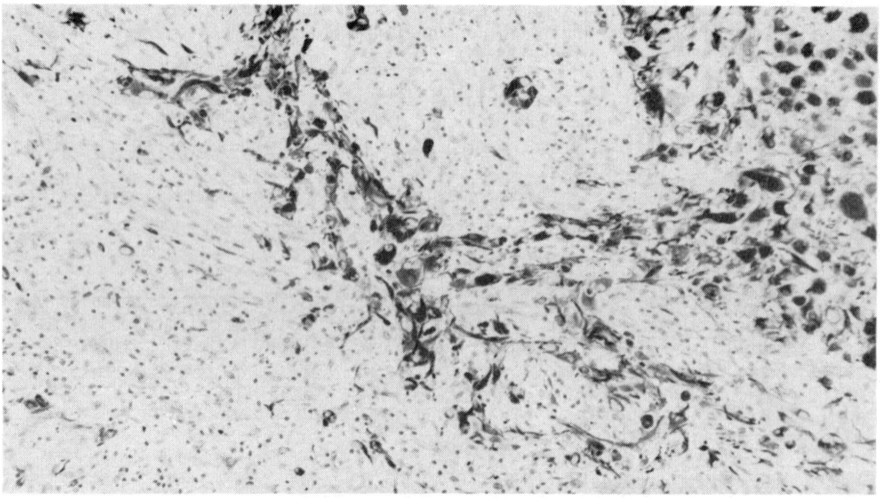

Fig. 3 : Variable GFAP-positivity in a glioblastoma multiforme : the fraction of GFAP-positive cells roughly corresponds with malignancy grade of glial tumor. In glioblastomas, GFAP positive cells are larger and have more abundant cytoplasm than GFAP-negative ones. (Polyclonal anti-GFAP, unlabelled antibody bridge technique, x 110).

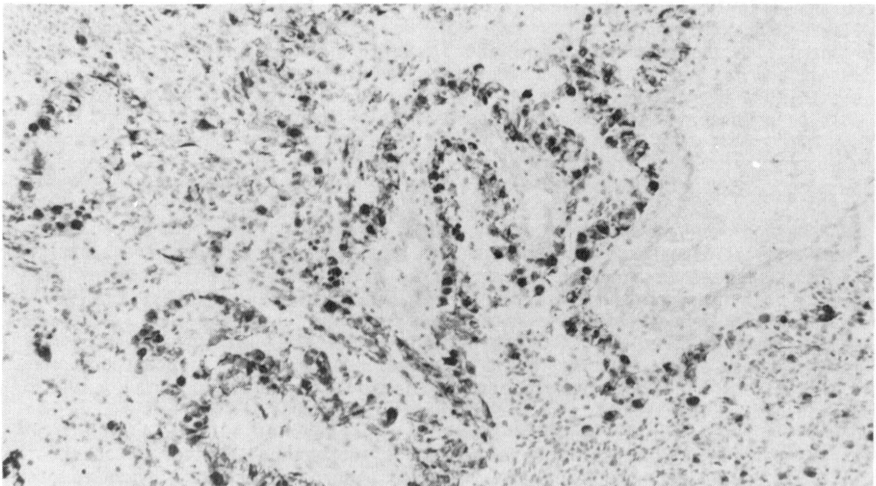

Fig. 4 : GFAP-positivity in grade IV glial tumors is mostly concentrated around the vessels. (Polyclonal anti GFAP, unlabelled antibody bridge technique haematoxylin counterstain, x 110).

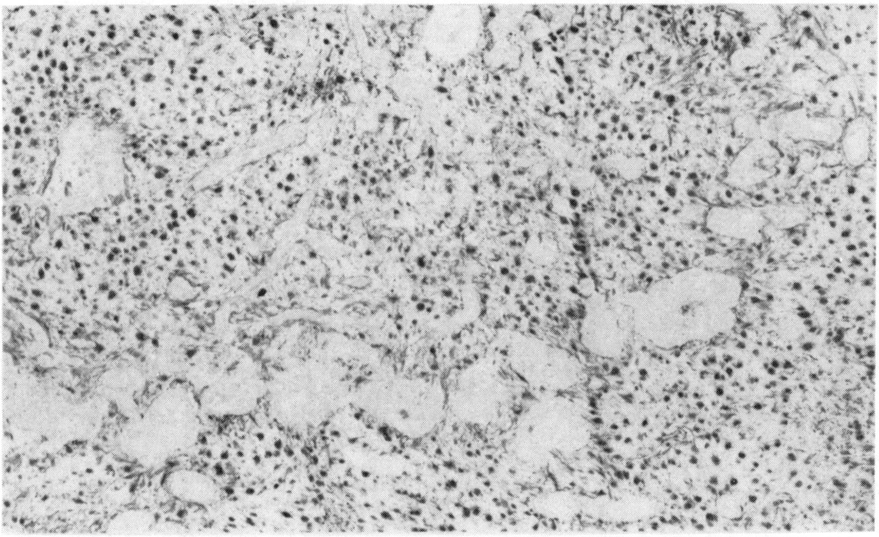

Fig. 5 : Vascular proliferation in a glioblastoma multiforme appears as GFAP negative areas. (G1 monoclonal antibody, ABC-method, haematoxylin counterstain, x 110).

positive than those at a distance. The reasons for this apparent differentiation are unclear.
Demonstration of GFAP has also been helpful in characterising tumor deposits in the leptomeninges (Cras et al., in press), and in extraneural sites (Haugen et al., 1984; Ojeda et al., 1984), as well as in the interpretation of cytological specimens (Collins, 1984)

QUANTITATIVE DETERMINATIONS OF GFAP IN BODY FLUIDS

GFAP concentrations were measured in cerebrospinal fluid (CSF) using a 2-site immunoradiometric assay (Noppe et al., 1979b). The mean GFAP concentration in the CSF from 142 patients without central nervous system (CNS) pathology was 3.2 ng/ml, with a standard deviation of 3.1 ng/ml (Table I). Therefore, GFAP concentrations above 9.4 ng/ml (mean + 2 standard deviations of controls) were considered pathological in subsequent analysis (Crols et al., 1983). The results of GFAP determinations in CSF in various neurological conditions are shown in Table II. These results can be divided in three groups (Crols et al., 1983). A first group represents those conditions in which high mean concentrations of GFAP are registered, and in which over 50% of cases have pathologically increased values. A second group represents those conditions in which the mean GFAP concentration is not spectacularly increased, but in which the range of concentrations is wide, and in which a fair percentage of cases have pathologically increased values. In a third group of conditions, increased GFAP concentrations in the CSF are encountered in less than 10% of cases. In contrast to an earlier report (Lowenthal et al., 1978), MS belongs to this last group.

It can be speculated that increased GFAP concentrations in the CSF either

Table I. GFAP concentration in CSF - control cases.

Diagnosis	Number of cases	GFAP concentration ng/ml mean	standard deviation
Nonorganic psychiatric	33	3.5	3.3
Disc herniation	75	3.2	3.4
Otovestibular, ophtalmol.	34	3.0	2.7
Total	142	3.2	3.1

result from massive CNS tissue destruction, or that they are related to pronounced astrocytic involvement in a disease process (e.g. reactive gliosis). GFAP concentrations were also measured in amniotic fluid samples. The first results indicated that determination of GFAP in amniotic fluid offered a reliable test for prenatal diagnosis of anencephaly, but that only half of spina bifida cases could be detected. Equally important were the low values of GFAP in those conditions often associated with high alpha-fetoprotein values, such as exomphalocoele. Although GFAP determination in amniotic fluid does not have the properties of a screening assay, it can be useful as an additional test in prenatal differential diagnosis (Van Regenmortel et al., in preparation).

DIFFERENT MOLECULAR WEIGHT FORMS OF GFAP.

Apart from its practical use as an astrocytic marker, there are some basic biological questions that can be asked on GFAP.
Relatively little is known of its function. GFAP is thought to be a major constituent of the intermediate filament of the astroglial cell, the astroglial fiber (Dahl et al., 1983). The protein exists in a water-soluble and a water-insoluble form, but the structural or chemical basis of this difference is not well understood yet (Eng, 1982). Facts are emerging though, on phosphorylation of GFAP, on structural homology with other filament proteins. One

Table II. GFAP concentration in CSF - pathological cases

Diagnosis	Number of cases	GFAP concentration (ng/ml) mean	range	% >9.4 ng/ml
Bacterial meningitis	13	19.0	1-74	76.9
Astrocytic tumors	12	43.3	2-151	66.7
Acute encephalitis	14	120.2	1-636	57.1
Syringomyelia	10	42.5	0-259	55.6
Vascular Parkinson syndrome	21	9.4	2-26	47.6
Dementia	46	8.4	0-74	32.6
Diabetes	31	6.7	0-30	32.3
Intoxications	33	9.2	0-46	27.3
Diffuse cerebrovascular	25	6.2	2-27	24.0
Acute stroke + TIA	82	16.1	0-342	20.7
Multiple sclerosis	79	6.4	0-242	8.9
Epilepsy	203	4.3	0-58	7.9
Amyotrophic lateral sclerosis	14	3.9	1-11	7.1
Parkinson's disease	6	2.2	1-6	0
Aseptic meningitis	10	2.1	1-6	0

controversial basic issue is the existence of different molecular weight (MW) forms of GFAP,that have different solubility properties (Eng 1982; Bigbee et al., 1983; Gheuens et al., 1984).

Immunoblots of various human brain extracts separated by Laemmli gel electrophoresis have been useful in the study of GFAP. Monoclonal (Gheuens et al., 1984) as well as polyclonal antibodies (Eng, 1982; Bigbee et al., 1983) identify different MW forms of GFAP in these blots. In addition it is clear that the band pattern differs from one cerebral region to another, and that in each case, relatively more of the lower forms are detected in the aqueous extracts, and relatively more of the higher MW forms in the detergent extracts (Gheuens et al., 1984). It therefore appears that the lower MW forms are relatively more water soluble than the higher MW forms. The significance of this is controversial. Lower MW forms of GFAP can result from post mortem or in vitro degradation (Dahl, 1976; Bigbee et al., 1983), and a calcium dependent proteinase active in this process has been identified (DeArmond et al., 1983). However, it was observed that the band pattern identified in these extracts by specific mono- or polyclonal antibody did not change after storage, that rat brain taken in the conditions of human autopsy contained only one high MW band, and also, that low MW forms could not be generated by deliberate attempts at degradation (Gheuens et al., 1984). It is therefore proposed as working hypothesis that the low MW forms of GFAP in human brain do not solely result from post mortem or in vitro degradation, that the MW is inversely correlated with the solubility of GFAP, and that low MW forms may have a biological role in vivo.

CONCLUSIONS

GFAP is mainly of use as an astrocytic marker. As such it is being used to identify cells in culture, and it is of use in general neuropathology and the pathology of brain tumors in particular. Though lagging behind the immunohistological studies of GFAP, quantitative determination of GFAP in body fluids seems to be a useful auxillary test in differential neurological diagnosis.

Apart from the fact that it is a constituent of an intermediate filament of astrocytes, little is known about the function of GFAP. A controversy surrounds the significance of the different MW forms of the protein. It is proposed that low MW forms of GFAP do exist *in vivo*, and that they play a biological role.

REFERENCES

Becker, L.E. and S. Takashima (1985). *Neuropediatrics* 16, 19-23.
Bigbee, J.W., D.D. Bigner, C. Pegrom and L.F. Eng (1983). *J. Neurochem.* 40, 460-467.
Bonnin, J.M., C.E. Pena and L.J. Rubinstein (1983). *J. Neuropathol. Exp. Neurol.* 42, 504-516.
Bonnin, J.M. and L.J. Rubinstein (1984). *J. Neurosurg.* 60, 1121-1133.
Brock, T.O. and D.L. McIlwain (1984). *J. Neuropathol. Exp. Neurol.* 43, 609-619.
Chamoles, N., D. Karcher, W. Zeman and A. Lowenthal (1970). *Brain Res.* 17, 315-324.
Collins, V.R. (1984). *Acta Cytol.* 28, 401-406.
Cras, P., J.J. Martin, M. Noppe and A. Neetens. *Proc. Joint World Meeting Neurophthalmology* (in press).
Crols, R., M. Noppe, J. Caers and A. Lowenthal (1983). *Neurochem. Pathol.* 1, 91-102.
Dahl, D. (1976). *Biochim. Biophys. Acta* 420, 142-154.
Dahl, D. and A. Bignami (1983). *Handbook of Neurochemistry* (Lajtha, A. ed.) 5, 127-151.

DeArmond, S.J., L.F. Eng and L.J. Rubinstein (1980). Path. Res. Pract. 168, 374-394.

DeArmond, S.J., M. Fajardo, S.A. Naughton and L.F. Eng (1983). Brain Res. 262, 275-282.

Eng, L.F. (1980) in Proteins of the Nervous System (Bradshaw R.A. and D.M. Schneider, eds.), 97-117. Raven Press, New York.

Eng, L.F. (1982). Scand. J. Immunol. 15 (suppl. 9), 41-51.

Eng, L.F. and L.J. Rubinstein (1978). J. Histochem. Cytochem. 26, 513-522.

Eng, L.F., J.J. Vanderhaeghen, A. Bignami and B. Gerstl (1971). Brain Res. 28, 351-354.

Gheuens, J., E. De Schutter, M. Noppe and A. Lowenthal (1984). J. Neurochem. 43, 964-970.

Gheuens, J., D. Karcher, M. Noppe and A. Lowenthal (1979). Acta Neurol. Belg. 79, 314-321.

Haugen, O.H. and C.R. Taylor (1984). Acta Pathol. Microbiol. Immunol. Scand. Sect. A 92, 9-14.

Herpers, M.J.H. and H. Budka (1984). Acta Neuropathol. 64, 265-272.

Herpers, M.J.H., H. Budka and D. McCormick (1984). Acta Neuropathol. 64, 333-338.

Kimura, T. and H. Budka (1984). Acta Neuropathol. 65, 46-52.

Kleinman, G.M., R.H. Young and R.E. Scully (1984). Hum.Pathol. 15, 632-638.

Lowenthal, A., M. Noppe, J. Gheuens and D. Karcher (1978). J. Neurol. 219, 87-91.

Ludwin, S.K., J.C. Kosek and L.F. Eng (1976). J. Comp. Neurol. 165, 197-207.

Mannoji, H., I. Takeshita, M. Fukin, M. Ohta and K. Kitamura (1981). Acta Neuropathol. 55, 63-69.

McComb, R.D., T.R. Jones, S.V. Pizzo and D.D. Bigner (1982). Acta Neuropathol. 56, 207-213.

McKeever, P.E., S. Laverson, P.L. Kornblith, R. Howard, E.A. Quindlen, B.H. Smith and B.M. Chronwall (1982). Acta Neuropath. 58, 69-72.

Noppe, M., J. Gheuens, D. Karcher and A. Lowenthal (1979a). in Protides of the Biological Fluids (Peeters, H., ed.) 27, 813-816, Pergamon Press, New York.

Noppe, M., A. Lowenthal, D. Karcher and J. Gheuens (1979b). J. Immunol. Meth. 27, 75-81.

Ojeda, V.J. and G.F. Sterrett (1984). Pathology 16, 217-221.

Palmer, J.O., A.G. Kasselberg and M.G. Netsky (1981). J. Neurosurg. 55, 161-169.

Rubinstein, L.J. (1982). Supplement to fascicle 6, 2nd Series American Forces Institute of Pathology (W.H. Hartmann, ed.) 21-26.

Rubinstein, L.J. and J.M. Brucher (1981) Acta Neuropathol. 53, 29-32.

Schiffer, D., M.T. Giordana, A. Mauro and A. Migheli (1984). Acta Neuropathol. 63, 108-116.

Szynas, J. and K.A. Mossmann (1984). Acta Neuropath. 62, 309-315.

Takashima, S., L. Becker, M. Nishimura and J. Tanoka (1984). Brain Develop. 6, 444-450.

Tascos, N.A., J. Parr, and N.K. Gonatas (1982). Hum.Pathol. 13, 454-458.

Trojanowski, J.Q. and V.M.-Y. Lee (1983). Hum. Pathol. 14, 281-285.

Vacca, L.L. and S.R.Nelson (1984) in Advances in cellular neurobiol. 5, 221-250, Academic Press, Orlando.

Molecular Cloning of cDNA for Bovine Glial Fibrillary Acidic Protein and Specific Localization of its mRNA in Astroglial Cells

S. Watanabe[†], Y. Endo[*,‡,#], T. Kitamura[†] and N. Sueoka[†]

[*]Shionogi Research Laboratory, Sionogi & Co. Ltd., Osaka 553, Japan
[†]Department of Pathology, Kyoto Prefectural University of Medicine, Kyoto 602, Japan
[‡]Department of MCDB, University of Colorado, Boulder, CO. 80309, USA

ABSTRACT

Total poly(A)-containing RNA from fetal spinal cord, which is an enriched source of the sequence for glial fibrillary acidic protein (GFAP), was used as a template for cDNA synthesis. Double-stranded cDNA was inserted into the Pst I site of pBR322 by hompolymeric dG-dC tailing. The recombinant plasmid with GFAP cDNA insert was examined by hybrid-selected mRNA translation, and one plasmid with a cDNA insert of approximately 600 base-pairs was positively selected. Using this cDNA insert, Northern blot analysis of bovine spinal cord mRNA revealed the presence of a 2.4 kilobase mRNA for bovine.

In situ cDNA-mRNA hybridization on the section of rat cerebrums disclosed that hybridization signals for the cDNA insert were located mainly on the cytoplasm of astroglial cells in the white matter and in the hippocampus. Moreover, when the *in situ* hybridization technique was applied on the retrograde degeneration of rat facial nuclei, which has been proved to be a GFAP-inducible system, intensive signals were detected on the damaged nuclei. These observations indicate that the 600-base pair insert is the cDNA fragment for bovine and rat GFAP.

KEYWORDS

GFAP; mRNA; cDNA, retrograde degeneration, facial nucleus, astroglial cells.

INTRODUCTION

Distinct types of intermediate filaments are expressed in the specific classes of higher eukaryotic cells: keratin filament in epithelial cells, glial filaments in astrocytes, neurofilaments in neurons, vimentin filaments in cells of mesenchymal origin and desmin filaments in muscle cells (for a review, see Lazarides, 1980). Glial fibrillary acidic protein (GFAP) is a component of the glial filament and seems to be an excellent marker of astroglial cells, although its function is hardly known (for review, see Eng, 1980). Molecular cloning of GFAP cDNA is needed to know the exact role of GFAP in the particular glial cells. We previously reported that fetal calf spinal cord is an excellent source of GFAP-mRNA (Watanabe and Sueoka, 1984). In this study, we isolated the clone with GFAP cDNA fragment.

Present Address: Department of Anatomy, Yamanashi Medical University, Tamahomura, Yamanashi 409-38, Japan.

MATERIALS AND METHODS

RNA preparation

Total poly(A)-containing RNA was isolated from fetal calf spinal cords as described elsewhere (Watanabe and Sueoka, 1984).

Construction of cDNA clones

Thirteen micrograms of total poly(A)-containing RNA was used to prepare cDNA essentially as described by Wickens et al. (1978) and Maniatis et al. (1982). The cDNA molecules were inserted into the Pst I site of pBR322 and transformed *Escherichia coli* HB101. Approximately 1200 clones which were sensitive to ampicillin but resistant to tetracycline were selected for further screening.

Colony hybridization

For the initial screening of clones, an *in situ* colony hybridization method (Maniatis et al., 1982) was employed using ^{32}P-cDNA probe made to total poly(A)-containing RNA of fetal calf spinal cord. Twenty-seven clones which gave strong hybridization signals were chosen for further screening.

Translation of hybrid-selected mRNA

Translation of plasmid-selected mRNA was carried out as described by Lewis et al. (1983). Plasmid DNA was prepared by a rapid boiling method (Holmes and Quigley, 1981) or centrifugation in CsCl gradients (Maniatis et al., 1982). *In vitro* translation and immunoprecipitation have been described by Watanabe and Sueoka (1984).

Northern blot hybridization

The cDNA insert was isolated after digestion of recombinant plasmid DNA with Pst I enzyme. ^{32}P-labeled nick-translated cDNA insert was hybridized to total mRNA by methods described previously (Thomas, 1980).

In situ cDNA-mRNA hybridization and immunohistochemical detection of GFAP

Four adult rats (Wister strain) were used. The left facial nerve of each animal was cut under the auricle, and the animals were sacrificed by decapitation at third or seventh day after the operation. Whole brains including brain stem were immersed immediately in the Bouin's fixative. The material was dehydrated and embedded in paraffin. Four-μm-thick sections containing facial nuclei or cerebrum were cut from the paraffin blocks. The dewaxed sections were treated with 0.2 N HCl for 10 min at room temperature, and then, immersed in 0.25% acetic anhydride in 0.1 M Tris-HCl buffer, pH 8.0, for 10 min at room temperature. They were rinsed in distilled water quickly, dehydrated in an ethanol series and air-dried. The hybridization was carried out in a solution containing ^{3}H-labeled cDNA (200 ng/ml), 50% formamide, 3 X SSC (450mM Nacl, 45mM sodium citrate), 1 X Denhardt' solution (0.02% each of polyvinyl pyrrolidone, bovine serum albumin and Ficoll) and sonicated *Escherichia coli* DNA (100 μg/ml). The mixture was boiled for 5 min, quickly cooled in ice water and applied on the sections. The sections were covered with glass coverslips and placed in a moist chamber for 17 hours at 37°C. After hybridization sections were rinsed several times in 2 X SSC, dehydrated and air-dried. They were then dipped in Kodak nuclear emulsion NTB2 diluted into 1:1 with water, and exposed at 4°C for 2 weeks. They were developed in Kodak D-19 developer diluted into 1:1 with water at 15°C for 3-5 min, fixed and washed in water. The sections were counter-stained with hematoxylin and eosin.
For immunohistochemistry the dewaxed sections were incubated with rabbit

antiserum against GFAP (kindly supplied by Dr. L.F. Eng) and the reaction was detected by indirect immunoperoxidase staining method.

RESULTS AND DISCUSSION

Total poly(A)-containing RNAs from fetal calf spinal cord were used as templates for cDNA synthesis. The double-stranded cDNA was inserted into the Pst I site of pBR322, and cloned in *Escherichia coli* HB101. As already reported, poly(A)-

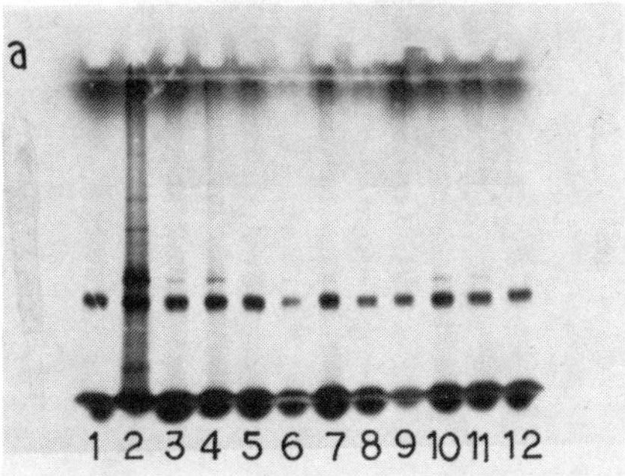

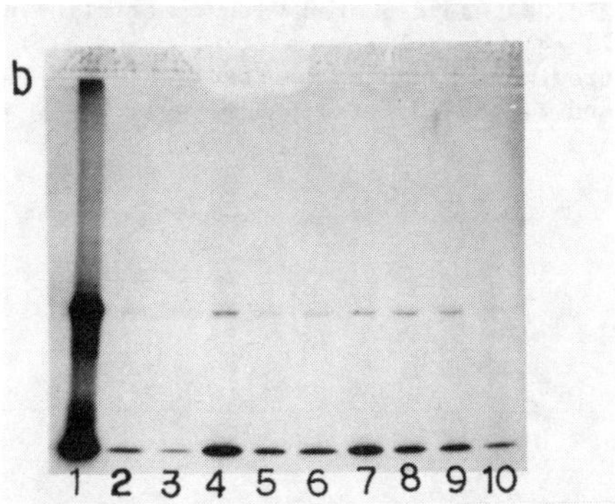

Fig. 1. Positive identification of GFAP cDNA clones by hydrid selection and *in vitro* translation.
Plasmid DNA immobilized on nitrocellulose filters was hybridized with poly(A)-containing RNA isolated from fetal spinal cords. The filters were washed, then the selected mRNA was eluted and rapidly cooled. These samples were used to direct *in vitro* protein synthesis in a reticulocyte lysate system. Analyses of the translation products by immunoprecipitation and/or SDS-polyacrylamide gel electrophoresis were performed as described in Materials and Methods. A is an autoradiogram displaying total ^{35}S-labeled polypeptides synthesized when the *in vitro* translation system was directed with no exogenous mRNA, lane 1; total poly(A)-containing RNA, lane 2; mRNA selected by plasmid DNA from pBR322, p411, p460, p507, p522, p528, p537, p540, p580 and pRSA13, lanes 3 to 12, respectively. B is an autoradiogram displaying ^{35}S-labeled polypeptide immunoprecipitated by anti-GFAP antiserum after *in vitro* translation had been directed with total poly(A)- containing RNA, lane 1; mRNA selected by plasmid DNA from pRSA13, pBR322, p411, p460, p507, p537, p540, p580 and pBR322, lanes 2 to 10, respectively.

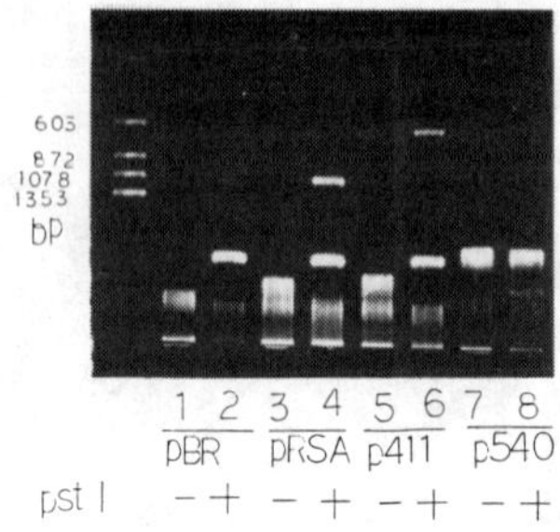

Fig. 2. Size determination of recombinant plasmid insert.
Purified plasmid DNA was digested with restriction enzyme Pst I and applied to 1.2% agarose gel (lanes 2, 4, 6 and 8). In lanes 1, 3, 5 and 7, Pst I treatment had been omitted. Plasmid DNAs used were: pBR322 for 1 and 2, pRSA13 for 3 and 4, p411 for 5 and 6, p540 for 7 and 8.

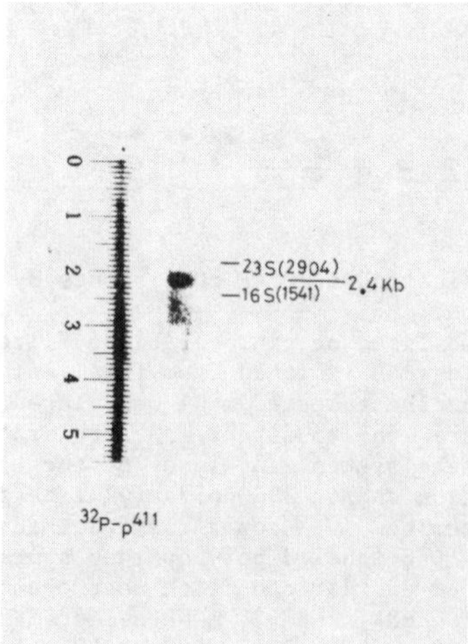

Fig. 3. Northern blot hybridization of mRNA.
Glyoxal-denatured poly(A)-containing RNAs from newborn-calf spinal cord was fractionated on a 1.4% agarose gel and transferred to nitro-cellulose paper, which was hybridized with the nick-translated cDNA insert (approximately 600 bases) from p411.

a

b

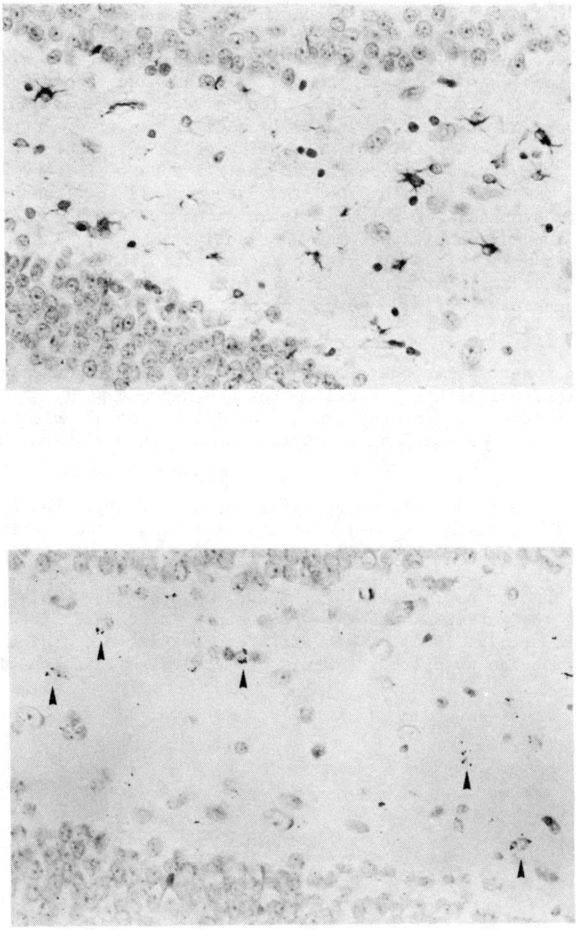

Fig. 4. Hippocampus of an adult rat.
a. Immunohistochemistry with anti-GFAP antiserum. Astroglial cells in this region show positive reaction in normal condition. X 400
b. *In situ* hybridization with the p411 cDNA insert. Silver grains (arrow head) are observed on the cytoplasm of astroglia. No grains are detected on the nerve cell layer. X 400

containing RNAs from fetal calf spinal cord could direct GFAP synthesis to have this protein comprise as much as about 1.4% of the total product (Watanabe and Sueoka, 1984). When ^{32}P-cDNA is used to make poly(A)-containing RNAs from fetal calf spinal cord, the recombinant clones with GFAP cDNA fragment are expected to give strong hybridization signals. Thus, in the *in situ* colony hybridization experiment, we obtained 27 clones with strong hybridization signals from approximately 1200 transormants, and candidate recombinant clones were analyzed by hybrid-selected translation. Plasmid hybrid-selected RNAs were translated in a cell-free system and translational products were immunoprecipitated with GFAP-antiserum. In several experiments, specific immunoprecipitants of products synthesized in response to selected RNAs by two recombinant plasmids, p411 and p540, always gave higher incorporation of ^{35}S-methionine than other recombinant plasmids or reference plasmids (pBR322, cloning vector, and pRSA13, rat serum albumin cDNA, Sargent et al. 1979 and 1981) used as negative controls (data not shown). Fig. 1A and B show SDS-polyacrylamide gel electrophoresis patterns of total *in vitro* translational products and their immunoprecipitants treated with GFAP antiserum. RNAs selected by two candidate plasmid (p411 and p540) DNAs could direct the synthesis of polypeptide which has a molecular weight of 50k daltons and is immunoprecipitated with GFAP-antiserum (Fig. 1A and 4 and 10, Fig. 1B lanes 4 and 8). Thus, these two recombinant clones (p411 and p540) seemed to contain cDNA inserts specifically hybridizing with GFAP mRNA. The recombinant clones p537 and p580 may also contain GFAP-cDNA, but these were not further characterized.

The size of the insert after the cleavage by Pst I digestion was determined on 1.2% (w/v) agarose gel electrophoresis. For one plasmid p411, it was approximately 600 bases long, but for the other plasmid, it was not cleaved by Pst I digestion (Fig. 2). The Pst I site of the latter plasmid DNA might be modified by mutation.

The length of mRNA hybridizable to the p411 insert was estimated by electrophoresis of bovine spinal cord poly(A)-containing RNAs on denaturing gel followed by transfer to nitrocellulose filter and hybridization with nick-translated p411 insert probes. Fig. 3 shows that ^{32}P-labeled probes hybridized to a major RNA band of approximately 2.4K nucleotides long. This mRNA is similar in size to GFAP-mRNAs described for mouse and chicken (Lewis et al., 1984; Capetanaki et al., 1984). When rat brain poly(A)-containing RNAs were used instead of bovine spinal cord RNAs, ^{32}P-labeled p411 probes also hybridized

a

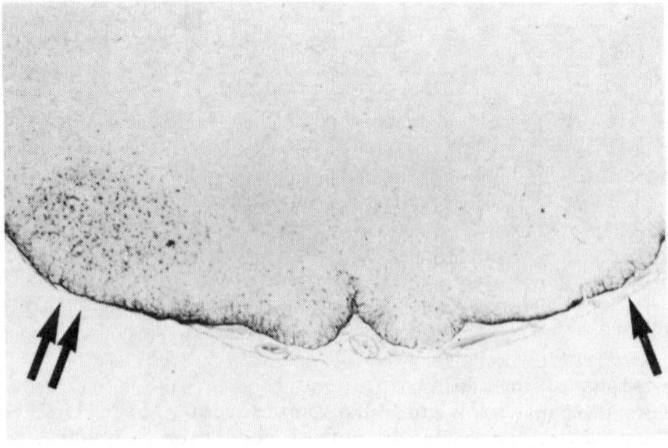

b

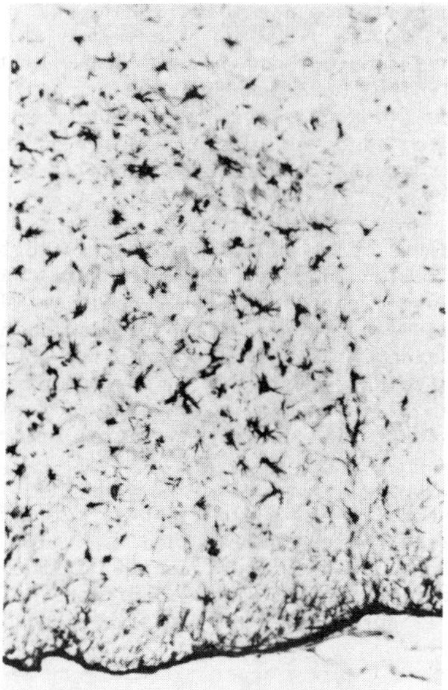

c

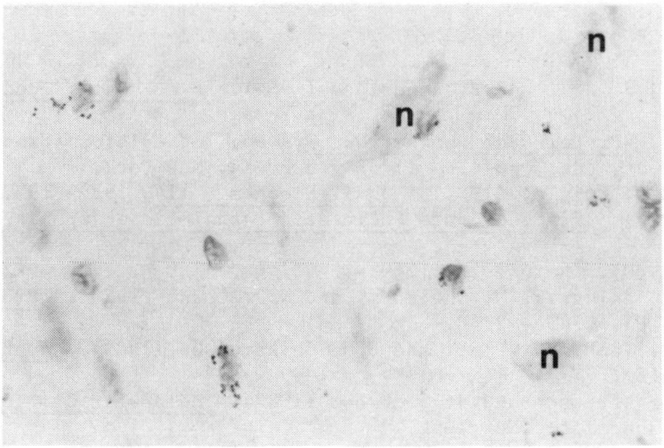

Fig. 5. Retrograde degeneration of rat facial nuclei.

a. Immunohistochemistry with the anti-GFAP antiserum. Seventh day after the operation. GFAP-positive cells are observed only in the area of the degenerated facial nucleus (double arrow). Few positive cells are detected on the control side (single arrow). X 20

b. High power view of the similar preparation as Figure 5a. Note the sharp boder of the GFAP-psotive zone. X 170

c. *In situ* hybridization with p411 cDNA insert. Astroglial cells in the degenerated nucleus are tagged with silver grains. Third day after the operation. "n" indicates nerve cells. X 800 to a few RNA bands of more than 2.2K bases long although long exposure was required to detect the bands (data not shown).

On the *in situ* hybridization using p411 insert, we could detect silver grains on the cytoplasm of astroglia in the hippocampus or in the white matter of rat cerebrums, where GFAP-positive cells are observed in normal condition (Fig. 4). No silver grains were detected on the nerve cells or on the mesenchymal cells. These findings indicate that the p411 insert reacts as the cDNA fragment for the GFAP of rats.

One of the authors, Kitamura, has pointed out that astroglia in the cerebral cortex or in the facial nucleus do not react with anti-GFAP antiserum in normal condition, but, when damaged, they become GFAP-positive (Kitamura, T., 1985). In the case of retrograde degeneration of the rat facial nucleus GFAP-positive astroglia were observed from the second day after the operation (Fig. 5a, b). On *in situ* hybridization using p411 insert cDNA a lot of silver grains were observed on the cytoplasm of astroglia in the damaged facial nuclei, whereas, only a few astroglia in the facial nuclei on control side were tagged with silver grains (Fig. 5c).

Generally GFAP is considered to be an excellent marker for astroglia. This may be true for the white matter of the brain, but in particular portions of the brain, such as cerebral cortices or facial nuclei, GFAP synthesis is induced in the astroglial cells under the pathological conditions. This may mean that the gene expression of GFAP is not always constitutive in all astroglial cells, but inducible in some.

AKNOWLEDGEMENT

The authors thank Dr. L. F. Eng for the supply of GFAP antiserum, and Drs H. Teraoka and S. Beckmann for helpful discussions.

REFERENCES

Capetanaki, Y.G., J. Ngai and E. Lazarides (1984). *Proc. Natl. Acad. Sci. USA, 81*, 6909 - 6913.

Eng, L.F. (1980). *Proteins of the Nervous System*, 2nd edition (Bradshaw R.A. and D.M. Schneider, eds), pp 85 - 117 Raven Press, New York.

Holmes, D.S. and M. Quigley (1981). *Anal. Biochem., 114*, 193 - 197.

Kitamura, T. (1985). *Transactiones Societatis Pathologicae Japoniae 74* (in press).

Lazarides, E. (1980). Nature 283, 249 - 256.

Lewis, S.A., J.M. Balcarek, M. Shelanski and N.J. Cowan (1984). *Proc. Natl. Acad. Sci. USA, 81*, 2743 - 2746.

Lewis, E.J., A.W. Tank, N. Weiner and D.M. Chikaraishi (1983). *J. Biol. Chem. 258*, 14632 - 14637.

Maniatis, T., E.F. Fritsch and J. Sambrook (1982). *Molecular Cloning*, Cold Spring Harbor Laboratory, New York.

Sargent, T.D., J-R. Wu, J.M. Sala-Trepat, R.B. Wallace, A.A. Reyes and J. Bonner (1979). *Proc. Natl. Acad. Sci. USA, 76*, 3256 - 3260.

Sargent, T.D., M. Yang and J. Bonner (1981). *Proc. Natl. Acad. Sci. 78*, 243 - 246.

Thomas, P.S. (1980). *Proc. Natl. Acad. Sci. USA, 77*, 5201- 5205.

Watanabe, S. and N. Sueoka (1984). *J. Neurochem. 42*, 1190 - 1193.

Development of Astrocytes: *in vivo* and *in vitro* Studies

A. J. Patel

MRC Developmental Neurobiology Unit, Institute of Neurology, 33 John's Mews, London, WC1N 2NS, UK

ABSTRACT

The major increase in the astrocytic proteins, glutamine synthetase (GS) and glial fibrillary acidic (GFA) protein, in three developmentally distinct regions in vivo is associated with maturation rather than with proliferation of astrocytes. Marked differences have been observed in the regional distribution of these proteins, the GS to GFA ratio varying 5-fold. This implies either that two types of astrocytes exist, one containing GS and the other GFA protein, or that all astroglial cells contain both proteins in varying amounts. Immunocytochemical observations on 14-day-old cultured cerebellar astrocytes were inconclusive, but those on forebrain astrocytes favoured the second possibility. Induction of GS by glucocorticoids was demonstrated in the brain in vivo, with the cerebellum showing the greatest response, followed by the olfactory bulbs and then the forebrain. Studies on astrocyte cultures showed that the regional differences were retained in vitro. These included the earlier development of GS and the lesser response of glucocorticoid treatment in the forebrain than in the cerebellum, although the hormonal induction could be elicited for a longer period in vitro than in vivo. Consistent with these findings, the forebrain or the forebrain astrocytes had lower concentrations of glucocorticoid receptors per unit GS activity than did the cerebellum or cerebellar astroglial cells, respectively. GS was also regulated by glutamine. The increases in GS activities, involving protein synthesis de novo and brought about by addition of dexamethasone and by removal of glutamine, were additive and therefore probably mediated by separate mechanisms. The observations with inhibitors which block protein synthesis at different stages indicated that the effect of glucocorticoids occurred at a transcriptional level, while that of glutamine removal was post-transcriptional, most probably by an increase in mRNA stability. Moreover, further differences in the biochemical properties of forebrain and cerebellar astrocytes have been observed. Finally, the distribution in neural cell types of enzymes and amino acids related to glutamate metabolism and of transport systems for acidic amino acids is in accord with our hypothesis on metabolic compartmentation of glutamate deduced from kinetic studies in vivo, and indicates that metabolic interaction occurs between astrocytes and neurones. Also, the comparison of the regional distribution of the astrocyte markers with that of certain neuronal indices has indicated that astrocytes in different parts of the brain have distinct biochemical properties which reflect the functional characteristics of the region.

Advances in the Biosciences Vol. 61

Printed in Great Britain

KEYWORDS

Regulation of astrocytic proteins; glutamine synthetase; glial fibrillary acidic protein; biochemical heterogeneity of astrocytes; glucocorticoid action; glutamine mediated regulation; ornithine decarboxylase; cerebellar astrocytes; forebrain astrocytes.

INTRODUCTION

In recent years our concept of the astrocyte as primarily a passive component separating neuronal elements has been radically changed. Our findings, together with those of others on the metabolism of labelled substrates, on the distribution of enzymes and free amino acids related to the metabolic compartmentation of glutamate, and on the uptake of the acidic amino acids in different cell types, have clearly demonstrated a role of astrocytes in the processing of amino acid neurotransmitters (for references see Balázs and Cremer, 1973; Patel, 1982; Gordon and Balázs, 1983; Patel and Hunt, 1985a). Astrocytes also possess a very efficient system for accumulation of K^+, which may be related to their part in ion homeostasis and tissue swelling (Walz and Hertz, 1983; Kuffler, et al, 1984). It has also been proposed that astrocytes provide the support for myelinated axons in the central nervous system (CNS; Peters, et al, 1976). Furthermore, in several genetic and non-genetic neurological disorders, the formation of reactive astrocytes has been observed most probably as a response to neural cell injury or death (Gall, et al, 1979; Davison and Thompson, 1981). On the other hand, the co-culturing of neural cells or the use of conditioned media have indicated that astroglial cells produce growth factors that enhance the survival and maturation of certain types of neurones (Lindsay, et al, 1982; Rudge, et al, 1985). Thus it is possible that glial cells themselves may be responsible for both the death of nerve cells and the subsequent reaction to their death. During the last few years therefore, we have carried out a systematic study to understand mechanisms involved in the regulation of the development of astroglial cells, and this chapter summarizes some of our findings.

IN VIVO DEVELOPMENT OF ASTROCYTE-ENRICHED PROTEINS

In the brain, a number of proteins, including glial fibrillary acidic (GFA) protein, glutamine synthetase (GS), S-100 protein and αα-enolase, are selectively enriched in astroglial cells (for references see Patel, et al, 1985b). In a search for appropriate markers of astrocyte development, we have estimated the ontogenetic changes in the amount of GS and GFA proteins in three asynchronously maturing regions of the rat brain (Table 1; Patel, et al, 1983a; Weir, et al, 1984). In the forebrain and olfactory bulbs the accretion of both GFA protein and GS activity occurred at a comparable age, and relative increases were more marked during the first 21 days after birth than subsequently. On the other hand, in the cerebellum, the developmental rise in GS activity and GFA protein was more or less similar up to 12 days after birth, but thereafter the amount of GFA protein increased at a much faster rate, reaching about 80% of the adult value by 20 days (Table 1). In contrast, GS activity increased at a much slower but linear rate up to 50 postnatal days, the activity at 20 days being only about 40% of the adult value. The acquisition of some of the astrocytic proteins at an early age may be related to the increase in astrocyte cell numbers, but the increase attributable to the proliferation must be rather small compared to that associated with the maturation of astrocytes. Moreover, since the accretion of both GS and GFA proteins continued well after the period when the rapid acquisition of cells, including astrocytes, in the brain has ceased (for references see Patel and Lewis, 1982), it would appear that the increases of these two proteins are associated with the differentiation rather than the acquisition of astrocytes (Patel, et al, 1983a; Weir, et al, 1984).

Table 1. In vivo and in vitro developmental patterns of astrocyte-enriched proteins.

The in vivo values are for whole brain regions of the rat, while in vitro results are for cultured astroglial cells derived from either newborn forebrain or 8-day-old cerebellum. The values for soluble GFA protein (μg/mg protein) are expressed as a percentage of the values at 65 days of age: these were 1.53 and 1.61 for the forebrain and cerebellum, respectively. The values on both in vivo and in vitro GS activity (μmol/h/mg protein) are expressed as a percentage of the in vivo value at 90 days of age: these were 2.41 and 3.07 for the forebrain and cerebellum, respectively.

Age in days	FOREBRAIN			CEREBELLUM		
	In vivo GFA protein	In vivo GS	In vitro GS	In vivo GFA protein	In vivo GS	In vitro GS
1	10	11	-	8	8	-
6	-	-	22	-	-	11
8	38	47	-	18	20	-
9	-	-	45	-	-	14
11	-	62	-	-	27	-
14	-	-	87	-	-	21
16	75	-	-	82	34	-
20	81	100	170	79	40	35
26	87	-	-	95	-	-
28	-	-	190	-	-	-
36	92	106	-	112	60	-
60	-	111	-	-	93	-
65	100	-	-	100	-	-
90	-	100	-	-	100	-

The values are taken from Patel, et al (1983a); Patel and Hunt (1985b) and Weir, et al (1984).

REGIONAL DISTRIBUTION OF GS AND GFA PROTEIN

The comparison of the adult concentrations of GS and GFA protein in the forebrain and the cerebellum indicated that the biochemical properties of astrocytes in these two brain regions may be different (Table 1; Patel, et al, 1983a; Weir, et al, 1984). This has now been confirmed in a detailed study, which has revealed marked differences in the distribution of GS activity and of GFA protein in 11 CNS regions (Fig. 1; Patel, et al, 1985b). The striatum and spinal cord contained the lowest concentration of GFA protein and GS activity respectively, while the olfactory bulbs had the highest level of both astrocytic proteins, the regional differences being up to 3-fold for GS and 4-fold for GFA protein. Immunochemical detection of GS and GFA proteins in whole homogenates of different regions confirmed that the variation of specific activities of GS and the concentration of soluble GFA protein were due to the differences in their absolute protein concentrations (Patel, et al, 1985b). For GFA protein, the regional variation in rat CNS also agreed with that previously reported in the human CNS (Palfreyman, et al, 1979). Furthermore, since intense immunostaining for both these marker proteins is seen in processes and the cell bodies of astrocytes (Bignami and Dahl, 1973; Ludwin, et al, 1976; Norenberg, 1979), we might conclude that there is a 3- to 4-fold variation in the relative volume occupied by the astroglial cells in different regions of the CNS. This would be in accord with the previous quantitative electron microscopic studies carried out by Wolff (1976) on the distribution of astrocyte volume in the neuropil of different brain areas.

One of the major findings in our study was the 5-fold variation in the GS to GFA protein ratio in different CNS regions. Expression of values as a GS to GFA protein ratio eliminates the variations in the proportion of the astrocytic volume in different brain parts. The findings provide direct evidence for the presence of more than one type of astroglial cell in terms of biochemical composition. The possibilities are either that there are two types of astrocytes, one containing GFA protein and the other containing GS, or that all astrocytes contain GS and GFA proteins but in varying amounts. Immunocytochemical examination indicated that in 7-day-old cultures of astrocytes derived from newborn forebrain more than 90% of cells were positive for GFA protein, while less than 30% of cells showed detectable immunofluorescence to GS antibody (Koppel, et al, 1985). However, the amount of GS increased 4-fold between 7- and 14-days in vitro (DIV; Table 1; Patel and Hunt, 1985b), and in 14-day-old cultures nearly all astroglial cells were positive for both GS and GFA proteins (Table 2). On the other hand, cerebellar astrocytes have half the activity of GS of forebrain astrocytes (Table 1) and in 14-day-old cerebellar culture only about 70% of GFA protein positive cells showed immunofluorescence to GS antibody (Table 2). Possible reasons for the absence of GS immunofluorescence in some young astrocytes might be: a very low amount of GS protein in immature astrocytes; the dispersal of GS, but not the cytoskeletal GFA protein, throughout the cell cytoplasm; and much lower titres of GS antisera compared to GFA protein antisera.

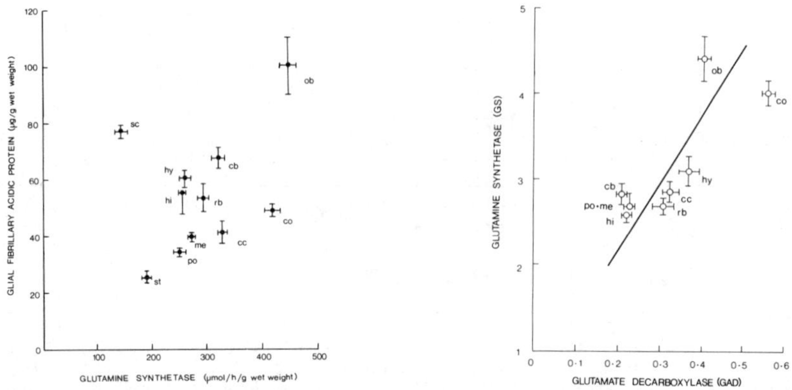

Fig. 1. Correlation of the activity of GS with either GFA protein (on left; n=33; r=0.33; not significant) or glutamate decarboxylase activity (on right and expressed as µmol/h/mg protein; n=24; r=0.78; p<0.001) in different regions of the adult rat CNS. Abbreviations: cc, cerebral cortex; cb, cerebellum; rb, residual brain; st, striatum; hi, hippocampus; co, colliculi; hy, hypothalamus; me, medulla; ob, olfactory bulbs; sc, spinal cord, po, pons.

COMPARISON BETWEEN IN VIVO AND IN VITRO DEVELOPMENT OF GS

In the brain, GS is mainly localized in the astroglial cells (Norenberg, 1979; Patel, et al, 1982); this allows a comparison between in vivo and in vitro developmental patterns of the enzyme. Since astrocytes may constitute a relatively small part of the brain volume (Wolff, 1976), it was anticipated that the specific activity in astrocyte cultures would be higher than that in vivo both for forebrain and cerebellum. As expected, in forebrain astrocytes the specific activity of GS after 14 DIV was always higher than at the corresponding age in vivo (Table 1; Patel and Hunt, 1985b). However, the cerebellar astrocyte cultures failed to show a similar phenomenon and at all ages the in vitro specific activity values were lower than the corresponding in vivo values (Table 1). These regional differences may relate to the relatively early maturation of GS in the forebrain compared to the cerebellum (Table 1; Patel, et al, 1983a), and this must be considered when comparing the in vivo and in vitro activities of the enzyme. Apart from these differences, there was considerable similarity between in vivo and in vitro developmental patterns for forebrain and cerebellar GS activity: the rapid increase in GS ceased after 20 days both in vivo and in vitro in forebrain, while GS in cerebellar cultures and in vivo increased in a more or less parallel fashion during the second and third weeks of age.

REGULATION OF GS

The regulation of GS is brought about by a variety of mechanisms (for references see Patel, 1982; Patel, et al, 1983b). In previous studies it has been shown that the level of GS in embryonic retina is controlled by glucocorticoids (Moscona and Piddington, 1966; Reif-Lehrer and Amos, 1968). More recently an increase of GS brought about by glucocorticoids has also been reported in cultured neural cells (Pishak and Phillips, 1980; Hallermayer, et al, 1981; Juurlink, et al, 1981), and in a variety of non-neural mammalian tissues (see Patel, et al, 1983b). In accord with these findings we have now shown that a glucocorticoid regulates the activity of GS in rat brain in vivo (Patel, et al, 1983b). The increase was more marked in young than in adult rats, and also varied considerably from one brain region to another. The cerebellum showed the greatest response, followed by the olfactory bulbs and then the forebrain, suggesting that the induction of GS depended on the maturational state of the region at the time of the hormone treatment. However, the relatively small response in the forebrain, particularly after a single dose of corticosterone, was surprising, and may suggest that forebrain astrocytes, irrespective of their developmental state, are less responsive to the hormone than astroglial cells of the cerebellum. In more recent studies, we have used primary astrocyte cultures, derived from forebrain or cerebellum, to elucidate in vivo regional differences in response to glucocorticoid treatment (Patel and Hunt, 1985b). In agreement with the in vivo findings (Patel, et al, 1983b), the steroid induction in forebrain astrocytes in vitro was significantly less than in cerebellar astroglial cells (Table 2). The extent of in vivo induction of GS activity due to glucocorticoid treatment decreased during maturation in both forebrain and cerebellum (Patel, et al, 1983b). However, a similar developmental decline was not observed in vitro, the induction of GS remained at a constant elevated level from 6 DIV to any time up to 28 DIV for both forebrain and cerebellar astrocyte cultures (Patel and Hunt, 1985b). The differences in the biochemical properties of these two types of astrocytes were intrinsic and were not related to the cell density, total protein content per cell or to the purity of the cultures (Patel and Hunt, 1985b).

Removal of glutamine from the culture medium also resulted in a marked increase in GS activity in cultures of a variety of cell types, including HeLa cells, Chinese hampster ovary cells, mouse L cells, hepatoma cells, neuroblastoma and C-6 glioma cells (for references see Lacoste, et al, 1982; Patel, 1982). In agreement with these observations on maintained cell lines, we have also found in both forebrain and cerebellar astroglial cells that replacement of a normal medium with one containing no added glutamine resulted in a rapid increase in the specific activity of GS during the first 24 hours, the higher level persisting throughout the period of glutamine deprivation (Patel, et al, 1985a). The relative increase (about 100%) in GS activity resulting from glutamine deprivation was similar in both types of astrocytes, even though the basal enzyme activity in the forebrain astrocytes was about 2.5-fold greater than in the cerebellar astroglial cells (Table 2). Furthermore, the increases caused by removal of glutamine and addition of dexamethasone were additive. They were also specific to GS, for neither treatment with glucocorticoid nor removal of

Table 2. Comparison of some of the biochemical properties of forebrain and cerebellar astroglial cells.

Primary cultures (14 DIV) enriched in normal astrocytes were obtained by plating dissociated cells from newborn forebrain and from either newborn or 8-day-old cerebellum of rats.

	Forebrain astrocytes	Cerebellar astrocytes	Forebrain / cerebellum ratios
I CELLULAR COMPOSITION (14 DIV)			
1. Immunocytochemically positive for GFAP (%)	>95	>90	
2. Immunocytochemically positive for GS (%)	>95	>60	
3. Contaminating fibroblast and endothelial cells (%)	<4	<7	
II GROWTH (14 DIV)			
1. Cell density at confluency (10^3 cells/cm^2)	310	210	1.48
2. Cell size distribution (% of total cells)			
5.5. - 10 µm	25	26	0.96
10 - 14.5 µm	46	22	2.09
> 14.5 µm	29	52	0.56
3. Protein/DNA (µg/ng atom-P)	8.9	12.5	0.71
III GLUTAMINE SYNTHETASE (GS; 14 DIV)			
1. GS activity (µmol/h/mg protein)	1.1-1.7	0.45-0.65	2.54
2. Dexamethasone treatment, 0.1 µM, 24 h (% induction)	50-100	130-180	0.48
3. Removal of glutamine, 48 h (% induction)	60-100	60-100	1.00
IV CYTOSOLIC GLUCOCORTICOID RECEPTORS (^{3}H-DEX)			
1. In vivo (14-day-old whole tissue)			
a. ^{3}H-DEX (fmol/mg protein)	199	438	0.45
b. ^{3}H-DEX/GS activity	100	447	0.22
2. In vitro (14 DIV astrocytes)			
a. ^{3}H-DEX (fmol/mg protein)	282	283	1.00
b. ^{3}H-DEX/GS activity	200	449	0.45
V ORNITHINE DECARBOXYLASE (ODC; 14 DIV)			
1. ODC activity (nmol/h/mg protein)	0.94	0.43	2.18
2. Removal of glutamine for 2 days (% of control)	13	44	0.30
3. Removal of glutamine for 7 days (% of controls)	90	51	1.76
VI FREE AMINO ACIDS (14 DIV)			
1. Glutamate (nmol/mg protein)	36	67	0.54
2. Aspartate (nmol/mg protein)	3.4	6.1	0.56
3. Glycine (nmol/mg protein)	12	60	0.20

Values are taken from Koppel, et al (1985); Patel and Hunt (1985a, 1985b); Patel, et al (1985a, 1985c).

glutamine, singly or together, had any appreciable effect on the cytosolic enzyme, lactate dehydrogenase, or on the protein content of the astroglial cells (Patel, et al, 1985a). These results suggest that the effects of glucocorticoid treatment and of removal of glutamine are mediated by different mechanisms.

POSSIBLE MECHANISMS OF GS REGULATION MEDIATED BY GLUCOCORTICOIDS OR BY GLUTAMINE REMOVAL

The mechanism of the action of glucocorticoid hormones has been extensively investigated (see Baxter and Rousseau, 1979). It involves the transport of hormone through the cell membrane and its binding to intracellular receptors; this complex in turn is translocated into nuclei where it associates with chromatin and elicits transcription of mRNA for a specific protein (e.g. GS), resulting in increased synthesis and accumulation of the protein. Therefore a close relationship between the amount of cytosolic glutocorticoid receptors and the physiological effects produced by the steroid in the cell is to be expected. The concentration of glucocorticoid receptors (specifically bound [^{3}H]dexamethasone per mg cytosol protein), as did GS activity, increased during development both in dissected brain tissue and in astrocyte cultures in vitro (A.J. Patel and A. Hunt unpublished observations; also see Pavlik and Buresova, 1984). Between 7 and 14 days, the increases in receptor binding were greater in cultured astrocytes from the forebrain (190%) and the cerebellum (170%) than the increases in vivo in the forebrain (160%) and the cerebellum (120%). In addition, both the concentration of glucocorticoid receptors per mg protein and the amount of receptors per unit GS activity were respectively 2.2- and 4.5-fold greater in cerebellum than in forebrain of 14-day-old rats (Table 2). Similarly, in cerebellar astrocytes, the concentration of steroid receptors was about 2.2-fold more than in forebrain astrocytes, when expressed as receptors per unit GS activity (Table 2). In comparison with whole forebrain or forebrain astrocytes, the relatively high level of glucocorticoid receptors per unit GS activity in the cerebellum both in vivo and in cultured cerebellar astrocytes was consistent with their greater responsiveness to the hormone (Table 2; Patel, et al, 1973b; Patel and Hunt, 1985b).

Comparatively little information is available on the mechanism of action of glutamine on GS activity. A possibility is that the removal of glutamine leads to the induction of a factor which directly activates or inhibits the enzyme (Bazzel, et al, 1976; Laitinen, et al, 1985). However, we have shown that the induction of GS activity is due to the increase in the amount of enzyme protein and not to activation of preformed enzyme (Patel, et al, 1983b). Also, the responses to both dexamethasone treatment and to removal of glutamine were blocked by the protein synthesis inhibitor, cycloheximide (Patel, et al, 1985a). These observations suggest that the primary mechanism of both these effects entails an increase in the availability of GS mRNA for enzyme synthesis, which could be due to increased transcription or to decreased breakdown of mRNA. Some indication as to whether the effect occurs at a transcriptional or at a post-transcriptional level has been obtained by using α-amanitin, an inhibitor of the DNA dependent RNA polymerase II catalyzing the synthesis of mRNA (Chambon, 1975). α-Amanitin markedly reduced the increase in GS activity after dexamethasone treatment, but not that following removal of glutamine from the growth medium (Patel, et al, 1985a). Thus the effect of glucocorticoid hormone appears to be transcriptional, while that of glutamine removal is post-transcriptional. It is possible that the latter effect may not be restricted to GS mRNA but could be a more general effect on total mRNA of the cell. This is supported by the effect of α-amanitin on ornithine decarboxylase (ODC), an enzyme with a very short half-life (Maudsley, 1971; Russel, 1980). In astrocytes grown in a medium containing glutamine, α-amanitin completely suppressed ODC activity, while about one-third of the original ODC activity remained after α-amanitin treatment in the cells grown in glutamine deficient medium (Patel, et al, 1985a). However, a direct experimental approach using a recombinant DNA probe for GS (Sanders and Wilson, 1984) is essential for final confirmation of our proposal that the removal of glutamine from the growth medium increases mRNA stability in primary cultures of astroglial cells. It is hoped that this approach will facilitate closer examination at the molecular level of the mechanisms regulating the expression of specific genes in astroglial cells.

REGULATION OF ODC IN TWO TYPES OF ASTROCYTES

ODC is an important enzyme in the biosynthesis of polyamines, which play a role in nucleic acid and protein synthesis and appears to be closely associated with both cellular growth and differentiation (Maudsley, 1979; Russell, 1980). Each region of the CNS seems to have a specific "ODC-polyamine" developmental pattern, and alterations in this system have been used as an index of perturbation of neuronal maturation (see Slotkin, 1979). Our recent findings show that, as with GS, forebrain astrocytes have greater ODC activity than do cerebellar astrocytes (Table 2) and the forebrain to cerebellum ratios for both GS and ODC activities of the astrocytes are very similar (Patel, et al, 1985a). However, in contrast to GS, when the cells were grown in glutamine deficient medium, effects on the ODC of these two types of astrocytes showed marked differences (Table 2; Patel, et al, 1985a). In forebrain astrocytes, ODC activity was only transiently reduced, while in cerebellar astrocytes the activity remained markedly decreased throughout the period of glutamine deprivation. Furthermore, contrary to general belief (see Slotkin, 1979), such a severe reduction in ODC activity had relatively little effect on cell numbers or protein content of the cerebellar astroglial cultures (Patel, et al, 1985a).

INTERACTION BETWEEN ASTROCYTES AND NEURONES

Kinetic studies on the oxidation of labelled substrates have indicated that the metabolism of glutamate and of the associated tricarboxylic acid (TCA) cycle intermediates are compartmented, and there seems to be an interaction between neurones and glial cells in the processing of amino acid neurotransmitters in the brain (for references see Balázs and Cremer, 1973; Patel, 1982). Briefly, it has been proposed that when transmitter amino acids, associated with the TCA cycle, are released from nerve terminals, they are largely taken up into astroglial cells. Here, they are converted into glutamine. Glutamine, in turn, is taken up, in part, after diffusion into the extra cellular space, by neuronal structures and hydrolysed to glutamate, thus balancing the loss of carbon units in the "cycle" that occurs as a result of release of the amino acids. In support of this concept, we have reported the selective distribution between neurones and astroglial cells of related free amino acids (Patel and Hunt, 1985a) and of their uptake (Gordon and Balázs, 1983), together with that of enzymes directly involved in the metabolic compartmentation of glutamate (Patel, et al, 1982). In neuronal cell cultures, the activity of glutaminase and the concentrations of glutamate, aspartate and GABA were several-fold greater than in astrocytes. In contrast, GS activity, amount of glutamine and uptake of glutamate were greater in astroglial cell cultures than in interneurone cultures. The observations were also in accordance with the immunocytochemical localization of GS in astrocytes (Norenberg, 1979) and of glutamate decarboxylase (McLaughlin, et al, 1974) and glutaminase (Cangro, et al, 1985) in neuronal structures. Furthermore, the finding that astrocytes from different CNS regions differ greatly in their GS content relative to GFA protein (Fig. 1; Patel, et al, 1985b) raised the possibility that a relationship may exist between the amount of GS and the number of cells and/or nerve terminals using GABA and glutamate neurotransmitters in a given region. The assessment of the distribution of a particular neurotransmitter system can be made by measuring the specific activity of the enzyme important in the synthesis of that transmitter (see Patel, et al, 1978; Gombos and Aunis, 1982). We have examined two such enzymes, glutamate decarboxylase and choline acetyltransferase, involved in the synthesis of GABA and acetylcholine, respectively. In agreement with the concept of a metabolic interaction between neurones and astrocytes in the processing of amino acid neurotransmitters, in different regions of the CNS the activity of GS was significantly correltated with that of glutamate decarboxylase (Fig. 1) but not with that of choline acetyltransferase (Patel, et al, 1985b). An apparent parallelism between immunohistochemically localized GS and suspected glutamatergic activity at a particular site has been noted previously (Norenberg, 1979). Also, it has been suggested that a relationship may exist between the proportion of glutamatergic innervation in a given brain region and the glutamate transport capacity of astroglial cells cultures from that region (Schousboe and Divac, 1979). On the other hand, in general, relatively high GFA protein to GS ratio in the area containing more white matter than grey matter (Patel, et al, 1985b), is consistent with astroglial cells providing the support to myelinated axons in the CNS (Peters, et al, 1976). Thus, it would appear that the astrocytes in different regions of the CNS have distinct biochemical properties, reflecting the functional characteristics of the neuronal pathway in that particular area.

ACKNOWLEDGEMENT

Gratitude is expressed to Robert Balázs for his interest. The author gratefully acknowledges the collaboration of Anthony Hunt, Hanno Koppel (Institute of Psychiatry), Caroline S.M. Tahourdin, David G.T. Thomas (Department of Neurological Surgery) and Mary D. Weir in these studies.

REFERENCES

Balázs, R., and J.E. Cremer, eds. (1973). Metabolic Compartmentation in the Brain, Macmillan, London.

Baxter, J.D., and G.G. Rousseau, eds. (1979). Glucocorticoid Hormone Action, Springer, Berlin.

Bazzel, K.L., G. Price, S. Tu, M. Griffin, R. Cox and N. Ghosh (1976). Europ. J. Biochem., 61, 493-499.

Bignami, A., and D. Dahl (1973). Brain Res., 49, 393-402.

Cangro, C.B., P.M. Sweetnam, J.R. Wrathall, W.B. Haser, N.P. Curthoys and J.H. Neale (1985). Brain Res., 336, 158-161.

Chambon, P. (1975). Ann. Rev. Biochem., 44, 613-638.

Davison, A.N., and R.H.S. Thompson, eds. (1981). The Molecular Basis of Neuropathology, Arnold, London.

Gall, C., G. Rose and G. Lynch (1979). J. Comp. Neurol., 183, 539-550.

Gombos, G., and D. Aunis (1982). Scand. J. Immunol. (Suppl. 9), 15, 317-338.

Gordon, R.D., and R. Balázs (1983). J. Neurochem., 40, 1090-1099.

Hallermayer, K., C. Harmening and B. Hamprecht (1981). J. Neurochem., 37, 43-52.

Juurlink, B.H.J., A. Schousboe, O.S. Jørgensen and L. Hertz (1981). J. Neurochem., 36, 136-142.

Koppel, H., A. Hunt and A.J. Patel (1985). Neuropath. Appl. Neurobiol., 11, 76.

Kuffler, S.W., J.G. Nicholls and A.R. Martin (1984). From Neuron to Brain, Sinauer, Sunderland.

Lacoste, L., K.D. Chaudhary and J. Lapointe (1982). J. Neurochem., 39, 78-85.

Laitinen, P.H., R.L. Huhtinen, O.A. Hietala and A.E.I. Pajunen (1985). J. Neurochem., 44, 1885-1891.

Lindsay, R.M., P.C. Barber, M.R.C. Sherwood, J. Zimmer and G. Raisman (1982). Brain Res., 243, 329-343.

Ludwin, S.K., J.C. Kosek and L.F. Eng (1976). J. Comp. Neurol., 165, 197-208.

Maudsley, D.V. (1979). Biochem. Pharmacol., 28, 153-161.

McLaughlin, B.J., J.G. Wood, K. Saito, R. Barber, J.E. Vaughn, E. Roberts and J.Y. Wu (1974). Brain Res., 76, 377-391.

Moscona, A.A. and R. Piddington (1966). Biochem. Biophys. Acta, 121, 409-411.

Norenberg, M.D. (1979). J. Histochem. Cytochem., 27, 756-762.

Palfreyman, J.W., D.G.T. Thomas, J.G. Ratcliffe and D.I. Graham (1979). J. Neurol. Sci., 41, 101-113.

Patel, A.J. (1982). Neurotransmitter: Interaction and Compartmentation (ed. H.F. Bradford), pp. 411-430, Plenum Press, New York.

Patel, A.J. and A. Hunt (1985a). J. Neurochem., 44, 1816-1821.

Patel, A.J. and A. Hunt (1985b). Develop. Brain Res., 18, 175-184.

Patel, A.J., and P.D. Lewis (1982). Mechanisms of Actions of Neurotoxic Substances (eds. K.N. Prasad and A. Vernadakis), pp. 181-218, Raven Press, New York.
Patel, A.J., M. del Vecchio and D.J. Atkinson (1978). Develop. Neurosci., 1, 41-53.
Patel, A.J., A. Hunt, R.D. Gordon and R. Balázs (1982). Develop. Brain Res., 4, 3-11.
Patel, A.J., A. Hunt and C.S.M. Tahourdin (1983a). Develop. Brain Res., 8, 31-37.
Patel, A.J., A. Hunt and C.S.M. Tahourdin (1983b). Develop. Brain Res., 10, 83-91.
Patel, A.J., A. Hunt and F. Faraji-Shadan (1985a). Develop. Brain Res., in press.
Patel, A.J., M.D. Weir, A. Hunt, C.S.M. Tahourdin and D.G.T. Thomas (1985b). Brain Res., 331, 1-9.
Patel, A.J., M.D. Weir, J.L. Darling, D.G.T. Thomas and A. Hunt (1985c). A Monograph Commemorating the Centenary of Rickman Godlee's Operation for Glioma (ed. M.D. Walker), Martinus Nijhoff, Boston, in press.
Pavlik, A., and M. Buresova (1984). Develop. Brain Res., 12, 13-20.
Peters, A., S.L. Palay and H.D. Webster (1976). The Fine Structure of the Nervous System: the Neurons and Supporting Cells, W.B. Saunders, Philadelphia.
Pishak, M.R., and A.T. Phillips (1980). J. Neurochem., 34, 866-872.
Reif-Lehrer, L., and H. Amos (1968). Biochem. J., 106, 425-430.
Rudge, J.S., M. Manthorpe and S. Varon (1985). Develop. Brain Res., 19, 161-172.
Russell, D. (1980). Pharmacology, 20, 117-129.
Sanders, P.G., and R.H. Wilson (1984). EMBO J., 3, 65-71.
Schousboe, A., and I. Divac (1979). Brain Res., 177, 407-409.
Slotkin, T.A. (1979). Life Sci., 24, 1623-1630.
Walz, W., and L. Hertz (1983). Prog. Neurobiol., 20, 133-183.
Weir, M.D., A.J. Patel, A. Hunt and D.G.T. Thomas (1984). Develop. Brain Res., 15, 147-154.
Wolff, J.R. (1976). Handbook of Electroencephalography and Clinical Neurophysiology (eds. A. Remond and O. Creutzfeldt), Vol. 2A, pp. 26-43, Elsevier, Amsterdam.

Glutamine Synthetase of the Chick Brain

G. Tholey, M. S. Ghandour, S. Bloch, M. Ledig and P. Mandel

Centre de Neurochimie di CNRS, 5 rue Blaise Pascal, 67084 Strasbourg CEDEX, France

ABSTRACT

We investigated immunocytochemically and biochemically the expression of glutamine synthetase (GS) in chick glial cells and neurons obtained in purified primary cultures and in tissue sections. The presence of GS is limited to astrocytes and their processes in tissue sections of chick cerebral cortex and cerebellum while the neurons and their structures do not express the enzyme. In contrast, neurons in pure primary cultures express a high level of GS as revealed by immunocytochemistry. The level of GS in primary cultures of astrocytes is modified by addition of fetal calf serum to the culture medium. Chick brain GS was purified and physicochemically characterized. Control of the enzyme activity by different divalent cations has been considered.

KEY WORDS

Glutamine synthetase; astrocytes; neurons; neurofilaments proteins; primary cultures; cellular localization; double immunostaining; physico-chemical characteristics; cation effect.

INTRODUCTION

Glutamine is involved in many synthesis of basic cellular components, like purines and pyrimidines nucleotides, nucleic acids, as well as the synthesis of the most abundant excitatory (glutamic acid) and inhibiting (GABA) amino acids.

The enzyme glutamine synthetase, GS (EC. 6.3.1.2.) which catalyzes the conversion of glutamate plus ammonium to glutamine, was purified and characterized from a great variety of prokaryotic and eukaryotic cells, as well as from mammalian brain (see for review Bloch et al, 1984). GS from these sources presents great structural and reactional similarities but none of these are physically or chemically identical. Moreover, discrepant conclusions concerning the regulatory effect of different metal ions on GS activity have been reported (Ip et al, 1983; Pishak et al, 1980). In order to characterize more precisely some physico-chemical properties of the chick brain GS, the enzyme was purified to apparent homogeneity by an original procedure and analysed.

Thus it is obvious that the synthesis of glutamine appears plays a critical role in brain function (Tapia, 1980). GS activity undergoes pronounced alterations during development and differentiation of the nervous tissue (Mahajan et al, 1982; Tholey et al, 1983) or in some neuropathological events, like Huntington's disease (Carter, 1983). Immunohistochemical studies on rat brain slices (Norenberg, 1979) showed that GS was exclusively located in astrocytes. By investigating immunocytochemically the cellular localization of GS in glial and neuronal cell lines, it has been observed that the enzyme may be present in both types of cell lines at low specific activity (Hallermayer et al, 1981). Moreover, an analysis of the distribution of GS in subcellular fractions of rat brain suggested a possible localization of the enzyme at the nerve terminal with the synaptic membrane (Dennis et al, 1980). The development of techniques to grow cerebral tissue in primary cultures has facilitated investigations on specific cerebral cell types. In the case of chick nerve tissue, glial cells (Sensenbrenner, 1977) and neurons (Pettmann et al, 1979) may be isolated in highly purified primary cultures, well characterized with respect to their morphological and biochemical maturation. Experiments were reported here in an attempt to investigate biochemically and immunocytochemically the cellular localization of GS in chick astrocytes and neurons in primary cultures and in chick brain slices.

RESULTS AND DISCUSSION

Purification of glutamine synthetase (Table 1)

Chick brains were homogenized in purification buffer 50 mM imidazole-HCl, containing 50 mM $MgCl_2$, 1 mM EDTA, 2 mM betamercaptoethanol and 0.1 mM phenylmethylsulfonyle fluoride (PMSF), pH 7.15. The crude homogenate was sonically disrupted, and centrifuged (106,000 g - 30 min). $(NH_4)SO_4$ was added to the supernatant. The precipitate obtained between 30 and 45% saturation was redissolved in the same buffer and dialyzed. The dialysate was loaded on an Affi Gel Blue column, washed with the purification buffer and eluted with 250 mM NaCl in the same buffer. Fractions with high enzyme activiy (more than 15 units/ml, determined as reported previousl, Bloch et al, 1984) were combined and the proteins precipitated by adding $(NH_4)_2SO_4$ until 45% saturation. The precipitate was redissolved in purification buffer, dialyzed against it and submitted to isoelectric focusing on 4% dextran gel containing 20 mM betamercaptoethanol and 2% ampholines (a mixture of 1.2% ampholines, pH 5.8 and 0.8% ampholines pH 7.9). The pH gradient finally obtained ranged between 4.2 and 7.7. After 19 hours electrofocusing at 8 W constant power, gel fractions were extruded and analysed for enzyme activity: higher than 10 units/ml fractions were combined and proteins precipitated by $(NH_4)_2SO_4$ adding until 45% saturation. The precipitate was redissolved in a minimum of purification buffer, and passed through a Sephadex-G150 column, equilibrated with purification buffer. Elution fractions with high enzyme activity (more than 7 units/ml) were combined and the proteins precipitated by $(NH_4)_2SO_4$ adding until 50% saturation. The precipitate was redissolved in purification buffer, stored at 0°C, and the characteristics of the purified enzyme preparation analysed.

Sedimentation analysis

Different molecular forms of the enzyme were separated by ultracentrifugation : aggregates which sediment very rapidly, and 3 forms of macromolecules, sedimenting separately with different velocities. Computer analysis of these results indicate that those forms correspond respectively to monomeric (23%), cubic tetrameric (less than 1%) and cubic octameric forms (77%).

TABLE I

PURIFICATION OF GLUTAMINE SYNTHETASE FROM CHICK BRAIN*

Step	Total protein (mg)	Total activity (units)	Specific activity (U/mg)	Purification	Yield %	Transferase activity/ Synthetase activity ratio
1. Crude extract (106,000 g supernatant)	559	1304	2.3	1	100	11.1
2. 30-45% $(NH_4)_2SO_4$ fractionation	182	1139	6.2	2.7	11.6	
3. Affi-Gel Blue column	10	343	34	14.6	26	
4. Isoelectric focusing	1.5	124	83	35	10	11.5
5. Sephadex G-150 column	0.7	64	91	39	5	11.3

*Specific activity is defined using the transferase reaction.

Electrophoresis analysis

Dissociation of the purified enzyme into monomeric forms by SDS analysis on polyacrylamide gels yielded only a single protein component, indicating that the subunits are identical. A similar subunit homogeneity of GS has also been reported for the enzyme of chick neurals retina (Sarkar et al, 1972), sheep brain (Ronzio et al, 1968), rat liver (Devel et al, 1978), plants (Mc Cormack et al, 1982) and bacterial enzymes (Adler et al, 1975). By comparison with standards, the molecular weight of the subunit was determined to be 45,000 ± 2,000, value similar to that estimated for other eucaryotic GS.

Molecular mass and subunit composition

Based on the results obtained from sedimentation equilibrium studies, a molecular mass of about 364,000 ± 12,000 has been determined. This value is close to those determined for pig brain: 370,000 (Stahl et al, 1972), chick retina: 340,000 (Sarkar et al, 1972), rat liver: 360,000 (Devel et al, 1978), sheep brain (Ronzio et al, 1969) and fit very well with the octameric structure proposed by Sarkar et al, 1972).

Regulation of the enzymatic activity

The optimal temperature of GS from chick brain was 45-50°C. Above these values, a rapid inactivation of the enzyme occurred.

The activity of the enzyme in the presence of different analogs of glutamate has been determined (Table II). Only D-glutamate and D-L α-methylglutamate, closely related to the normal substrate showed slight activity: the purified enzyme is highly specific with L-glutamate as substrate.

TABLE II

SUBSTRATE SPECIFICITY OF CHICK BRAIN GS.

Substrate (50 mM)	GS activity %
L-glutamate	100
D-glutamate	37,4
DL-α-methylglutamate	44
L-glutamine	2,6
D-glutamine	2,3
L-aspartate	0,9
D-aspartate	0,6
GABA	0,3
γ-OH-butyrate	0,3

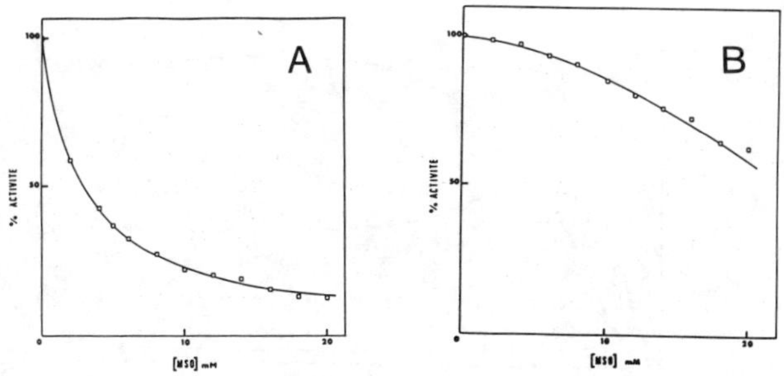

Fig. 1. Effect of L-methionine-DL-sulfoximine (MSO) on the synthetase (A) and transferase (B) activity of chick brain GS. In the presence of 1 mM MSO, a decrease of 26% of the initial activity occurs; 50% of the initial activity remains in the presence of 2.9 mM MSO. With similar concentrations of this inhibitor, the transferase activity is only reduced by 0.7 and 1.7% respectively.

The presence, simultaneously, of a synthetase and a transferase activity site on the same molecule was confirmed by determining the ratio of both activities during the different steps of the preparation. This value was not modified and always very close to 11. These results are in agreement with those of Juurlink (1982) who observed no significant variation in the GT/GS ratio of cultured mouse astroglial cells by cortisol and diB-cAMP treatment.

The inhibitory effect of methionine sulfoximine (MSO) toward GS and GT activity was determined: synthetase activity is approximately 10 fold more sensitive than transferase activity to MSO (Fig. 1).

A decrease of the ATP/ADP ratio drastically reduces the activity of the enzyme. Maximum level of activity is expressed in the presence of ATP concentration at least 5 fold higher than ADP value.

Concerning the metal ions requirements of chick brain GS, it was observed that among the various cations studied (Ca^{++}, Co^{++}, Fe^{++}, Li^{+}, Mg^{++}, Mn^{++}, Ni^{++}, Zn^{++}) only Mn^{++}, Mg^{++} and Co^{++} stimulated the synthetase activity. In the presence of 1 mM ATP, the order of activating ability of these cations was: $Mg^{++} > Co^{++} > Mn^{++}$ (Fig. 2). Moreover, it appears that the activatory effect of Mn^{++} concerns only a very sharp range of concentration (between 0 and 0.5 mM); 1 mM Mn^{++} inhibits already very strongly GS activity. In the range of cation concentration examined (0 - 20 mM) a slight inhibitory effect is also observed with Co^{++} (at a concentration superior to 5 mM). Nothing similar is observed with Mg^{++}. Activatory effects of Mg^{++} and Mn^{++} toward GS activity have previously been reported by Wedler et al (1982) for the ovine brain enzyme. Concerning an activatory effect of Co^{++}, a similar effect has been reported for GS of Anabaena (Orr et al, 1981) and that of bovine retina (Pahuja and Reid, 1985). Concerning chick brain GS, no effect of Co^{++} has ever been reported.

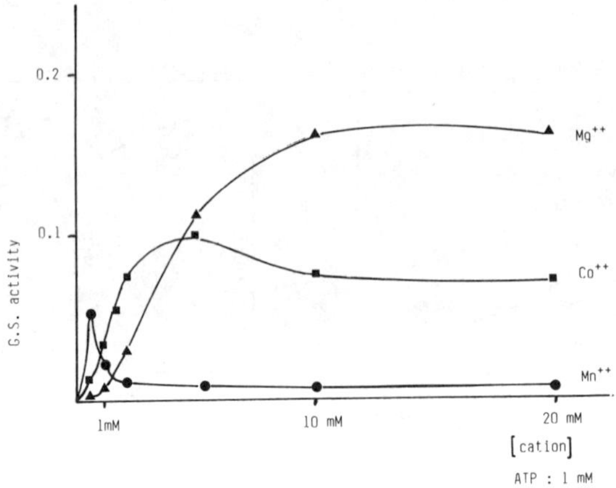

Fig.2. Effect of Mn^{++}, Mg^{++} and Co^{++} on GS activity in the presence of 1 mM ATP. Activity is expressed as OD 540 nm in 10 min. The highest GS activity is observed in the presence of 10 mM Mg^{++}. The optimum for Mn^{++} and Co^{++} concentrations were respectively 0.5 and 5 mM, supporting respectively 30 and 61% of the maximum Mg^{++} dependant activity.

Immunocytochemical procedures : Cerebral hemispheres and cerebellar sections were prepared as previously described (Ghandour et al, 1979, 1980). Briefly, chick embryos (14 and 19 days) were fixed by trans-cardiac perfusion with 4% paraformaldehyde and 0.1% glutaraldehyde. The removed tissues were postfixed for 1 hour at 4°C in the same fixative. 100 μM thick vibratome sections were obtained and then preincubated for overnight in phosphate buffer saline (PBS) containing 3% BSA. For cultures, the cells were washed twice with PBS and then fixed for 1h at 4°C. After washes, the cells were treated with 5% acetic acid in 96% ethanol for 10 min at -20°C. For immunostaining, tissue sections and cultivated cells were incubated for 1h at room temperature in 1/100 dilution GS rabbit immune serum and then with fluosceine isothiocyanate (FITC) or horseradish peroxidase (PO) - conjugated sheep antirabbit IgG (Pasteur Production Paris). The peroxidase activity was revealed with 0.01% 3,3' diaminobenzidine 4-HCl (DAB) and 0.001% H_2O_2 for 10 min. For the double labeling technique in cultivated neurons, the cells fixed as described above were incubated simultaneously with mouse monoclonal antibodies to neurofilaments at 1/200 ascite fluid dilution and with rabbit immune serum against GS at 1/100 dilution (1h, room temperature). After washes, the cells were incubated for 1h with both FITC - conjugated sheep anti-mouse IgG and tetramethyl rhodamine isothiocyanate (TRITC) - conjugated sheep anti-rabbit IgG. Tissue sections and cells in tissue culture were examined with a LEITZ orthoplan microscope equipped with epifluorescence attachments.

Immunocytochemical localization of GS

The cellular localization of GS in cerebellar tissue sections and in cerebral cortex from 19 day old chick embryos is limited to astrocytes and their processes (Fig. 3 and 4). The Bergmann fibers, their end feet on Pia mater and their cell bodies in Purkinje cell layer are strongly fluorescent (Fig. 3). Other protoplasmic astrocytes in the granular layer express the GS. Purkinge and granule cells, other neuronal structures and the newly formed interneurons in the external granular layer are not stained. In cerebral cortex (Fig. 4) the staining is limited to astrocytes, their processes and their feet surrounding the blood vessels. Similar results were obtained from 14 day-old embryos (not shown). Mainly the protoplasmic astrocytes are stained, while the fibrous astrocytes in the white matter are less stained. The present results are in accordance with the previous data concerning GS localization in the rat CNS (Norenberg, 1979; Norenberg and Martinez-Hernandez, 1979). Neurons in pure primary culture, derived from 7 day-old brain embryos are able to express GS as shown by immunochemistry (Fig. 5). The identity of those neurons that contain GS is confirmed by the double staining experiment with neurofilament proteins NF (Fig. 6). All the cells NF^+ are GS^+. The staining of neurons in primary cultures for GS suggests that those cells are able to express the enzyme when they are growing alone, in an artificial system deprivated from astrocytes influence.

Astrocytes in primary cultures contain GS as well as in vivo (Fig. 7 and 8). Addition of fetal calf serum to the culture medium decreases the level of GS in the cells : this can be observed immunochemically (Fig. 7 and 8) and biochemically by measuring GS activity of the cells (data not shown).

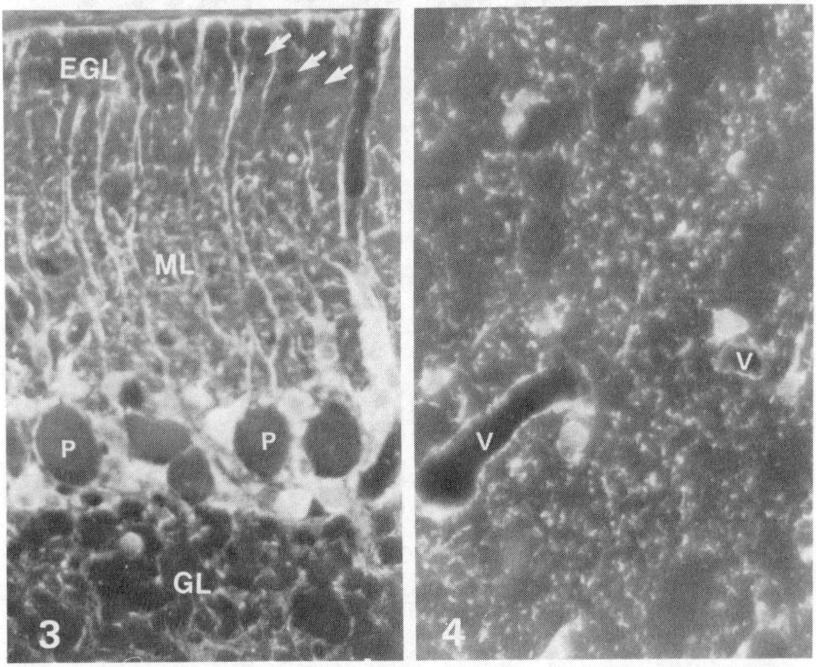

Fig.3 and 4. Immunocytochemical localization of GS in 19 day old chick embryo cerebellum (Fig. 3) and cerebral cortex (Fig. 4); EGL : external granular layer; ML, molecular layer; P, Purkinje cell layer; GL, granular layer; V, blood vessel. In the Fig. 3, the fluorescence is limited to astrocytic Bergmann fibers in the molecular layer and to their cell bodies in Purkinje cell layer. Other astrocytes in the granular layer are also labeled. The neuroblasts and the migrating immature granul cells in the EGL and ML are negatively stained (arrows). In Fig. 4. a number of cortical astrocytes and their end feet on blood vessels are fluorescent (x 400).

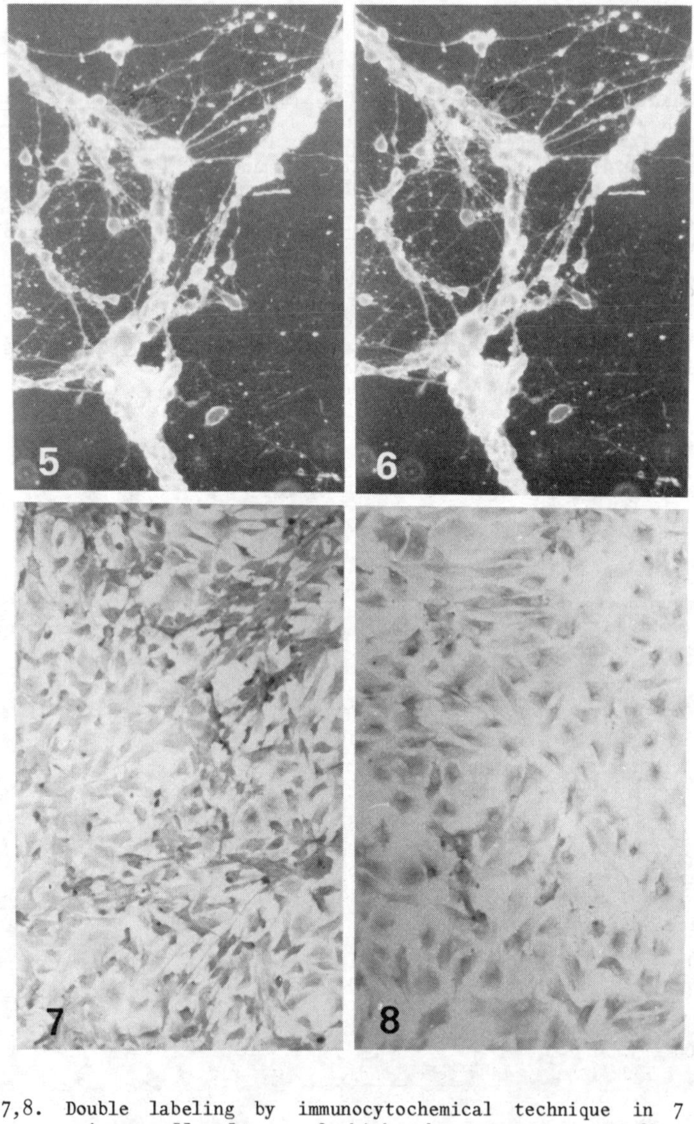

Fig.5,6,7,8. Double labeling by immunocytochemical technique in 7 day old primary cell cultures of chick embryo neurones. In fig. 5 immuno fluorescent neurones stained for neurofilaments (NF) using monoclonal antibodies and then FITC-conjugated sheep anti-mouse IgG (x200). In Fig. 6, the same cells as above stained for GS with polyclonal rabbit immune serum and then with TRITC-conjugated sheep anti rabbit IgG. Al the cells NF^{+} are GS^{+} (x200). Fig. 7 and 8 : immunocytochemistry of GS in 14 day old astrocytes primary cell cultures derived from chick embryo, in Sato defined medium (Fig. 7) and in the presence of 10% fetal calf serum added (Fig. 8). Astrocytes in the presence of serum are less stained (x200).

CONCLUSION

The present data concerning the presence of GS in both chick neurons and astrocytes in primary cultures is in accordance with the observation of the presence, at low specific activity, of the enzyme in glial and neuronal cell lines (Hallermayer *et al*, 1981). The double staining methods for GS and various cell markers applicated to purified primary cultures of chick glial cells and neurons clearly indicated the presence of the enzyme in both types of nerve cells. However, the basal activity in these cultures is far below the activity in the brain of the newborn chicken. Moreover, immunohistochemical studies on slices of chick brain embryo showed that GS is indeed located exclusively in astrocytes. These contradictory results concerning GS in neurons in primary culture may indicate that regulation phenomena occur *in vivo*, inhibiting the expression of the enzyme in this type of cells, and do not occur in culture. More investigations are necessary in order to determine the nature of this type of regulation.

Concerning the purified enzyme, it is yet impossible to elucidate the specific function of Mg^{++}, Mn^{++} and Co^{++} on the regulation of chick brain GS. It is probable that the enzyme possesses more than one type of cation binding and regulatory site(s) to which the cation may bind for activatory or (and) inhibitory effects. Whether regulatory cations interfere preferentially with one or more of these sites remains to be elucidated.

ACKNOWLEDGMENTS

We wish to thank Dr. N. Sternberger, Center for Brain Research, Rochester University, USA for the generous gift of monoclonal antibodies directed against neurofilaments. Mrs. M.A. Laluet and J. Ledig are thanked for their expert technical assistance and Mrs. S. Ott for typing the manuscript.

REFERENCES

Adler, S.P., D. Purich, and E.R. Stadtman (1975). J. Biol. Chem., 250, 6264-6272.
Bloch, S., M. Ledig, P. Mandel and G. Tholey (1984). Comptes Rendus Acad. Sc. Paris, 298, 127-130.
Carter, C.V. (1983). Lif Sci., 32, 1949-1955.
Dennis, S.C., J.C.K. Lai and J.B. Clark (1980). Brain Res., 197, 469-475.
Devel, T.F., M. Louie, A. and Lerner (1978). J. Biol. Chem., 253, 6111- 6118.
Ghandour, M.S., O.K. Langley, G. Vincendon and G. Gombos (1979). J. Histochem. Cytochem., 27, 1634-1637.
Ghandour, M.S., Langley, O.K., Vincendon, G., Gombos, G., Filippi D., Limizin N., Dalmasso C. and Laurent G. (1980). Neuroscience 5, 559-571.
Hallermayer, K., C. Harmening and B. Hamprecht, B. (1981). J. Neurochem., 37, 43-52.
Ip, S.M., P. Rowell and W.D.P. Stewart (1983). Biochem. Biophys. Res. Commun. 114, 206-213.
Juurlink, B.H.J. (1982). Neurochem. Res., 7, 905-910.
Mahajan, R.G., S. Mandal, A.K. Sinma and K.L. Mukherjee (19???). J. Neurochem., 38, 356-359.
Mc Cormack, D.K., K.J.F., Farnden and M.J. Boland (1982). Arch. Biochem. Biophys., 218, 561-571.
Norenberg, M.D. (1979). J. Histochem. Cytochem., 27, 756-762.
Norenberg, M.D. and A. Martinez-Hernandez (1979). Brain Res., 161, 303- 310.
Orr, J., L.M. Keefer, P. Keim, T.D. Nguyen, T. Wellems, R.L. Heinrikson, and R. Haselkorn (1981). J. Biol. Chem., 256, 13091-13098.
Pamuja, S.L. and T.W. Reid (1985). Exp. Eye Res., 40, 75-83.
Pettmann, B., J.C. Louis and M. Sensenbrenner (1979). Nature (Lond), 281, 378-380.
Pishak, M.R. and A.T. Phillips (1980). J. Neurochem., 34, 866-872.

Ronzio, R.A., Rowe, W.B., Wilk, S. and A. Meister (1969). Biochemistry, 8, 2674-2680.

Sarkar, P.K., D.A. Fischman, E. Goldwasser and A.A. Moscona (1972). J. Biol. Chem., 247, 7743-7749.

Sensenbrenner, M. (1977). Cell, Tissue and Organ Cultures, S. Fedoroff, L. Hertz, ed. Academic Press, New York, 191-213.

Stahl, J. and L. Jaenicke (1972). Eur. J. Biochem., 29, 401-407.

Tapia, R. (1980). in "Glutamine: metabolism, enzymology and regulation", Mora, J. and R. Palacios, ed. Academic Press, New York, 285-297.

Tholey, G., M. Ledig, S. Bloch, P. Mandel (1983). Neurochem. Res. 8, 1233-1243.

Wedler, F.C., R.B. Denman, W.G. Roby (1982). Biochemistry, 24, 6389-6396.

Carbonic Anhydrase in Young and Mature Cultured Rat Glial Cells

J. P. Delaunoy, D. Langui, S. Ghandour and M. Sesenbrenner

Centre de Neurochimie du CNRS, 5 rue Blaise Pascal, 67084 Strasbourg, France

ABSTRACT

The generation of monoclonal antibodies against carbonic anhydrase II (CAII) and of polyclonal antibodies against glial fibrillary acidic protein (GFAP) has made it possible to demonstrate the existence of bipotential glial cells, which could be precursors of oligodendrocytes or astrocytes. Moreover, we have confirmed that CAII in mature glial cells is present only in oligodendroglial cells. Finally, we have shown that a soluble extract from pure chick neurons in culture or neuronal conditioned media have an effect on the morphological maturation of oligodendroglial cells.

KEY WORDS

Carbonic anhydrase; glial fibrillary acidic protein; monoclonal antibodies; oligodendroglial cell; astrocyte; immunocytochemistry.

INTRODUCTION

In the central nervous system (CNS) the isoenzyme II of carbonic anhydrase (CAII), which catalyzes the hydration of carbon dioxide, is present in particulate and soluble form (Karler and Woodbury, 1960 ; Dodd, et al., 1981). Our immunohistochemical studies with polyclonal antibodies have shown in the CNS of the rat, mouse and human (Delaunoy, et al., 1977 ; Borrelli, et al., 1982) as well as in primary glial cell cultures of mouse and rat (Delaunoy, et al., 1980 ; Sarliève, et al., 1981) that CAII is localized essentially in oligodendroglial cells and the myelin sheath. This exclusive localization indicates that oligodendroglial cells and myelin are implicated in ionic exchanges by regulating the concentration of bicarbonate ion and suggests that astrocytes cannot assure the rapid reaction of carbon dioxide hydration. Although our results on the localization of CAII were confirmed by Ghandour, et al. (1979) and Kumpulainen and Korhonen (1982), other authors claimed the presence of CAII also in protoplasmic astrocytes (Parthe, 1981).

We have reexamined this aspect and applied double immunocytochemical staining to distinguish oligodendroglial cells from astroglial cells. For this purpose, we have developed monoclonal antibodies against CAII in mice and polyclonal

Printed in Great Britain

antibodies against GFAP in rabbit (Langui, et al., 1985). Moreover, with these two antibodies we have been able to demonstrate the presence of CAII and GFAP simultaneously in the same glial cell and to follow the development of such cells with time in culture.

MATERIALS AND METHODS

Primary cultures of glial cells were derived from dissociated brain hemispheres of Wistar newborn rat (Booher and Sensenbrenner, 1972). Meninges were carefully discarded and brain hemispheres passed through a nylon sieve (82 µm pore size) in Waymouth's medium. The cells were seeded onto Falcon petri dishes (1-2 x 10' cells per cm^2). The nutrient culture medium was composed of Waymouth's medium MD 705/1 (Flow) supplemented with 10 % calf serum (Gibco) containing antibiotics.
Pure neuronal cultures from 8-day-old chick embryos were prepared by the method of Pettmann, et al. (1979); but after 24 hours the serum-containing medium was replaced by a serum-free defined medium consisting of Waymouth's medium plus 5 µg/ml insulin (Sigma), 50 µg/ml transferrin (Sigma), 20 nM progesterone (Sigma), 100 nM putrescine dihydrochloride (Sigma) and 30 nM sodium selenite (Merck). After 5, 7 and 9 days in culture, media conditioned by neurons were removed, centrifuged and kept frozen until use. At the same culture times soluble extracts were obtained by homogenizing the neurons in phosphate saline buffer followed by a centrifugation at 10.000 g for 1 hour. These soluble extracts contained about 200 µg/ml proteins, measured by the method of Lowry, et al. (1951). The conditioned culture medium as well as the neuronal extract, were added to glial cell cultures 48 hours before performing the experiments, in the proportion of 50 % (v/v) and 20 % (v/v), respectively.
Polyclonal antibodies against GFAP were obtained in rabbit by injecting a mixture of Freund's complete adjuvant and GFAP purified from cultures of astrocytes by the method of Rasmussen, et al. (1981). Polyclonal antibodies against CAII were obtained as described by Delaunoy, et al. (1977) and monoclonal antibodies (MABs) as described by Langui, et al. (1985). For double immunocytochemical localization we have used the following method : cells were fixed with F.E.A. (formaldehyde 5, ethanol 69, acetic acid 5, water 22) as described by Cammer, et al. (1985). After fixation and washing, cells were incubated with polyclonal antibodies against GFAP (1:50) in culture medium containing MABs against CAII. After washing, a second incubation was made with a solution containing sheep anti-rabbit globulin bound to FITC and sheep anti-mouse globulin bound to TRITC (Sigma). Preparations were observed in a Zeiss fluorescence microscope. For single labelling we have used goat anti-mouse or anti-rabbit globulin (Biosys) coupled to peroxidase.

RESULTS

Similar results were obtained by using either polyclonal or monoclonal antibodies against CAII (Fig. 1 a, b) : only small cells, morphologically characteristic of oligodendrocytes, were stained in the cytoplasm and processes.
Our observations after double immunocytochemical staining and during the development of the cultures led us to consider three distinct phases. A first one, from culture initiation to day 5, containing cells that stained neither for CAII nor for GFAP. A second stage, from days 5 to 13, where GFAP$^+$ and CAII$^+$ cells were present as well as some cells stained simultaneously for the two proteins (GFAP$^+$, CAII$^+$). A third one, after 13 days in culture, in which cells were stained either for GFAP or CAII, but never simultaneously for both markers. These results indicate that the two markers appeared in glial cells around 5 days in culture and that between day 5 and 13 some double stained cells were present which contained both CAII and GFAP. The number of double stained cells was about 1 or 2 percent of the total cell population or 5 to 10 percent of the CAII positive cells (Figs 2 a, b, c). Morphologically these cells resemble oligodendrocytes and GFAP is present in the soma but never in the long processes.

Figure 1

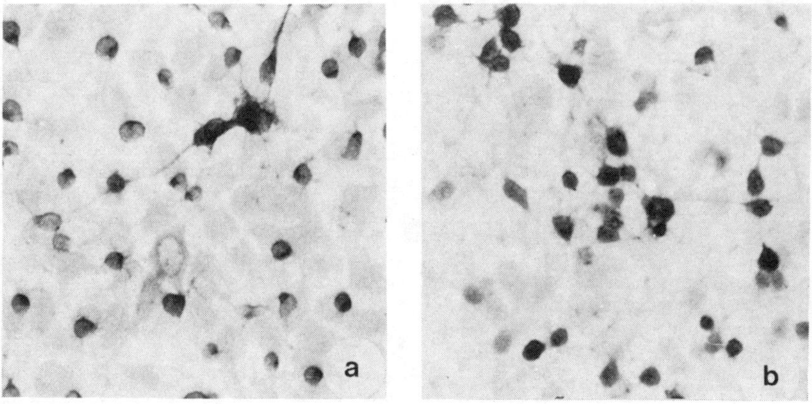

Figure 3

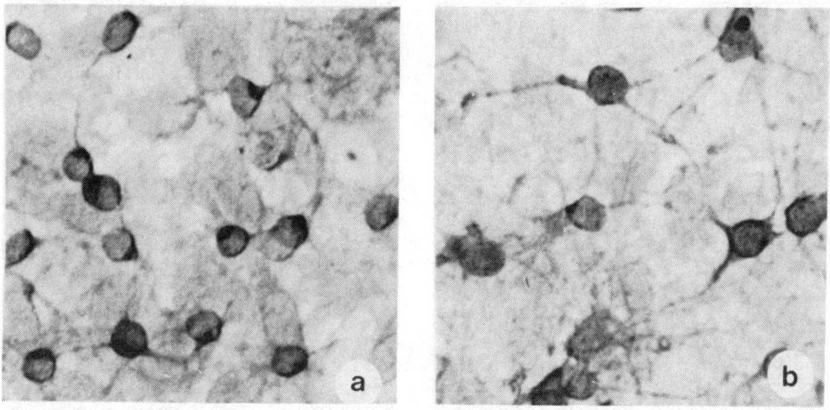

Figure 2

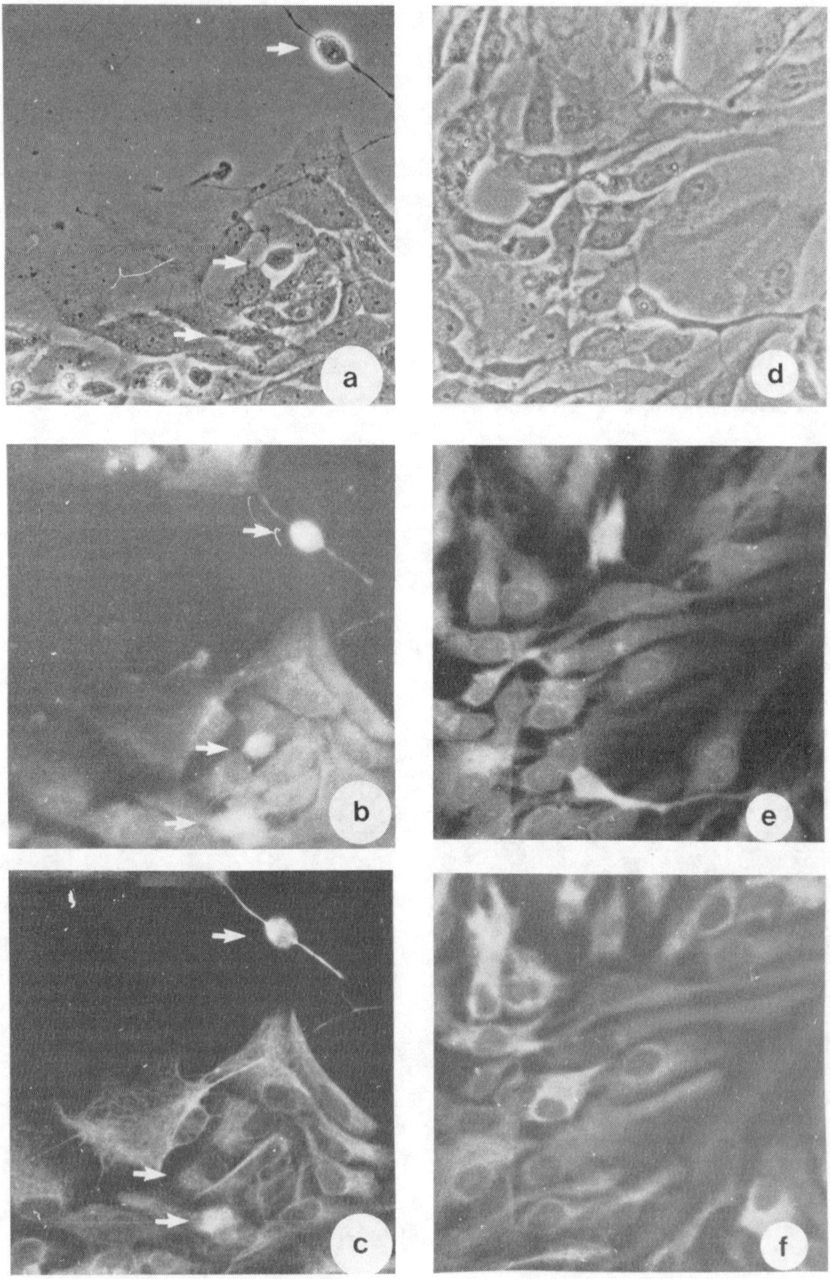

Fig. 1. Immunocytochemical localization of CAII in primary cultures of rat glial cells; 20 days in culture.
a) First incubation with polyclonal antibodies (1:50) and second incubation with anti-rabbit globulins bound to peroxidase.
b) First incubation with monoclonal antibodies and second incubation with anti-mouse globulins bound to peroxidase.
Only the cytoplasm and processes of small cells with a typical morphology of oligodendrocytes are stained with polyclonal and monoclonal antibodies. x 250.

Fig. 2. Double immunocytochemical localization of CAII and GFAPin primary cultures of glial cells
a,b,c: 5 days in culture.
a) Phase contrast micrograph.
b) TRITC-fluorescence staining for CAII: three cells are stained (arrows).
c) FITC-fluorescence staining for GFAP: most flat cells are stained with a typical picture of GFAP staining, the three cells seen to be CAII positive in b are also stained for GFAP.
d,e,f: 15 days in culture.
d) Phase contrast micrograph.
e) Staining for CAII (TRITC).
f) Staining for GFAP (FITC).
Cells stained for CAII are not for GFAP and cells stained for GFAP are not CAII positive. x 700.

Fig. 3. Immunocytochemical localization of CAII in primary glial cell cultures; 20 days in culture.
First incubation with monoclonal antibodies against CAII, second incubation with antimouse globulins bound to peroxidase.
a) Control culture.
b) Cells treated 48 hours before staining with soluble chick neuronal extract.
x 1000.

Double staining experiments made after 13 days in culture showed (Fig. 2 d, e, f) astrocytic-like cells stained for GFAP and oligodendroglial-like cells stained for CAII, but we never observed cells which contained simultaneously the two proteins.

Recent and preliminary experiments have shown that in the presence of conditioned media from chick neurons or soluble extracts of these neurons, oligodendrocytes which contain CAII remained well dispersed and extended more and longer processes compared to control cultures (Fig. 3 a, b).

DISCUSSION

Immunocytochemical studies with specific monoclonal antibodies confirmed our previous results on the presence of CAII in mature oligodendrocytes (Delaunoy, et al., 1977 ; Delaunoy, et al., 1980). In our immunochemical experiments we have never detected CAII in mature astrocytes. The simultaneous presence of the two glial markers, GFAP and CAII, in immature glial cells could be interpreted according to different ways. Our suggestion, based on the present morphological observations, is that the cells might be precursors of oligodendroglial cells and contained soluble GFAP not included in the cytoskeleton. The presence of GFAP in young oligodendroblasts in vivo and in vitro has been already described by Choi and Kim (1984) and Raff, et al. (1983). However, it is not possible from the present results, to affirm if these cells become oligodendrocytes, astrocytes or both cell types.

We have shown that extracts from cultured chick neurons exert a stimulatory effect on the morphology of oligodendroglial cells, indicating that these extracts contain active factors. Further investigations will be needed to search if these factors can influence the number of bipotential cells and can affect the orientation of these cells into astrocytes or oligodendrocytes.

CONCLUSION

We have confirmed that carbonic anhydrase II is present in oligodendroglial cells and not detected in astrocytes when these cells have reached a sufficient degree of maturity. We have further demonstrated that precursors of glial cells were still present after the apparition of CAII and GFAP. Finally, we have shown that soluble extract from neurons or neuronal conditioned media influence the morphology of oligodendroglial cells. Now, we have to investigate if the soluble extract modulates the differentiation of glial precursors toward oligodendrocyte or astrocyte.

REFERENCES

Booher, J., and M. Sensenbrenner (1972). Neurobiol., 2, 97-105.

Borrelli, E., O.K. Langley, M.S. Ghandour, J.P. Delaunoy and G. Gombos (1982). Neurosci. Letters, 32, 321-27.

Cammer, N., R. Sachi, and V. Sapirstein (1985). J. Histochem. Cytochem., 33, 45-54.

Choi, B.H., and R.C. Kim (1984). Science, 223, 407-408.

Delaunoy, J.P., G. Roussel, and P. Mandel (1977). C.R. Acad. Sci., 285, 801-803.

Delaunoy, J.P., F. Hog, G. Devilliers, M. Bansart, P. Mandel, and M. Sensenbrenner (1980). Cell. Mol. Biol., 26, 235-240.

Dodd, P.R., J.A. Hardy, A.E. Oakley, J.A. Edwardson, E.K. Perry, and J.P. Delaunoy (1981). Brain Res., 226, 107-118.

Ghandour, M.S., O.K. Langley, G. Vincendon, and G. Gombos (1979). J. Histochem. Cytochem., 27, 1634-1637.

Karler, R., and D.M. Woodbury (1960). Biochem. J., 75, 538-543.

Kumpulainen, T., and K.L. Korhonen (1982). J. Histochem.Cytochem., 30, 283-292.

Langui, D., J.P. Delaunoy, M.S. Ghandour, and M. Sensenbrenner (1985). Neurosci. Letters, in press.

Lowry, J., N.JH. Rosebrough, A.L. Farr, and R.J. Randall (1951). J.Biol. Chem., 193, 265-275.

Parthe, V. (1981). J. Neurosci. Res., 6, 119-131.

Pettmann, B., J.C. Louis, and M. Sensenbrenner (1979). Nature, 281, 378-380.

Raff, M.C., R.H. Miller, and M. Noble (1983). Nature, 303, 390-396.

Rasmussen, S., M. Sensenbrenner, G. Devilliers, A. Schousboe, and E. Bock (1981). Neurosci. Letters, 25, 119-124.

Sarliève, L.L., J.P. Delaunoy, A. Dierich, A. Ebel, M. Fabre, P. Mandel, G. Rebel, G. Vincendon, M. Wintzerith, and A.N.K. Yusufi (1981). J. Neurosci. Res., 6, 659-683.

Glycolipid Markers in Glial Cells

N. Baumann, B. Zalc, A. Baron-Van Evercooren,
M. Gumpel and C. Jacque

Laboratoire de Neurochimie INSERM I 134, Hôpital de la
Salpêtrière, 76751 Paris CEDEX 13, France

ABSTRACT.

Many of the antigenic determinants of glial cell surface molecules are carbohydrate in nature and include epitopes of glycolipids. Biochemical investigations immunocharacterization on thin-layer plates and immunolocalization are complementary techniques for their characterization. Galactosylceramide is characteristic of myelinating oligodendrocytes but also of satellite ones, and of myelinating and non myelinating Schwann cells (in situ). Sulfogalactosylceramide is present on myelinating cells but also detected on ciliated ependymal cells and subpial astrocytic processes. Seminolipid appears to be a Schwann cell marker detected by monoclonal antibody 224-58. Gangliosides such as GD3, GM1 or those revealed by monoclonal antibody A2B5, although present on other cell types in the nervous system, may be also considered as glial antigen markers at certain stages of development.

KEYWORDS

Glycosphingolipids; galactosylceramide ; sulfogalactosylceramide ; seminolipid ; ganglioside GD3 ; ganglioside GM1 ; A2B5 monoclonal antibody.

INTRODUCTION

Glycosphingolipids are ubiquitous membrane components, the majority of which are assumed to be present at the outer leaflets of plasma membrane. Changes in cell surface molecules are found in the processes of embryonic development and cellular differentiation. Many of the antigenic determinants at the plasma membrane are carbohydrate in nature and include epitopes of glycosphingolipids. There is increasing evidence that numerous surface antigens detected by monoclonal antibodies are glycolipids which may be minor components of a cell. Ideally, to be considered as a marker, a specific surface antigen must be characterized biochemically and the immunological specificity of the antibody used for the determination of its localization demonstrated. In contrast to many protein antigens, the glycolipid antigen epitope structure of a monoclonal antibody can be well defined. In this respect, the association of biochemical and immunological techniques has appeared very useful, especially the immunocharacterization of glycolipids separated on thin-layer plates incubated first with a specific antibody and second with an ^{125}I-antibody directed against the first antibody (Magnani et al, 1981, 1982)or a peroxidase-labeled anti-Ig antibody (Harpin et al, 1985). Ideally also, a characteristic glycolipid surface antigen must be detected by immunohistochemical techniques both on tissue sections and in tissue culture. Interpretation of tissue culture data alone may be difficult as glycosphingolipids (not glycoproteins), either added (O'Keefe et al, 1977) or present in a culture medium containing fetal calf serum (Kemp and Stoolmiller, 1976) may be incorporated in a membrane modifying its composition, and function as well as those synthesized in situ (Hakomori, 1981). In the nervous system, several glial glycosphingolipid markers have been identified, which are specific of a glial cell type, of a differentiation stage, of a functional subclass or linked to pathological conditions.

GALACTOSYLCERAMIDE.

Galactosylceramide (GalC) is an excellent marker of oligodendrocyte and Schwann cells. It has been localized on these myelinating cells with the use of polyclonal and monoclonal (Ranscht et al, 1982)antibodies. Use of GalC as an oligodendrocyte marker in culture has first been pointed out by Raff et al (1978).On myelinated axon preparation, it is present at the surface of the myelin sheath (Dupouey et al, 1979). On brain tissue sections, Zalc et al (1981) have reported that GalC is present only in myelin and oligodendrocytes. In the myelinated tracts, isolated oligodendrocytes and intrafascicular oligodendroglial cells are GalC positive. Outside myelinated tracts, satellite oligodendrocytes are also GalC positive. Interestingly enough they have found that subependymal cells,in the close vicinity to corpus callosum are also positive for GalC. This suggests that, in adult animals, immature oligodendrocytes start synthetizing GalC before their migration in the corpus callosum. In culture, GalC seems to be the first known myelinic marker synthetized by oligodendrocytes (Abney et al, 1981). During development, in dissociated cell preparations, comparing different myelin constituents, oligodendrocytes first express GalC, then Wolfgramm protein W1, then myelin basic protein (MBP) and proteolipid protein (PLP). This sequence is always respected, whatever brain area is explored. The absolute timing of appearance of GalC differs from one area to another, respecting the caudo-rostral timing of myelination (Monge et al, in preparation). Although GalC has been thought to be a characteristic of myelinating Schwann cells, Jessen et al,(1985) have shown that non myelinating Schwann cells in situ do form significant amount of this glycolipid, i.e. when in contact with axons. The rapid loss of expression of this myelin lipid by neonatal rat Schwann cells that are cultured in the absence of neurons contrasts sharply with the long-term expression of this component by cultured oligodendroglia (Raff et al, 1978; Lisak et al, 1981)

This antigen appears to be absolutely required for myelination to occur or myelin to be maintained. Studies on in vitro and in vivo antibody-mediated immunopathologic reactions of the CNS and PNS, with anti-GalC antibodies have recently been reviewed by Lisak (1984). Anti-GalC antibodies can cause demyelination in vitro, in CNS or PNS cultures, and in vivo after intraneural injection in the optic nerve or the sciatic nerve.

SULFOGALACTOSYLCERAMIDE.

Sulfogalactosylceramide (SGC) has been localized on myelin sheaths and the intrafascicular oligodendroglia by Zalc et al (1981). It has also been localized in other structures not belonging to the oligodendroglial cell line ; thus this glycolipid does not appear to be as good an oligodendroglial marker as GalC. They have also shown that ependymal cells are highly labeled by an anti-sulfatide antibody as are some astrocyte fiber processes in contact with these cells or with the pia arachnoid, especially the Bergmann fibers in the cerebellum.

On astroglial structures SGC-positive, an antiserum raised against human Tamm-Horsfall (TH) protein was found to react (Zalc et al, 1984). Indirect immunofluorescence on rat brain tissue allowed us to localize this brain TH-cross-reacting material to ependymal cells, astrocytic processes such as the Bergmann fibers or astrocytic feet in contact with either the blood vessels or the meninges. In these structures, as in the kidney (luminal face of the thick ascending limb of the loop of Henle), the localization of TH and SGC are strictly superimposable. Since TH and SGC in the kidney are localized on a membrane that possesses an electrogenic Cl pump, we propose that the astroglial structures which contain these two molecules are also the site of a Cl^- transport system. SGC on the cell surface appears to be an inducible marker in C6 glioma cells. Normally absent on this cell line (Robert et al, 1977), it can be synthetized and appears on the cell surface when cells are cultured in the presence of psychotropic drugs such as desipramine (Tocqué et al, 1985).

SEMINOLIPID.

Recently Goujet-Zalc et al(1986)and Guerci et al (1986)have demonstrated by indirect immunofluorescence the cellular localization of a monoclonal antibody (mAB 224-58), produced after immunization of a mouse with human central nervous system (CNS) myelin. Serologically, monoclonal antibody 224-58 was found to be specific for 3'sulfomonogalactosylglycolipids,namely sulfogalactosylceramide (SGC) and 3'sulfogalactosyl 1-O-alkyl ether 2-O-acyl glycerol (seminolipid). This monoclonal antibody did not bind to sulfogalactosylceramide containing tissues such as kidney, liver, spleen or brain, nor to muscle. However monoclonal antibody 224-58 did stain positively mouse, rat and human peripheral nerve sections. In these latter sections, monoclonal antibody 224-58 is bound to Schwann cell bodies and processes. The specificity of monoclonal antibody 224-58 for Schwann cells was ascertained on teased rat sciatic nerves and rat Schwann cell cultures. Cells positive for monoclonal antibody 224-58 are also positive for laminin and negative for Thy 1-1 antigens both teased fibers and Schwann cell culture. In addition, in teased nerve preparations, monoclonal antibody 224-58 positive cells are also galactosylceramide and sulfogalactosylceramide positive. Isolated Schwann cells also express 224-58 antigen, even after prolonged time in culture. On testis sections, which contain both sulfogalactosylceramide and 3'sulfogalactosyl 1-O-alkyl ether 2-O acylglycerol, the sulfogalactosylceramide positive cells, i.e. the spermatogonia, are always 224-58 negative. But the other germinal cells are 224-58 positive. This suggests that while monoclonal antibody 224-58 does not discriminate between sulfogalactosylceramide and 3'sulfo galactosyl 1-O-alkyl ether 2-O-acyl glycerol in serological tests, these lipids in their naturally occuring membrane acquire a spatial configuration which make them distinguishable for their respective antibody.

GANGLIOSIDE GD3.

From determination of ganglioside pattern in relation to brain development, it has been assumed that a predominance of GD3 in brain may characterize mitotically active cells of the nervous system, central (Dreyfus et al 1975, Yavin and Yavin 1979; Irwin et al, 1980) and peripheral (Yates and Thompson, 1978; Chou et al, 1982). Studies on the cellular localization of this glycolipid have been performed using a mouse monoclonal antibody that recognize GD3, i.e. R24 (Pukel et al,1982). Goldman et al (1984)have shown in rat tissue sections that the antibody binds to cells in several areas known to contain immature neuroectodermal populations. In adult rat CNS, they could not detect any binding on neurons or glia, except for scattered cells in the adult suventricular zone. Thus they believe this ganglioside to be characteristic of both neuronal and glial precursors. Rösner et al (1985) in the chicken consider also that it is only localized on mitotically active cells. In the adult human brain, Graus et al (1984) found positive staining with R24 antibody in the surrounding neuropil of many neuronal groups, in ependymal cells and glial processes around blood vessels and in the subependymal region. Thus, there may be species differences for mammalian brains.

In the search of an in vivo and in vitro model to study the modulation of ganglioside composition in relation to different stages of development, we have focussed our interest on a neurological mutant, the Trembler mouse. This mutant is characterized by an hypertrophic peripheral neuropathy with segmental dys- and demyelination and Schwann cell proliferation (Perkins et al, 1981). Ganglioside isolation and analysis of this class of glycolipids on thin-layer chromatography followed by a densitometric analysis after resorcinol reagent, evidenced several modifications of the ganglioside profile in the Trembler mouse and revealed the detection only in the mutant of a ganglioside migrating with the Rf of GD3 (Harpin et al, 1982). Further studies were performed using the monoclonal antibody R24 so as to immunolocalize GD3 in the peripheral nerve. The staining of this antibody seemed to be analogous to laminin visualized with an antibody against laminin (from H. Kleinman) staining being observed at the level of Schwann cells and of their basal lamina in both Trembler and control nerves (Baron-Van Evercooren unpublished). Thus it did not seem to be correlated with Schwann cell proliferation in Trembler mice. In view of the discrepancy observed between biochemical determinations and immunohistochemistry, further characterization of the antibody was undertaken by immunocharacterization on thin layer plates. R 24 was shown to reveal several gangliosides of CNS ganglioside mixture. Preliminary studies indicate that R 24 detects several ganglioside antigens on mouse peripheral nerves (Harpin et al, in preparation).

GANGLIOSIDES REACTING WITH A2B5 MONOCLONAL ANTIBODY

Monoclonal antibody A2B5 (Eisenbarth et al, 1979) reacts with many gangliosides in neuronal tissue (Kundu et al, 1983) possibly minor ones (Fredman et al, 1984). Immunodetection on thin-layer plates of separated gangliosides shows that the antibody bind to many antigens (Fredman et al, 1984). We have also shown by the same technique that several gangliosides, mostly minor bind to A2B5) in neuroblastoma human tumors (unpublished). In the chicken retina, this antigen can be localized on most neuronal cells (Eisenbarth et al, 1979). Berg and Schachner (1982) and Schnitzer and Schachner (1982) have reported that A2B5 bind to a small proportion of astrocytes and oligodendrocytes (as well as the majority of neurons) in cultures of a variety of vertebrate CNS tissues. In the rat, Raff et al (1983a and b) have shown that a bipotential glial progenitor cell in the developing rat optic nerve differentiates into a fibrous astrocyte (type 2-astrocyte) if cultured in the presence of 10% fetal calf serum and into an oligodendrocyte if cultured in serum-free medium. This has been confirmed when the 0-2A progenitor cells are grown as isolated cells in microculture (Temple and Raff, 1985). Both the progenitor cell and the type-2 astrocyte express antigens reacting with A2B5 monoclonal antibody on their cell surface. They mention that fetal calf serum acts directly on the precursor cell to induce it to differentiate into type-2 astrocytes. Although Raff et al (1983a) have shown that gangliosides can be synthesized in the culture conditions that they used, it is noteworthy that fetal calf serum contains also gangliosides reacting with A2B5 monoclonal antibody (Harpin et al, in preparation) when immunodetected on thin layer plates.

GANGLIOSIDE GM1.

Immunocytochemical methods with cholera toxin detect GM1 on PNS and CNS neurons (Raff, 1979). GM1 although present in these structures is characteristic of isolated CNS myelin (Suzuki et al, 1967; Ledeen et al, 1973)and considerably reduced in the dysmyelinating quaking mutant (Iwamori et al, 1985) in which their is a defect in myelin compaction. In PNS, GM1 is a minor component (Harpin et al, 1982; Chou et al, 1982); it is primarily localized on specialized Schwann cell structures in the nodal region ; low amount are also found on the internodal surfaces particularly of the smallest myelinated fibers (Ganser et al 1983).

CONCLUSION.

The glycosphingolipids have a unique molecular character, particularly their carbohydrate structure,that is subject to changes under both genomic and extragenomic (environmental conditions). This dual nature (Nakakuma et al, 1984) suggests that these molecules are well qualified to be important in a variety of cell biological functions (Hakomori, 1981) such as cell recognition, growth, receptor function and other cell social activities in addition to malignant cell transformation in which characteristic changes in glycosphingolipid compositions and metabolism have been reported. No doubt that the nervous system, in which glycolipids are so important quantitatively and qualitatively, is a special field of investigation to determine the roles of these cellular markers during development and in the different types of functional interactions involving neurons and glial cells.

REFERENCES

Abney E.R., P.P. Bartlett and M.C. Raff (1981). Develop. Biol.83, 301-310.

Berg G.J. and M. Schachner (1982), Cell Tissue Res., 224, 625-636.

Chou K.H., C.E. Nolan, and F.B. Jungalwala (1982), J. Neurochem., 39, 1547-1558.

Dreyfus H.P., P.F. Urban, S. Edel-Harth, and P. Mandel (1975), J. Neurochem 10, 429-440.

Dupouey P., B. Zalc, M. Lefroit-Joly, and D. Gomes(1979),Cell Mol. Biol. 11, 269-272.

Eisenbarth G.S., F.S. Walsh and M. Nirenberg (1979), Proc. Natl. Acad. Sci (USA) 75, 4913-4917.

Fredman P., J.L. Magnani, M. Nirenberg, V. Ginsburg (1984), Arch Biochem. Biophys. 233, 661-666.

Ganser A.L., D.A. Kirschner and M. Willinger (1983) J. Neurocytol. 12, 921-938.

Goldman J.E., M.Hirano, R.K. Yu and T.N. Seyfried (1984), J. Neuroimmunol. 7, 179-192.

Goujet-Zalc C., A. Guerci, G. Dubois and B. Zalc. (1986) J. Neurochem (in press).

Graus F., C. Cordon-Cardo, A.N. Houghton, M.R. Melamed, and L.J. Old (1984), Brain Res. 324, 190-194.

Guerci A., M. Monge, A. Baron-Van Evercooren, S. Dancea, J.M. Boutry, C. Goujet-Zalc and B. Zalc. (1986). J. Neurochem (in press).

Hakomori S. (1981) Annu. Rev. Biochem.50, 733-764.

Harpin M.L., J. Portoukalian and N. Baumann (1982), Neurochem. Res.7, 1367-1373.

Harpin M.L., M.J. Coulon-Morelec, P. Yeni, F. Danon and N. Baumann. (1985), J. Immunol. Meth. 78, 135-141

Irwin L.N., D.B. Michael,and C.C. Irwin (1980), J. Neurochem. 34, 1527-1530

Iwamori M., M.L. Harpin, F. Lachapelle and N. Baumann (1985), J. Neurochem.45 73-78.

Jessen K., L. Morgan, M. Brammer (1985), J. Cell. Biol. 101, 1134-1143.

Kemp S.F., A.C. Stoolmiller (1976), J. Biol. Chem. 251, 7626-7631.

Kundu S.K., M.A. Pleatman, W.A. Redwine, A.E. Boyd and D.M. Marcus (1983), Biochem. Biophys. Res. Commun. 116, 836-841.

Ledeen R.W., R.K. Yu and L. Eng. (1973), J. Neurochem 21, 829-839.

Lisak R.P., D.E. Pleasure, D.H.Silberberg, M.C. Manning, T. Saida (1981) Brain Res. 223, 107-122.

Lisak R.P.(1984)Neuroimmunology,Raven Press New York (pp 167-177).

Magnani J.L., M. Brockhaus, D.F. Smith, M. Blaszczyck, K.F. Mitchell,Z.Steplewski and H. Koprowski (1981), Science 212, 55-56.

Magnani J.LL, B. Nilsson, M. Brockhaus, D. Zopg, Z. Steplewski, H. Koprowski and V. Ginsburg (1982), Biol. Chem. 257, 14365-14369.

Nakakukama H., Y. Sanai, K., Shiroki, and Y. Nagai (1984) J. Biochem 96, 1471-1480

O'Keefe E. and P. Cuastrecasas. (1977), Life Sci. 21, 1649-1654.

Perkins C.S., A.J. Aguayo and G.M. Bray (1981), Neuropath. Appl. Neurobiol. 7 115-126.

Pukel C.S., K.O. Lloyd, L.R. Travassos, W.G. Dippold, H.F., Oettgen and L.J. Old, (1982), J. Exp. Med.155, 1133-1147.

Raff M.C., R. Mirsky,K. Fields, R.P. Lisak, S. Dortman, D.H. Silberberg, N.A. Gregson, S. Leibowitz and M.C. Kennedy.(1978), Nature 274, 813-816.

Raff M.C., K.L. Fields, S. Hakomori R. Mirsky, R.M. Pruss and J. Winter (1979) Nature 273, 672-673.

Raff M.C., E.R. Abney, J. Cohen, R. Lindsay and M. Noble. (1983a) J. Neurosci.3 1289-1300.

Raff M.C., R.H. Miller and M. Noble. (1983), Nature 303, 390-396.

Rantsch B., P.A. Clapshaw, J. Price, M. Noble and W. Seifert (1982) Proc. Natl. Acad. Sci. (USA)79, 2709-2713.

Robert J., G. Rebel and P. Mandel (1977) J. Lip. Res. 18, 517-520

Rösner H., M. Al-Aqtum, and S. Henke-Fahle. (1985) Dev. Brain Res. 18, 85-95

Schnitzer J. and M. Schachner (1982) Cell Tissue Res. 224, 625-636.

Suzuki K.,S.E. Poduslo and W.T. Norton(1967) Biochem. Biophys. Acta 144, 375-381.

Temple S. and M.C. Raff (1985) Nature 313, 223-225

Tocqué B., S. Albouz, J.M. Boutry, F. Le Saux, J.J. Hauw, J.J. , R. Bourdon, N. Baumann and B. Zalc.(1984), J. Neurochem. 42, 1101-1106.

Yates A.J. and D.K. Thompson (1978) J. Neurochem. 30, 1649-1651.

Yavin E. and Z. Yavin (1979), Dev. Neurosci. 2, 25-37.

Zalc B., M. Monge, P. Dupouey, J.J. Hauw, and N. Baumann (1981) Brain Res 211, 341-354.

Zalc B., A. Collet, M. Monge, M.P. Ollier-Hartmann, C. Jacque, L. Hartmann and N. Baumann. (1984) Brain Res 291, 182-187.

Expression of Histocompatibility Antigens on the Oligodendrocyte Surface

A. Suzumura and D. H. Silberberg

Department of Neurology, University of Pennsylvania School of Medicine, Philadelphia, Pennsylvania, USA

ABSTRACT

The major histocompatibility complex (MHC) class I antigens are reportedly expressed on the cells of most tissues. However, the level of H-2 antigens on brain cells has been shown to be extremely low, by immunohistochemical studies, and H-2 antigen has not been demonstrated on unstimulated oligodendrocytes.

In this study, we found that the expression of mouse MHC class I antigen (H-2) on isolated cultured mouse oligodendrocytes is induced by: 1) soluble factors produced by concanavalin A-activated T-lymphocytes, and 2) cloned gamma-interferon. H-2 antigens were demonstrated by indirect immunofluorescence and radioimmunoassay. Surface expression of Ia antigens was not induced by these factors. Autoradiographic studies indicate that this induction of H-2 expression is not accompanied by proliferation of oligodendrocytes. The phenomenon of induction of expression of surface H-2 antigen on oligodendrocytes by products of immune reactions may have a role in the pathogenesis of multiple sclerosis or other disorders of the central nervous system.

KEY WORDS

Oligodendrocytes; major histocompatibility complex (MHC) antigens; H-2 antigen; Ia antigen; T cell; lymphokine; multiple sclerosis; demyelination.

INTRODUCTION

The recognition of major histocompatibility complex (MHC) class I antigens (H-2D and K subregions in mouse, HLA-A and B in man) by cytotoxic T-cells responding to virus infected cells, tumor cells and chemically modified cells is necessary for their interaction (reviewed in Zinkernagel and Doherty, 1979). The cytotoxic T-cells can lyse only antigen-bearing target cells which are of the same H-2 (or HLA) haplotype as the host. H-2 antigen is reportedly expressed on cells of most tissues. However, the level of H-2 antigens in brain has been shown to be extremely low by absorption studies (Pizarro, et al, 1967, Schachner and Sidman, 1973). H-2 antigen was undetectable in histological sections of mouse cerebellum, although some fibroblast-like cells and some astrocytes in dissociated cerebellum culture of 6 day old mouse expressed H-2 antigen on their surface (Schnitzer and Schachner, 1981).

On the other hand, MHC class II antigens (Ia antigens) are found primarily on B-lymphocytes and on activated T-cells. These are also found on antigen presenting cells such as dendritic cells and cells of macrophage-monocyte lineage including Kupfer cells and Langerhans cells. Class II antigens are key elements in the control of the immune response to antigen, functioning in the recognition of antigen by regulatory T-lymphocytes (reviewed in Kaufman, et al, 1984). Although there were several conflicting reports about Ia expression on brain cells (Natali, et al, 1981, Ting, et al, 1981), recent studies by Hart et al (1981), Lisak et al (1983) and Hirsch et al (1983) indicated that oligodendrocytes and non-stimulated astrocytes were Ia-negative. H-2 (or HLA) antigen has not been demonstrated on unstimulated oligodendrocytes, which produce and maintain myelin in the central nervous system, and are the possible target in multiple sclerosis (MS). Although the etiology and pathogenesis of MS are still unclear, immunological mechanisms and/or viral infection probably play a critical role in the development of the MS lesion (reviewed in Waksman, 1981).

Recently, it has been shown that soluble factors from activated T-cells, or interferon alone, enhanced the expression of MHC antigen in several cell types (Wallach, et al, 1982, Basham and Merigan, 1983, Kim, et al, 1983, Pober, et al, 1983). More recently, these immune factors have been shown to induce H-2 expression on most neural cells (Wong, et al, 1984), and Ia expression on some astrocytes (Hirsch, et al, 1983, Fontana, et al, 1984, Wong, et al, 1984).

In this study, we show the induction of H-2 antigen in isolated mouse oligodendrocytes, in which H-2 antigen cannot be detected before stimulation, with soluble factors from concanavalin A (Con A) activated T-cells or gamma interferon (γ-IFN). These materials, however, do not induce Ia expression on isolated cultured oligodendrocytes. This finding of H-2 induction on oligodendrocytes may provide the next step in approaching the interaction between oligodendrocytes and the immune system.

MATERIALS & METHODS

Oligodendrocytes from 3 different strains; AKR (H-2D^k, K^k), C57BL/6 (H-2D^b, K^b) and BALB/c (H-2D^d, K^d) mice, were isolated from primary dissociated brain cell cultures of newborn mice at 14 days *in vitro* as described previously (Suzumura, et al, 1984). The purity of oligodendrocytes at the time of the binding assay (3 days after isolation) was 80-95% by indirect immunofluorescence with anti-galactocerebroside antibody. Monoclonal antibodies against mouse H-2; anti-H-2D^k, K^k, anti-H-2D^b, K^b and anti-H-2D^d, K^d and against mouse Ia antigen; anti-I-A^k, anti-I-A^b and anti-I-A^d antibody were obtained from Bionetics Laboratory (Charleston, SC). Supernatant media of non-producing mouse myeloma cell line (SP2/0) to which 100 ug/ml of mouse IgG was added (Cappel) was employed as the control for these monoclonal antibodies. Soluble factors from activated T-cells (Con A sup) were prepared by stimulating mouse spleen cells from each strain with Con A (final concentration 4 ug/ml) for 3 days. Cloned murine γ-interferon was kindly provided by Dr. S.M. Phillips. Supernatant from non-stimulated spleen cell cultures (NS sup) and Con A alone were used for the controls. Oligodendrocytes were cultured with or without Con A sup (10% in final concentration), γ-IFN (10 u/ml), NS sup (10% in final concentration) or Con A alone (0.4 ug/ml in final concentration) for 2 days, starting 1 day after isolation. The expression of H-2 antigen on oligodendrocytes was then assessed by indirect immunofluorescence and radioimmunoassay using unfixed viable cells, as detailed elsewhere (Suzumura and Silberberg 1985).

In order to examine whether the H-2 induction was accompanied by simultaneous proliferation of oligodendrocytes, autoradiography combined with anti-GalC immuno-staining was carried out using oligodendrocytes from AKR mice with or without stimulation with various concentration of Con A sup or γ-IFN (Suzumura, et al, 1984, Suzumura and Silberberg, 1985). The percentage of GalC positive cells with more than 5 grains over the nucleus was enumerated by scoring at least 200 GalC positive cells per coverslip.

RESULTS

As shown in Table I, oligodendrocytes did not express either H-2 antigen or Ia antigen on their surface before stimulation and after stimulation with Con A alone or NS sup. However, if they were stimulated with Con A sup or γ-IFN, the corresponding H-2 antigen, but not Ia antigen, became detectable on their surface (Figure 1). Because of cross reactivity of anti-H-2D^d, K^d with H-2D^b, K^b antigen, weak staining of C57BL/6 oligodendrocytes with anti-H-2D^d, K^d was observed.

The data shown above were obtained using Con A sup of the same strain as the oligodendrocytes. We also stimulated oligodendrocytes with Con A sup from different strains (AKR oligodendrocytes with C57BL/6 Con A sup or with BALB/c Con A sup) or from different species (C57BL/6 oligodendrocytes with Sprague Dawley rat Con A sup), which demonstrated the same H-2 induction when examined by indirect immunofluorescence. GFAP positive astrocytes and fibronectin positive firbroblasts also expressed the appropriate H-2 antigen only after stimulation with Con A sup or γ-IFN. Macrophage-microglia, identified by their property of phagocytosing latex particles exhibited non-specific staining including staining of endocytosed IgG. Minimum requirement for the induction of H-2 expression was 2.0% Con A sup or 0.2 u/ml γ-IFN for 24 h.

In order to confirm this H-2 induction, solid phase radioimmunoassay was carried out. Again, only oligodendrocytes stimulated with Con A sup and γ-IFN showed significantly higher binding with the corresponding anti-H-2 antibody (Table II). These results also demonstrated the induction of H-2 antigen on their surface after stimulation with Con A sup and γ-IFN. However, the same stimulation did not induce Ia antigen on oligodendrocytes.

The autoradiography showed no significant increase of ^{3}H-thymidine incorporation into oligodendrocytes after Con A sup or γ-IFN stimulation (Table III).

Table I. H-2 Expression on Oligodendrocytes: Indirect Immunofluorescence*

Cultures	Monoclonal antibody against					
	H-2D^k^,K^k^	H-2D^b^,K^b^	H-2D^d^,K^d^	Ia^k^	Ia^b^	Ia^d^
AKR (H-2D^k^,K^k^)						
oligo	-	-	-	-	-	-
oligo + Con A	-	-	-	-	-	-
oligo + NS sup	-	-	-	-	-	-
oligo + Con A sup	+	-	-	-	-	-
oligo + γ-IFN	+	-	-	-	-	-
C57BL/6 (H-2D^b^,K^b^)						
oligo	-	-	-	-	-	-
oligo + Con A	-	-	-	-	-	-
oligo + NS sup	-	-	-	-	-	-
oligo + Con A sup	-	+	±	-	-	-
oligo + γ-IFN	-	+	±	-	-	-
BALB/c (H-2D^d^,K^d^)						
oligo	-	-	-	-	-	-
oligo + Con A	-	-	-	-	-	-
oligo + NS sup	-	-	-	-	-	-
oligo + Con A sup	-	-	+	-	-	-
oligo + γ-IFN	-	-	+	-	-	-

*The results were expressed as-: negative; ±: weakly positive; and +: positive

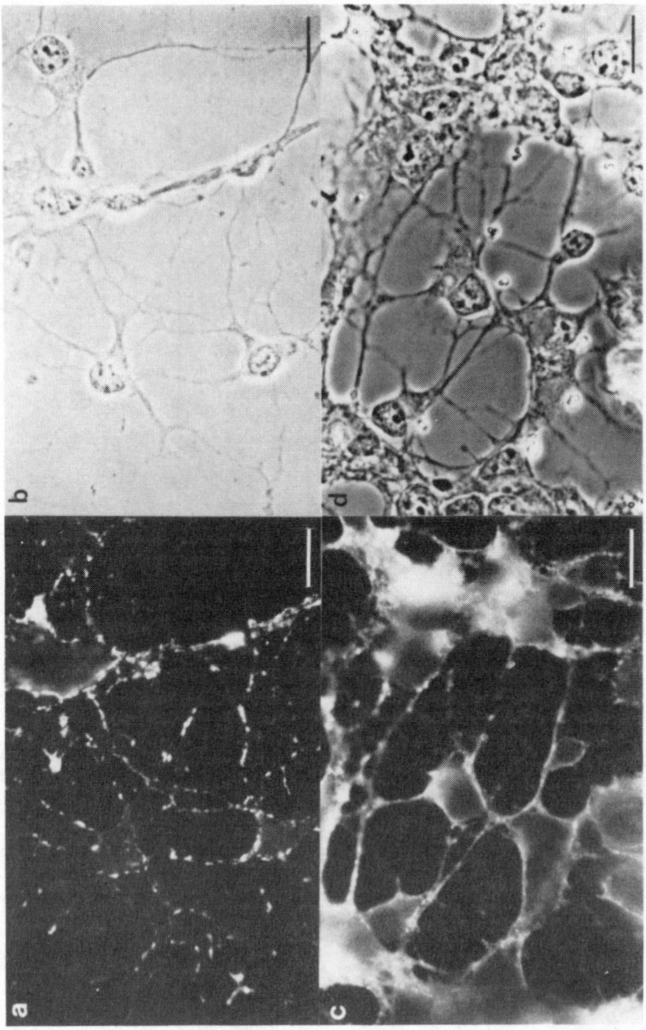

Figure 1 H-2 expresion on oligodendrocytes.

C57BL/6 oligodendrocytes stimulated with Con A sup, viewed with a; rhodamine (H-$2D^b$,K^b) and b; phase microscopy. Double labelling of unstimulated cultures with anti-GalC and anti-H-$2D^b$,K^b showed that most of the cells were GalC positive (c; fluorescein, d; phase) and they did not express H-$2D^b$,K^b (not shown). Bar=15 um.

Table II. Expression of MHC on oligodendrocytes: Radioimmunoassay

Culture	$H\text{-}2^d$	Ia^d	$H\text{-}2^k$	Ia^k	normal serum	$SP^2/0$	GalC**
BALB/c ($H\text{-}2^d$, Ia^d)							
oligo	519±69	542±95	583±68	481±75	550±32	528±67	720±42*
oligo + Con A	572±67	588±72	467±72	521±39	551±45	498±63	686±48*
oligo + NS sup	586±62	NT	499±47	NT	557±64	NT	771±31*
oligo + Con A sup	1698±154*	629±39	661±74	628±72	661±68	690±55	913±60*
oligo + γ-IFN	1209±193*	699±79	661±74	628±72	661±68	690±255	NT

The data represent M±SD (cpm) from triplicate coverslips.

*$P<0.001$ when compared with negative controls and other uncorresponding anti-MHC antibodies by a T test. NT; not tested.

**Monoclonal antibody to galactocerebroside (Rostami et al. 1984).

Table III. Effect of Con A Sup and γ-IFN on Oligodendrocyte Proliferation

		^3H -labelled cell/GalC positive cell (%)*	
Con A sup	0.%	0.3 ± 0.5	(0 - 0.9)
	5%	0.7 ± 0.2	(0.5 - 0.9)
	10%	0.1 ± 0.2	(0 - 0.5)
	25%	0.7 ± 0.6	(0 - 1.3)
	50%	0.5 ± 0.4	(0 - 0.8)
γ-IFN	0	0.3 ± 0.6	(0 - 1.5)
	0.2 u/ml	0.3 ± 0.8	(0 - 1.9)
	2.0 u/ml	0.6 ± 1.0	(0 - 2.4)
	10 u/ml	0.7 ± 0.6	(0 - 1.4)
	50 u/ml	0.2 ± 0.5	(0 - 1.1)

*The results were obtained from a minimum of triplicate and quadricate coverslips. The data represent mean ± standard deviation (and ranges).

DISCUSSION

From the data presented here, it is clear that oligodendrocytes express H-2 antigen, but not Ia, <u>in vitro</u>, under certain experimental conditions. However, the mechanism of this induction remains to be solved. Con A alone or NS sup did not stimulate the H-2 induction. In addition, the H-2 induction was not inhibited by the addition of α-methyl-D-mannoside (Sigma, 10 ug/ml in final concentration) with Con A sup which blocks the binding of Con A to cells. Cloned γ-IFN alone induced H-2 antigen expression on oligodendrocytes. These observations indicated that the H-2 induction was mediated by soluble factors from Con A activated spleen cells including γ-IFN, but not by Con A itself. Reportedly, supernatant of mitogen activated T-cells enhances the proliferation of glial cells (Fontana, <u>et al</u>, 1980), including oligodendrocytes (Merrill, <u>et al</u>, 1984). However, we could not confirm the proliferation of oligodendrocytes by Con A sup or γ-IFN in our experimental conditions.

Lymphokines from actived T-cells or γ-IFN alone reportedly enhanced the expression of Ia antigen, as well as H-2 antigen, on several cell types (Wong, <u>et al</u>, 1982, Basham and Merigan, 1983, Kim, <u>et al</u>, 1983, Pober, <u>et al</u>, 1983, Roehan, <u>et al</u>, 1984). However, in our experiments, Con A sup or γ-IFN failed to induce Ia antigen on oligodendrocytes. We also stimulated oligodendrocytes with these materials in various concentrations (Con A sup 2.0-50%, γ-IFN 0.2-100 u/ml) for up to 14 days, but we could not detect Ia antigen expression on oligodendrocytes. Normally, oligodendrocytes as well as other brain cells are probably isolated from such T-cell soluble factors or cytokines by the blood-brain-barrier (BBB), and may not express H-2 antigen <u>in vivo</u>. However, it is possible that the impairment of the BBB as observed in MS (Aita, <u>et al</u>, 1978, Takeoka, <u>et al</u>, 1983) may result in the induction of H-2 antigen, which then render oligodendrocytes possible target cells for cytotoxic T-cells.

ACKNOWLEDGEMENTS

This work was supported by grants from NIH, NS11037, and from the Multiple Sclerosis Society. We thank Dr. Robert P. Lisak for his suggestions, and Lori M. Buell for typing the manuscript.

REFERENCES

Aita, J.F., Bennet, D.R., Anderson, R.E. and Ziter, F. (1978). Neurology, 28, 251-255.

Basham, T.Y. and Merigan, T.C. (1983). J. Immunol., 130, 1492-1294.

Fontana, A., Grieder, A., Arrenbrecht, S. and Grob, P.J. (1980). J. Neurol. Sci., 46, 55-62.

Fontana, A., Fierz, W. and Wekerle, H. (1984). Nature, 307, 273-276.

Hart, D.N.J., and Fabre, J.W. (1981). J. Exp. Med., 154, 347-361.

Hirsch, M.R., Wietzerbin, J., Pierres, M. and Goridis C. (1983). Neurosci. Lett. 41, 199-204.

Kaufman, J.F., Auffray, C., Korman, A.J., Schackelford A. and Strominger, J. (1984). Cell, 36, 1-13.

Kim, K.J., Chaouat, G., Leiseron, W.M., King, J. and de Maeyer, E. (983). Cell Immunol., 76, 253-267.

Lisak, R.P., Hirayama, M., Kuchmy, D., Rosenzweig, A., Kim, S.U., Pleasure, D.E. and Silberberg, D.H. (983). Brain Res., 289, 285-292.

Merrill, J.E., Kutsunai, S., Mohlstrom, C., Hofman, F., Groopman, J. and Golde, D.W. (1984). Science, 224, 1428-1430.

Natali, P.G., DeMartino C., Quaranta, V., Nicotra, M.R., Frezza, F., Pellegrino, A.M. and Ferrone, S. (1981). Transplantation, 31, 75-78.

Pizarro, O., Hoecker, G., Rubinstein, P. and Ramos, A. (1967). Proc. Natl. Acad. Sci. (Wash.), 47, 1900-1907.

Pober, J.S., Gimbrone, M.D., Cotran, R.S., Reiss, C.S., Brackoff, S.J., Fiers, W. and Ault, K.A. (1983). J. Exp. Med., 157, 1339-1353.

Roehan, N.W., Leibson, H.J., Zlotnik, A., Kappler, J., Marrack, P. and Cambrier, J.C. (1984). J. Exp. Med. 160, 679-694.

Rostami, A., Eccleston, P.A., Silberberg, D.H., Hirayama, M., Lisak, R.P., Pleasure, D.E. and Phillips, S.M. (984). Brain Res., 298, 203-208.

Schachner, M. and Sidman, R.L. (1973). Brain Res., 60, 191-198.

Schnitzer, J. and Schachner, M. (1981). J. Neuroimmunol., 1, 429-456.

Suzumura, A., Bhat, S., Eccleston, P.A., Lisak, R.P. and Silberberg, D.H. (984). Brain Res., 324, 379-383.

Suzumura, A. and Silberberg, D.H. (1985). Brain Res., 336, 171-175.

Takeoka, T., Shinohara, Y., Furumi, K. and Mori, K. (1983). J. Neurochem., 41, 1102-1108.

Waksman, B.H. (981). Immunology Today, 2, 87-93.

Wallach, D., Fellous, M. and Revel, M. (1982). Nature, 299, 833-836.

Wong, G.H.W., Clark-Lewis, I., McKimm-Breschkin, J.L., Schrader, J.W. (1982). Proc. Natl. Acad. Sci., U.S.A., 79, 6899-6993.

Wong, G.H.W., Bartlett, P.F., Clark-Lewis, I., Battye, F. and Schrader J.W. (1984). Nature, 310, 688-691.

Zinkernagel, R.M. and Doherty, P.C. (1979). Adv. Immunol., 27, 52-177.

Monoclonal Antibodies Directed Against Oligodenrocytes

M. Dumas*, S. Szuchet[†] and B. G. W. Arnason[†]

*Department of Neurology, University of Berne, Inselspital CH 3010 Berne, Switzerland

[†]Department of Neurology, University of Chicago, School of Medicine, Chicago, Il 60637, USA

ABSTRACT

In this study, we document the binding of monoclonal antibodies (Mab) to ovine oligodendrocytes (ODC), in situ and in vitro.

KEYWORDS

Oligodendrocytes; cell culture ; tissue sections; electron microscopy ; monoclonal antibodies ; immunocytochemistry.

In Multiple Sclerosis ODC may be the target of an autoimmune attack (Szuchet and Dumas, 1983). The purpose of this study is to investigate antigens which are specifically expressed by ODC.
We have generated a battery of Mabs to ODC (for details cf. Dumas, et al, 1985) Among those, Mab I 6G1 stains the surface of oligodendroglial cell bodies and processes (Fig. 1a) whereas labelling with Mab VIII 1E9 is confined to the cytoplasm (Fig. 1b). The electron-microscopic examination of an unstained, but contrasted, culture reveals typical features of ODC, e.g. electron-dense vesicles, microtubules (Fig. 1c) and extensive formation of membranes which are decorated by Mab I 6G1 (Fig. id). When applied to frozen sections Mab 1 6G1 labels cellular perikarya aligned in rows (Fig. 2), as interfascicular ODC do (Rio-Hortega, 1928). We are currently investigating whether these Mabs may distinguish the different classes of ODC (Rio-Hortega, 1928) on the basis of their antigen expression.

ACKNOWLEDGEMENT

Supported by the Swiss and American (Grant No. RG-1223-B3) Multiple Sclerosis Societies.

REFERENCES

Dumas, M., A., Zurbriggen, M. Vandevelde, S.H. Yim, B.G.W. Arnason, S. Szuchet and C. Meier (1985). J. Neuroimmunol., 9, 55-67.
Rio-Hortega, P. del. (1928). Real.Soc. espanola de Historia Natural, 14, 14-122
Szuchet, S. and M. Dumas (1983). Neurol. Clinics, 1, 729-755.

Printed in Great Britain

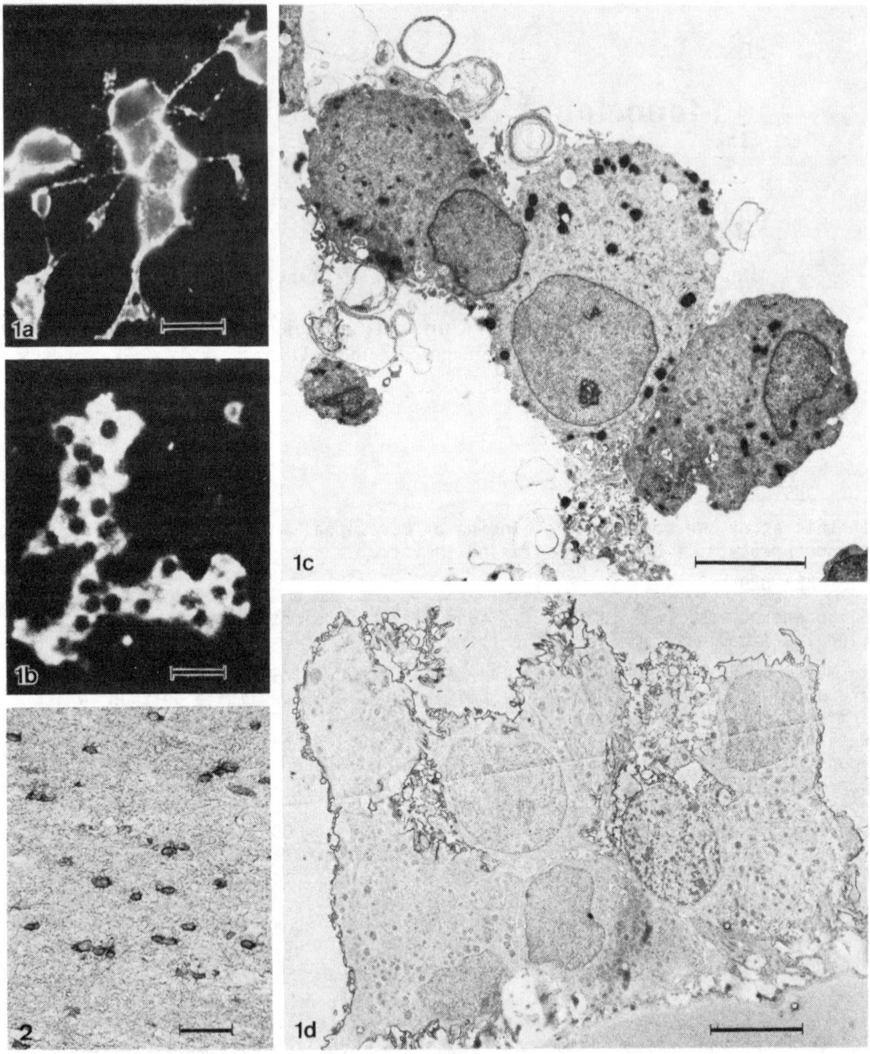

Fig. 1. Binding studies on isolated oligodendrocytes from lamb white matter, 30 days old culture, unlabelled PAP staining.
a: Membrane staining unfixed culture, Mab I 6G1. Bar 15 µm.
b: Intracytoplasmic staining, fixed culture, Mab VIII 1E9. Bar 15 µm.
c: Electron micrograph, not immunostained, conventional contrasting. Bar 5 µm.
d: Immuno-electron mocrograph, Mab 1 6G1. Bar 5 µm.
Fig. 2. Cryostat section from lamb medulla oblongata, Mab I 6G1. Bar 50 µm.

Immunohistochemical Localization of Prostaglandin D Synthetase in Oligodendrocytes in the Central Nervous System of Adult Rat

Y. Urade*, T. Kaneko†, N. Fujimoto*,
A. Konishi†, N. Mizuno† and O. Hayaishi*

*Hayaishi Bioinformation Transfer Project, Research Development Corporation of Japan, Kyoto 601, Japan
†Department of Anatomy (1st Division), Kyoto University Faculty of Medicine, Kyoto 606, Japan

ABSTRACT

In the central nervous system of adult rats, prostaglandin D synthetase-like immunoreactivity is localized in oligodendrocytes.

KEY WORDS : Prostaglandin D_2; immunohistochemistry.

INTRODUCTION

Prostaglandin D_2 is a major prostaglandin in the central nervous system of the rat (Narumiya, et al., 1982) and human (Ogorochi, et al., 1984), and the enzyme responsible for the biosynthesis of prostaglandin D_2 is localized in the central nervous system (Abdel-Halim, et al. and Sun, et al., 1977). In this paper, to clarify which types of cells produce prostaglandin D_2, we examined the immunohistochemical localization of prostaglandin D synthetase in the central nervous system of adult rats.

RESULTS AND DISCUSSION

Prostaglandin D synthetase was purified from rat brains to apparent homogeneity and the antibody was raised against the enzyme in a rabbit (Urade, et al., 1985a). The antibody absorbed more than 90% of the enzyme activity in crude extracts of the rat brain under the condition of antibody excess, and showed only one positive band in the Western blotting of the crude extracts at the same position as that of the purified enzyme. These results indicate that the antibody can bind almost all of the enzyme in the central nervous system, and that the antibody bind only the enzyme among various proteins in the central nervous system.

In the immunoperoxidase staining of the brain and spinal cord of adult rats (> 8 weeks old), the immunoreactivity was found in small non-neuronal cells with a soma diameter of about 10 μm (Urade, et al., 1985b). The small immunoreactive cells were widely distributed in the central nervous system and more enriched in the white matter than in the gray matter (Fig. 1). The regional distribution of the immunoreactive cells was similar to that of structures with immunoreactivity for myelin basic protein. The immunoreactive cells did not contain the immunoreactivity for glial fibrillary acidic protein. By electron microscopy, the cells had scanty cytoplasm and contained a small oval nucleus. The immunoreactive cells were, therefore, considered to be oligodendrocytes. These results indicate that prostaglandin D_2 is synthesized in oligodendrocytes in the central nervous system of adult rats.

Advances in the Biosciences Vol. 61

Printed in Great Britain

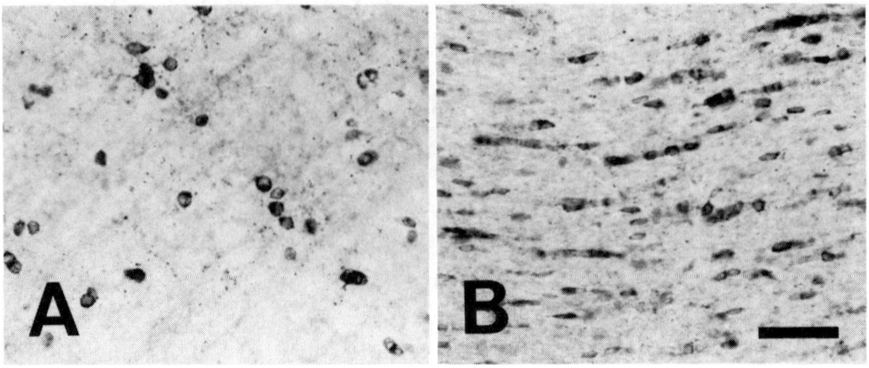

Fig. 1. Prostaglandin D synthetase-immunoreactive oligodendrocytes in the cerebral cortex (A) and the corpus callosum (B) of 12 week-old rat. Bar: 50 μm.

However, the immunoreactivity of this enzyme was not detected in oligodendrocytes of neonatal rats (2 weeks old) (Urade, et al., 1985b), at which age most oligodendroglial markers are reportedly expressed. In the adult rat, the immunoreactivity was observed in the cell body of oligodendrocytes but not in myelin and also not in Schwann cells of the peripheral nervous system. Thus, prostaglandin D synthetase is a new useful marker specific for the cell body of mature oligodendrocytes, in which prostaglandin D_2 may function on the maintenance of myelin or on the neuron-glia interaction.

ACKNOWLEDGMENT

We are grateful to Dr. T. Saida, Utano National Hospital, for providing the anti-(myelin basic protein)-antiserum. We thank Drs. Y. Goh, H. Hayashi, S. Ito, R. Ueno and Y. Watanabe of our project for kind suggestions and discussion. We also thank K. Amano, N. Eguchi, M. Tokumoto, Y. Ueha and Y. Yamamoto for their technical assistance.

REFERENCES

Abdel-Halim, M.S., M. Hamberg, B. Sjöguist and E. Ånggård (1977). Prostaglandins, 14, 633-643.

Narumiya. S., T. Ogorochi, K. Nakao and O. Hayaishi (1982). Life Sci., 31, 2093-2103.

Ogorochi, T., S. Narumiya, N. Mizuno and O. Hayaishi (1984). J. Neurochem., 43, 71-82.

Sun, F.F., J.P. Chapman and J.C. McGuire (1977). Prostaglandins, 14, 1055-1074.

Urade, Y., N. Fujimoto and O. Hayaishi (1985a). J. Biol. Chem., in press.

Urade, Y., T. Kaneko, N. Fujimoto, Y. Watanabe, N. Mizuno and O. Hayaishi (1985b). Adv. Prostaglandin, Thromboxane, Leukotriene Res., 15, 549-551.

Synthesis and Phosphorylation of Myelin Basic Protein in Oligodendrocyte Cultures: Regulatory Events

T. Vartanian*,†, S. H. Yim*, S. Szuchet* and G. Dawson*,‡

Departments of *Neurology, †Biochemistry, ‡Pediatrics, and The Brain Research Institute, The University of Chicago, Chicago, Illinois 60637, USA

KEYWORDS

Myelination; myelin proteins; glial cells; post-translational modification.

Myelin basic proteins (MBP) are found in the cytoplasm of oligodendrocytes (OLG) during myelinogenesis; they are expressed maximally when myelination begins but subside to barely detectable levels thereafter (Sternberger, 1984). Neither the signal nor the mechanism of induction of MBP synthesis are known. MBP undergoes two types of post-translational modifications: phosphorylation and methylation. The functional role of these modifications is not understood. Having shown that mature OLG, in long term cultures, re-express MBP (Szuchet, et al, 1980), we used such cultures to investigate the events that regulate the synthesis and phosphorylation of MBP. Our results are consistent with the hypothesis that: a) an external signal is required for the phosphorylation and increased synthesis of MBP; b) in vitro, cell-substratum adhesion mediates these events, i.e., it acts as a surrogate for the putative signal and c) there is increased synthesis and phosphorylation of MBP with time in culture.

OLG are isolated from lamb white matter and cultured as described (Szuchet and Yim, 1984). We keep OLG in two physical states: as floating clusters (B3.f) and as attached cells (B3.fA). The transition from B3.f to B3.fA is marked by important physical and biological changes.

B3.f cells synthesize some MBP (Fig.1a), but do not phosphorylate it within detectable levels (Fig.1b). Routinely B3.f cells are maintained for 4 days; prolonging this time to 8 or more days does not result in increased synthesis or phosphorylation of MBP (Fig.2). This is observed at a time when there is active synthesis and phosphorylation of other OLG proteins.

During the first 24 hours following cell attachment to a polylysine substratum, there is phosphorylation of MBP as well as a distinct increase in the synthesis of MBP (Fig.1). Both events continue to rise throughout the first five weeks in culture but level off in the subsequent three weeks (Figs.1a&b). Immunocytochemical detection of MBP, which is equivocal for B3.f cells (Szuchet, et al, 1980), is strongly positive in B3.fA cells.

The increased incorporation of $^{32}PO_4$ into MBP with time may result from either an increase in phosphorylation concomitant with synthesis or from the turnover of phosphate in existing MBP. Conceivably, both events take place. In myelin, turnover of MBP phosphate occurs much more rapidly than turnover of the peptide backbone (Des Jordins and Morell, 1983).

The mechanism whereby cell adhesion triggers the phosphorylation of MBP is unknown. Protein kinase C (PK-C), the enzyme responsible for MBP phosphorylation in myelin (Turner, et al, 1982), is active when in a quaternary complex with phosphatidylserine (PS), Ca^{++} and diacylglycerol (DAG; Nishizuka, 1984). DAG is thought to be the rate limiting factor in PK-C activation and its action is mimicked by tumor promoting phorbol esters (TPA). The fact that PK-C activity from cell homogenates (assayed using exogenous histone and added PS, Ca^{++}, and DAG) does not change following cell attachment along with the evidence that TPA causes rapid phosphorylation of MBP in B3.f cells may be taken as an indication that subsequent to cell attachment there is a release of DAG. We are thus investigating the induction of phospholipase C-linked receptors and the role of the second messenger DAG in mediating the phosphorylation of MBP in OLG. The alternative possibility that MBP and PK-C are compartmentalized separately in B3.f cells but brought together in B.3fA cells is also being considered.

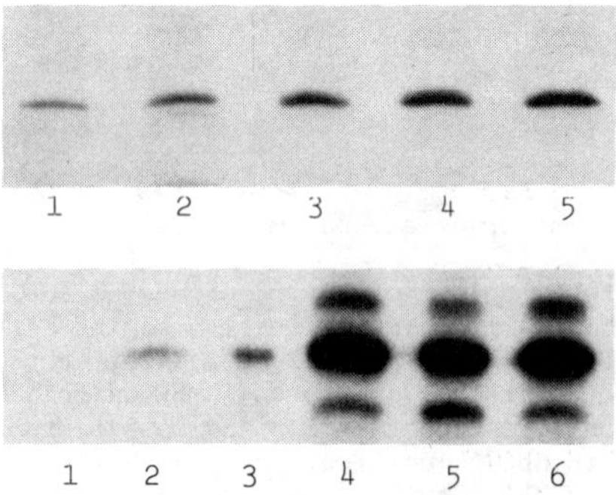

Figure 1: Synthesis and phosphorylation of MBP in OLG cultures. Cells were kept in DMEM plus 20% horse serum and 2μM L-gln; labeled with either 5μCi of ^{14}C-leu for 20 hr or 50μCi of $^{32}PO_4$ for 30 min. MBP were immunoprecipitated using rabbit polyclonal anti-MBP (gift from A.T. Campagnoni), resolved on 12% SDS-PAGE and visualized by autoradiography.
a) ^{14}C-polypeptides. Lanes: 1,B3.f; 2-5,B3.fA attached for 1,26,35 and 58d, respectively. Prior to adhesion, cells are left for 3-4d as floaters.
b) ^{32}P-polypeptides. Lanes 1,B3.f; 2-6,B3.fA attached for 1,14,27,42 and 58d, respectively.

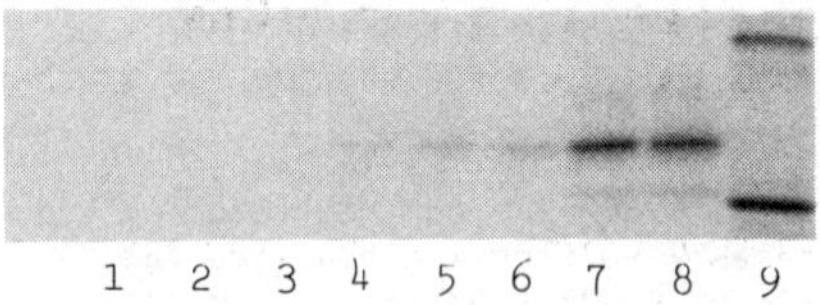

Figure 2: Autoradiogram of ^{32}P-MBP. MBP were immunoprecipitated as above and resolved by 7-25% linear gradient SDS-PAGE. Lanes: 1-3,B.3f after 4,5, and 8d; 4-8,B3.fA, after 1,1,5,10, and 22d, respectively. Lanes 4 and 5 are from cells kept as floaters for 4 and 7d, respectively, prior to 1d of attachment. Lane 9,^{14}C-stds.

ACKNOWLEDGEMENT

One of us (TKV) was the recipient of a Telfeyan Scholarship. Supported by grants RG-1223-C4 (Natl. MS Soc.) and HD 06426 (NICHD).

REFERENCES

Des Jordins KC and P Morell (1983). J Cell Biol 97, 438-446.
Nishizuka Y (1984). Nature 308, 693-698.
Sternberger N (1984). Oligodendroglia, Academic Press, New York.
Szuchet S, K Stefansson, RL Wollmann, G Dawson and BGW Arnason (1980). Brain Res 200, 151-164.
Szuchet S and SH Yim (1984). Neurosci Res 11, 131-144.
Turner RS, CH Jen Chou, RF Kibler and JF Kuo (1982). J Neurochem 39, 1397-1404.

Session III

NEURON-GLIA MOLECULAR RELATIONSHIPS

A. ION HOMEOSTATIS
B. DEVELOPMENTAL ASPECTS

Co-chair : L. HERTZ (*Saskatchewan, Canada*)
G. FRANCK (*Liège, Belgium*)

C. METABOLISM AND NEUROTRANSMISSION
D. MYELINATION

Co-chair : S. VARON (*San Diego, USA*)
W. T. NORTON (*New York, USA*)

Mechanisms of K^+ Regulation in Brain Passive of Active Process?

G. Franck*, T. Grisar† and G. Moonen*

*Centre Hospitalier Universitaire di Sart Tilman, Service de Neurologie, 4000 Liège, Belgium
†Laboratoire de Biochimie Générale et Comparée, Place Delcour, 17, 4020 Liège, Belgium

INTRODUCTION

Repetitive action potentials result in accumulation of potassium ions in the extracellular space adjacent to neurons (Franck et al., 1978; Sykova, 1983). From a resting level of 3mM, K^+ can increase up to 10 of 20mM during intense nerve activity whereas values as high as 30, 50 or 80mM have been seen in spreading depression, anoxia or hypoglycemia (Franck et al., 1978; Somjen, 1980; Sykova, 1983).

Increases of $(K^+)_o$ would have an excitatory effect on neurons since K^+ depolarizes neuronal membranes and evokes repetitive firing of isolated axons. Also, the amplitude of spike potentials and excitatory postsynaptic potentials of motor neurons shows a linear relationship to the resting potential versus external potassium.

Evidences supporting the view that K^+ is involved in the so-called "primary afferent depolarization" and in transmitter release from primary afferent terminals have also been reviewed by Sykova (1983). However, increased $(K^+)_o$ can have dual effect according to the level of this elevation : low increases in general facilitate neuronal excitability and synaptic transmission while higher elevations (greater than 6-8mM) could lead to neuronal depression (Sykova, 1983). Moreover, Weight and Erulkar (1976) illustrated how the modulation of transmitter release by K^+ accumulation induced by postsynaptic impulse activity can function as a direct negative feedback, supporting many results showing that the frequency of impulses in some firing neurons increases first a "burst" and then the neurons stop firing (Weight and Erulkar, 1976).
Therefore, the mechanisms by which accumulated K^+ is cleared from the ECS are of major importance when discussing the control systems of neuronal excitability both under normal and pathological conditions, although many observations support the conclusion that excessive changes in $(K^+)_o$ are probably compensated by other transports across glial membranes, bloodbrain barrier and diffusion in the extracellular space.

The possible involvement of glial cells in this regulatory process has been proposed by several authors (Franck et al., 1978; Franck et al., 1973) but opinions remained divided about the nature (i.e.

Advances in the Biosciences Vol. 61

Printed in Great Britain

active or passive) by which glial cells (mainly astrocytes) could limit the rise of K^+. Electrophysiological results in general favor the concept of passive spatial buffering system as proposed by Orkand (1969) whereas biochemical observations supported the idea of an active accumulation of K^+ mediated by the (Na^+, K^+)-ATPase (Franck *et al*., 1978; Franck *et al*., 1983).

K^+ transport in glial cells

According to the "spatial buffer mechanism" (Orkand, 1969), K^+ enters the glial cells in the region with a high $(K^+)_o$ and leaves it elsewhere in regions where the $(K^+)_o$ is lower. This local build-up of extracellular K^+ and the consequent glial cell membrane depolarization, causes current to flow through the so-called "glial syncitium". This purely passive process only requires that the local glial cell membrane is permeable to K^+. Two facts from electrophysiological studies were the ingredients of this theory :

(a) Glial cells function as highly selective K^+ electrodes, membrane potential varying with changes of $(K^+)_o$, according to the Nernst Law ("Nernstian Behavior"). Moreover, only single K^+ channel has been recently recorded with oligodendroglial membrane patches (Kettenmann *et al*, 1983). Accordingly, most investigators agree that glial membranes are almost exclusively permeable to K^+ (see also Ward, 1978).

(b) Glial cells are electrically coupled both *in vivo* (Kuffler and Nicholls, 1966; Ransom and Goldring, 1973) and *in vitro* (Moonen and Nelson, 1978; Moonen *et al*., 1980) with evidence of the existence of low-resistance interconnections between astrocytes (gap junctions).

An opposite school to the "spatial buffering function" of glial cells contends that extracellular K^+ is cleared by active transport mechanisms and that glial membranes, in fact, do not behave as a perfect K^+ electrode. Although this may be due to a non-negligible Na^+ and Cl^- conductance (Walz and Hertz, 1984); this non-Nernstian behavior of glial cells is more likely the consequence of increased intracellular K^+ concentration when the extracellular K^+ is raised (Coles and Tsacopoulos, 1979).

In fact, several investigators found K^+ accumulation in glial structures when exposed to elevated external K^+ concentrations (Franck *et al*., 1978; Franck *et al*., 1983; Grisar and Franck, 1981). Early slice experiments by Franck (Franck *et al*., 1978; Grisar and Franck, 1981) have clearly shown that an increase of the external potassium from 3 to 20mM led to a step increase in intracellular K^+ concentration in a cellular compartment identified as the glial cells (Grisar and Franck, 1981).

The ability of isolated glial cells to accumulate potassium efficiently was then demonstrated with either bulk isolated (Henn *et al*. 1972) or cultured astrocytes (Hertz, 1979). Moreover, faster accumulation of labelled K^+ into glial compartment of brain slices (Franck, 1970) as well as into astrocytes of primary cultures (Hertz, 1979; Moonen *et al*., 1980) than into corresponding neuronal structures also suggested that the main uptake of K^+ occurs in glia cells. The correlation between rates of ^{43}K uptake and potassium concentration indicated a Km value of 10-15mM for K^+ transport into cultured astrocytes (Franck *at al*., 1983) suggesting a major activation of this process when the extracellular concentration is increased whell above the resting level (Franck *et al*., 1978; Futamachi and Pedlely, 1976).

Walz <u>et al</u>. (1984) have recently confirmed that membrane potential characteristics of astrocytes in primary culture reveal exclusive potassium conductance and potassium accumulator properties.

Several mechanisms have been proposed to explain this specific glial K^+ accumulation :

(1) the passive entrance of the "spatial buffer mechanism",

(2) a K^+-Cl^- exchange,

(3) a net active uptake of K^+, and

(4) a Na^+-K^+ exchange through the Na^+, K^+ pump.

As recalled by Gardner-Medwin (1983), the spatial buffer mechanism is a purely passive process, a form of facilitated K^+ diffusion. This transport, however, would require that K^+ passively cross the glial membrane against an important electrochemical gradient in such a manner that $(K\ )_i$ does not vary significantly. This is no longer supported by experiments cited above and recent evidence shows that a large change in intraglial K^+ concentration does exist in "glia elements" of the retina of the drone honeybee when light is flashed to activate surrounding photoreceptors ("neuronal elements") (Coles and Tsacopoulos, 1979).

Moreover, one has to keep in mind that the cellular membrane is a phospholipid bilayer with embedded active proteins, on constant modifications of their tridimensional structures. Energy sources like ATP and other phosphate donors are constantly used, maintaining the membraneous system far away from the thermodynamic equilibrium. In other words, the cellular membranes have to be thought in terms of "open thermodynamic systems". This is particularly true as far as cation movements are concerned, since (Na^+, K^+)-ATPase exists and is a specific energy dependent molecular machinery for cation transport (see below). Hence, from this theoretical point of view, we feel that uptake of K^+ within the glial cells like any other cells, must be an active process.

In fact, a potassium induced stimulation of oxygen uptake in glial cells has been illustrated in brain slices (Hertz, 1978), in isolated glial cells and in subventricular white matter which is known to be highly enriched in glial cells. Its virtual absence in jimpy mice (Hertz, 1977 and 1978), which are deficient in glial cells, also supports this notion. Similarly, a potassium induced reduction of ATP content observed in brain slices seems to occur predominantly in astrocytes (Hertz, 1978). Although these experiments on K^+ enhanced O_2 uptake only provide indirect arguments and were most often observed in high potassium media (+/- 50mM), these observations give further support to the concept that the metabolic effects of increased extracellular potassium concentrations are exerted mainly on glial cells.

A K^+-dependent uptake of chloride into glia has also been described under circumstances requiring elevated (and pathological) extracellular levels of potassium (Bourke and Tower, 1966; Bourke and Nelson, 1972 and Bourke <u>et al</u>., 1978). This involves both passive and carrier mediated transport of cations with Cl^-, and transfer of an osmotic equivalent of water. Carbonic anhydrase was shown to be implicated in this process (Bourke and Nelson, 1972). As discussed by us elsewhere (Grisar and Franck, 1981), a crucial K^+ concentration, i.e. +/- 20mM, seems de demarcate the border between the so-called "physiological" and "pathological" biochemistry events observed in brain tissue incubated <u>in vitro</u>.

In our opinion, this Cl^-, K^+-dependent water uptake probably corresponds to a model of brain oedema, as recently discussed by Kimelberg et al.1979 and Walz and Hertz (1984). An ouabain-resistant net uptake of potassium into cultured astrocytes over and above that caused by diffusion has been shown to be much more intensive than in neurons, a mechanism that appears to be stimulated by elevated potassium concentrations (Walz and Hertz, 1982). Finally, these glial cells contain a potassium-dependent, ouabain-sensitive, hyperpolarizing electrogenic pump, has also been shown be several electrophysiological studies (Moonen and Nelson, 1978; Moonen et al. 1980). Its significance in K^+ homeostasis versus axonal pump has been questioned, however (Somjen, 1980). Walz et al. (1984) also claimed that an ouabain effect on the glial membrane potential does not rule out the possibility of an impairment of the only K^+ accumulation, rather than an impairment in the electrogenic exchange of Na^+ and K^+.

The potassium-sodium exchange observed in brain slices, when external potassium increases from 3.7 to 20mM, is presumably mediated by the Na^+, K^+ pump (Franck et al., 1978, Franck et al., 1983). Because this exchange occurs essentially in a glial compartment, the concept of a specific activation of the glial Na^+, K^+ pump in K^+ enriched media has recently been favored (Franck et al., 1983). Until now, the key experiments, calculating simultaneous fluxes (both outfluxes and influxes) of Na^+ and K^+ in isolated glial versus neuronal cells in progressive K^+ enriched media, both in vitro and in vivo, are still missing.

The recent discovery of a specific K^+ activated glial (Na^+, K^+)-ATPase has supported the idea of a special role of glial cells in an active control of external K^+ concentration mediated by the Na^+, K^+ pump (Grisar et al., 1979; Grisar et al., 1978).

Brain (Na^+, K^+)-ATPase

Another way to illustrate a key role of astrocytes in active control of extracellular K^+ in the CNS would be to demonstrate that these cells, in fact, do have a specific molecular machinery to actively pump K^+, in K^+ enriched media. This approach, however, is very often neglected by most electrophysiological studies which underestimate the significance of the molecular pathways used by almost all the cellular membranes along the phylogenetic tree from *E. coli* to human brain, to carry K^+ inside the cells, i.e. the Na^+, K^+ pump, enzymatically controlled by the (Na^+, K^+)-ATPase.

In 1957, Skou (1957) demonstrated that a microsomial fraction of crab (*Carcinus maenas*) leg nerve was able to hydrolyse ATP in the presence of Mg^{2+}, Na^+ and K^+ ions. K^+ activated ATP hydrolysis only in the presence of both Mg^2 and Na^+. That this system was in fact the enzymatic support of the Na^+, K^+ pump was further demonstrated by the fact that the same low concentrations of cardiac glycosides (ouabain-strophantidin) inhibited both the activity of this microsomial fraction and Na^+ outflux (Albers, 1976; Glynn and Karlish, 1975).

This sodium and potassium ion stimulated adenosine-triphosphatase, a membrane-embedded protein, was then shown to be the enzyme responsible for the active transport of Na^+ and K^+ ions in most animal cells (Albers, 1976). Since this time, several reviews and symposia have been dedicated to this fundamental enzyme (Albers, 1976; Glynn and Karlish, 1975).

The evolutionary significance of this enzyme is to maintain an osmotic equilibrium in the intracellular compartment and it is important to remember that up to 30 % of the total energy supplied by the mitochondrial respiratory chain is devoted to this Na^+, K^+ pump in a large variety of mammalian cells (Albers, 1976).

Our early experiments (Franck, 1970; Franck et al., 1978; Grisar and Franck, 1981) showed a sodium-potassium exchange in mammalian brain slices when increasing the external potassium concentrations from 3 to 20mM. We concluded that this was probably due to the activation of the above-described Na^+, K^+ pump. These conclusions prompted us to investigate the catalytic properties of both glial and neuronal (Na^+, K^+)-ATPase. However, when we started to investigate the relative contribution of glial cells and neurons in the clearance of external K^+, studies on the K^+ activation mechanisms of the enzyme within neuronal, synaptic and glial fractions provided more information on their membraneous physiological characteristics than the absolute levels of (Na^+, K^+)-ATPase.

We utilized tissue fractionation on sucrose-ficoll gradients for preparing whole glial or neuronal cells, synaptosomes and plasma membranes, from different mammalian species such as rabbitt, cat and man. Obviously, the purity and integrity of these different fractions is an important factor in determining the validity of conclusions based on results obtained in this type of study. Since 1969, this problem has been extensively discussed elsewhere (Blomstrand and Hamberger, 1969; Grisar et al., 1979; Grisar et al., 1980; Grisar et al., 1978; Hamberger, 1971; Henn, Haljamme and Hamberger, 1972; Hertz, 1977).

ATPase activities were first determined by a conventional technique (detecting the amount of inorganic phosphate produced by the enzymatic reaction) and later by a kinetic method determining the initial velocity (measuring the rate of liberation of protons in low buffered media) (see Grisar et al., 1979 and Grisar et al. 1978 for review).

Sensitivity to potassium differed between glial and neuronal ensymes supporting data previously shown by Henn et al. (1972). Neuronal enzymes reached maximal activities at 3-5mM K^+. Similar results were observed with the $K^\pm$pNPPase (Franck et al., 1983; Grisar et al., 1978). This was further confirmed by Moonen and Franck (1977) on pure mature cultured astrocytes. Like the Na^+, K^+ exchange observed in rat brain slices, this glial enzyme characteristic is dependent upon a process of maturation since it cannot be observed either in bulk isolated glial cells from young animals (Franck et al., 1978; Grisar and Franck, 1975) or in immature cultured astrocytes (Moonen and Franck, 1977) where neuronal and glial enzyme exhibited similar kinetic in response to K^+. More recently, Atterwill et al. (1984) showed that optimal activity by K^+ ions was observed in 1mM K^+ media in immature cultured astrocytes from rat cerebella, while it was detected only at 8-10mM K^+ in corresponding neuronal cultures. Although no statistical conclusions can be drawn from this latter study, these discrepancies (as otherwise pointed out by the authors) might be related to the immaturity of the preparation.

More interestingly, our kinetic approach measures an accurate initial velocity of the enzymatic reaction (Grisar et al., 1978), as demonstrated from Linewearver-Burke diagrams. The apparent affinity of the glial enzyme for the substrate (ATP. Mg^2) is considerably increased by high K^+ ions concentrations (i.e. 20mM).

Accordingly, the so-called "physiological efficiency" (i.e. Vmax/Km app ratio) of the glial (Na^+, K^+)-ATPase is specifically enhanced in K^+ enriched media while this phenomenon is absent in neuronal preparations (Grisar et al., 1979; Grisar et al., 1980; Grisar et al., 1978). Regardless of the kinetic models used to explain (Na^+, K^+)-ATPase activities, glial and neuronal ATPases mus be two different molecular entities based on their different affinities to (K^+). This only occured in adult or mature neural preparations.

Sweadner (1979) confirmed the presence of two molecular forms of brain (Na^+, K^+)-ATPase by resolving on gel electrophoresis in sodium dodecylsulfate two subspecies of the enzyme catalytic alpha subunits, so-called $alpha^+$ and alpha subunits. The pure form of $alpha^+$ was only seen in rat brain axolemma, whereas alpha was isolated in pure cultured astrocytes. Hence, it was hypothesized that alpha $^+$ might be neuronal form while alpha would be more specific for glial cells.

In brief, we feel that the relative proportion of $alpha^+$ and alpha in a neural membrane could determine both ouabain and K^+ sensitivity of (Na^+, K^+)-ATPase and accordingly of the Na^+, K^+ pump. For example, high concentrations of the alpha subunit in mature glial cells, could explain their specific abilities to pump external K^+ when it reaches higher concentration (up to 20mM). If only $alpha^+$ or higher concentrations of $alpha^=$ is present like in axo lemma (Sweadner, 1979), the membrane would pump K^+_o on higher yield but in the presence of more physiological variations of K^+_o, i.e. between 3-6mM.

Experiments comparing ion fluxes measurements and relative proportion of $alpha^+$ and alpha (Na^+, K^+)-ATPase subspecies in both glial and neuronal structures are needed to verify this hypothesis.

ACKNOWLEDGMENTS

I express my gratitude to Professors G. Franck, E. Schoffeniels and A.V. Delgado-Escueta for their constant enthusiasm and scientific support.

I thank A. Minet and M. Vergniolles-Burette for their excellent technical work and I am grateful to V. Louis for secretarial and word-processing assistance.

This work was also supported by the Belgian Rotary Foundation and Department of Health and Human Services contract N01-NS-0-2332.

REFERENCES

Albers, R.W. (1976) : The (sodium plus potassium)-transport ATPase. In : The enzymes of Biological Membranes, (eds.) Martonosi AN, Plenum Press, New York, pp. 283-295.

Blomstrand, C.; Hemberger, A. (1969) : Protein turnover in cell-enriched fractions from rabbit brain. J. Neurochem. 16 : 1401-1407.

Bourke, R.S. and Tower, D.B. (1966) : Fluid compartmentation and electrolytes of cat cerebral cortex in vitro. II. Sodium, potassium and chloride of nature cerebral cortex. J. Neurochem. 13 : 1099.

Bourke, R.S. and Nelson, K.M. (1972) : Further studies on the K^+ dependent transport of chloride. J. Neurochem. 19 : 663-675.

Bourke, R.S.; Daze, M.A. and Kimelberg, H.K. (1978) : Chloride transport in mammalian astroglia. In : Dynamic Properties of

Glial Cells, (eds.) Schoffeniels, E.; Franck, G.; Hertz, L. and Tower, D.B., Pergamon Press, Oxford, pp. 337.

Coles, J.A. and Tsacopoulos, M. (1979) : Potassium activity in photoreceptor, glial cells and extracellular space in the drone retina. Changes during photostimulation. J. Physiol. (London) 290 : 525-549.

Franck, G. (1970) : Echanges cationiques au niveau des neurones et des cellules gliales du cerveau. Arch. Internat. Physiol. Biochem. 78 : 613-720.

Franck, G.; Grisar, T.; Moonen, G. and Schoffeniels, E. (1978) : Potassium transport in mammalian astroglia. In : Dynamic Properties of Glial Cells (eds.) Schoffeniels, E.; Franck, G.; Hertz, L. and Tower, D.B., Pergamon Press, Oxford, pp. 315-325.

Franck, G.; Grisar, T. and Moonen, G. (1983) : Glial and neuronal Na^+, K^+ pump. Adv. Cell Neurobiol. 4 : 133-159.

Futamachi, K.J. and Pedley, T.A. (1976) : Glial cells and extracellular potassium : their relationship in mammalian cortex. Brain Res. 109 : 311-322.

Gardner-Medwin, A.R. (1983) : A study of the mechanisms by which potassium moves through brain tissue in the rat. J. Physiol. 335 : 353-374.

Glynn, I.M. and Karlish, S.J.D. (1975) : The sodium pump. Ann. Rev. Physiol. 37 : 13-80.

Grisar, T.; Frère, J-M.; Franck, G. (1979) : Effects of K^+ ions on kinetic properties of the Na^+K^+ATPase (EC 3.6.1.3.) of bulk brain cortex. Brain Research 165 : 87-103.

Grisar, T.; Franck, G. and Schoffeniels, E. (1980) : Glial control of neuronal excitability in mammals. II. Enzymatic evidence : Two molecular forms of the Na^+K^+-ATPase in brain. Neurochem. Internat. 2 : 311-320.

Grisar, T.; Frère, J-M.; Charlier-Grisar, J.; Franck, G. and Schoffeniels, E. (1978) : Synaptosomal Na^+K^+-ATPase is a hysteretic enzyme. Febs. Letters 89 : 173-176.

Grisar, T.; Franck, G. (1975) : Glial cell control of extra-cellular K^+ content in the CNS. Abstr. of the 5th Int. Mtg. of the Int. Soc. for Neurochem. (Barcelona) 160 : 228.

Grisar, T.; Franck, G. and Schoffeniels, E. (1978) : K^+ activation mechanisms of the Na^+K^+-ATPase of bulk isolated glial cells and neurons. In : Dynamic Properties of Glial Cells, (eds.) Schoffeniels, E.; Franck, G.; Hertz, L. and Tower, D.B., Pergamon Press, Oxford, pp. 353-363.

Grisar, T. and Franck, G. (1981) : Effect to changing K^+ ion on rat cerebral cortex slices in vitro. A study development. J. Neurochem. 36 : 1853.

Hamberger, A. (1971) : Amino acid uptake in neuronal and glial cell fractions from rabbit cerebral cortex. Brain Res. 31 : 169-178.

Henn, F.A.; Haljamme, M. and Hamberger, A. (1972) : Glial cell function : active control of extracellular K^+ concentration. Brain Res. 43 : 437-443.

Hertz, L. (1977) : Biochemistry of glial cells. In : Cell, Tissue and Organ Cultures in Neurobiology (eds.) Fedoroff, S. and Hertz, L., Academic Press, New York, pp. 39-71.

Hertz, L. (1978) : Energy metabolism in glial cells. In : Dynamic Properties of Glial Cells, (eds.) Schoffeniels, E.; Franck, G.; Hertz, L. and Tower, D.B. - Pergamon Press, Oxford, pp. 121-131.

Hertz, L. (1979) : Functional Interactions between neurons and astrocytes. II. Potassium homeostasis at the cellular level. Prog. Neurobiol. 13 : 324-368.

Kettenmann, H.; Sonnhof, U. and Schachner, M. (1983) : Exclusive potassium dependence of the membrane potential in cultured mouse oligodendrocytes. J. Neurosci. 3 : 500-505.

Kimelberg, H.K.; Biddlecome, S. and Bourke, R.S. (1979) : SITS inhibitable Cl^- transport and Na^+ dependent H^- production in primary astroglial cultures. Brain Res. 173 : 111.

Kukes, G.; Elul, R. and De Vellis, J. (1976) : The ionic basis of the membrane potential in a rat glial cell line. Brain Res. 104 : 71-92.

Kuffler, S.W. and Nicholls, J.G. (1966) : The physiology of neuroglia cells. Ergeb. Physiol. 57 : 1-90.

Matsuura, S. (1969) : Effects of changes in extracellular K^+ concentration on resting and action potentials of the toad spinal Motoneuro. Osaka City Med. J. 15 : 29-45.

Moonen, G. and Franck, G. (1977) : Potassium effect on (Na^+, K^+)-ATPase activity of cultured new-born rat astroblasts during differentiation. Neurosci. Letters 4 : 263.

Moonen, G. and Nelson, P.G. (1978) : Some physiological properties in astrocytes in primary cultures. In : Dynamic Properties of Glial Cells, (eds.) Schoffeniels, E.; Franck, G.; Tower, D.B.; Hertz, L., Pergamon Press, New York, pp. 389-393.

Moonen, G.; Franck, G. and Schoffeniels, E. (1980) : Glial control of neuronal excitability in mammals : 1. Electrophysiological and isotopic evidence in culture. Neurochem. Intern. 2 : 299.

Orkand, R.K. (1969) : Neuroglial-neuronal interactions. In : Basic Mechanisms of the Epilepsies, (eds.) Jasper, H.H.; Ward, A. and Zope, A., Little Brown, Boston, Massachusetts, pp. 737-746.

Ransom, B.R. and Goldring, S. (1973) : Ionic Determinants of membrane potential of cells presumed to be glia in cerebral cortex of cat. J. Neurophysiol. 36 : 855-868.

Somjen, G.G. (1980) : Influence of potassium and neuroglia in the generation of seizures and their treatment. In : Antiepileptic Drugs : Mechanisms of action, Eds. Glaser, G.H. et al., Raven Press, New York - pp. 155-167.

Sweadner, K.J. (1979) : Two molecular forms of (Na^+, K^+)-stimulated ATPase in brain : separation and difference in affinity for strophantidin. J. Biol. Chem. 254 - 6060-6067.

Sykova, E. (1983) : Extracellular K^+ accumulation in the central nervous system. Prog. Biophys. Mol. Biol. 42 : 135-189.

Walz, W. and Hertz, L. (1982) : Ouabain-sensitiv and ouabain resistant net uptake of potassium into astrocytes and neurons in primary cultures. J. Neurochem. 39 : 70-77.

Walz, W. and Hertz, L. (1984) : Comparison between fluxes of potassium and of chloride in astrocytes in primary cultures. Brain Research, in press.

Walz, W.; Wuttke, W. and Hertz, L. (1984) : Astrocytes in primary cultures : membrane potential characteristics reveal exclusive potassium conductance and potassium accumulator properties. Brain Research, in press.

Potassium Channels and Carriers in Glial Cell Membranes

W. Walz

Department ofg Physiology, College of Medicine, University of Saskatchewan, Saskatoon, Saskatchewan, S7N 0W0 Canada

ABSTRACT

Resting permeability for K^+ through channels and carrier-mediated K^+ accumulation in response to increased external K^+ were investigated in rat, mouse and chick astrocytes in primary cultures. Carrier-mediated K^+ transport was found to be identical in rat and mouse astrocytes, chick astrocytes did not respond with a net K^+ accumulation when external $[K^+]$ was increased. Mouse astrocytes (but not rat and chick astrocytes) had in addition a barium-sensitive K^+ channel, which mediated the transport of K^+ with a velocity 50 times higher than the carrier. With physiologically increased external $[K^+]$ mouse and rat astrocytes exhibited increased Na^+, K^+-ATPase-mediated K^+ uptake and the KCl carrier used the Na^+ driving force to increase internal K^+ and Cl^- content. This increased the internal water content without delay. Na^+ ions were cycled across the membrane by Na^+,K^+-ATPase and KCl carrier. At pathologically increased external $[K^+]$ the Na^+,K^+-ATPase activity was not additionally increased and the driving force for the Na^+ ion is reduced, decreasing the uptake through the KCl carrier. The pathologically increased $[K^+]$ and water content was accomplished by passive K^+ movements through channels probably due to Donnan forces.

KEYWORDS

Barium; cell swelling; furosemide; glial cells; ouabain; potassium; sodium.

INTRODUCTION

The main focus of investigations of glial cell function has been and still is their role in extracellular transmitter and K^+ removal. Transmitter substances and K^+ ions are released by active neurons and they in turn interact with neuronal function. Their concentrations in the neuronal microenvironment must therefore be controlled by uptake processes in which glial cells take part (for reviews see Walz and Hertz, 1983; Syková, 1983; Kimelberg, 1983). This review concentrates on what is known about K^+ uptake systems of glial cells and on some obvious methodological problems associated with these studies. Almost all the information available about glial cell function comes from astrocytes or oligodendrocytes in primary cultures from newborn rat or mouse brain. A few other model systems are used, all not of mammalian origin: the leech CNS, the honey bee retina and the optic nerve of the mudpuppy. The methodological problems associated with the extensive use of primary culture model systems are the intensive coupling of astrocytes by gap junctions and dramatic differences in the quantity and quality of K^+ channels present in rat astrocytes on one hand and mouse astrocytes on the other hand. The aim of this review is therefore to draw attention on these problems for future interpretation of the results by authors using tissue culture model systems.

UNIDIRECTIONAL INFLUX AND NET ACCUMULATION OF POTASSIUM

For an understanding of the experimental strategy, it is necessary to distinguish clearly between unidirectional influx of K^+ and net accumulation of K^+. During equilibrium state with respect to ion distribution, the cells do not experience any change, because the total concentration of ions present is constant and it is assumed that the cells cannot distinguish between different isotopes of one ion. However, because the amount of extracellular radioactivity is known (f.e. in Bq/g of K^+), it is easy to calculate the ion exchange of the external with the internal compartment. The velocity with which the isotope is taken up into the cell is a measure of its steady-state turnover and therefore for its permeability. The higher the permeability (or turnover rate measured as the rate constant in min^{-1}, expressed as the time in min for the total content of the cell to be exchanged with the external solution), the faster an equilibrated distribution of the isotope is reached. The isotope is then distributed with the same specific radioactivity (Bq ^{42}K/g ^{39}K) inside and outside the cell. The permeability for K^+ is so high, that despite its very high internal content it reaches equilibrium within seconds, in contrast to the Cl^- ion which takes minutes to reach equilibrium with its internal Cl^- content, which is only 15% of that of K^+. The rate constant for K^+ is 2.7 min^{-1}, i.e. within 1 min the internal K^+ exchanges 2.7 times with the outside. This is a resting turnover and means that efflux and influx are equal, so that the content does not change. The unidirectional influx rate is measured by using the linear part of the radioisotope accumulation curve (10 s

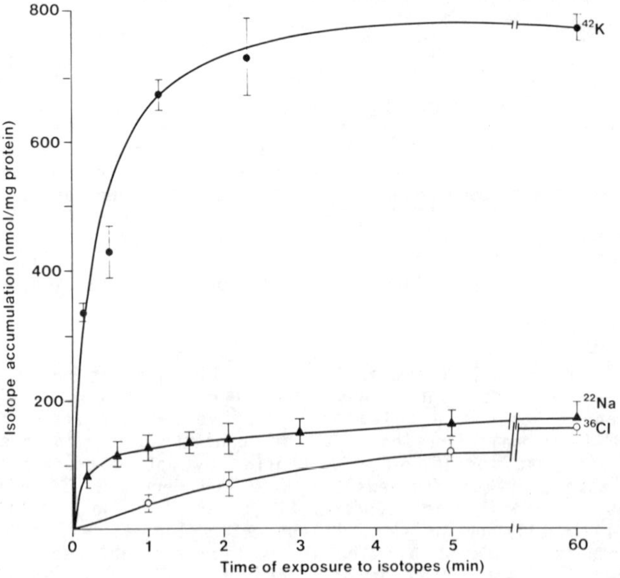

Fig. 1 Accumulation of isotopes into mouse astrocytes (35 days old, DBcAMP-treated) during steady-state conditions with respect to ion concentrations.

for K^+) and is expressed in units of nmol K^+ $xmg^{-1} \cdot xmin^{-1}$. This high and exclusive K^+ permeability of astrocytes is reflected by their membrane electrical properties. The membrane input resistance is very low (2MΩ), the membrane time constant is very short (<400 us) and the membrane potential is an exclusive Nernst potential for K^+. A high membrane permeability is the result of ion movements through channels. Only a very small part (<5%) of these K^+ transmembrane movements is mediated by ion carriers. Thus, through these channels the turnover rate of K^+ (or ^{42}K) is kept very high. However, it is a symmetrical exchange (unidirectional influx=unidirectional efflux) and this does not change the content and therefore there is no net accumulation or net loss of K^+. It should be noted that according to the spatial buffer theory (Orkand et al, 1966) the glial cells use this high permeability to remove excess external K^+, however, this is not accomplished by net accumulation, but by redistribution of K^+ along the glial network. Spatial buffering uses passive K^+ ion carried currents across cell membranes, which are channel-mediated and involve no net accumulation. In order for glial cells to store K^+ temporarily, they have to net accumulate it and this has to be accomplished by carriers. Such a net accumulation is measured by loading the cells under steady-state conditions with ^{42}K until equilibrium of the radioisotope is reached. Then, external $[K^+]$ is increased, but at the same time the specific radioactivity of the external medium maintained (constant amount of Bq ^{42}K/g ^{39}K). Now, any net change in the K^+ content of the cell is reflected by a proportional net accumulation of ^{42}K.

We used monolayer cultures of primary cortical astrocytes from newborn embryos (Hertz et al., 1982). These monolayers cultures are homogenous and are well suited for ion transport analysis, because the extracellular space is small and not separated by diffusional barriers from the bulk fluid of the cell culture dish. Washing procedures do not present any problems (Walz and Hertz, 1983b). In the following part of the article I shall present an up to date view about K^+ transport in these astrocytic cultures through channels and carriers during resting conditions and under physiologically and pathologically elevated external K^+.

THE IMPACT OF SPECIES ORIGIN ON ION TRANSPORT OF CULTURED ASTROCYTES

The three laboratories so far principally involved in ion transport studies of cultured astrocytes report differences of several orders of magnitude in crucial parameters of K^+ transport. Values for the unidirectional fluxes of K^+ of 2025 have been reported for mouse (Hertz, 1978; Walz and Hertz, 1983a) and 35 (Kimelberg et al, 1979) and 26 (Moonen et al, 1980) nmol K^+ xmg^{-1} protein $xmin^{-1}$ for rat cultures, respectively. These differences seem to concern generally cation metabolism, since Kimelberg et al (1979) reported a unidirectional Na^+ flux of 8 nmol xmg^{-1} $xmin^{-1}$ and Walz and Hertz (1984) found Na^+ fluxes between 500 and 550 nmolxmg^{-1}. In contast both groups (Kimelberg, 1981; Walz and Hertz, 1983) agree on a unidirectional Cl^- influx of around 35 nmol$xmg^{-1}$$xmin^{-1}$. Table 1 gives a comparison of cation transport in mouse, rat (as studied by Walz and Kimelberg, 1985) and chick astrocytes.

The abilities of astrocytes in accumulating K^+ when they are exposed to increased extracellular $[K^+]$ (12 mM) shows clear and dramatic differences between all three species. 50 uM $BaCl_2$ is used to distinguish channel-mediated and carrier-mediated unidirectional K^+ transport (Walz et al.,1984). Only mouse astrocytes have a very high K^+ turnover component mediated by a barium-sensitive K^+ channel. This is about 50 times higher than the carrier-mediated transport component. This carrier-mediated component is about the same in mouse and rat astrocytes and is in both stimulated by 12 mM K^+ with the same time course. The channel-mediated component in rat astrocytes is smaller than their carrier-mediated component and 100 times smaller than the channel-mediated component in mouse astrocytes with respect to K^+ transport. The chick astrocytes differ completely from rodent astrocytes: 1) They exhibit an insignificant channel-mediated component. 2) Their carrier-mediated transport is very low. 3) This carrier-mediated component is not activated by increases in extracellular [K^+].

Table 1. Comparison of cation transport in rodent and chick astrocytes. All cultures were 30-40 days old and not treated with DBcAMP.

Parameter	Unit	Mouse	Rat	Chick
Content:	nmol x mg^{-1}			
K^+		715	541	641
Na^+		161	120	119
Influx:	nmol x mg^{-1} x min^{-1}			
K^+ (Ba^{2+}-sensitive)		2420	23	1
K^+ (Ba^{2+}-insensitive)		55	72	17
Na^+		488	415	319
K^+ Net Accumulation	nmol x mg^{-1}			
(within 2 min following an external increase of K^+ to 12 mM)		350	350	0

Chick astrocytes: Components of unidirectional K^+ flux.
Chick astrocytes were analyzed in detail, because their low flux rates gave a chance to get a complete overview about all transport systems involved. There are three transport systems which completely acount for all transmembrane fluxes of K^+: 1) the K^+ channel blocker barium blocked 1.3 nmol x mg^{-1} x min^{-1} (8%). 2) The Na^+, K^+-ATPase blocker ouabain blocked 12.9 nmol x mg^{-1} x min^{-1} (75%). 3) The Na^+, Cl^-, K^+ cotransport inhibitor furosemide blocked 4.1 nmol x mg^{-1} x min^{-1} (17%). The total fluxes were found to be 18.3 nmol x mg^{-1} x min^{-1}, which is the same as the sum of all drug-inhibitable fluxes. These experiments confirm that barium, ouabain and furosemide block in steady-state conditions all the K^+ fluxes. As we shall see later, this is still true during activation of K^+ transport at higher K^+ concentrations with rodent astrocytes.

Importance of these findings. From these results it has to be expected that the membrane potential of astrocyctes from mouse brain is clamped at the K^+ equilibrium potential due to their very high K^+ permeability. That means, there are no contributions to the membrane conductance from other ions or any electrogenic components. Indeed, Bowman and Kimelberg (1984) found an electrogenic component in rat astrocytes, whereas Walz and Hertz (1984) did not with mouse astrocytes. Only changes in K^+ distribution across the cell membrane would affect the membrane potential of mouse astrocytes, unless the large K^+ permeability is blocked for example by barium. Indeed, in terms of K^+ transport, mouse astrocytes can be "transformed" into rat astrocytes through barium treatment (Walz and Kimelberg, 1985).

Influence of cell culture environment on K^+ channels. In some studies we (Walz and Kimelberg, 1985; Walz, 1986) looked at the influence of cell culture conditions on the amount of K^+ transported through the barium-sensitive K^+ channel in mouse astrocytes. We found that age or time in cultures had negligible impact on K^+ channels. Treatment with 0.25 mM DBcAMP from day 14 on suppressed about 25% of the channel-mediated flux. The highest impact, however, was caused by replacing the growth medium with physiological saline. This doubled the fluxes through the channels (see Walz, 1986). This probably was due to some or one component(s) in the growth medium, which inhibited fluxes through the barium-sensitive channel.

Which species-derived culture reflects the in vivo situation? Astrocytic cultures are used as a model for the *in vivo* situation and such a dramatic difference such as that between the channel-mediated K^+ fluxes (K^+ permeability) in cultures derived from different species has to be dealt with. There is an observation that suggests that the properties of mouse cultures come closest to the *in vivo* situation. The unidirectional K^+ fluxes of primary cultures of mouse cortical neurons are 85 nmol x mg^{-1} x min^{-1}, the same as rat astrocytes, but 25-30 times lower than mouse astrocytes. The work of Trachtenberg and Pollen (1970) as well as that of Hounsgaard and Nicholson (1983) suggests that the resting K^+ permeability of glial cells *in situ* resembles that of neurons during the peak K^+ conductance phase of the action potential. It is much higher that the neuronal resting permeability and this is one of the reasons which enable glial cells to act as spatial buffers. The observation that rat astrocytes in culture have the same magnitude of K^+ resting permeability than cultured neurons, could mean that they lack K^+ channels which are normally present.

COMPARISON OF MEMBRANE PROPERTIES OF CULTURED MOUSE ASTROCYTES WITH THOSE MEASURED IN GLIAL CELLS *IN SITU*

As just explained in the previous paragraph, there are a few observations that cultured mouse astrocytes might come close to mammalian glial cells *in situ*, with respect to their membrane properties. Table 2 summarizes some basic properties and compares them with measured physiological parameters of glial cell studies *in situ*. One problem is that the only mammalian glial cell studies *in situ* so far measured only membrane potential, input resistance, time constant and ion dependence of the resisting potential. Information about changes in intracellular ion concentration are only available from invertebrate glial cells. As can be seen from Table 2 all the

Table 2. Comparison of membrane properties of cultured mouse astrocytes with those measured in glial cells *in situ*.

Parameter	In situ	Mouse astrocytes
Membrane potential (mV)	-92 1)	-92
Membrane input resistance (mΩ)	10 1) 2 2)	2
Membrane time constant (μs)	385 1) 200 2)	<400
Relative ion permeability	exclusive K^+ 3)	exclusive K^+
K^+ accumulation in response to elevated K^+	yes 4)	yes
Concomittant decrease in Na^+	no 4)	no
Active Cl^- accumulation	yes 4)	yes

References: 1) glial cells in the cat cortex (Trachtenberg and Pollen, 1970); 2) glial cells from CA1 region of guinea pig hippocampal slices (Schwartzkroin and Prince, 1979); 3) for review see Walz and Hertz, 1983; 4) leech CNS, Walz and Schlue, 1982; Schlue and Wuttke, 1983; W. Wuttke, personal communication; retina of the honey-bee: Coles and Tsacopoulos, 1979; Coles and Orkand, 1985.

physiological membrane parameters are the same as for mammalian glial cells *in situ*. Some basic active transport properties of cultured mouse astrocytes are the same in glial cells of the honey-bee retina and leech CNS. Therefore, the use of mouse astrocytes as a model for mammalian glial cells *in situ* is strengthened by this comparision and there are no obvious problems to anticipate.

Barium and the membrane input resistance. As measured by Walz (1986) the membrane input resistance of 3.4 ± 1.1 MΩ is increased during 50 μM $BaCl_2$ application to 7.4 ± 2.4 M Ω (118%). Application of 5mM barium increased the input resistance by 167% (Fig. 2). However, this high barium concentration interferes with the carrier-mediated transport and decreases intracellular $[K^+]$ (Walz *et al*., 1984). 50 μM barium did not change the resting potential of astrocytes, but led to spontaneous depolarizations (Fig. 2) due to a voltage-dependent Ca^{2+} conductance (MacVicar, 1984). 5mM barium depolarized astrocytes directly (Fig. 2), a similar barium concentration depolarized glial cells *in situ* (Walz and MacVicar, 1985). The increase of the input resistance by 118% when 50 μM $BaCl_2$ was introduced was much less than expected by a block of unidirectional influx through a barium-sensitive channel, which accounts for 95% of all transmembrane influxes. The explanation is that in cultured astrocytes all cells are connected by gap junctions and they represent therefore multidimensional systems. In these systems it is nearly impossible to determine membrane resistance changes from changes in input resistance, because the input resistance is highly insensitive to changes in membrane resistance (Jack *et al*., 1975). Barium (50 μM) is the drug that should affect glial cell K^+ permeability the most. However, it has only small effects on membrane input resistance. One can therefore expect that drugs and other substances that modulate glial cell K^+ permeability have only a very small

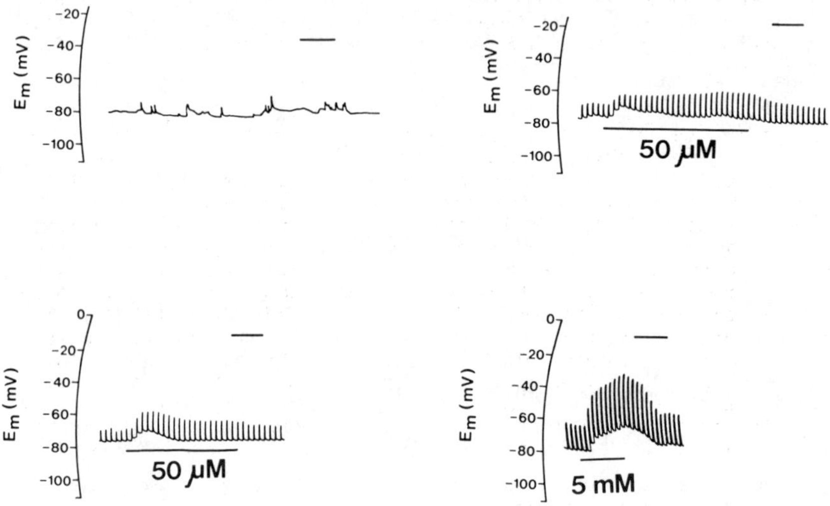

Fig. 2. Intracellular recordings from mouse astrocytes with dual-channel electrodes. UL: Membrane potential during exposure of the cell to 50 μM $BaCl_2$. The resting membrane potential was not changed, the membrane, however, exhibited spontaneous depolarizations. LL and UR: The cell was exposed for a short time (bar) with 50 μM $BaCl_2$ and the membrane input resistance was measured by the aid of constant current pulses through the second channel of the electrode. LR: The cell was for 3 min perfused with 5 mM $BaCl_2$ and the membrane input resistance measured. In 5mM barium the cell depolarized in two phases. The bar on the right corner of each registration represents 2 min.

effect or no effect at all on membrane input resistance. It might be also the case that the membrane input resistance of glial cells with different density in culture (i.e. degree of electronic coupling with other glial cells) or of glial cells in explant cultures with mixed cell types has a different sensitiity to K^+ channel modulating agents.

POTASSIUM CLEARANCE DURING PHYSIOLOGICALLY ELEVATED POTASSIUM

An increase of external $[K^+]$ from 3 to 12 mM K^+ was selected as a situation that mimics the physiological situation during neuronal activiy (Walz and Hinks, 1983, Walz and Hertz, 1985). 12mM is the maximum ceiling level during physiological stimulation in the CNS (Heinemann and Lux, 1977). The intracellular K^+ content of cultured mouse astrocytes increased by 63% within 50s of such a change to reach a new steady-state value. Barium application did not change this accumulation, ruling out the involvement of K^+ channels. Ouabain and furosemide together inhibit this net accumulation. There is a small barium-sensitive K^+ accumulation present, but only if the carriers are blocked by ouabain and furosemide. This is probably due to an increase by Donnan forces, which only plays a role when the carriers are inactive. The furosemide and ouabain-sensitive components of the net accumulation are equally high (250 nmol x mg^{-1} each), however, the total accumulation (400 nmol x mg^{-1}) is less than their sum (500 nmol x mg^{-1}). Another problem is that there is no or only a slight concomitant decrease in Na^+ content, despite the partial ouabain-sensitivity of the K^+ accumulation (Walz and Hertz, 1983; Coles and Orkand, 1983;1985). A concomittant furosemide sensitive Cl^- accumulation was found, however (Walz and Hertz, 1983a). The furosemide-sensitive K^+ accumulation was sensitive to Cl^- and Na^+ removal. It is abolished by a reversal in the Na^+ gradient, in the presence of a physiological external Na^+ concentration. It was found (Walz and Hertz, 1984a) that in the presence of furosemide there was a decrease in Na^+ influx together with a tendency towards a decrease in Na^+ content, whereas Na^+ influx was not altered.

A transmembrane Na^+ cycle. The above observations, including the partial overlap of the ouabain and furosemide effects, can be explained by the existence of a transmembrane Na^+ cycle (Walz and Hinks, 1986). This postulate is supported by experiments with mouse astrocytes which show that an increase in internal Na^+ is stimulating the turnover rate of the Na^+, K^+-ATPase. The furosemide-sensitive KCl carrier is using the Na^+ driving force as an energy source, thereby transporting cations into the cell. An increase in external $[K^+]$ would stimulate both the Na^+, K^+-ATPase and the Na^+, K^+,Cl^- carrier. This would increase internal $[Na^+]$, which in turn would stimulate the Na^+, K^+-ATPase, keeping the internal $[Na^+]$ low. This low internal Na^+ concentration would prevent a decline in the inward Na^+ driving force and keep the carrier's energy source intact. It would explain a partial ouabain-sensitivity of the KCL carrier by a decrease of the Na^+ gradient and a partial furosemide-sensitivity of the Na^+,K^+-ATPase by a decline in internal $[Na^+]$ and a subsequent inhibiting effect on the Na^+,K^+-ATPase as a result of the internal $[Na^+]$ decline. Thus, the Na^+ ion would couple and synchronize both pumps.

Astrocytic swelling. A calculation shows that an activation of this cycle has to increase internal osmolarity. During the 50s accumulation of K^+ following an increase to 12 mM external K^+, the cells gain a total of 400 nmol x mg^{-1} K^+ (200 each from the Na^+,K^+-ATPase and Na^+,K^+,Cl^- carrier). The Na^+,K^+,Cl^- carrier is electroneutral, transporting Na^+,K^+ and 2 Cl^- (Geck et al., 1980; Kimelberg and Hirata, 1981; Walz et al., 1984). That means an additional 400 nmol xmg^{-1} Cl^- ions are accumulated. The Na^+ content does not change, as measured. This is a net accumulation of 800 nmol ions xmg^{-1}. We found (Walz and Hinks, 1985) that the astrocytes have a high water permeability and the water content increased without delay to ion accumulation by 2.82 µl x mg^{-1}.

800 nmol ions x mg^{-1} divided by the water increase of 2.82 μl x mg^{-1} gives 284 mOsm of the additional water space. The osmolarity of the growth medium and physiological saline is 300 mOsm, i.e. identical with the calculated osmolarity of the additional water space.

This concept involves of course, that glial cell *in vivo* swell also during normal physiological activity. This is a new postulation. Dietzel *et al.* (1982) found a reduction of the ECS in cat cortex during physiological stimulation and Orkand *et al.* (1984) found the same for the retina of the drone during light stimulation. These observations might be explained by glial cell swelling. Moreover, Ransom *et al* (1985) found an activitydependent shrinkage of the ECS in rat optic nerve, which only occured in the presence of glial cells and which was furosemide-sensitive. These results are in perfect agreement with the K^+ clearance properties of cultured mouse astrocytes.

POTASSIUM CLEARANCE DURING PATHOLOGICALLY ELEVATED POTASSIUM

Under normal conditions K^+ change in the ECS is limited to a ceiling level of 12 mM. However, pathological conditions such a hypoxia, hypoglycemia, trauma, and ischemia lead to a rapid increase in the extracellular $[K^+]$ to values which may reach 30 to 100 mM (for a review see Walz and Hertz, 1983c). These increases are rapidly reversible, provided the pathological state does not last too long and lead to irreversible damage. We selected an increase to 54 mM K^+ as a representative value of an increase of external $[K^+]$ to pathological values and analyzed the K^+ net accumulation into astrocytes in detail. Ouabain application had no effect on the net accumulation as compared with ouabain-treated cells at normal K^+. Furosemide had a small inhibitory effect, much less than at 12 mM K^+. Greater inhibition is exerted by 50 μM barium, even in the absence of both other inhibitors. The presence of all three inhibitors (barium, ouabain, furosemide) blocks net accumulation completely. Thus, there is a difference to the physiological situation, not only in the amount of accumulated K^+: The ouabain-sensitive component seems reduced, but this is probably due to a decrease in K^+ loss at high K^+. Inhibition by furosemide is clearly much less (150 at 54 mMK^+ compared with 250 nmol xmg^{-1} at 12 mMK^+). The barium-sensitive component, i.e. the channel-mediated component is clearly the most important one. This is probably due to a K^+ accumulation (probably with Cl^-) by Donnan forces, which uses the K^+ channel. This high K^+ concentration induced swelling of astrocytes by about 4 μl water xmg^{-1} protein (or 38% of the normal volume) compared with 2.82 μl xmg^{-1} at 12 mMK^+. The water accumulation at pathological levels was as fast as the K^+ accumulation, showing that there was no change in water permeability.

<u>Intense furosemide-sensitive K^+ accumulation at high extracellular K^+ levels.</u> Walz and Hertz (1984b) postulated a dramatic increase of furosemide-sensitive K^+ accumulation in astrocytes at high K^+ levels. This postulation was based on measurements of the initial unidirectional K^+ influx following a K^+ increase. The measurements of K^+ net accumulation reported here suggest the opposite. A furosemide-sensitive net accumulation of 150 nmol xmg^{-1} following a change to 54 mMK^+ was also reported by Walz and Hertz (1984b). If the furosemide-sensitive KCl carrier is using the Na^+ gradient as an energy source (as suggested in the previous chapter), then the KCl net accumulation would be, indeed, decreased during high external K^+: The Na^+ driving force would be diminished by about half of its value. The Na^+ equilibrium potential is at about +54 mV (190 nmol xmg^{-1} at 10.5 μl xmg^{-1} cell water results in 18 mM internal Na+ concentration, external Na^+ is 144 mM) and the membrane potential depolarizes at 54 mMK^+ from -92 mV to -30 mV. This decrease in net accumulation of the furosemide-sensitive component was found by Walz and Hertz (1984b) and in the experiments reported here. The increase in the furosemide-sensitive unidirectional K^+ influx might be the result of a furosemide-sensitive homoexchange (see Tupper, 1975). A dramatic and biphasic increase in total unidirectional K^+ influx at high K^+ as reported by Hertz (1978) could

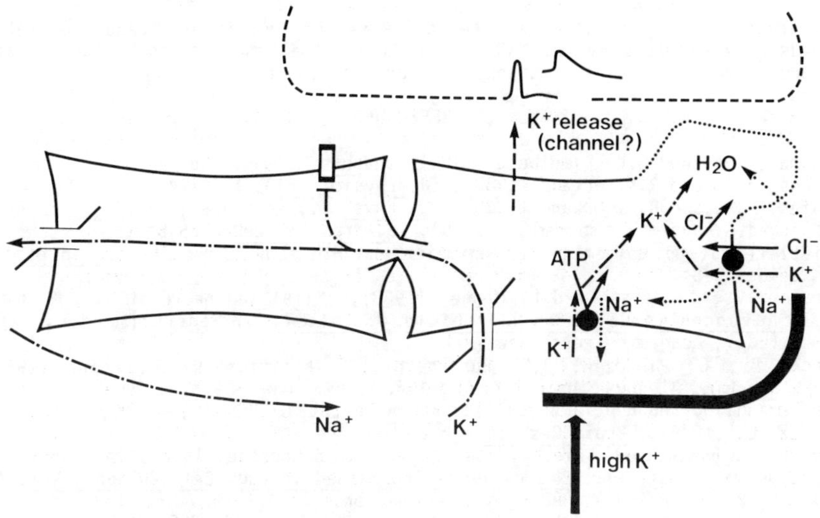

Fig. 3. Schematic model of K^+ homeostatic mechanisms of glial cells as concluded from data presented here. For details see text.

not be reproduced in this laboratory and is also not present in the results reported by Walz and Hertz (1984b). The Na^+, K^+-ATPase activity is not increased by K^+ increases beyond 12 mMK^+ (Franck <u>et al</u>., 1983).

CONCLUSIONS

A hypothetical model is descirbed in this paragraph and is schematically drawn in Fig. 3 to explain K^+ clearance mechanisms. It involves competing K^+ removal systems, evoked by increased external K^+: spatial buffer K^+ currents through channels and K^+ removal by the Na^+ cycle (Na^+, K^+-ATPase amd KCl carrier). K^+ current pathways cross cell limits through gap junctions and can be guided by inhibitors of K^+ channels. The external return current is mainly carried by Na^+ (and Cl^-), but is negligible in terms of relative concentration changes. The Na^+ cycle results in swelling of glial cells. The net accumulation of K^+ by the Na^+ cycle must be subsequently reversible, because it is not to expect that the cells store K^+ indefinitely. Indeed, we found a K^+ release system, evoked by decreases of external K^+ and probably using a K^+ channel, although the existence of the necessary driving force is yet unclear. Glial cells could modulate neuronal function via increased extracellular K^+ (for K^+-neuronal excitation interactions see Walz and Hertz, 1983c): 1) The swelling of astrocytes reduces the ECS and thus increases K^+ concentration even in areas where there is originally no increased external K^+. 2) The K^+ release by astrocytes might increase the external K^+ locally. 3) K^+ spatial buffer currents could be guided by a combination of open and closed channels to areas where they increase external K^+. At high external K^+ the spatial buffer currents and K^+ movements through channels by Donnan forces might be increased. The Na^+ cycle is disturbed and the KCl carrier activity reduced to about one half.

ACKNOWLEDGEMENT

The author is a scholar of the Medical Research Council of Canada. Operating funds were provided by the MRC of Canada and the Saskatchewan Health Research Board. Bruce Van Vliet is thanked for critically reviewing the text.

REFERENCES

Bowman, C.L. and H.K. Kimelberg (1984). Nature 311, 656-659.
Coles, J.A. and R.K. Orkand (1983). J. Physiol. 340, 157-174.
Coles, J.A. and R.K. Orkand (1985). J. Physiol., in press.
Coles, J.A. and M. Tsacopoulos (1979). J. Physiol. 290, 525-549.
Dietzel, I., U. Heinemann, G. Hofmeier and H.D. Lux (1982). Exp. Brain Res. 46, 73-84.
Franck, G., T. Grisar and G. Moonen (1983). Glial and neuronal Na^+, K^+ pump. Advances in Cellular Neurobiology, Vol. 4, S. Fedoroff and L. Hertz (Eds), Academic Press, New York.
Geck, P., C. Pietrzyk, B.C. Burckhardt, B. Pfeiffer and E. Heinz (1980). Biochim. Biophys. Acta 600, 432-447.
Heinemann, U. and H.D. Lux (1977). Brain Res. 120, 231-249.
Hertz, L. (1978). Brain Res. 145, 202-208.
Hertz, L. B.H.J. Juurlink, H. Fosmark and A. Schousboe (1982). Astrocytes in primary cultures. Neuroscience Approached Through Cell Culture, Vol. I., S. Pfeiffer (ed.), CRC Press, Boca Raton.
Hounsgaard, J. and C. Nicholson (1983). J. Physiol. 340, 359-388
Jack, J.J.B., D. Noble and R.W. Tsien (1975). Electric Current Flow in Excitable Cells, Clarendon Press, Oxford.
Kimelberg, H.K. (1981). Biochim. Biophys. Acta 646, 179-184.
Kimelberg, H.K. (1983). Cell. Molec. Neurobiol. 3, 1-16.
Kimelberg, H.K., C. Bowman, S. Biddlecome and R.S. Bourke (1979). Brain Res. 177, 533-550.
Kimelberg, H.K. and H. Hirata (1981). Soc. Neurosci. Abstr. 7, 698.
MacVicar, B.A. (1984). Science 226, 1345-1347.
Moonen, G., G. Franck and E. Schoffeniels (1980). Neurochem. Int. 2, 299-310.
Orkand, R.K., I. Dietzel, and J.A. Coles (1984). Neurosci. Lett. 45, 273-278.
Orkand, R.K., J.G. Nicholls and S.W. Kuffler (1966). J. Neurophysiol. 29, 788-806.
Ransom, B.R., C.L. Yamate and B.W. Connors (1985). J. Neurosci. 5, 532-535.
Schwartzkroin, P.A. and D.A. Prince (1979). Brain Res. 161, 533-538.
Sykova, E. (1983). Prog. Biophys. Molec. Biol. 42, 135-189.
Trachtenberg, M.A. and D.A. Pollen (1970). Science 167, 1248-1253.
Tupper, J.T. (1975). Biochim. Biophys. Acta 394, 586-596.
Walz, W. (1986). Ann. N.Y. Acad. Sci., in press.
Walz. W., E. Hertz and L. Hertz (1983). Prog. Neuro-Psychopharmacol. and Biol. Psychiat. 7, 697-702.
Walz, W. and L. Hertz (1983a). Brain Res. 277, 321-328.
Walz, W. and L. Hertz (1983b). J. Neurosci. Res. 10, 411-423.
Walz, W. and L. Hertz (1983c). Prog. Neurobiol. 20, 133-183.
Walz, W. and L. Hertz (1984a). J. Neurosci. Res. 11, 231-239.
Walz, W. and L. Hertz (1984b). J. Cereb. Blood Flow Metab. 4, 301-304.
Walz, W. and E.C. Hinks (1985). Brain Res., in press.
Walz, W. and E.C. Hinks (1986). Brain Res., in press.
Walz, W. and H.K. Kimelberg (1985). Brain Res., in press.
Walz, W. and B.A. MacVicar (1985). Soc. Neurosci. Abstr., in press.
Walz, W. and W.R. Schlue (1982). Brain Res. 239, 119-138.
Walz, W., M. Shargool and L. Hertz (1984). Neuroscience 13, 945-949.
Walz, W., W. Wuttke and L. Hertz (1984). Brain Res. 292, 367-374.

Oligodenrocytes Control Extracellular Potassium by Active Uptake and Spatial Buffering

H. Kettenmann

Department of Neurobiology, University of Heidelberg, Im Neuenheimer Feld 504, 6900-Heidelberf, FRG

ABSTRACT

The membrane potential of oligodendrocytes identified by cell-type specific antibodies is dominated by the K^+-conductance. Exposed to elevated extracellular K^+-concentration ($[K^+]_o$) the membrane potential depolarizes and the intracellular K^+-concentration ($[K^+]_i$) increases. This uptake of K^+ into the oligodendrocytes depends on homogeneous $[K^+]_o$ elevation and is therefore not mediated by spatial buffer currents.

Spatial buffer currents, however, are generated when the membrane potential differs from the K^+-equilibrium potential by an unequal distribution of $[K^+]_o$ along the cell membrane. In a model system based on passive electrical membrane properties determined from a single oligodendrocyte, local elevation of $[K^+]_o$ resulted in considerable differences between membrane potential and K^+-equilibrium potential. By current injection into an oligodendrocyte the membrane potential was displaced and thereby a difference was generated between membrane potential and K^+-equilibrium potential. The resulting changes in $[K^+]_i$ were in the range of several mM and show that oligodendrocytes have a great capacity to buffer K^+ by passive current movement.

It is therefore concluded that oligodendrocytes have two powerful mechanisms to compensate the variations in extracellular $[K^+]$, active uptake and redistribution by spatial buffer currents.

KEYWORDS

oligodendrocyte - cell culture - membrane properties - K^+ regulation - spatial buffering

INTRODUCTION

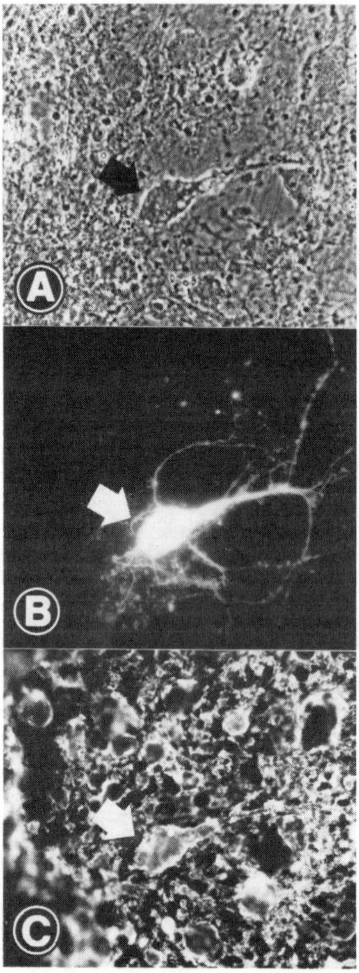

Regulation of K^+ homeostasis in the brain is important for the function of the nervous system. Glial cells participate in the clearance of K^+ in regions where K^+ accumulates due to neuronal activity (Houndsgaard and Nicholson, 1982). The mechanisms involved are thought to be either active uptake or/and redistribution by spatial buffer currents (Orkand, 1977). An accumulation of K^+ triggered by an elevation of $[K^+]_o$ has been shown for invertebrate glial cells (Schlue and Wuttke, 1982; Coles and Tsacopoulos, 1979). An active uptake of K^+ was demonstrated for cultured astrocytes (Walz et al., 1984). Since astrocytes form a large syncytium (Kettenmann et al., 1983a), they fulfil the requirements for the generation of spatial buffer currents induced by an uneven distribution of $[K^+]_o$. The occurrence of these spatial buffer currents has been directly shown (Gardner-Medwin, 1983a,b). The participation of oligodendrocytes in K^+-clearance, however, has not yet been investigated. To approach this problem we have chosen a culture system to carry out electrophysiological investigations on identified oligodendrocytes. We first studied the basic membrane properties of these cells in order to understand the membrane responses induced by $[K^+]_o$ alterations. Based on the results of these studies we proceeded to approach the problem of whether oligodendrocytes are capable of participating in the clearance of extracellular K^+.

Fig. 1 Identification of oligodendrocytes. A cell shown in phase contrast (A) was injected with Lucifer Yellow (B) after recording. Subsequently the culture was stained with antibody against the cell surface antigen O4 to identify oligodendroglial membrane structures (C).

MATERIALS AND METHODS

Oligodendrocytes were obtained from embryonic mouse spinal cord and maintained in culture for 3 to 6 weeks (Kettenmann et al., 1983a). During recording with standard electrophysiological equipment (Kettenmann et al., 1983a) the culture was perfused with a salt solution containing in mM: NaCl 116; KCl 5.4; $CaCl_2$ 1.8; $NaHCO_3$ 26; NaH_2PO_4 1; $MgSO_4$ 0.8; glucose 5.5. Electrodes contained 3 M KCl for membrane potential recording, 4% Lucifer Yellow for intracellular staining, and the bathing salt solution for patch-clamp studies (Kettenmann et al., 1982, 1984a; Kettenmann et al., 1983a,b). K^+-sensitive microelectrodes were fabricated as described (Kettenmann et al., 1983a) using Corning 477317 as exchange resin. Immunocytochemical techniques were used as described by Sommer and Schachner (1981).

RESULTS

Identification of oligodendrocytes
Cells were identified by the monoclonal antibody against O4, which reacts specifically with the surface of mature and more immature oligodendrocytes (Schachner et al., 1981; Sommer and Schachner, 1981). Physiologically characterized cells were injected with Lucifer Yellow to mark their location in the culture. Subsequently the entire culture was immunolabelled with O4 to identify oligodendroglial membrane structures (Fig. 1). This procedure allowed an unequivocal correlation of the physiological data with a subpopulation of glial cells, the oligodendrocytes (Kettenmann et al., 1983a).

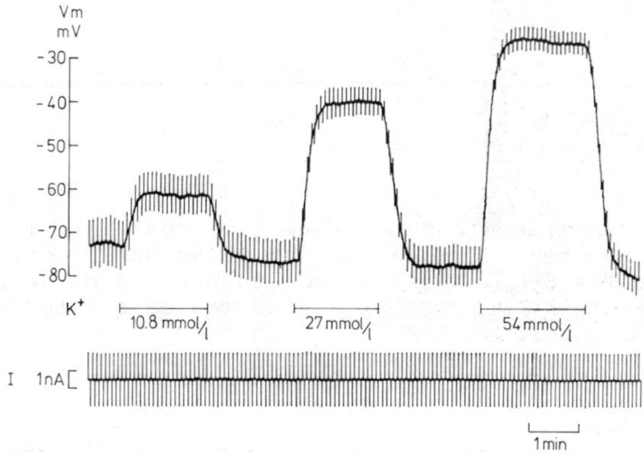

Fig. 2 Response of an oligodendrocyte to elevation of $[K^+]_o$. During measurement of membrane potential (V_m) constant current pulses were injected with a separate electrode to monitor input resistance. $[K^+]_o$ was elevated in the bath (10.8, 27, 54 mM, respectively). The membrane potential is depolarized while input resistance decreases.

Ionic dependence of the membrane potential

The membrane potential (average -67 mV) depended strongly on $[K^+]_o$. A tenfold increase of $[K^+]_o$ resulted in a 52 mV depolarization, less than predicted by the Nernst equation assuming constant $[K^+]_o$ (Fig. 2). Short termed alterations of $[Na^+]$, $[Cl^-]$ and $[Ca^{++}]$ did not change the membrane potential (Kettenmann et al., 1983a).

Passive electrical membrane properties

The time constant and input resistance of the oligodendrocytic membrane determined by injection of current pulses and analyzing the resulting voltage changes (Kettenmann et al., 1984b) (Fig. 3) was on average 3.3 MΩ and 3.5 msec, respectively, and decreased when $[K^+]_o$ was elevated (Fig. 2). By combining these results with the determination of total membrane area of the characterized cell the specific membrane resistance could be calculated ($1.3 \times 10^3 \Omega cm^2$) (Kettenmann et al., 1984b).

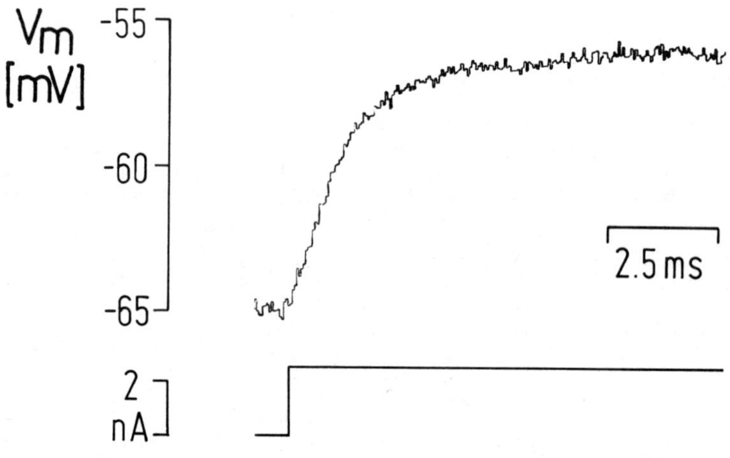

Fig. 3 Input resistance and membrane time constant. The response of the membrane potential to a current pulse was recorded. The input resistance was 3 MΩ. The time constant could be determined from the derivation of the voltage slope and was calculated to be 3 ms.

Membrane channels

The patch-clamp technique was used to study the channel system(s) responsible for the electrical behaviour of the oligodendrocytic membrane. Membrane patches were isolated and the current through the patch was studied at different membrane potentials (Fig. 4). Single channels could be distinguished having a conductance of 6-200 pS (Kettenmann et al., 1982, 1984a). The opening and closure behaviour was voltage dependent. The average open time increased with depolarization. The membrane patch could be isolated from the cell in the inside-out configuration (Hamill et al., 1981) and the still functional channel could be studied. Under these conditions the ionic environment was altered to test the selectivity of the channel for K^+-ions. The channel current was affected by changes of K^+ and the current voltage curve shifted as expected from the Nernst equation for an exclusive K^+ conductance. These highly selective K^+ channels were the only type which were found in the membrane of cultured oligodendrocytes (Kettenmann et al., 1984a).

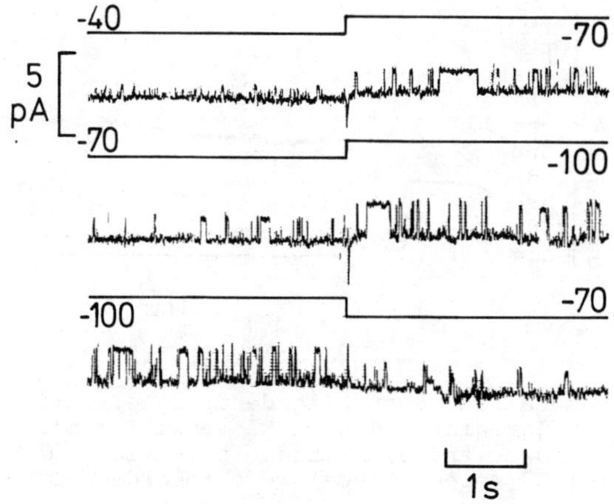

Fig. 4 Single channel recording. The current through a membrane patch was recorded at different membrane potentials as indicated. Generation of outward current by channel opening is displayed in the upward direction. The membrane potential was changed as indicated by bars. The average open time of the channel can be seen to increase with depolarization.

Uptake of K^+

The capacity of oligodendrocytes to take up K^+ during extracellular elevation was tested. Cells were therefore penetrated with K^+-sensitive microelectrodes and $[K^+]_o$ was elevated. The rise in intracellular $[K^+]$ was dependent on $[K^+]_o$. The uptake of K^+ into oligodendrocytes amounted to several mM (Fig. 5). Since there was no gradient of $[K^+]_o$ along the membrane a contribution of spatial buffering could be excluded. Elevation of $[K^+]_o$ therefore triggered an uptake mechanism in oligodendrocytes. The increase in $[K^+]_i$ during $[K^+]_o$ elevation also explains the previous observation that the membrane potential of oligodendrocytes depends not only on $[K^+]_o$, but also on $[K^+]_i$ (Kettenmann et al., 1983a).

The driving potential of the spatial buffer current

During spatial buffering a K^+-current is flowing across the glial membrane driven by the difference between membrane potential and K^+-equilibrium potential. This difference is generated by an uneven distribution of $[K^+]_o$ along the cell membrane. The response of an idealized cell to local elevated $[K^+]_o$ in a model system is shown (Fig. 6). The $[K^+]_o$ increase was assumed to occur in the somatic region of the schematic cell. This results in a depolarization of the cell which is highest in the somatic region.

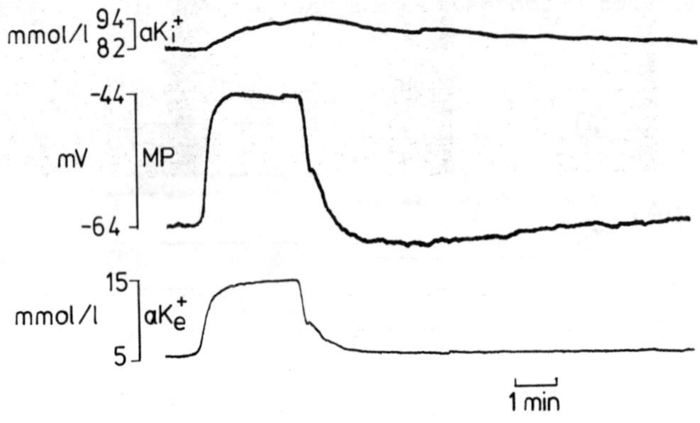

Fig. 5 $[K^+]_i$-accumulation by an oligodendrocyte. During elevation of extracellular $[K^+]$ registered by a K^+-sensitive microelectrode close to the cell (lower trace) membrane potential (middle trace) and intracellular $[K^+]$ (upper trace) were recorded. The cell accumulates K^+.

Intracellular $[K^+]$ changes induced by transmembrane currents
By injection of current carried by ions with a low membrane permeability such as Li^+, TMA^+, Na^+ or Cl^- the membrane potential was displaced a few mV from its resting value. Positive current reduces $[K^+]_i$ within a few minutes by several mM, while injection of negative current results in an increase of $[K^+]_i$. As shown in Fig. 7 a depolarization of 3 mV from the resting value leads to a decrease of $[K^+]_i$ by more than 5 mM within 1 minute.

DISCUSSION

Membrane properties of oligodendrocytes in relation to astrocytes
The present data are the first recordings from identified oligodendrocytes and can only be compared to those from astrocytes or unidentified non-excitable cells. The deviation of the membrane potential from the expected Nernst slope is similar to that seen for cultured astrocytes and unidentified mammalian glial cells in vivo (Moonen and Nelson, 1978; Ransom and Goldring, 1973). As shown for oligodendrocytes, the deviation is due to uptake of K^+ and not to the influence of other conductances. It is therefore likely that the membranes of all CNS macroglia, namely astrocytes and oligodendrocytes are highly permeable to K^+.

Mechanism of K^+ uptake
We have shown that oligodendrocytes have the capability to accumulate K^+ when $[K^+]_o$ was homogeneously raised. This excludes a contribution of spatial buffer currents. The uptake could either be due to an increase of pump activity or/and to the reduction of the Na^+

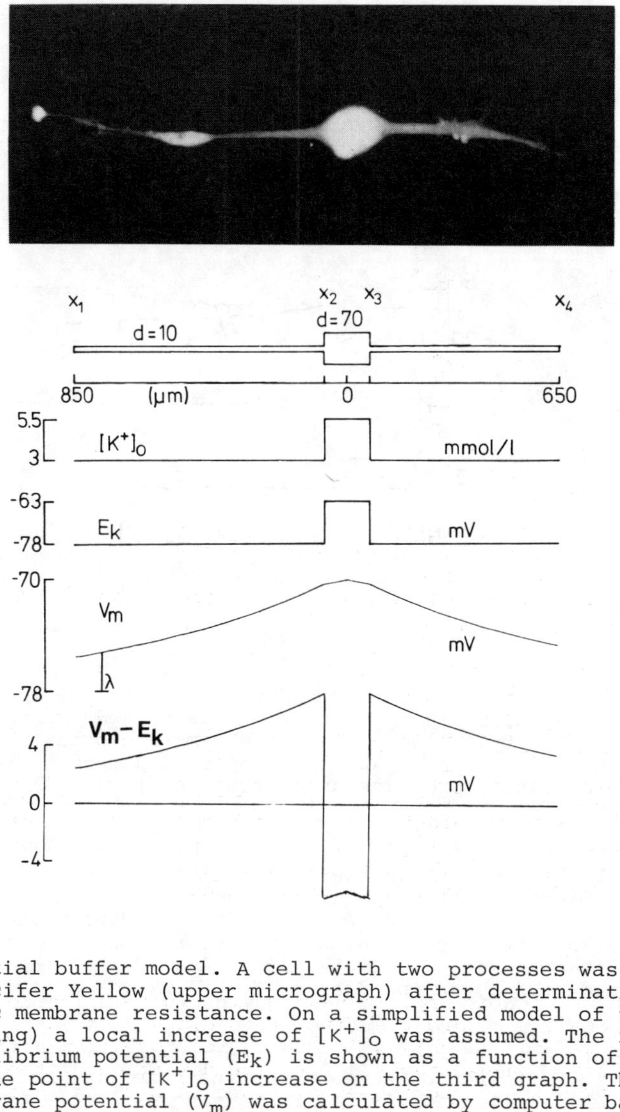

Fig. 6 Spatial buffer model. A cell with two processes was injected with Lucifer Yellow (upper micrograph) after determination of the specific membrane resistance. On a simplified model of the cell (upper drawing) a local increase of $[K^+]_o$ was assumed. The resulting K^+-equilibrium potential (E_k) is shown as a function of distance from the point of $[K^+]_o$ increase on the third graph. The slope of the membrane potential (V_m) was calculated by computer based on the specific membrane resistance in relation to distance. The length constant (λ) is indicated by bar. The difference between K^+-equilibrium potential and membrane potential (V_m-E_k) is given on the bottom graph. The slope of the depolarization along the cell is determined by the specific membrane resistance. It results in a difference between membrane potential and K^+-equilibrium potential. This potential difference induces a current driving K^+ into the cell at regions where $[K^+]_o$ is high. In the periphery with resting $[K^+]_o$, K^+ leaves the cell.

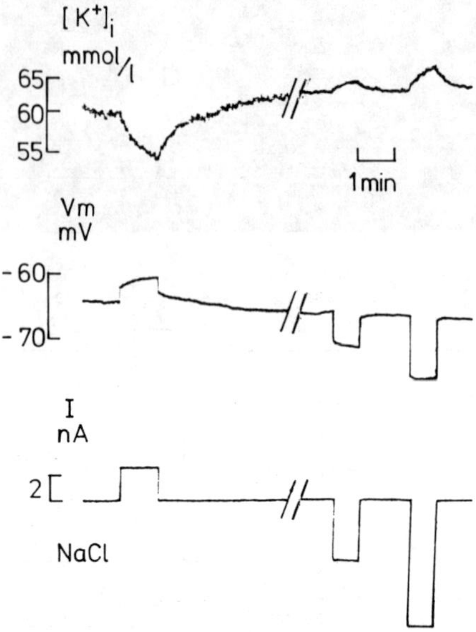

Fig. 7 During recording of membrane potential (V_m) and $[K^+]_i$ with a K^+-sensitive microelectrode current was passed through a separate, NaCl-filled electrode to displace the membrane potential from its resting value. A depolarization of the cell provokes a decrease and a hyperpolarization provokes an increase of $[K^+]_i$.

leakage. It has been shown that the activity of the Na^+/K^+-ATPase of bulk isolated glial cells is increased by elevated $[K^+]$ (Grisar et al., 1979). Since the K^+ gradient in oligodendrocytes is maintained by the activity of the Na^+/K^+-ATPase (Kettenmann et al., 1983a; Kettenmann et al., 1985), it is likely that the K^+ accumulation is mediated by an increase of Na^+/K^+-ATPase activity.

Spatial buffering of K^+

The driving force for spatial buffer currents across a glial membrane is the difference between membrane potential and K^+-equilibrium potential (Orkand, 1977). For a single oligodendrocyte we could give an estimation of the expected potential difference based on measurements of the specific membrane resistance. When a similar potential difference was experimentally installed by injection of current, substantial changes in $[K^+]_i$ could be observed. We have, however, no experimental evidence for the potential distribution in a glial syncytium. We have shown that oligodendrocytes form gap junctions, but to a much smaller extent than astrocytes and that it is also possible to find electrically isolated oligodendrocytes in culture (Kettenmann et al., 1983b). We conclude that at least uncoupled oligodendrocytes have the capacity to efficiently regulate $[K^+]_o$ by spatial buffer currents.

Contribution of oligodendrocytes to K^+ clearance in the brain
In the culture system cells face an infinite extracellular space. In vivo, the extracellular space is smaller than the glial volume. Since oligodendrocytes are capable of moving K^+ across the membrane to such an extent that marked changes in the intracellular concentration can be observed, the resulting changes in the extracellular space must be even larger. In the intact nervous system oligodendrocytes are therefore potent regulators of the extracellular K^+-milieu both by spatial buffering and active uptake processes.

Acknowledgement
The author thanks Drs. H.D. Lux, R.K. Orkand, U. Sonnhof and M. Schachner for helpful discussion and Hermann and Lilly Schilling Stiftung for support.

REFERENCES

Coles, J.A. and M. Tsacopoulos (1979). J. Physiol., 290, 525-549.
Gardner-Medwin, A.R. (1983a). J. Physiol., 335, 353-374.
Gardner-Medwin, A.R. (1983b). J. Physiol., 335, 393-426.
Grisar, T., J.-M. Frere and G. Franck (1979). Brain Res., 165, 87-103.
Hamill, D.P., A. Marty, E. Neher, B. Sakmann and F.J. Sigworth (1981). Pflügers Arch., 391, 85-100.
Houndsgaard, J. and C. Nicholson (1982). J. Physiol., 340, 359-388.
Kettenmann, H., R.K. Orkand, H.D. Lux and M. Schachner (1982). Neurosci. Lett., 32, 41-46.
Kettenmann, H., U. Sonnhof and M. Schachner (1983a). J. Neurosci., 3, 500-505.
Kettenmann, H., R.K. Orkand and M. Schachner (1983b). J. Neurosci., 3, 506-516.
Kettenmann, H., R.K. Orkand and H.D. Lux (1984a). Pflügers Arch., 400, 215-221.
Kettenmann, H., U. Sonnhof, H. Camerer, S. Kuhlmann, R.K. Orkand and M. Schachner (1984b). Pflügers Arch., 401, 3324-3332.
Kettenmann, H., E. Sykova, R.K. Orkand and M. Schachner (1985). Submitted for publication.
Moonen, G. and P.G. Nelson (1978). Dynamic Properties of Glial Cells, Pergamon Press, London.
Orkand, R.K. (1977). Cellular Biology of Neurons, Williams & Wilkins, Baltimore.
Ransom, B. and S. Goldring (1973). J. Neurophysiol., 36, 855-868.
Schachner, M., S.U. Kim and R. Zehnle (1981). Dev. Biol., 833, 3328-3338.
Schlue, W.R. and W. Wuttke (1983). Brain Res., 270, 368-372.
Sommer, I. and M. Schachner (1981). Dev. Biol., 833, 311-327.
Walz, W., W. Wuttke and L. Hertz (1984). Brain Res., 292, 367-374.

Recent Advances in Brain (Na^+, K^+)-ATPases

D. Guillaume*,†, T. Grisar* and
A. V. Delgado-Escueta†

*Institut de Biochimie Générale et Comparée, Laboratoire de Neurochimie, Université de Liège, Place Delcour, 17, 4020 Liège, Belgium
†Molecular Neurosciences Laboratories, Comprehensive Epilepsy Program, Reed Neurological Research Center, University of California at Los Angeles, Los Angeles, CA 90024, USA

ABSTRACT

Based on kinetic and thermodynamic data, glial and neuronal (Na^+, K^+)-ATPases have been shown to be two different molecular entities characterized by their different affinities to $[K+]_o$. On the other hand, the presence of two catalytic subunits of brain (Na^+, K^+)-ATPase on SDS - polyacrylamide gel electrophoresis has also been demonstrated. We have isolated these subunits from mouse, cat and human brain and examined their affinity for K^+ ions by studying their phosphorylation level. Preliminary results show slightly different affinity constants for K^+ between the 2 subunits. The hypothesis of a cellular specificity is proposed : the high molecular weight form of catalytic subunit, with a lower affinity constant for K^+, and the low molecular weight form, with a higher affinity constant for K^+, are thought to be the neuronal and glial forms of (Na^+, K^+)-ATPase, respectively. We also demonstrated that phenytoin, a potent antiepileptic drug, is acting on the (Na^+, K^+)-ATPase by dephosphorylation of its E_2-P intermediate.

KEY WORDS

(Na^+, K^+)-ATPase, phosphorylation, K^+, phenytoin.

INTRODUCTION

The maintenance of ionic gradients across cell membranes is characteristic of almost all mammalian cells. The physiological mechanism for creating and maintaining cytoplasmic concentrations of Na^+ below and K^+ above those in the extracellular fluid is the $[Na^+, K^+]$ pump. In nervous tissues, the importance of the $[Na^+, K^+]$ pump is obvious. More specifically, Na^+ and K^+ gradients that are necessary for initiation of the action potential are dependent on that enzymatic mechanism (Siegel et al., 1981). The mechanism by which K^+ is cleared from the extracellular space after neuronal discharges is of great interest. Extracellular K^+ may increase up to 3mM after neuronal firing in normal conditions, but during epileptic episodes it may reach 10-15mM. Elevated extracellular K^+ increases neuronal excitability by modification of neurotransmitter release and

partial depolarization of neurons, so control of K^+ in the extracellular space of the central nervous system is of critical importance. Neurons reaccumulate K^+ by active transport mediated by the neuronal (Na^+, K^+)-ATPase. As discussed above by Franck, (Na^+, K^+)-ATPase in glial cells may also be involved in clearing extracellular K^+ after neuronal discharge, although this mechanism remains controversial (Grisar et al., 1978; Franck et al., 1983).
Several authors are favoring the spatial buffering mechanism hypothesis (Kuffler, 1967; Lux, 1974), referring to a passive transport of K^+ ions into the glial cells. However, in the last few years, several pieces of evidence have been added to that controversy, suggesting that not only (Na^+, K^+)-ATPases in neurons and glia are different molecular entities (see below), but also that glial (Na^+, K^+)-ATPase is of importance in the regulation of extracellular K^+ (Franck et al., 1983).
In this paper, we present the most recent advances in brain (Na^+, K^+)-ATPase from several laboratories, including ours, related to the existence of two molecular forms and to interactions with phenytoin.

Two molecular forms of (Na^+, K^+)-ATPase : neuronal and glial forms ?

In 1978, based on kinetic studies, Grisar et al. had already proposed that the glial and neuronal (Na^+, K^+)-ATPases could be two molecular entities different by their K^+ sensitivity. The glial enzyme is specifically sensitive to high K^+ concentrations (20mM), whereas the neuronal enzyme has a maximal activity in presence of 3-5mM K^+.
In 1978 also, Marks and Seeds were the first to propose that the inhibition of brain (Na^+, K^+)-ATPase by ouabain is complex and may involve two binding sites. One year later, Sweadner showed the same results on brain (Na^+, K^+)-ATPases. On the other hand, only one inhibition site is described in non nervous tissues. More recently (1984), Atterwill et al. have confirmed those results. Astrocytes in culture "contain predominantly, but not exclusively, a (Na^+, K^+)-ATPase with low affinity for ouabain (73 % of the total) that is similar to the single enzyme form in the kidney. This form constituted a significantly smaller proportion of the (Na^+,K^+)-ATPase in the cultured neuronal preparations (55 %)". They suggest that both forms are present in neuronal and glial cells, but in different proportions.

But, the first direct evidence for 2 molecular forms of (Na^+, K^+)-ATPase has come from Sweadner's work. After SDS-gel electrophoresis of (Na^+, K^+)-ATPase enriched preparations from rat brain, gels demonstrate the existence of two distinct subspecies of catalytic subunit : alpha (+) and alpha (-), which can be identified by Na^+ dependent, K^+ sensitive phosphorylation.
Both forms exist in brain tissue, while only the lower molecular weight form is present in muscular, cardiac and renal tissue. She proposed that the two forms are located in different kinds of brain cells : one is the only (Na^+, K^+)-ATPase of astrocytes, alpha (-), while the other is the only (Na^+, K^+)-ATPase of axolemma, alpha (+).

Since that first report, several authors have described the existence of two molecular forms of (Na^+, K^+)-ATPase on SDS-gel electrophoresis. Peterson and Hokin (1981) have also shown that the (Na^+, K^+)-ATPase purified from the brine shrimp nauplii reveals two large subunits. Their relative proportions appear to vary during development (Peterson et al., 1982). In 1984, Churchill presented data indicating that cardioactive steroids have higher affinity for the higher molecular weight form alpha (+). The incorporation of radioactive phosphate is greater in the higher molecular weight form alpha (+) (Churchill et al., 1984), suggesting that ATP and ouabain have a higher affinity for that form.

Recently, Matsuda and Cooper (1983) demonstrated that pyrithiamin inhibits the activity of (Na^+, K^+)-ATPase derived from neuronal tissue but had no effect on non neuronal tissue. Matsuda et al. (1984) showed afterwards that the treatment of the enzyme with pyrithiamin decreased the phosphorylation of only the higher molecular weight form, suggesting cautiously that alpha (+) and alpha (-) could be assigned to neuron and glia respectively.
Recent experiments, using a monoclonal antibody, have shown that there are antigenically distinct (Na^+, K^+)-ATPases in different tissues of the chicken (Fambrough and Bayne, 1983).
In the brine shrimp again, the catalytic subunit precursor appears to be synthesized on membrane bound polysomes and is present as two separate peptides on SDS gels, that suggests that the two forms of catalytic subunit seen in mature brine shrimp are products of two distinct messenger RNAs (Fisher et al., 1984). The existence of two different molecular forms has been further supported by the work of Morohashi and Kawamura (1984), analysing the NH_2- terminal amino acid sequence of the catalytic subunit of (Na^+, K^+)-ATPase. At the third position, a small but distinct amount of lysine was found in addition to glycine, suggesting the possible presence of two isoforms of (Na^+, K^+)-ATPase different from each other at least at the third residue. Shull et al. (1985) and Kawakami et al. (1985) have been able to determine the complete amino acid sequence of the catalytic subunit in sheep kidney and Torpedo Californica electroplax, respectively. They showed the considerable homology between the N-terminal sequences of sheep and dog kidney, duck nasal salt gland, crustacean (brine shrimp) and Torpedo Californica electroplax (Na^+, K^+)-ATPase.

In summary, at least two forms of (Na^+, K^+)-ATPase with different affinities for ouabain and ATP, different molecular weights and apparently different amino acid sequences exist in different tissues and especially in the brain. However, the issue of a cellular specificity - glial versus neuronal - in the brain remains an open question.

PERSONAL RESULTS

Because our previous results had shown a differential effect of K^+ on the kinetic of glial and neuronal (Na^+, K^+)-ATPases (Grisar et al., 1978), we wanted to study the effect of K^+ on the phosphorylation of (Na^+, K^+)-ATPase catalytic subunits.

Under various conditions, (Na^+, K^+)-ATPase is catalyzing different reactions that may be viewed as portions of a cycle leading to ATP hydrolysis. The sequence of the enzymatic reaction can then be written as a phosphorylation - dephosphorylation cycle described by Albers et al. (1963) and Post et al. (1965). Many kinetic experiments support the basic assumption of two phosphorylated and two dephosphorylated enzyme conformations : E_1 and E_2 correspond to two different structural conformations of the enzyme.

$$E_1 + ATP \rightleftharpoons E_1\ ATP \underset{Mg^{++}}{\overset{Na^+}{\rightleftharpoons}} ADP + E_1\text{-}P \underset{Mg^{++}}{\rightleftharpoons} E_2\text{-}P \overset{K^+}{\rightleftharpoons} E_2 + P_i \rightleftharpoons E_1$$

According to the model, the reaction involves a phosphorylation step in presence of Na^+ and Mg^{2+} ions producing an E_1-P intermediate, immediately converted into an E_2-P intermediate. The dephosphorylation step induced by K+ completes the cycle.

We used C57/Bl mice and sacrified them at the age of 19 to 22 days. According to published procedures (Delgado-Escueta <u>et al.</u>, 1974), cat cortex was removed from normal adult cats, after control of the electrical activity by electrocorticography. Human cortex samples were obtained from patients operated on for a non epileptogenic deeply lying tumor. As previously described (Guillaume <u>et al.</u>, 1985), all samples were processed through several centrifugation steps up to a microsomal fraction.
That fraction was incubated in presence of SDS and centrifuged on the top of a continuous sucrose gradient. This procedure provided us with an enzyme of specific activity around 120umoles of Pi/mg prot/hr. The enzymatic preparation was then phosphorylated by [^{32} P-ATP] and the samples were analyzed by SDS-gel electrophoresis according to the procedure of Laemmli (1970). The silver stained gels were submitted to autoradiography.

Sodium-stimulated phosphorylation and potassium-induced dephosphorylation are unique characteristics of the large alpha catalytic subunit of (Na^+, K^+)-ATPase. Its molecular weight was around 100.000 daltons for every species we studied. In 6% acrylamide gels, we were able to resolve the (Na^+, K^+)-ATPase catalytic subunit into its two molecular forms : alpha (+), the higher molecular weight form, and alpha (-), the lower molecular weight form. However, the difference in molecular weight and therefore the resolution between the two subspecies appeared greater in humans compared to cats and mice, allowing an easier study of their relative phosphorylation level in humans.

1) <u>K^+ and phosphorylation of (Na^+, K^+)-ATPase</u>

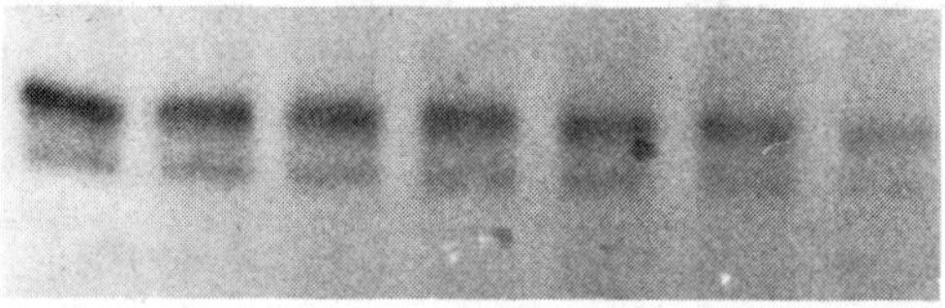

<u>Figure 1</u> : Effect of K^+ on the phosphorylation of human (Na^+, K^+) -ATPase catalytic subunit. The partially purified enzyme was phosphorylated in presence of 140mM Na^+ and varying concentrations of K^+ : from left to right : 0, 0.01, 0.05, 0.1, 0.3, 0.5 and 1mM K^+.
In human brain, the two catalytic subunit subspecies, alpha (+) and alpha (-) are well resolved in 6 % SDS-polyacrylamide gels. Both subspecies are phosphorylated in presence of Na^+ and progressively dephosphorylated in presence of increasing concentrations of K^+.

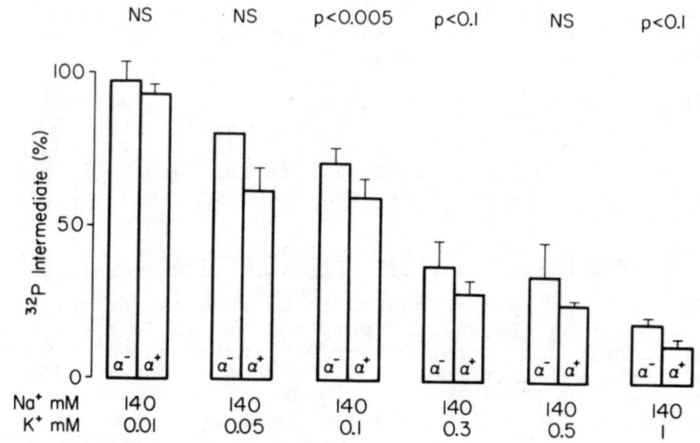

Figure 2 : Effect of K^+ on the absolute level of phosphorylation of (Na^+, K^+)-ATPase catalytic subunits ($alpha^+$ and $alpha^-$) purified from C57/Bl mouse brains. The results are expressed as a percentage of the absolute reference level, measured in presence of 5μM ATP, 3mM Mg Cl_2, 30mM Tris HCl, pH 7.1 and 140 mM NaCl. Data are mean ± SD values (n = 3). Statistical significance has been determined by t-paired Student test of absolute values of densitometric units for alpha (+) and alpha (-).

Increasing concentrations of K^+ have a progressive dephosphorylation effect on both subunits (Figure 1). However, preliminary results obtained from C57/Bl mice indicate that the rate of dephosphorylation is slightly different for alpha (+) and alpha (-). Any concentration of K^+ below 1mM is slightly more effective on the phosphorylation level of alpha (+), suggesting that alpha (+) has a slightly higher affinity for K^+ (Figure 2). More experiments are presently in progress in order to validate those differences.

Table 1 : Affinity constants of alpha (+) and alpha (-) for K^+ in human brain (In μM K^+).

	Alpha (+)	Alpha (-)	Difference in %
1	138	166	+ 20
2	79	95	+ 20
3	154	180	+ 17

The three human cortex samples were obtained from three different patients operated on for a non epileptogenic deeply lying tumor. The affinity constants were calculated from phosphorylation levels at three to five different K^+ concentrations.

In normal human cortex, where the resolution of the two catalytic subunit subspecies is best, data are clearer. In all experiments (n = 3), the affinity constants for K^+ are around 20 % higher for alpha (-) compared to alpha (+) (Table 1). Because of the limited availability of normal human tissue, those results are still preliminary and need further confirmation.
However, they may suggest that alpha (+), which is already known to have a higher affinity for ATP and ouabain, has also a higher affinity for K^+ compared to alpha (-). Those results, together with the differential effect of K^+ on the kinetic of glial and neuronal (Na^+, K^+)-ATPases, could further support the hypothesis that alpha (+) is the neuronal form, or at least the form in higher proportion in neurons and that alpha (-) is the glial or non-neuronal form.

2) Phenytoin and phosphorylation of (Na^+, K^+)-ATPase.

Phenytoin (PHT) has been suggested to work as an antiepileptic drug by stimulating the Na^+, K^+ pump and the (Na^+, K^+)-ATPase. In 1955, Woodbury demonstrated that PHT increased Na^+ turnover in whole brain, while it prevented the rise in intracellular Na^+ caused by maximal electroshock seizures.
These experiments were the basis for Woodbury's hypothesis that the anticonvulsant effects of PHT were a result of its enhancement of active Na^+, K^+ transport across the cell membranes. This concept found support in the results of Delgado-Escueta et al. (1974, 1980), who studied the ion transport properties of synaptosomes isolated from normal and epileptic animals. Those studies showed that PHT stimulates (Na^+, K^+)-ATPase activity and synaptic potassium uptake in ionic conditions simulating a depolarized state.
More recently, White et al. (1985) showed that therapeutic concentrations of PHT increased permeability of cultured glial cells to K^+, resulting in hyperpolarization. However, as far back as 1968, Rawson and Pincus questioned Woodbury's hypothesis. They proposed the Na^+ blockade hypothesis based on observations that PHT inhibited brain microsomial (Na^+, K^+)-ATPase and the passive inward movements of Na^+ ions in lobster nerves, brain slices and squid axons (Lipicky et al., 1972). Studies by Swanson and Crane (1972) examining the effects of PHT on ion transport and ATP levels in brain slices have shown that the decrease of intracellular K^+ and ATP, the increase of intracellular Na^+ normally elicited by electrical stimulation are reduced by PHT. More recently, Catterall (1981) reported that PHT blocks sodium channel activation by allosteric competition. Those studies may suggest that PHT acts by decreasing the passive inward movement of Na^+ and not by stimulating the (Na^+, K^+)-ATPase. But it is conceivable that two or more mechanisms exist; which one is related to the antiepileptic action of phenytoin is however not clear.

Using the same technical procedure, we studied the effect of PHT on the phosphorylation of partially purified (Na^+, K^+)-ATPase. The study has been mainly performed on (Na^+, K^+)-ATPase purified from C57/Bl mouse brains.
As mentioned earlier, the imperfect resolution of the two catalytic subunit subspecies on the autoradiograph did not allow us to study the influence of PHT on each one's level of phosphorylation. Hence, results presented in this section are only dealing with the effect of phenytoin on net phosphorylation of the entire catalytic subunit without regard to differential effects on alpha (-) and alpha (+) subspecies. Present experiments are examining that aspect on (Na^+, K^+)-ATPase purified from cat and human cortex.

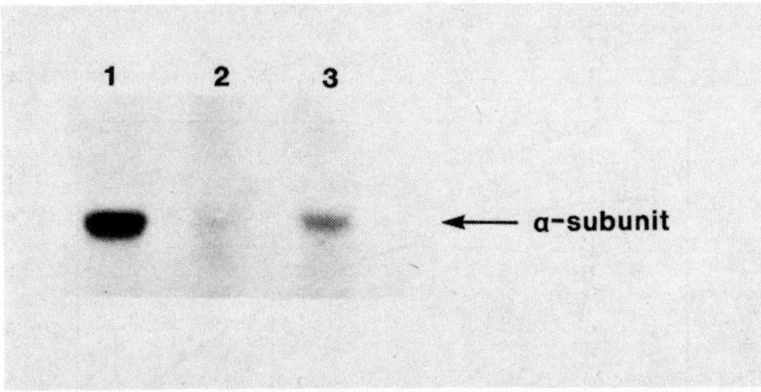

Figure 3 : Effect of phenytoin on the phosphorylation of C57/Bl mouse brains (Na^+, K^+)-ATPase.
The partially purified mouse brain (Na^+, K^+)-ATPase was phosphorylated by $[^{32}P]$ -ATP in reference conditions (3mM Mg Cl_2, 3mM Tris-HCl, pH 7.1, 140mM NaCl) (Lane 1), with 3mM KCl (Lane 2), and with 0.1mM phenytoin (Lane 3), subjected to SDS-Page and prepared for autoradiography.

Table 2 : Effect of 3mM K^+ and 0.1mM PHT on the net phosphorylation level of (Na^+, K^+)-ATPase catalytic subunit in C57/Bl mouse, cat and human brains.

Na^+ mM K^+ mM PHT mM	140 0 0	140 3 0	140 0 0.1
C57/Bl mice	100 %	11 ± 11 % n = 10 $p < 0.005$	46 ± 22 % n = 8 $p < 0.005$
Cats	100 %	10 ± 4 % n = 3 $p < 0.005$	49 ± 26 % n = 3 $p < 0.01$
Humans	100 %	9 ± 3 % n = 3 $p < 0.005$	66 ± 10 % n = 3 $p < 0.01$

Results are expressed as percentage of the reference level measured in presence of 3mM Mg Cl_2, 30mM Tris-HCl, pH 7.1, 140mM NaCl and 5µM $[^{32}P]$ -ATP. Data are mean ± SD values. Statistical significance has been determined by t-paired Student test of absolute values of densitometric units.

At 10^{-4}M, phenytoin decreased the net phosphorylation level of the alpha subunit in the presence of 140mM Na^+ (Figure 3). Table 2 gives a quantification of that effect : 10^{-4} M phenytoin decreased the alpha subunit net phosphorylation level by about 50% in C57/Bl mice ($p < 0.005$) and cats ($p < 0.01$) and by 35 % in humans ($p < 0.01$). By comparison, 3mM K^+ decreased that level by about 90 % ($p < 0.005$) in the three species. The dose dependence of that effect is shown by Figure 4. In C57/Bl mice, it disappears at phenytoin concentrations below 10^{-7}M.

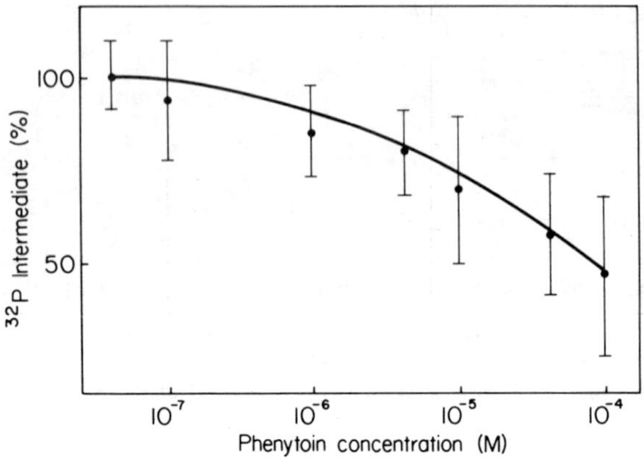

Figure 4 : Effect of phenytoin (5.10^{-8} to 10^{-4} M) on the phosphorylation level of mouse brain (Na^+, K^+)-ATPase catalytic subunit. The partially purified enzyme was preincubated with different concentrations of phenytoin and phosphorylated as previously described. Data are expressed as a percentage of control values (without phenytoin) and represent mean $\pm$ SD values of at least 3 different experiments.

Oligomycin and N-ethyl-maleimide (NEM) are substances known to block the conversion between the two phosphorylated intermediates, E_1-P and E_2-P (Post et al., 1963; Albers, 1967). E_1-P is a high energy phosphate intermediate, which can be decomposed in presence of ADP, while E_2-P does not react with ADP, but with K^+. After incubation of the enzymatic preparation with oligomycin (50 µg/ml) or N-ethyl-maleimide (10^{-3}M), the sensitivities of the phosphorylated intermediates to K^+, ADP and phenytoin were studied (Figure 5A-C). The native mouse brain enzyme phosphorylated intermediate was very sensitive to K^+ (90 % dephosphorylation) and less to ADP (65 % dephosphorylation). This suggests that, as proposed by Post et al. (1965), in normal conditions (presence of Na^+ and Mg^{++} ions) the E_1-P intermediate was transitory and had immediately converted to the E_2-P intermediate. In presence of oligomycin or NEM, the phosphorylated intermediate was more sensitive to ADP (75% and 83% dephosphorylation, respectively) than to K^+ (55% and 60% dephosphorylation, respectively), suggesting a significant but not total blockade of E_1-P conversion to E_2-P. After incubation with oligomycin or NEM, PHT did not affect the phosphorylation of the (Na^+, K^+)-ATPase catalytic subunit (+ 2% and -10% respectively). Consequently, that effect on the native brain enzyme has to be specifically on the E_2-P intermediate by inducing its dephosphorylation, as K^+ does.

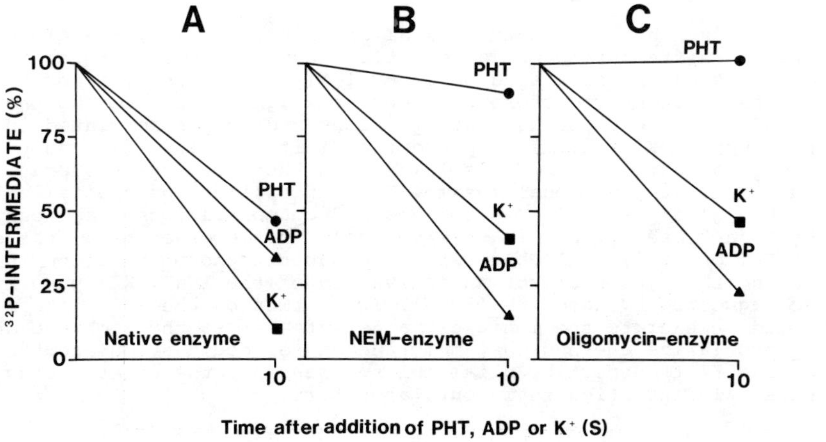

Figure 5 : Dephosphorylation of the phosphorylated intermediate of C57/Bl mouse brains (Na^+, K^+)-ATPase. The phosphorylated enzyme was formed with [^{32}P]-ATP in the absence (A) or presence of 10^{-3}M N-ethylmaleimide (B) or 50μg/ml oligomycin (C).

All values, except the two concerning the effect of PHT in presence of NEM or oligomycin, show a significant decrease ($p < 0.005$) of the phosphorylation level compared to references conditions.

$p < 0.01$ when the effect of ADP or PHT is compared to the effect of K^+ on the native enzyme.

$p < 0.05$ when the effect of ADP is compared to the effect of K^+ on the NEM or oligomycin treated enzyme.

Sen et al. (1969) proposed that the K^+ sensitive phosphorylated form of (Na^+, K^+)-ATPase, E_2-P, is the active intermediate that binds with ouabain. Thus, phenytoin, K^+ and ouabain appear to have a similar site of action, specifically at the external side of the membrane, the E_2-P form of the (Na^+, K^+)-ATPase.
However, the issue remains, whether or not our observations on the catalytic subunit of (Na^+, K^+)-ATPase are related to the antiepileptic properties of PHT or to its toxic side effects. It is difficult to differentiate PHT's regulating effects on convulsive activity from its toxic side effects, such as decline in cognitive powers, ataxia and slowing of EEG rhythms. Our experiments were performed on normal brains of C57/Bl mice, cats and humans and PHT, at therapeutic concentrations (10^{-6} to 10^{-4} M), decreased net phosphorylation level of the alpha subunit of (Na^+, K^+)-ATPase by 20 to 50%. At lower PHT concentrations (10^{-7} to 10^{-10} M), the effects were negligible. These observations do not support arguments for either antiepileptic or toxic effects. PHT effects were observed at therapeutic concentrations and similar, if not more pronounced, effects could as easily occur with toxic concentrations. What is needed is to correlate the in vitro biochemical actions of therapeutic concentrations of PHT with the in vivo suppression of epileptic activities as measured clinically and electroencephalographically. These experiments are now in progress.

CONCLUSION

Recent data have shown that glial and neuronal (Na^+, K^+)-ATPases are two different molecular entities. We propose that those two entities can be separated on SDS-gel electrophoresis, according to their slightly different molecular weight. Both are phosphorylated in presence of Na^+ and Mg^{++} ions and dephosphorylated in presence of K^+ ions. The apparent affinity of alpha (-), supposedly the glial enzyme, for K^+ is lower than the affinity of alpha (+), supposedly the neuronal enzyme. That underlines the specialization of the glial Na^+, K^+ pump for high K^+ concentrations (see chapter by Franck and Grisar, in this volume). On the other hand, the effect of PHT on (Na^+,K^+)-ATPase catalytic subunit phosphorylation is favoring the hypothesis of an activation of the (Na^+, K^+) pump by PHT, as suggested by Woodbury (1955). Its action on the E_2-P intermediate could accelerate the ionic exchange rate across the membrane and hyperpolarize the neuronal membrane and/or could stimulate the active K^+ transport by glial cells. These hypotheses are being presently investigated in our laboratory.

ACKNOWLEDGEMENTS

This investigation has been supported by a research grant from the Epilepsy Foundation of America and, at least, in part by Federal Funds from the Department of Health and Human Services, under contract number N01-NS-2332.

The contents of this publication do not necessarily reflect the view or policies of the Department of Health and Human Services, nor mention any trade names, commercial products, or organization imply endorsement by the U.S. government.

D. Guillaume is also supported by the National Fund for Scientific Research, Belgium (F.N.R.S.).

REFERENCES

Albers, R.W., Fahn, S. and Koval, G.J. (1963). The role of sodium ions in the activation of Electrophorus electric organ adenosine triphosphatase. Proc. Natl. Acad. Sci. U.S.A., 50 : 474 - 481.

Albers, R.W. (1967). Biochemical aspects of active transport. Ann. Rev. Biochem., 36 : 727 - 756.

Atterwill, C.K., Cunningham, V.J. and Balazs, R. (1984). Characterization of (Na^+, K^+)-ATPase in cultured and separated neuronal and glial cells from rat cerebellum. J. Neurochem., 43 : 8 - 18.

Catterall, W.A. (1981). Inhibition of voltage-sensitive sodium channels in neuroblastoma cells by antiarrhytmic drugs. Mol. Pharm., 20 : 356 - 362.

Churchill, L.; Hall, C.C.; Peterson, G.L.; Ruoho, A.E. and Hokin, L.E. (1984). Photoaffinity labeling of the ouabain binding site in (Na^+, K^+)-ATPase in developing Brine Shrimp. J. Exp. Zool., 231 : 343 - 350.

Churchill, L. (1984). Differences in phosphorylation of the two large subunits of Brine Shrimp (Na^+, K^+)-ATPase. J. Exp. Zool. 231 : 335 - 341.

Delgado-Escueta, A.V., Davidson, D., Hartwig, G. and Reilly, E. (1974). The freezing lesion II : Potassium transport within nerve terminals isolated from the epileptogenic foci. Brain Res., 78 : 223 - 237.

Delgado-Escueta, A.V. and Davidson, D. (1974). Freezing lesions III : The effects of diphenylhydantoin on potassium transport within nerve terminals from the primary foci. Brain Res., 86 : 85 - 96.

Delgado-Escueta, A.V. and Horan, M.P. (1980). Phenytoin : Biochemical membrane studies, in Antiepileptic Drugs, Mechanisms of Action, edited by Glaser, G.H., Penry, J.K., Woodbury, D.M., 377 - 398. Raven Press, New York.

Fambrough, D.M. and Bayne, E.K. (1983). Multiple forms of (Na^+, K^+) -ATPase in the chicken. J. Biol. Chem., 258 : 3926 - 3935.

Fisher, J.A.; Baxter-Lowe, L.A. and Hokin, L.E. (1984). Site of synthesis of the alpha and beta subunits of the (Na^+, K^+)-ATPase in Brine Shrimp Nauplii. J. Biol. Chem., 259 : 14217 - 14221.

Franck, G.; Grisar, T. and Moonen, G. (1983). Glial and neuronal Na^+, K^+ pump. In : Advances in Cellular Neurobiology, edited by S. Fedoroff and L. Hertz, Vol. 4, 133 - 159.

Grisar, T.; Franck, G. and Schoffeniels, E. (1978). K^+ activation mechanisms of the (Na^+, K^+)-ATPase of bulk isolated glia cells and neurons. In : Dynamic properties of glia cells, edited by E. Schoffeniels, G. Franck, L. Hertz and D.B. Tower, 353 - 363. Pergamon Press, New York.

Guillaume, D.; Grisar, T. and Delgado-Escueta, A.V. (1985). Phenytoin dephosphorylates the catalytic subunit of the (Na^+, K^+)-ATPase in C57/Bl mice. J. Neurochem. Submitted.

Kawakami, K.; Noguchi, S.; Noda, M.; Takahashi, H; Ohta, T.; Kawamura, M.; Nojima, H.; Nagano, K.; Hirosa, T.; Inayama, S.; Hayashida, H.; Miyata, T. and Numa, S. (1985) : Primary structure of the alpha subunit of Torpedo Californica (Na^+, K^+)-ATPase deduced from cDNA sequence. Nature., 316 : 733-736.

Kuffler, S.W. (1967) : Neuroglial cells : Physiological properties and a potassium mediated effect of neuronal activity on the glial cell membrane potential. Proc. R. Soc. London , Ser. B., 168 : 1 - 21.

Laemmli, U.K. (1970). Cleavage of structural proteins during the assembly of the head of bacteriophage T4. Nature. 227 : 680 - 685.

Lipicky, R.J.; Gilbert, D.L. and Stillman, I.M. (1972). Diphenylhydantoin inhibition of sodium conductance in the squid giant axon, Proc. Natl. Acad. Sci. U.S.A. 69 : 1758 - 1760.

Lux, H.D. (1974). The kinetics of extracellular potassium. Relation to epileptogenesis. Epilepsia., 15 : 375 - 393.

Marks, M.J. and Seeds, N.W. (1978). A heterogeneous ouabain -ATPase interaction in mouse brain. Life Sci., 23 : 2735 - 2744.

Matsuda, T. and Cooper, J.R. (1983). Inhibition of neuronal sodium and potassium activated adenosine triphosphoatase by pyrithiamin. Biochem. 22 : 2209-2213.

Matsuda, T.; Iwata, H. and Cooper J.R. (1984). Specific inactivation of $alpha^+$ molecular form of (Na^+, K^+)-ATPase by pyrithiamin. J. Biol. Chem., 259 : 3858 - 3863.

Morohashi, M and Kawamura, M. (1984). Solubilization and purification of Artemia salina (Na,K)-activated ATPase and NH_2-terminal amino acid sequence of its larger subunit. J. Biol. Chem., 259, 14928 - 14934.

Peterson, G.L. and Hokin, L.E. (1981). Molecular weight and stoichiometry of the sodium and potassium activated adenosine triphosphatase subunits. J. Biol. Chem., 256 : 3751 - 3761.

Peterson, G.L.; Churchill, L.; Fisher, J.A. and Hokin, L.E. (1982). Structural and biosynthetic studies on the two molecular forms of the (Na^+, K^+)-activated adenosine triphosphatase large subunit in Artemia salina nauplii. J. Exp. Zool. - 221 : 291 - 308.

Post, R.L.; Sen, A.K. and Rosenthal A.S. (1965). A phosphorylated intermediate in adenosine triphosphate-dependent sodium and potassium transport across Kidney membranes. J. Biol. Chem. 240 : 1437 - 1445.

Post, R.L.; Kume, S.; Tobin, T.; Orcutt, B. and Sen, A.K. (1965). Flexibility of an active center in sodium-plus-potassium adenosine triphosphatase. J. Gen. Physiol., 54 : 306 - 326.

Rawson, M.D. and Pincus, J.H. (1968). The effect of diphenylhydantoin on sodium, potassium, magnesium-activated adenosine triphosphatase in microsomal fractions of rat and guinea pig brain and on whole homogenates of human brain. Biochem. Parmacol., 17 : 573 - 579.

Siegel, S.J.; Stahl, W.L. and Swanson, P.D. (1981). Ion transport. In : Basic Neurochemistry, edited by G.J. Siegel; R.W. Albers, R. Katzman and B.W. Agranoff, 107 - 143. Little Brown, Boston.

Shull, G.E.; Schwartz, A. and Lingrel, J.B. (1985). Amino acid sequence of the catalytic subunit of the (Na^+, K^+)-ATPase deduced from a complementary DNA. Nature, 316 : 691 - 695.

Swanson, P.D. and Crane, P.O. (1972). Diphenylhydantoin and movement of radioactive sodium into electrically stimulated cerebral slices. Biochem. Pharm., 21 : 2899 - 2905.

Sweadner, M.J. (1979). Two molecular forms of (Na^+, K^+)-stimulated ATPase in brain. Separation and difference affinity for strophanthidin. J. Biol. Chem., 254, 6060 - 6067.

White, H.S.; Yen-Chow, Y.C.; Chow, S.Y.; Kemp, J.W. and Woodbury, D.M. (1965). Effects of phenytoin on primary glial cell cultures. Epilepsia, 26 : 58 - 68.

Woodbury, D.M. (1955). Effects of diphenylhydantoin on electrolytes and radiosodium turnover in brain and other tissues of normal hyponatremic and post-ictal rats. J. Pharmacol. Exp. Ther., 115 : 74 - 95.

Volume Regulation in Primary Astrocyte Cultures

H. K. Kimelberg*,† and M. V. Frangakis*

*Division of Neurosurgery, †Departments of Anatomy and Biochemistry, Albany Medical College, Albany, New York 12208, USA

ABSTRACT

Using radiolabelled markers of the intracellular space, we found that astrocytes in primary monolayer culture prepared from neonatal rat cerebral cortex swelled or shrunk in response to changes im media osmolality. They also regulated their volume back to normal under hypotonic conditions (regulatory volume decrease or RVD), but did not volume regulate after shrinkage in response to hyperosmotic conditions. They did show regulatory volume increase (RVI) after shrinking due to being placed in isotonic media after a previous exposure to hypotonic media. This RVI was inhibited by furosemide and bumetanide implicating the cation plus chloride co-transport system in this process. Under isotonic conditions the cells swelled in response to 0.1mM L-glutamate, as also did rat cerebrocortical slices. The cells also swelled in response to increased media [K^+] and upon addition of 10mM lactic acid. In contrast they shrunk in response to 0.1mM l-norepinephrine.

KEYWORDS

Astrocyte cultures; co-transport; furosemide; bumetanide; volume regulation; glutamate; potassium; lactic acid; norepinephrine.

INTRODUCTION

Mammalian astroglia swell readily in a number of pathological states. Thus studies on their volume control and ion transport are of fundamental importance to an understanding of brain pathology, as well as normal brain functions. We have used rat astrocytes in primary monolayer cultures as a model for studying astrocyte properties (Kimelberg, 1983).

The major electrical membrane property of the cells in these cultures is determined by K^+ conductance, since these cells have large negative membrane potentials which are very sensitive to changes in medium [K^+]. There may be minor conductances to other ions (Kimelberg, et al, 1979b). However, in addition to a major K^+ conductance permeability, there are a number of electrically silent co- and exchange transport systems. In such cultures we have shown the presence of a SITS-sensitive anion exchange system transporting Cl^- and HCO_3^- (Kimelberg, et al, 1979a; Kimelberg, 1981), an amiloride-sensitive Na^+/H^+ exchange system (Kimelberg, et al, 1979a; Kimelberg and Ricard, 1982) and a furosemide and bumetanide sensitive KCl co-transport system (Kimelberg and Hirata, 1981). This co-transport system may also transport Na^+ (Kimelberg and Frangakis, 1985), presumably as the driving force for uptake of both K^+ and Cl^- against their respective electrochemical gradients (Kinne, 1985). These transport systems are depicted in figure 1 and the sites of action of the different inhibitors are also shown. If the furosemide co-transport system also transports Na^+ inwards then this could lead to active accumulation of Cl^- inside the cell. This would then be expected to drive the SITS-sensitive anion exchange system in the direction shown, with HCO_3^- entering the cell. This then buffers the cell by associating with H^+ and recycling via membrane permeable CO_2. Thus this system, plus the amiloride-sensitive $Na^+ \leftrightarrow H^+$ exchange system serves to buffer pH_i. Na^+ is then pumped out of the cell via the (Na^+,K^+) pump (not shown). With high CO_2 the anion exchange system should be driven in the opposite direction, and with Na^+ taken up on the $Na^+ \leftrightarrow H^+$ exchange system can lead to accumulation of NaCl with cell swelling (Cala, 1980; Kimelberg, et al, 1979a).

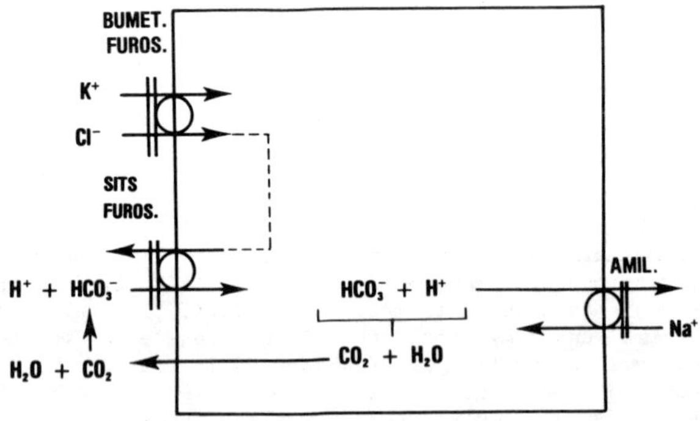

Fig. 1. Model of co- and exchange transport processes found in astrocytes in primary culture. Furos, furosemide; bumet, bumetanide; SITS, 4-acetamido-4'-isothiocyanostilbene-2,2' disulfonic acid; amil, amiloride. See text for further details.

These exchange and co-transport processes have been shown to be involved in volume regulation in a number of different cell types, such as red blood cells (Cala, 1980; Kregenow, 1981; Lowe and Lambert, 1983; Ueberschar and Bakker-Grunwald, 1983), Ehrlich ascites tumor cells (Geck, et al, 1980; Hoffman, et al, 1983), lymphocytes (Grinstein, et al, 1984), lymphoblasts (Rosenberg, 1972) and epithelial cells (Fisher, et al, 1981; Ussing, 1982). We have therefore examined the involvement of these processes in primary astrocyte cultures in volume regulation. This regulation involves regulatory volume decrease (RVD) and regulatory volume increase (RVI), when cells regulate their volume back towards control levels after an initial swelling or shrinkage when exposed to hypotonic or hypertonic medium respectively. We have studied some of the contributions of the ion co- and exchange transport processes to volume control caused by exposure of the cultured astrocytes to hyper- and hypo-osmotic media. We have also begun to study the effects of a number of compounds, known to be released in pathological states where astrocytic swelling is common, on the volume of the cultured cells.

EXPERIMENTAL METHODS

Cell culture

Cultures were prepared from the cerebral cortices of 1-2 day old Sprague-Dawley rats after careful removal of the meninges. The cell suspension prior to plating was prepared by sequential dissociation of the tissue by means of the bacterial protease Dispase, as previously described (Frangakis and Kimelberg, 1984). The cells were grown in Eagles BME + 10% fetal calf serum with added supplements and antibiotics (Frangakis and Kimelberg, 1984) in 12 well trays (Costar) and were used when 3 to 5 weeks old. The cultures were at least 95% astrocytic as judged by staining for GFAP using a 1:200 dilution of a monoclonal antibody (Amersham) to human GFAP, and rhodamine conjugated rabbit anti mouse IgG (Accurate Chemicals), as previously described (Frangakis and Kimelberg, 1984).

Ion transport

These studies were performed essentially as previously described (Kimelberg, 1981; Kimelberg, et al, 1979a,b). Growth medium was removed and the cells washed three times with HCO_3^- or in some cases HEPES buffered balanced salt solution (Na^+, 145; K^+, 4.5; Mg^{2+}, 0.4; Ca^{2+}, 1.3; Cl^-, 123; glucose, 10; SO_4^{2-}, 1.2; phosphate, 1.2; HCO_3^-, 25 bubbled with 5% CO_2/95% air to pH 7.4 or HEPES 25 adjusted to pH 7.4 with NaOH. All values refer to mmol/liter). The cells were then incubated in this solution at 37°C for up to 30 minutes, and also in a 5% CO_2/95% air atmosphere for the HCO_3^- buffered solution. Inhibitors were added during this period when preincubation was desired. This solution was then removed and 0.5 to 1.0ml of the same media, warmed to 37°C and containing ^{14}C labelled urea or 3-0-methyl glucose with or without inhibitors was added to each well. After varying times the media was rapidly aspirated off and each well washed 6 or 7 times with 1ml ice-cold 0.32M sucrose containing 10mM Tris-Cl (pH 7.4) and 0.4mM Ca $(NO_3)_2$, by rapid addition with an automatic pipette attached to a reservoir and immediate aspiration with a tube attached to a vacuum pump (total washing time was approximately 10-15 seconds). A zero time uptake of ^{14}C urea or ^{14}C 3-0-methyl glucose when the isotope was added and the cells immediately washed, or a ^{14}C sucrose or ^{14}C inulin space, was always subtracted to correct for extracellular, adherent medium (Kimelberg, et al, 1979a). The cell monolayer from each well was then solubilized in 0.5ml 1N NaOH at room temperature for about 15-30 minutes. This was removed and the wells washed with 0.5ml distilled water which was added to the NaOH digest. Aliquots were taken for scintillation counting and protein determination by the Lowry-Folin method.

RESULTS AND DISCUSSION

In figure 2 we show that the cultured astrocytes exhibit regulatory volume decrease (RVD) when exposed to hypotonic media. The cell volume was measured using ^{14}C labelled urea. Uptake of ^{14}C labelled urea was first allowed to proceed for 30 minutes. The media was then changed to a hypotonic one and the cell content of ^{14}C urea was then measured at the times shown. After an initial rapid

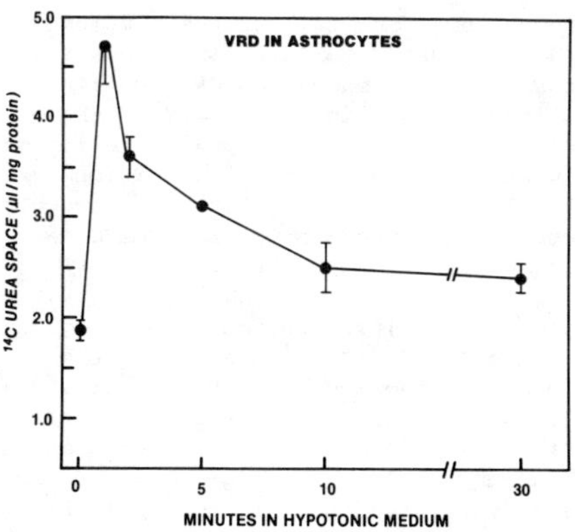

Fig. 2. Regulatory volume decrease in primary astrocyte cultures. Cultures were incubated with ^{14}C urea for 30 min in HCO_3^- buffered media (see Methods). At 0 time on the graph medium was changed to a hypotonic medium (NaCl 43mM instead of 140mM) with ^{14}C urea at the same specific activity. At times shown the cells were washed and ^{14}C urea remaining in cells determined. The urea space was calculated based on the specific activity of ^{14}C urea in the medium in cpm/µl. n = 3 ± S.D. for each point.

swelling phase it can be seen that these cells then volume regulate reaching close to the pre-hypotonic exposure levels by 10 minutes.

We found no significant effect of furosemide or SITS on RVD and this may be because the loss of osmoles principally involves loss of KCl through separate K and Cl conductance channels, as recently reported for lymphocytes (Grinstein, 1984). Furthermore, when the astrocytes shrink in response to exposure to medium made hypertonic by addition of 200mM mannitol, they do not show RVI but remain at this shrunken volume for up to at least 90 minutes (Kimelberg and Frangakis, 1985). Volume regulatory increase after exposure to hypotonic media and then subsequent shrinkage upon re-exposure to isotonic media has been reported (Grinstein, et al, 1984; Hoffman, et al, 1983). This has been suggested to be due to loss of KCl from the cell so that when the cells are re-exposed to isotonic media there is increased driving force for inward transport of KCl on the co-transport uptake system (see fig. 1). We have found the same shrinkage of our cells after exposure to isotonic medium following a previous exposure to hypotonic medium and also a re-swelling back to control levels. This is shown in figure 3 where it is also shown that this re-swelling is inhibited by furosemide or bumetanide. This suggests that the co-transport system is involved here. Amiloride, which inhibits the $Na^+ \leftrightarrow H^+$ exchange system in our cells (Kimelberg and Ricard, 1982) had no effect , suggesting that $Na^+ \leftrightarrow H^+$ plus $Cl^- \leftrightarrow HCO_3^-$ uptake leading to a gain of NaCl by the cells, as has been reported to occur for RVI in amphiuma red blood cells (Cala, 1980), is not involved.

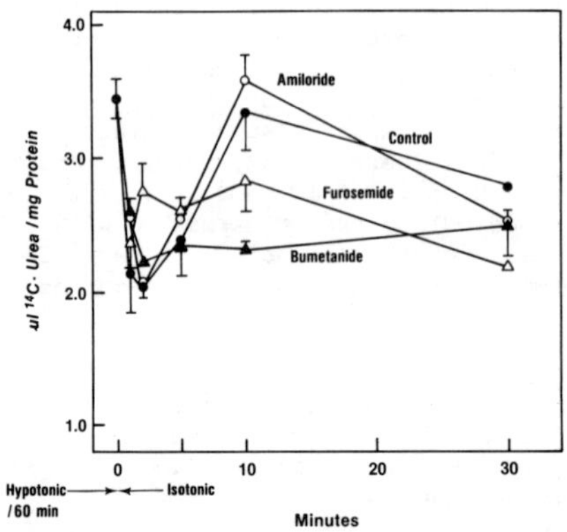

Fig. 3. Effects of transport inhibitors on volume changes as indicated by changes in ^{14}C urea space (expressed as µl/mg protein) following RVD in hypotonic medium (see fig. 2) followed by exposure to isotonic medium at 0 time on graph (indicated below the abscissa). The inhibitors were added at 0 time when the cells were exposed to isotonic medium after the initial 60 min exposure to hypotonic medium HCO_3^- buffered medium used throughout. The inhibitor concentrations used were; amiloride and furosemide 1mM; bumetanid 0.01mM. n = 3 wells $\pm$ SEM.

The preceding data show that astrocytes in culture can regulate their volume in response to changing osmolality of the media and under certain conditions this uses the KCl (+ Na^+?) co-transport mechanism. In the mammalian CNS swelling of astrocytes occurs under conditions where large changes in osmolality of the extracellular fluid are probably not occurring, such as ischemia, hypoxia and trauma (see Kimelberg, et al, 1982). Some likely effectors under these conditions are raised extracellular $[K^+]$, transmitters and lactic acid. Swelling of astrocytes in response to such agents may involve some of the transport mechanisms discussed above or perhaps other mechanisms.

The data in figure 4 show that increasing medium $[K^+]$ from 4.5mM to 50mM by replacement with Na^+ causes an increase of the cellular ^{14}C urea space. Here the entire time course of ^{14}C urea uptake is measured showing that the ^{14}C urea space reaches a reasonably constant level at 10 minutes. This effect of raised $[K^+]_o$ could be due either to activation of the KCl (+ Na^+?) co-transport system or, since we have raised $[K^+]_o$ without reducing $[Cl^-]_o$, swelling due to the increased outside Donnan ratio. The co-transport system does not appear to be involved since neither furosemide or bumetanide had any effect, as shown in figure

5. As expected SITS also had no effect since there is no reason to expect that raising $[K^+]_o$ would directly activate the anion exchange system.

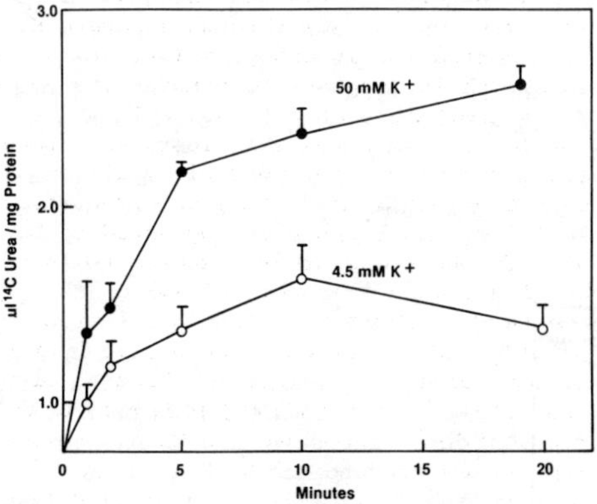

Fig. 4. Increase of ^{14}C urea space in µl/mg protein with time at normal (4.5mM) and raised (50mM) K^+. K^+ replaced Na^+. K^+ raised at t = 0 on graph when ^{14}C urea was added after an initial 15 min pre-incubation in normal HCO_3^- buffered medium. n = 3 ± S.D.

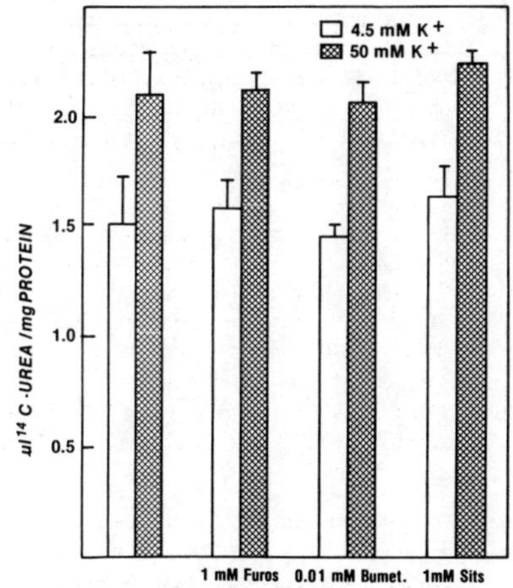

Fig. 5. Effect of different inhibitors on K^+ induced swelling in primary astrocyte cultures. ^{14}C urea uptake measured for 10 min (see fig. 4). The cells were incubated in regular HCO_3^- buffered medium for 15 min before adding ^{14}C urea (1 µCi/well/0.5ml) in normal or raised $[K^+]_o$. Uptake was then measured for 10 min. Furosemide and bumetanide were added at same time as ^{14}C urea. SITS was preincubated with cells for 15 min. Concentrations shown on graph. n = 3 ± SEM.

Based on the results shown in figure 5, K^+ induced swelling appears likely to be mediated via Donnan effects with K^+ entering through K^+ channels. However, the K^+ induced increase in the ^{14}C urea space does not seem to involve Ba^{2+} sensitive K^+ channels, which are more common in mouse than rat astrocytes (Walz and Kimelberg, 1985; Walz, et al, 1984), since 1mM Ba^{2+} did not affect the K^+ induced increase in the $[^{14}C]$ urea space (fig. 6). Also, 1mM amiloride, which also partly blocks steady state K^+ transport (Walz and Kimelberg, 1985), had no effect. A Further test for K^+ moving through diffusional channels under Donnan forces would be to show the effect of raised $[K^+]_o$ in Cl^- free media or with media Cl^- reduced to keep a constant K x Cl product.

We have found that L- but not D-glutamate at 0.01 to 1mM will increase the ^{14}C urea space of these cells. This is shown in figure 7. 10^{-4}M L-glutamate also stimulates uptake of $^{22}Na^+$ in the presence of ouabain but this uptake is insensitive to the co-transport inhibitors so that it might involve uptake of Na glutamate which is known to be very intense in these cells. For example, L-glu uptake shows about a 100-fold higher Vmax than uptake of GABA and a comparable Km of 50μM (Hertz, 1979). Also swelling is only seen in the presence of ouabain, which by inhibiting the (Na + K) ATPase presumably prevents the pumping out of the accumulated Na^+. Since we have found (Bowman and Kimelberg, 1984) that L- but not D-glutamate depolarizes astrocytes in culture, and if this is due to the same mechanism as the L-glutamate induced swelling, this suggests that such uptake is electrogenic involving more than one Na^+ per glutamate. However, uptake of glutamate into rat astrocyte cultures showed only a 1:1 Na^+ to glutamate ratio (Drejer, et al, 1982). Alternatively L-glu could be stimulating Na^+ uptake, perhaps due to a conductance increase, via action on an L-glutamate receptor. In agreement with this, kainic acid also causes depolarization of astrocytes (Bowman and Kimelberg, 1984).

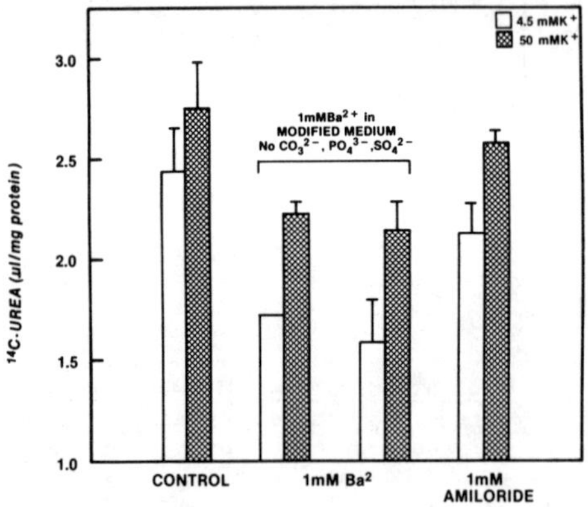

Fig. 6. Lack of effect of 1mM Ba^{2+} and 1mM amiloride on K^+-induced increase in ^{14}C urea space. Other conditions as in figs. 4 and 5. The divalent anions shown were omitted from the media to prevent precipitation of Ba^{2+}. This by itself reduced the ^{14}C urea space. HEPES buffered media as described in Materials and Methods was used, except it contained no PO_4 and Ca^{2+} gluconate instead of $CaSO_4$. n = 3 $\pm$ SEM.

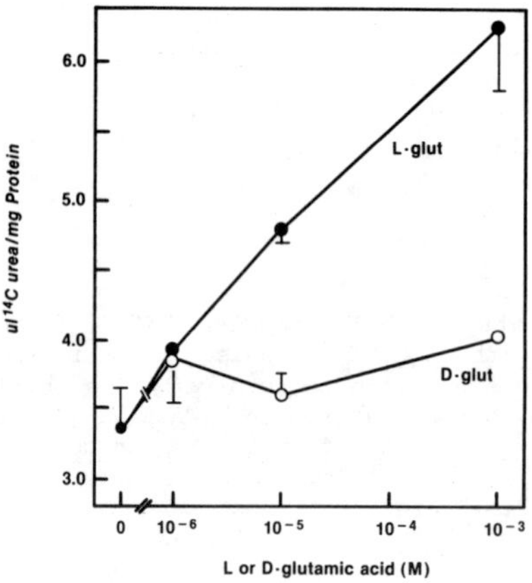

Fig. 7. Effect of L- or D-glutamic acid on ^{14}C urea space of primary astrocyte cultures. Cells preincubated for 15 min and then D- or L-glutamic acid, plus 1mM ouabain and ^{14}C urea, were added to the final concentrations shown and the uptake of ^{14}C urea measured for 15 min. n = 3 $\pm$ SEM where indicated.

We have also found that exposure to 10mM lactic acid, which is known to rise to 15-20mM in ischemia and hypoxia (Rehncrona, et al, 1981; Siesjo, 1984) also causes swelling of the cells, as measured by the ^{14}C 3-O-methyl glucose space (Kletzien, et al, 1975) as shown in figure 8. This is in spite of increased osmolality due to the added lactic acid. In contrast, 0.01mM norepinephrine caused shrinkage of the cells.

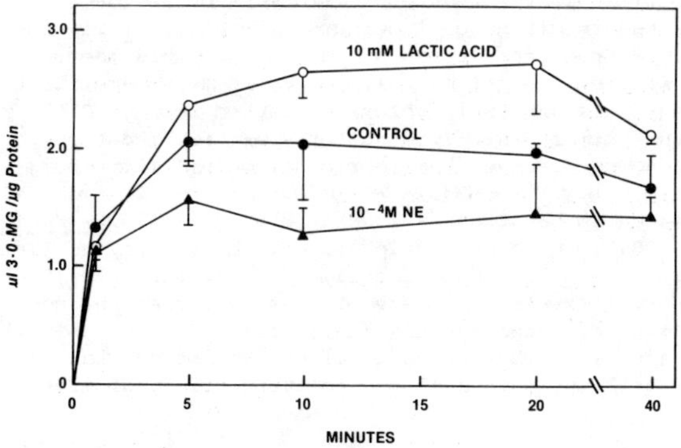

Fig. 8. Effects of 10mM lactic acid (O) or 0.1 mM l-norepinephrine (▲), on time course of uptake of [^{14}C] 3-0-methyl D-glucose. Cells preincubated for 20 min in HCO_3^- buffered medium and then [^{14}C] 3-OMG with or without lactic acid or NE was added and the time course of [^{14}C] 3-OMG uptake measured.

This effect by NE was unexpected since in cat brain we have observed swelling by NE, as well as by histamine and adenosine (Bourke, et al, 1983). Glutamate is also known to cause swelling of rat and cat brain slices at concentrations of 10mM (Bourke and Tower, 1966; Lund-Andersen and Hertz, 1970). We have also observed swelling of rat cerebrocortical slices by 1mM glutamate as shown in Table 1. In this case the swelling was mainly associated with increased uptake of Na^+ in HEPES buffered media. In HCO_3^- buffered media there was greater swelling associated with increased uptake of Cl^- as well as Na^+

Table 1: Increased swelling and ion content of rat brain cerebrocortical slices due to added 1mM L-glutamate

Medium	L-glutamate induced increase in wet weight (Δ percent initial weight) or ion content (µmoles/g initial wet weight)			
	Δwet weight	ΔCl^-	ΔNa^+	ΔK^+
HEPES buffered	4.5***	2.3**	17.9***	2.9***
HCO_3^-	7.2***	9.1***	24.1***	1.7*

Rat cerebrocortical slices from adult rats were incubated for a total of 1 hr in HEPES (see Materials and Methods) buffered media with or without 10mM glutamate. $NaHCO_3$ (10mM) when present was added after 20 minutes. Swelling and ion content were expressed and other experimental conditions were as previously described for cat cerebrocortical slices (Bourke, et al, 1983). Number of slices for each buffer condition and ± glutamate was 15. All increases were significantly different from control slices to which glutamate was not added at the *** < 0.001, ** < 0.01, *<0.02 level by Student t-test.

DISCUSSION

The data presented in this paper show that rat astrocytes in primary culture will swell in response to exposure to hypotonic medium, and then show regulatory volume decrease (RVD)in the continued presence of hypotonic medium. This is similar to the behavior of a number of other cell types including lymphoblasts and lymphocytes (Rosenberg, 1972; Grinstein, 1984), avian erythrocytes (Kregenow, 1981), frog epithelial cells (Ussing, 1982), Ehrlich ascites tumor cells (Hoffman, et al, 1983) and proximal convoluted rabbit kidney tubules (Gagnon, et al, 1982). RVD has been reported in glioma cell cultures (Kempski, et al, 1983, 1984), and also in rat primary astrocyte cultures by Holtzman at this meeting (see this volume). In these studies the glioma cells or astrocytes were first detached from their growing surfaces and their volumes measured using a Coulter counter. In contrast, when the primary astrocyte cultures are exposed to hypertonic medium they shrink, but do not recover their original volume (Kimelberg and Frangakis, 1985). Thus they only show RVI after previous RVD (fig. 3). Lack of hypertonic-induced RVI is seen in hepatocytes (Bakker-Grunwald, 1983), C_6 cells (Kempski, et al, 1983, 1984), and human peripheral blood lymphocytes (Grinstein, et al, 1984). However, some cells, specifically avian erythrocytes, restore their volumes towards normal levels by uptake of ions by either the co-transport or anion plus $Na^+ \leftrightarrow H^+$ exchange systems (Cala, 1980; Kregenow, 1981). Thus the cultured astrocytes show some, but not all, of the volume regulatory properties exhibited by other cell types in response to changes in medium osmolality.

The ability of the astrocyte to show RVD under hypoosmotic conditions may contribute to volume regulation of the brain under hyponatremia or other hypoosmotic conditions (Chan and Fishman, 1985). However, under the pathological conditions where astroglia are usually found to be swollen, large changes in extracellular osmolality do not seem likely. Thus it seemed reasonable to also look for effectors of astrocytic swelling under isosmotic conditions. We have clearly identified one such agent, L-glutamate, which also causes swelling of brain slices (Lund-Andersen and Hertz, 1970; Bourke and Tower, 1966) and of astrocytic cells *in situ* (Van Harreveld and Fifkova, 1971). Also glutamate is extensively released in edema plus ischemic conditions (Maier-Hauff, et al, 1984). In the astrocyte cultures we found that L-glu will increase the [^{14}C] urea space, but only in the presence of ouabain. This is presumably due to L-glu stimulating uptake of Na^+ which is then immediately pumped out by the astroglial (Na^+, K^+) pump, since we have found that 10^{-4}M L-glu causes a Na^+ dependent depolarization of these cells which, in the presence of ouabain, is followed by a marked and transient hyperpolarization when ouabain is washed out. This is presumably due to marked activation of the Na,K pump, pumping in the electrogenic 3Na/2K mode, due to the increased intracellular Na^+ load (Bowman and Kimelberg, 1984). This, however, assumes that the swelling and depolarization response to L-glu involves the same mechanisms. Glutamate has also been reported to increase the volume of C_6 glioma cells by 8% (Kempski, et al, 1982). A relatively high concentration of glutamate (15mM) was used, perhaps because ouabain or some other means to inhibit the (Na^+, K^+) pump was not used in these experiments. Decreased activity of the (Na^+, K^+) pump is likely to be found under the pathological states in which astrocytes are swollen because of depletion of substrate or O_2, and thus there will be a decreased ability to pump Na^+ out. Two likely possibilities for the mechanism of the L-glutamate-induced increased Na^+ influx are Na^+-dependent uptake of L-glutamate or glutamate-activated changes in Na^+ conductance. Further work on the effects of different inhibitors of exchange and co-transport systems and L-glu receptor antagonists on L-glu-induced swelling, ion transport and depolarization should help to resolve this problem. The lack of swelling by l-norepinephrine was surprising since this transmitter, as well as histamine and adenosine, causes swelling of cat brain slices *in vitro*, which is associated with astrocytic swelling (Bourke, et al, 1983). It suggests that indirect effects of l-NE on neurons, such as increased neuronal activity and/or metabolism, are involved in the astrocytic swelling seen in brain tissue. Lactic acid acidosis is a common correlate of conditions in which astrocytic swelling is seen (Siesjo, 1984), and thus the swelling of astrocytes in culture when lactic acid is added suggests a direct effect.

In conclusion the volume of astrocytes growing in monolayer culture can be easily measured by tracer techniques and they respond in a conventional way to changes in media osmolality. These cells also show an ability to volume regulate, showing RVD in hypotonic media. They also show a capacity for RVI after an initial hypotonic induced period of swelling and RVD, utilizing the furosemide sensitive KCl co-transport system. The cells in culture also swell in response to L-glu, raised $[K^+]_o$ and lactic acid under isosmotic conditions, but the mechanisms for these processes have not yet been elucidated. However, swelling due to increased $[K^+]_o$ may be partly due to activation of the co-transport system since swelling of mouse astrocyte cultures due to increased $[K^+]_o$ was inhibited by furosemide and by omission of media Cl^- (see Walz, this volume and Walz and Hertz, 1984). Continued studies with primary astrocyte cultures, perhaps in conjunction with studies on brain slices and intact brain tissue, should provide important information on the likely mechanisms of astrocytic swelling with implications for the normal and pathological roles of these cells in the mammalian central nervous system.

ACKNOWLEDGEMENTS

Supported by grant BNS 8213873 from NSF and a grant from Merck Sharp & Dohme. We thank J. Rose for doing the experiments reported in Table 1 and E.P. Graham for typing this manuscript.

REFERENCES

Bakker-Grunwald, T. (1983). Biochim. Biophys. Acta 731, 239-242.
Bourke, R.S., H.K. Kimelberg, H. Daze and G. Church (1983). Neurochem. Res. 8, 5-24.
Bourke, R.S. and D.B. Tower (1966). J. Neurochem. 13, 1079-1117.
Bowman, C.L. and H.K. Kimelberg (1984). Nature 311, 656-659.
Cala, P.M. (1980). J. Gen. Physiol. 76, 683-708.
Chan, P.H. and R.A. Fishman (1985). Handbook of Neurochemistry, Plenum Publish. Corp., pp. 153-174.
Drejer, J., O.M. Larsson and A. Schousboe (1982). Exp. Brain Res. 47, 259-269.
Fisher, R.A., B.-E. Persson and K.R. Spring (1981). Science 214, 1357-1359.
Frangakis, M.V. and H.K. Kimelberg (1984). Neurochem. Res. 9, 1685-1694.
Gagnon, J., D. Ouimet, H. Nguyen, R. Laprade, C. LeGrimellec, S. Carriere and J. Cardinal (1982). Amer. J. Physiol. 243, F408-F415.
Geck, P., C. Pietrzyk, B.-C. Burckhardt, B. Pfeiffer and E. Heinz (1980). Biochim. Biophys. Acta 600, 432-447.
Grinstein, S., A. Rothstein, B. Sarkadi and E.W. Gelfand (1984). Am. J. Physiol. 246, C204-C215.
Hertz, L. (1979). Progress in Neurobiology, 13, Pergamon Press, Great Britain, p. 300.
Hoffman, E.D., C. Sjoholm and L.O. Simonsen (1983). J. Membrane Biol. 76, 269-280.
Kempski, O., L. Chaussy, U. Gross, M. Zimmer and A. Baethmann (1983). Brain Res. 279, 217-228.
Kempski, O., U. Gross and A. Baethmann (1982). Advances in Neurosurgery, 10, Springer-Verlag, Berlin, pp. 254-258.
Kempski, O., M. Zimmer, L. Chaussy and A. Baethmann (1984). Recent Progress in the Study and Therapy of Brain Edema, Plenum Publish. Corp. pp. 151-158.
Kimelberg, H.K. (1981). Biochim. Biophys. Acta 646, 179-184.
Kimelberg, H.K. (1983). Cell. Molec. Neurobiol. 3, 1-16.
Kimelberg, H.K., S. Biddlecome and R.S. Bourke (1979a). Brain Res. 173, 111-124.
Kimelberg, H.K., R.S. Bourke, P.E. Stieg, K.D. Barron, H. Hirata, E.W. Pelton and L.R. Nelson (1982). Head Injury: Basic and Clinical Aspects (eds. R.G. Grossman and P.L. Gildenberg), Raven Press, pp. 31-44.
Kimelberg, H.K., C. Bowman, S. Biddlecome and R.S. Bourke (1979b). Brain Res. 177, 533-330.
Kimelberg, H.K. and M.V. Frangakis (1985). Brain Res. 361, 125-134.
Kimelberg, H.K. and H. Hirata (1981). Soc. Neurosci. Abst. 7, 698.

Kimelberg, H.K. and C. Ricard (1982). Trans. Amer. Soc. Neurochem. 13, 112.
Kinne, R. (1984). Ann. N.Y. Acad. Sci. 435, 39-46.
Kletzien, R.F., M.W. Pariza, J.E. Becker and V.R. Potter (1975). Anal. Biochim. 68, 537-544.
Kregenow, F.M. (1981). Ann. Rev. Physiol. 43, 493-505.
Lowe, C. and A. Lambert (1983). Biochim. Biophys. Acta 694, 353-374.
Lund-Andersen, H. and L. Hertz (1970). Exp. Brain Res. 11, 199-212.
Maier-Hauff, K., M. Lange, L. Schurer, Ch. Guggenbichler, W. Vogt, K. Jacob and A. Baethmann (1984). Recent Progress in the Study and Therapy of Brain Edema, Plenum Publish. Corp., pp. 183-192.
Rehncrona, S., I. Rosen and B.K. Siesjo (1981). J. Cerebral Blood Flow and Metabolism, 1, 297-311.
Rosenberg, H.M., B.B. Shank and E.C. Gregg (1972). J. Cell. Physiol. 80, 23-32.
Siesjo, B.K. (1984). J. Neurosurg. 60, 883-908.
Ueberschar, S. and T. Bakker-Grunwald (1983). Biochim. Biophys. Acta 731, 243-250.
Ussing, H.H. (1982). Acta Physiol. Scand. 114, 363-369.
Van Harreveld, A. and E. Fifkova (1971). Exp. & Molec. Path. 15, 61-81.
Walz, W. and L. Hertz (1984). J. Cerebral Blood Flow and Metabolism, 4, 201-304.
Walz, W. and H.K. Kimelberg (1985). Brain Res. 340, 333-340.
Walz, W., M. Shargool and L. Hertz (1984). Neuroscience 13, 945-949.

Energy-dependent Volume Regulation in Primary Cultured Astrocytes: Relevance of Cytotoxic Edema

D. Holtzman, J. E. Olson and R. Sankar

Divisiion of Pediatric Neurology, Department of Psychiatry and Neurology, Tulane University School of Medicine, 1415 Tulane Avenue, New Orleans, Louisiana 70112, USA

ABSTRACT

In hypo-osmotic NaCl medium, primary cultured astrocytes swell initially more than predicted from osmotic theory. Cell volumes then decrease about 30%. This volume decrease results from a decrease in intracellular osmoles by a mechanism tightly coupled to (Na,K)-ATPase. Inhibition of this volume decrease by octanoic acid, but not by ammonium, may contribute to brain cellular edema in Reye's Syndrome.
Keywords: astrocytes, ATPase, edema, Reye's Syndrome

INTRODUCTION

The astrocyte is thought to be important in the control of water volume and ion concentrations in brain extracellular space (Kimelberg and Bourke, 1982). Relevant to this critical cell function, we reported that astrocytes from primary cell culture swell more in hypo-osmotic NaCl medium than predicted by osmotic theory (Olson and Holtzman, 1982). We now describe, in these cells, a regulatory volume decrease (RVD) following swelling in hypo-osmotic medium.

METHODS

Cerebral astrocyte cultures, prepared from 3 day old rat pups (Olson and Holtzman, 1980), were grown to confluence for 2-4 weeks. Cells were removed from the dishes with 0.25% trypsin plus 1 mM EDTA and suspended in iso-osmotic phosphate buffered saline (PBS) containing 137 mM NaCl. Using a Coulter Counter, cell volumes were measured 1 min, 15 min, and 30 min after resuspending in hypo-osmotic (68.5 mM NaCl) or iso-osmotic PBS (Olson and Holtzman, 1982). Metabolic inhibitors were added when the cells were resuspended. Respiratory rates were measured polarographically (Olson and Holtzman, 1982).

RESULTS AND DISCUSSION

After the initial swelling, astrocytes in hypo-osmotic PBS shrink about 40-50% more than cells suspended in iso-osmotic PBS (Olson et al., in preparation). When resuspended in iso-osmotic PBS after 30 min, cells are 10% smaller than cells continuously suspended in iso-osmotic medium. Thus, astrocyte RVD involves loss of intracellular osmoles.

Effects of metabolic inhibitors on astrocyte RVD are shown in Fig. 1. Their effects on astrocyte respiration are described separately (Olson and Holtzman, 1980; Olson et al., in preparation). Uncoupling oxidative phosphorylation (dinitrophenol 50 uM) has no effect on RVD. Inhibitors of cell respiration (antimycin-a, not shown; DNP 1 mM) and of oxidative phosphorylation (oligomycin) decrease RVD by 50%. Ouabain, an inhibitor of (Na,K)-ATPase, completely blocks RVD. Ouabain also inhibits cell respiration more in hypo-osmotic PBS than in the iso-osmotic PBS. Thus, RVD depends on energy metabolism and, more specifically, on (Na,K)-ATPase activity, as described also in lymphoblasts (Rosenberg et al., 1972). In contrast, RVD in a glioma cell line is only partly inhibited by ouabain (Kempski et al., 1984).

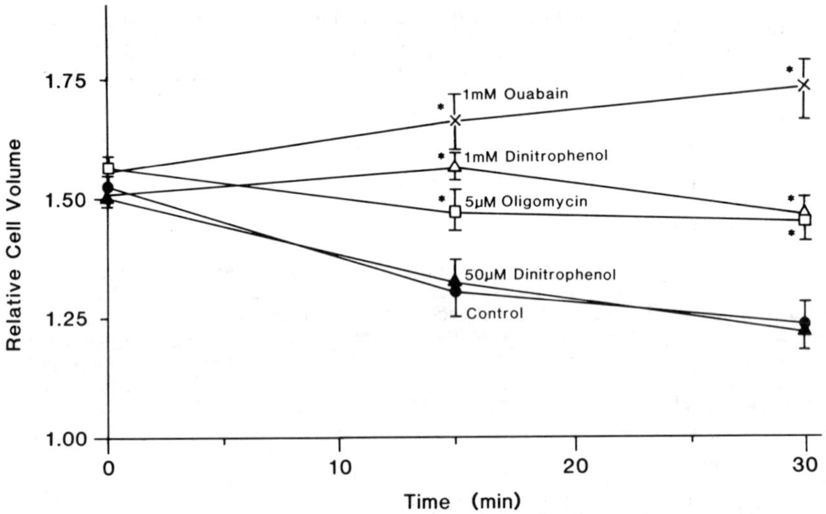

Figure 1. Volume of astrocytes suspended in hypo-osmolar PBS (mean ± SEM of at least 4 studies) expressed relative to the volume of astrocytes simultaneously suspended in iso-osmotic PBS. For control cells (closed circles), both hypo-osmotic and iso-osmotic media contained no inhibitor. For experimental cells, the hypo-osmotic medium contained a metabolic inhibitor while the iso-osmotic PBS did not. * indicates that the mean relative volume of experimental cells was significantly different ($p < .001$) from the mean relative volume of matched control cells determined by Student's t-test for paired samples.

We also measured the effects on RVD and on cell respiration of agents thought to be important in the pathogenesis of brain cellular edema in Reye's Syndrome (Table 1). In this acute childhood illness, astrocyte swelling has been attributed to elevated blood levels of octanoate, ammonia or salicylates (Delong and Glick, 1982). Octanoate, like ouabain, completely inhibits RVD in astrocytes. Neither very large concentrations of NH_4Cl nor salicylate affect RVD. Salicylate (5 mM) is a very effective uncoupler of cellular oxidative phosphorylation (oligomycin-inhibited respiration is stimulated 420 ± 58%). In contrast, octanoate (.4 mM) both partially reverses oligomycin inhibition and inhibits basal respiration. The observation that octanoate completely blocks RVD, unlike the limited effects seen with other inhibitors or uncouplers of cell respiration, suggests that inhibition of RVD may be due to inhibition of (Na,K)-ATPase by octanoate (Dahl, 1968; Trauner, 1980). The inhibition of astrocyte

RVD supports the proposed role of octanoate, but not salicylate or ammonia, in the pathogenesis of brain cellular edema in Reye's Syndrome.

TABLE 1. Effects of agents, which are associated with Reye's Syndrome, on astrocyte RVD. Relative cell volume is defined in Fig. 1.

Agent		Relative Cell Volumes	
		0 min	30 min
None		1.45 ± 0.03 (21)[a]	1.24 ± 0.03 (21)
Octanoate	0.1 mM	1.59 ± 0.03 (5)	1.22 ± 0.03 (5)
	1.0 mM	1.30 ± 0.09 (14)	1.36 ± 0.10 (11)[b]
	4.0 mM	1.45 ± 0.03 (6)	1.82 ± 0.08 (6)[b]
Salicylate	5 mM	1.37 ± 0.03 (6)	1.26 ± 0.04 (8)
	10 mM	1.38 ± 0.03 (6)	1.29 ± 0.04 (5)
NH_4Cl	5 mM	1.44 ± 0.03 (4)	1.06 ± 0.05 (4)

[a] Values in parentheses are the number of experiments comprising each mean value.

[b] Significantly greater than the mean of matched control cells ($p < .001$).

The responses of primary cultured astrocyte to decreased osmolarity support the proposal that this cell, *in vivo*, is important in control of brain extracellular water. The astrocyte swells more than predicted by osmotic theory when placed in a NaCl-containing medium of low osmolarity (Olson and Holtzman, 1982). This would facilitate water uptake under conditions of decreased extracellular Na. We now describe a regulatory volume decrease in which water and solute are transported out of cells. If, *in situ*, these processes result in net transport of water and solute from extracellular space to capillary surface, the astrocyte could provide the mechanism for energy-dependent osmoregulation in the brain. The proposed role for inhibition of astrocyte RVD and (Na,K)-ATPase activity in the pathogenesis of cerebral cellular edema must be studied further *in vitro*, in animal models, and in man.

ACKNOWLEDGEMENT. This work was supported by the Robert Katz Research Foundation and the Research Development Fund of the Dept. of Psychiatry and Neurology.

REFERENCES

1. Dahl DR (1968). *J Neurochem* 15, 815-820.
2. DeLong GR and Glick (1982). *Pediatrics* 69, 53-63.
3. Kempski O, Zimmer M, Chaussy L, and Baethmann A (1984). In Go KG and Baethmann E (eds): *Recent Progress in the Study and Therapy of Brain Edema*. New York, Plenum Press. pp. 151-158.
4. Kimelberg HK and Bourke AS (1982). In Lajtha A (ed.) *Handbook of Neurochemistry*, Vol. 1 (second edition). New York, Plenum Press. pp. 31-67.
5. Olson JE and Holtzman D (1980). *J Neurosci Res* 5, 497-506.
6. Olson J and Holtzman D (1982). *Brain Res* 246, 273-279.
7. Rosenberg HM, Shank BB, and Gregg EC (1972). *J Gen Physiol* 80, 23-32.
8. Trauner DA (1980). *Pediatr Res* 14, 844-845.

pH Control of K^+ Channels in Astrocytes

W. Walz

Department of Physiology, College of Medicine, University of Saskatchewan, Saskatoon, Saskatchewan, S7N 0W0 Canada

Normal and abnormal neuronal activity in the brain cortex is usually followed by a long-lasting acidic phase of the extracellular space (Kraig, et al., 1983). During normal neuronal activity the interstitial pH of 7.4 changed to 7.0 and this pH decrease was related to frequency, intensity and duration of the stimulus used to excite neurons. During ischemia and spreading depression the acidic pH shift reached values of 6.6.

Glial cells are known to remove excess extracellular K^+ by passive currents through their K^+ channels (spatial buffering) and by carrier-mediated uptake through Na^+-K^+-ATPase and a furosemide-sensitive KCl carrier. Mouse astrocytes in primary cultures were used to investigate the influence of different pH changes on both K^+ clearance pathways. Radiotracer methodology was used to analyze these changes. Prior to the experiments the cells were incubated in Hepes buffered salt solution (pH 7.4). In this situation 95% of the influx rates are mediated by barium (50 µM)-sensitive K^+ channels and the remaining 5% by ion carrier systems. Exposure of the cells to Hepes BSS with a pH of 6.6 for 30 min blocked the K^+ influx through the barium-sensitive K^+ channels by 41%. The barium-insensitive carrier-mediated influx was not influenced in this experimental situation. An analysis of the dependence of this effect upon the extracellular hydrogen ion concentration revealed the following: A decrease of the pH from 7.4 to 7.0 did not change the unidirectional K^+ flux rate, a further drop to 6.8 blocked about 15% of the K^+ channels. A pH of 6.6 had an even more drastic effect as described above (41% inhibition). In another experiment the time necessary for extracellular pH changes to act before exerting their inhibitory action of the permeability of the K^+ channel pathway was investigated. If the cells are exposed to a BSS with pH 6.6, 4-5 min are necessary to exert the maximum effect on the K^+ channels. A shorter exposure time has no effect at all, a longer exposure time (up to 30 min) did not increase or weaken the inhibition of channel-mediated K^+ influx. When astrocytes are exposed to a sudden increase of the external K^+ concentration from 3 to 12 mM they accumulate intracellular K^+ ions by about 50% of their normal content within 50s. This increase was not changed in Hepes BSS as compared with MEM medium, although the time course was somewhat increased in BSS. In BSS with a pH 6.6 (30 min exposure) the net accumulation was only 29% of the controls and this block of net accumulation was not based on slowing down of the time needed to accumulate the K^+.

These results demonstrate a biphasic effect of pathological pH changes. The K^+ channels of glial cells are by 41% closed and the net accumulation of K^+ is

Advances in the Biosciences Vol. 61

inhibited by 71%. The ion carriers are obviously only inhibited when they are stimulated by external K^+ increases, but not in equilibrium. The block of the K^+ channels is only relevant during pathological pH decreases, physiological pH decreases do not change the K^+ channel permeability. From the time needed to close the pH-sensitive K^+ channels (4-5 min) after a change in pH, it can be concluded that an intracellular and not an extracellular acidification is necessary, and that the hydrogen ion mediated block is from an intracellular site of the channel. Amiloride (1 mM), an inhibitor of the Na^+-H^+ exchange, lowers the intracellular pH of astrocytes (Kimelberg and Ricard, 1982). Exposure of the cells to amiloride for different times suppressed the barium-sensitive K^+ flux independent of the time exposure. 20 min of amiloride application resulted in a 53% inhibition of the channel-mediated K^+ flux. This inhibitory amiloride-effect is still present in addition to the channel block directly by solutions with low pH, however it is reduced in amplitude. This could be a sign that such a Na^+-H^+ exchange is involved in pH regulation. Low internal pH affects K^+ clearance through channels and carriers.

These results might be helpful in understanding the spread of abnormal neuronal activity which involves excess extracellular K^+ like in spreading depression and epileptic seizures, where conventional synaptic transmission and voltage-dependent neuronal Na^+ channels are not involved.

ACKNOWLEDGEMENT

The author is a scholar of the Medical Research Council of Canada. Operating funds were provided by the MRC of Canada.

REFERENCES

Kimelberg, H.K. and C. Ricard (1982). Trans. Am. Soc. Neurochem. 13, 112.

Kraig, R.P., C.R. Ferreira-Filho and C. Nicholson (1983). J. Neurophysiol., 49, 831-850.

Plasminogen Activators in Brain Development

I. Selak, B. Rogister, P. P. Lefebvre, P. Leprince and G. Moonen

Secteur de Neurologie, Institut de Médicine, et Département de Cytologie et Histologie, Université de Liège, Rue de Pitteurs 20, 4030 Liège, Belgium

ABSTRACT

In this communication, we report the identification of plasminogen activators in brain tissue : the urokinase-type and the tissue-type. We show that these activators are involved in at least two steps of neuro-ontogenesis, neuronal migration and astrocytic proliferation. About neuronal migration, we also demonstrate that a glycoprotein of the extracellular matrix, laminin, is able to alter the migratory behaviour of cultured neurons and that it is synthetized and secreted by astrocytes.

KEYWORDS

Plasminogen activators; urokinase; neuronal migration; astrocytes proliferation; laminin.

INTRODUCTION

Neutral extracellular proteolysis has been since long time associated with highly plastic events such as cell migration or tissue remodeling both in normal and tumoral material. For instance ovulation has been shown to involve a proteolytic step which is under hormonal control (Beers *et al.*, 1975). Activated macrophages are known to release plasminogen activators and that release is inhibited by corticosteroids (Vassali *et al.*, 1976). Cancer cells are known to release plasminogen activators activity and this has been related to the invasion and metastasis processes. Nervous system ontogenesis involves a highly ordered sequence of events including cell proliferation, migration, process formation... The purpose of the present communication is to summarize the data obtained in our laboratory concerning the possible role of the plasminogen activators system in neuronal migration and astrocytes proliferation.

I. CHARACTERIZATION OF PLASMINOGEN ACTIVATORS IN THE NERVOUS SYSTEM.

The plasminogen activators have been characterized using zymography after SDS-PAGE of medium conditioned by developing cerebellum. The plasminogen activators activity resolved into two bands with apparent molecular weight of 48 and 75 Kd (Fig. 1). Notice that no proteolytic activity has been observed in the absence of plasminogen in the casein underlay. Those bands were further characterized using immunological methods and could be related to the urokinase (UK)-type and

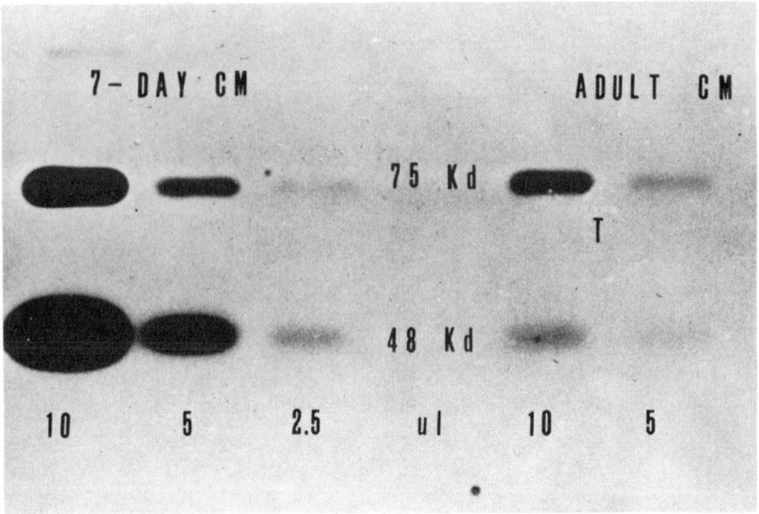

Fig. 1 SDS-PAGE of medium conditioned by 7-day old or adult rat cerebellum followed by casein underlay (Moonen et al., 1985). Equivalent quantity of conditioned material has been utilized in both conditions.

the tissue-type plasminogen activator (TPA) (Moonen et al., 1985). Interestingly enough much lower activity could be found in media conditioned by adult material. It is important to point out that the conditioning material is heterogeneous containing neurons, glial cells, vascular and mesenchymal structures. However, using astrocytes primary cultures (more than 90 % of the cells are GFA-P positive astrocytes) only the UK-type was found (Leprince et al., in preparation).

II. NEURONAL MIGRATION.

Neuronal migration in the developing cerebellum follows an outside-inside radial pathway (Rakic, 1972) involving a close relationship between the Bergmann fibers and the migrating neurons. The molecular mechanism underlying such interactions is largely unknown. We have previously reported that plasminogen activator activity is involved since proteases inhibitors can block neuronal migration (Moonen et al., 1982) (Fig. 2). However, the substrate on which such proteases could act is unknown as well as their site of action. That led us to investigate the possible involvment of a well known glycoprotein, laminin. Indeed; it has been shown that laminin can modulate neuritic elongation and is released by several nervous system derived cell types (Baron Van Evercooren et al., 1982; Carbonetto, 1984; Lander et al., 1985; Manthorpe et al., 1983; Sanes, 1985).

When newborn rat cerebellar microexplants are cultured in serum-free conditions on a polyornithine substratum, a massive outgrowth of neural processes is observed with few cells coming out the microexplants (Fig. 3A). On laminin precoated tissue culture dishes on the other hand, the culture appearance is clearly different (Fig. 3B). Indeed, a massive migration of phase bright, process-bearing cell occurs, while few flat cells can be seen. No such effect could be obtained using either fibronectin or type IV collagen which are other well known constituent of the extracellular matrix. A more detailed view of these events can be obtained when the glial fibrillary acidic protein (GFA-P) is

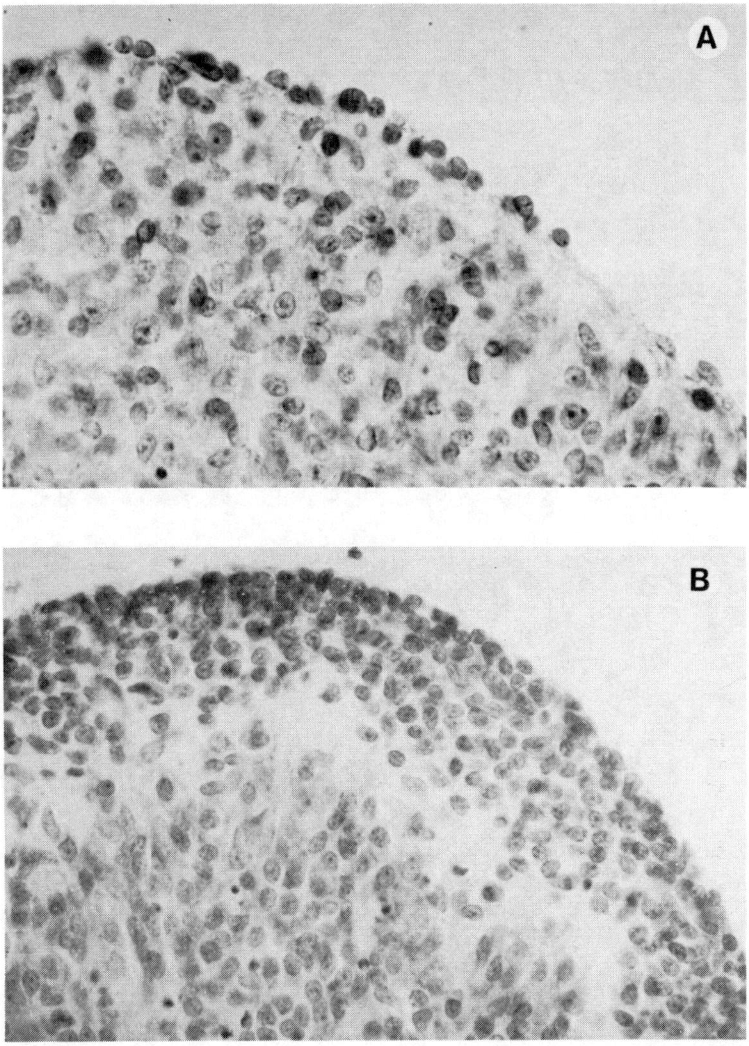

Fig. 2 Sections through 7-day old rat paraflocculus cultured for 48 hrs on a giratory shaker (Moonen et al., 1982) in MEM medium supplemented with 10 % horse serum (A) or in the presence of diisopropylfluorophosphate (DFP) 10^{-4} M (B). (Magn. x 625).

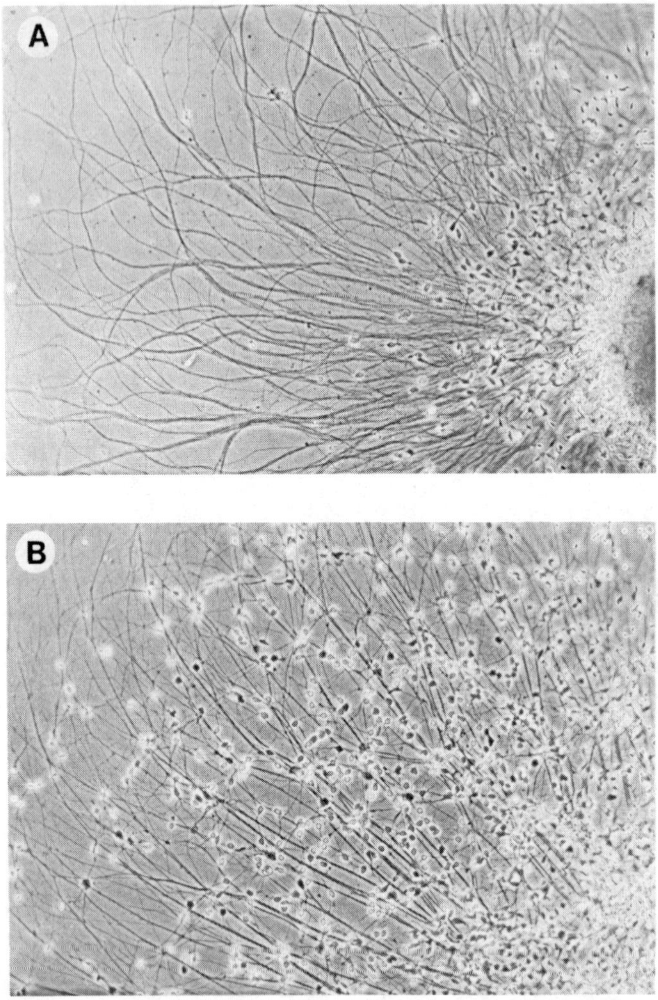

Fig. 3 Phase contrast micrographs of newborn rat cerebellar microexplants cultured for 4 days on a polyornithine substratum (A) or on laminin-coated dishes (B) in serum-free conditions (Selak et al., 1985).
(Magn. x 150).

revealed using immunocytochemistry on cultured microexplants. In the control condition (on polyornithine) GFA-P positive processes can be identified with a radial distribution around the microexplants (Fig. 4A). In those conditions, the phase-bright, process-bearing cells, identified as neurons (after neurofilament immunostaining) are in close relationship with the GFA-P positive processes. On laminin (Fig. 4B), on the other hand, while a few neuronal cells are still in the vicinity of astroglial processes, many of them are isolated and migrate randomly directly on the substratum.

Furthermore, we have been able to show that laminin is synthetized and released by newborn rat cerebellar microexplants and that can be related to astrocytic component of the culture since more laminin is synthesized if astrocytic proliferation is induced in the microexplants and since it is found in the medium and cell extracts of primary astrocytes cultures (Selak et al., 1985). That effect of laminin on the migratory behaviour of neurons (a so far unreported effect of that glycoprotein) and the demonstration of laminin synthesis in the developing cerebellum strongly support the idea of a key role of laminin in the control of that developmental event. The relationship between laminin and plasminogen activators, if any, is not known. The tissue-type activator but not the UK-type has been reported to have an affinity for laminin. The Ng cell adhesion molecules (Ng-CAM) have also been suggested to be required for granule cells migration along radial glia in the developing cerebellum since that migration is reduced by mono or polyclonal antibodies to L 1 (a glycoprotein initially identified with monoclonal antibodies) and later shown to have common antigenic determinant to Ng-CAM. The control of the fate of a post-migratory neurons leaving the external granular layer and going through the molecular and the Purkinje cell layer to the internal granular layer appears then rather complex but clarifying those mechanisms will undoubtly help us to understand how cytoarchitectonic is build and possibly impaired (such as in the mice mutants).

III. GLIAL PROLIFERATION.

Proliferation of astrocytes in the developing brain occur mainly after birth. The study of glial proliferation led to the identification of several mitogens such as the astrocyte growth factor (Pettman et al., 1980), the glial growth factor (Brckes et al., 1980) and the glia maturation factor (Lim and O'Connel, 1982). We have recently reported that medium conditioned by microexplants of either newborn rat cerebral cortex or cerebellum is mitogenic for astrocytes using a serum-free in vitro assay (Moonen et al., 1985). That mitogenic effect could be related to the plasminogen activators of those media since it is suppressed by inhibitors and mimicked by a semi-purified preparation of urokinase (Fig. 5).

It is worth mentioning again that no evidence could be obtained for an endogenous plasmin generating system and that the mitogenic effect of human urokinase is observed in the absence of added plasmin generating system or in the presence of plasmin inhibitors (Moonen et al., 1985). All these data were obtained using an heterogeneous (neurons plus glia) conditioning tissue as well as an heterogeneous bio-assay (microexplants). More recently, we have been able to reproduce a similar mitogenic effect of urokinase using primary cultures of newborn rat cerebral cortex as well as purified Schwann cells. Since it appears that the medium conditioned by astrocytes themselves contains urokinase it is possible that urokinase could act as an autocrine growth promoting substance.

The mechanism of action of urokinase remains unknown although different sets of data are relevant to that question. The effect of urokinase and insulin are additive which suggest that the mechanism of action of urokinase is different from that of insulin and related insulin-like growth factors. Secondly, the urokinase mitogenic effect seems to involve a binding of urokinase to the cells (Vassali et al., 1985) since the 33 Kd form, which lacks the cell binding site but not the proteolytic site, is uneffective. Finally, the urokinase release

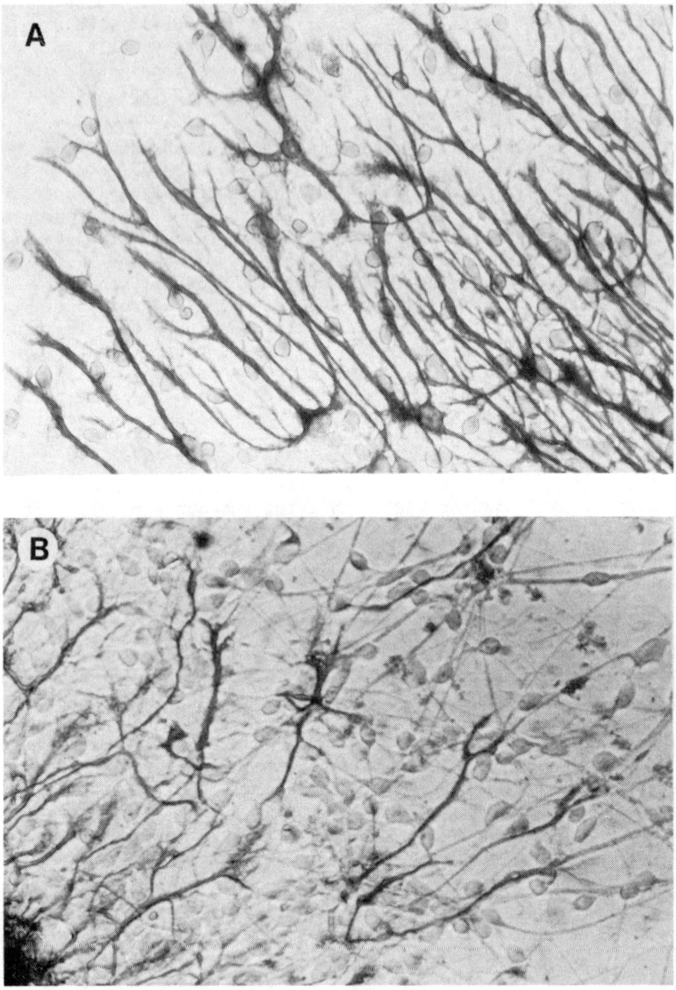

Fig. 4 Peroxydase-antiperoxydase GFA-P immunostaining of newborn rat cerebellar microexplants cultured for 4 days on either polyornithine (A) or laminin substratum (B). (Magn. x 225).

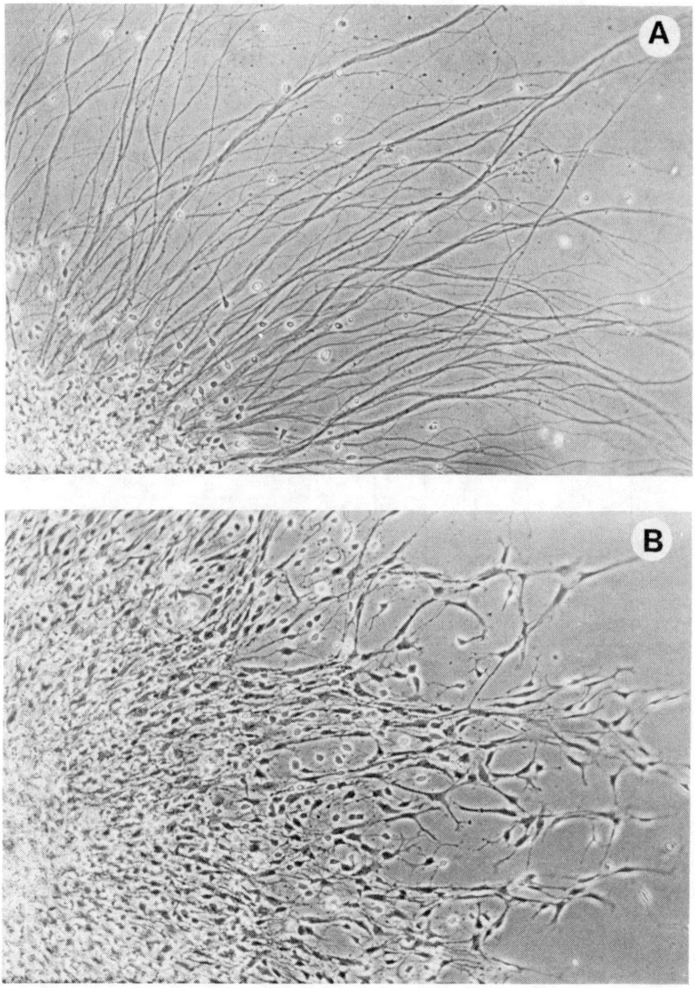

Fig. 5 Phase contrast micrographs of newborn rat cerebellar microexplants cultured for 4 days on polyornithin in serum-free conditions with only insulin (A) or in the presence of insulin and semi-purified preparation of human urokinase (15 U/ml) (Moonen et al., 1985). (Magn. x 150).

could well be one of the consequences of a mitogenic stimulation since it appears to be modulated by the proliferation status of the culture (Fig. 6).

DISCUSSION

Although not specific for the nervous system, the plasminogen activators appear to play a significant developmental role in the nervous system that we are only at the beginning of understanding. The modulation of such proteolytic activity is likely to be complex, involving control of release, possible control of pro-enzyme to enzyme conversion and existence of endogenous inhibitors which have been reported both for the nervous system and other systems (Knauer and Cunningham, 1984; Grau-Wagemans et al., 1984; Low et al., 1982). Since those proteases appear to be involved both in cellular proliferation and migration the question arises of a possible biological step that would be common to both events. Our current hypothesis is that that common step could be the modulation of cell to substrate or cell to cell interaction.

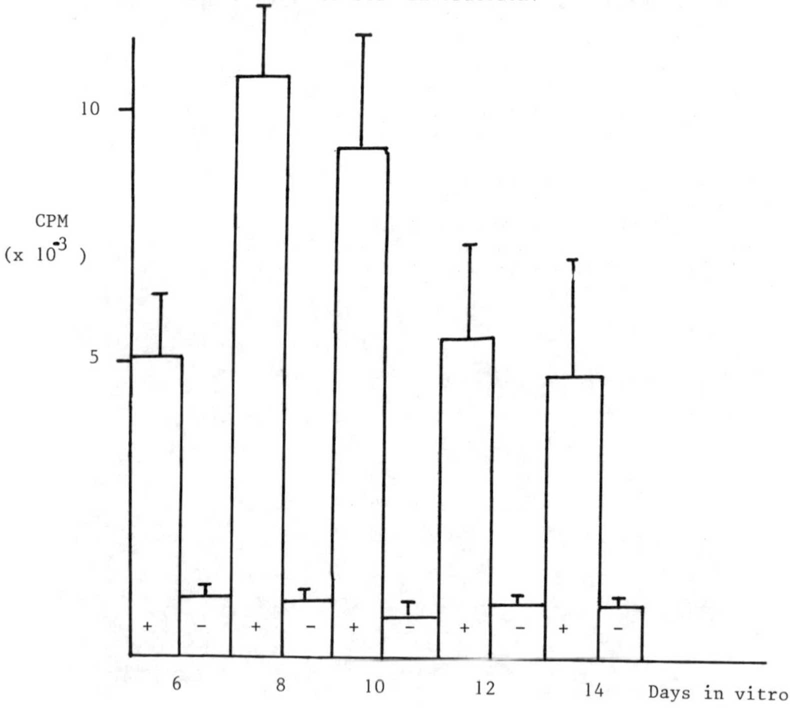

Fig. 6 Plasminogen activators activity (CPM) produced by primary cultures of newborn rat cerebral cortex astrocytes as function of time of culture. Plasminogen activators are measured using a ^{125}I-fibrin assay in the presence (+) or absence (-) of plasminogen.

REFERENCES

Baron van Evercooren, A., H.K. Kleinman, S. Ohno, P. Marangos, J.P. Schwartz, and M.E. Dubois-Dalcq (1982). J.Neurosc.Res. 8, 179-194.

Beers, W., S. Stricklan, and E. Reich (1975). Cell 6, 387-394.

Brockes, J.P., G.E. Lemke, and D.R. Balzer (1980). J.Biol.Chem. 255, 8374-8377.

Carbonetto, S. (1984). Trends in Neurosci. 10, 382-387.

Grau-Wagemans, M.-P., I. Selak, P.P. Lefebvre, and G. Moonen (1984). Develop. Brain Res. 15, 11-19.

Knauer, D.J. and D.D. Cunningham (1984). Trends in Bioch.Sci., 231-233.

Lander, A.D., D.K. Fujii, and L.F. Reichardt (1985). Proc.Natl Acad.Sci. U.S.A. 82, 2183-2187.

Lim, R. and M.M. O'Connell (1982). Develop. Brain Res. 5, 29-39.

Low, D.A., R.W. Scott, J.B. Baker, and D.D. Cunningham (1982). Nature 298, 476-478.

Manthorpe, M., E. Engvall, E. Ruoslathi, F.M. Longo, G. Davis, and S. Varon (1983). J.Cell Biol. 97, 1882-1890.

Moonen, G., M.-P. Grau-Wagemans, and I. Selak (1982). Nature 298, 753-755.

Moonen, G., M.-P. Grau-Wagemans, I. Selak, P.P. Lefebvre, B. Rogister, J.D. Vassalli, and D. Belin (1985). Develop. Brain Res. 20, 41-48.

Pettman, B., M. Sensenbrenner, and G. Labourdette (1980). FEBS Lett. 118, 195-199.

Rakic, P. (1972). J.Comp.Neurol. 145, 61-84.

Sanes, J.R. (1985). Nature 315, 714-715.

Selak, I., J.M. Foidart, and G. Moonen (1985). Submitted.

Vassalli, J.D., J. Hamilton, and E. Reich (1976). Cell 38, 271-281.

Vassalli, J.D., D. Baccino, and D. Belin (1985). J.Cell Biol. 100, 86-92.

Astroglia Production of Neuronotrophic and Neurite-promoting Agents

S. Varon, J. S. Rudge, G. E. Davis, S. D. Skaper and M. Manthorpe

Department of Biology and School of Medicine, University of California, San Diego, La Jolla, California 92093, USA

Neuroglial cells, during development as well as in response to traumatic or pathological insults in the adult nervous system, are in a position to present their neighboring neurons with critical supplies of extrinsic agents modulating neuronal survival and function, axonal extension, and repair processes (Varon and Somjen, 1979; Manthorpe et al., 1985a). Cultures of normal Schwann cells and their tumoral counterparts present and/or release at least two classes of agents beneficial to neurons from peripheral (PNS) and central (CNS) nervous systems (reviewed in Varon and Manthorpe, 1982).

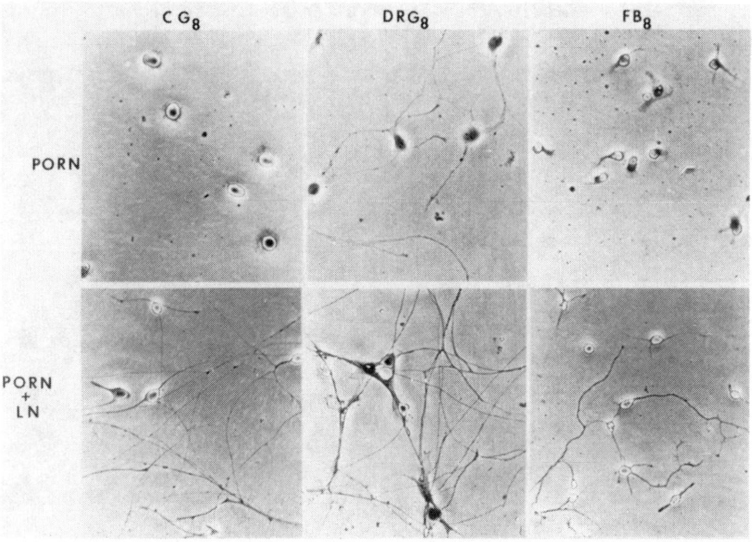

Fig. 1. Neuritic response to bound laminin (PORN+LN) by trophic supported PNS and CNS neurons.

Members of one class, designated as neuronotrophic agents, support neuronal survival and general growth capability while those of the other class, designated as neurite-promoting agents, specifically stimulate the outgrowth of neuronal processes. Both types of agents can be readily detected and measured by use of dissociated neuronal cell cultures, illustrated in Figure 1 (Varon et al., 1983). In turn, purified glial cell cultures can be obtained from both PNS and CNS tissues, and used to investigate the action of glia-modulating agents capable of altering morphological, biochemical and secretory properties of the glial cells themselves (see Varon and Manthorpe, 1982; Manthorpe et al., 1985a). We shall review here the ability of cultured astroglial cells to release into their medium both neuronotrophic and neurite promoting agents.

1. The astroglial microculture system

Thousands of replicate microcultures of rat astroglial cells can be set up and maintained with a chemically defined medium and a defined substratum (Rudge et al., 1985). Neonatal rat cerebra are dissociated and the cells grown for 10 days on culture plastic in serum-supplemented Eagle's Basal Medium (EBM, enriched with bicarbonate, glucose and glutamine). This permits a reproducible collection of 6 x 10^6 primary cells/cerebrum, more than 95% of which display the glial fibrillary acidic protein (GFAP) antigen. The harvested cells are re-seeded at 16,000 cells per 6 mm microwell on a polyornithine/fibronectin substratum in serum-free EBM. One can thus generate nearly 400 microcultures from one original cerebrum. Of the 16,000 cells seeded per well, about 12,000 attach in 2 hr and form a nearly confluent, flat cell culture as shown in Figure 2,A. Nearly all of these flat cells are GFAP-positive and can be made to assume a stellate morphology (Fig. 2,B) by specific treatments which raise intracellular cyclic AMP levels (Manthorpe et al., 1979; Magistretti et al., 1983; Skaper et al., 1985). The secondary microcultures can be maintained at a constant cell number for 7 days without re-feeding or for weeks with medium changes every 3 days -- and can be retriggered into proliferation by presentation of serum or other agents (Rudge et al., 1985; Katoh-Semba et al., 1985). All cultures receive a medium change 2 hr after seeding in order to minimize serum or cell lysate carryovers from the primary cell harvest.

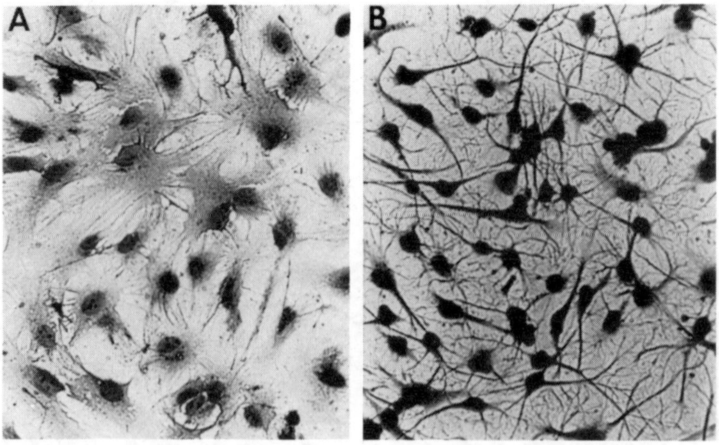

Fig. 2. Rat astroglial cells cultured without (A) or with (B) 1 mM dibutyryl cAMP.

2. Output of neuroactive substances by untreated astroglial microcultures

Media conditioned by individual microcultures for various times after the 2 hr attachment period can be collected and tested for neuronotrophic and neurite-promoting activities (Rudge et al., 1985; Manthorpe et al., 1985a). The test chick neurons used in most experiments (cf. Fig. 1) are from: (1) E8 (embryonic day 8) dorsal root ganglia for the measurement of Nerve Growth Factor (NGF); (2) E8 ciliary ganglia for Ciliary Neuronotrophic Factor (CNTF) activity; E8 forebrain for trophic activity addressing CNS neurons (CNS-TA); and E8 ciliary ganglia again for determinations of substratum-binding Neurite Promoting Factors (NPFs).

Figure 3 shows that the astroglial microcultures release at least four distinct types of agents addressing nerve cells. The four activities accumulate and/or decline in the astroglia-conditioned media with reproducible but distinct time patterns. NGF increases gradually to day 3 and declines thereafter. CNTF is already highest at the earliest time tested, maintains a plateau for about 3 days, and then declines (more slowly than does NGF). CNS trophic activity reaches a plateau already by day 2 and maintains it at least until day 5. NPF activity accumulates gradually and linearly with time over (and beyond) the 5 day period. These temporal differences must reflect independent regulations of production, release and/or stability of the corresponding molecules. Future work will aim at i) analyzing the events underlying each time pattern, ii) using the microcultures to screen for biochemical or cellular agents that would enhance or inhibit the astroglial output of any one of the four activities, and iii) investigating further such "glia-modulating" agent(s) to identify which output step (synthesis, secretion, stabilization) is altered.

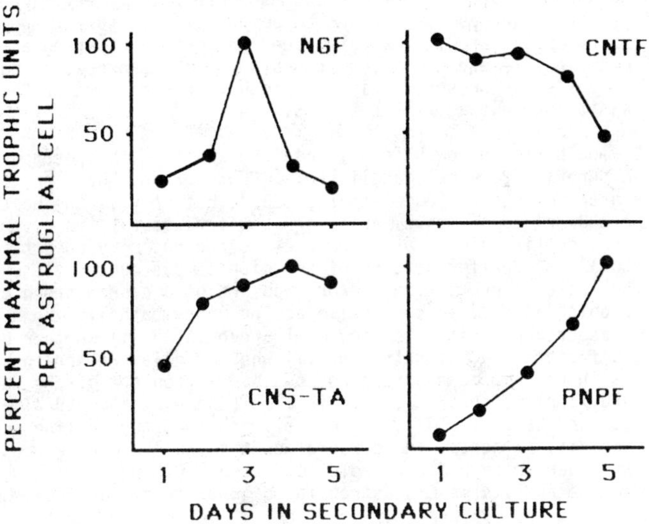

Fig. 3. Astroglial output of neuronotrophic and neurite-promoting agents.

3. Implications of an astroglial involvement in neuroactive agents

The preceding observations demonstrate an ability of CNS astroglial cells to generate trophic and neurite-promoting agents for a variety of neurons. It is intriguing that at least two such agents (NGF, CNTF) are traditionally known to address PNS rather than CNS neurons. Two possibilities deserve attention: (1) PNS-addressing agents may also be relevant to CNS neurons, and/or (2) both PNS and CNS glia can produce different agents for either or both central and peripheral neurons, but will do so in vivo only under selective influences from their microenvironment. These concepts, which are increasingly attracting the interest of neuroscientists, have been reviewed recently (Varon et al., 1984a; Varon 1985). Only some aspects are selected for discussion here.

a. Nerve Growth Factor. NGF is known to act mainly on dorsal root and sympathetic ganglionic neurons. Dependence on NGF for neuronal survival, (the basis for in vitro neuronotrophic assays) declines with embryonic age of chick dorsal root ganglion neurons, and other neuronotrophic factors such as CNTF (Skaper and Varon, 1985) or brain-derived growth factor (Barde et al., 1982) can increasingly replace NGF in that role. E8 chick dorsal root ganglionic neurons are the test neurons of choice for NGF, since NGF is the only known trophic factor which can support them. Further verification of an NGF identity can be obtained by ascertaining whether the unknown activity is blocked or immunosequestered by antibodies from the appropriate animal species -- subject to adequate cross-species immunoreactivity.

Other investigators have recognized the production of NGF by normal (e.g. Varon et al., 1974; Lindsay et al., 1982) and tumoral (e.g. Schwartz et al., 1977; Perez-Polo et al., 1977) glial cell cultures, and its modulation by specific chemicals. In two such studies (Varon et al., 1974; Lindsay et al., 1982), the test neurons were actually seeded in the concurrent presence of glial cells such that neuron-glia interactions could occur. Very recent studies have demonstrated an ability of NGF to perform from a surface-anchored state (Gundersen, 1985), as CNTF has also been shown to do (Carnow et al., 1985; see further on). These new observations raise the provocative possibility that glial cells may present their NGF product on their own surface rather than releasing it into extracellular spaces.

In vivo, a "sympathetic ingrowth" can invade the adult rat hippocampus from the rat's own superior cervical ganglia (e.g. Crutcher and Chandler, 1984) or from ganglia grafted near it (Gage et al., 1984), but only if the host hippocampus has been deprived of its fimbria-fornix afferents (including the septum-generated cholinergic input). The view is increasingly entertained (see also further on) that deafferentation of the hippocampus causes elements of this tissue to increase production and/or exposure of a neuronotrophic factor exerting a "tropic" as well as supportive action on sympathetic axons. It remains to be ascertained, if this view will prove valid, i) whether or not the postulated factor is NGF itself, and ii) whether glial rather than neuronal elements in the hippocampus are responsible for output of the factor. Also recent are findings that NGF, both in vivo (Ghahn et al., 1983) and in vitro (Honegger and Lenoir, 1982), selectively enhances the choline acetyltransferase activity of septal and cerebral cortex cholinergic neurons and, therefore, may indeed play a direct role for certain CNS neurons -- which may add meaning to the ability of CNS astroglia to generate and deliver NGF.

b. Ciliary Neuronotrophic Factors. CNTFs have been purified from the intrinsic muscle tissue of chick embryo eye (Barbin et al., 1984) and more recently from the sciatic nerve of adult rats (Manthorpe et al., 1985b). Both chick eye and rat nerve CNTFs also support the survival of dorsal root and sympathetic ganglionic neurons (at embryonic ages slightly older than those already responsive to NGF). In vivo CNS studies have revealed that CNTF activity i) occurs in rat CNS tissues at relatively low levels, and ii) increases several-fold in both tissue and extracellular fluids in the vicinity of either a mechanical or a chemical (kainic acid) lesion (Manthorpe et al., 1983a; Nieto-Sampedro et al., 1983). The temporal increase of CNTF activity in injured entorhinal cortex correlates closely with the successful survival of cholinergic neurons in embryonic rat CNS grafts placed within the lesion. The high degree of reactive gliosis which characterizes both mechanical and chemical lesions has encouraged the speculation that astroglial cells of the adult host CNS are responsible for the trauma-triggered accumulation of CNTF activity.

A procedure has been developed in this laboratory which allows the visualization of CNTF polypeptide bands without prior purification (Carnow et al., 1985). E8 chick ciliary ganglionic neurons, seeded on nitrocellulose paper, will survive if CNTF is present in the culture medium, and they can be visualized 24 hr later by their ability to take up and reduce a vital dye. Neuronal survival will occur equally well if CNTF is present in a nitrocellulose-bound form rather than in the medium, suggesting that trophic factors may indeed work when anchored to a surface -- as already noted for NGF -- and survival will be restricted to the nitrocellulose regions carrying the anchored CNTF. Crude CNTF source materials (e.g. tissue extracts) can be submitted to sodium dodecyl sulfate polyacrylamide gel electrophoresis (SDS-PAGE), the polypeptide bands "blotted" from the electrophoretic gel to nitrocellulose, and the CNTF band position "probed" by culturing ciliary ganglionic neurons on the nitrocellulose blot in a CNTF-free medium. After dye staining and fixation, the blot can be scanned longitudinally to recognize the neuronal band, with neurons counted across the band width for quantitative CNTF evaluations. Using this blot-and-culture technique, chick eye, rat nerve and rat CNS wound fluid were shown to contain CNTFs with molecular weights of approximately 21 and 24 kD, respectively (Carnow et al., 1985). For the chick eye and rat nerve CNTFs, these same molecular weights correspond to those of the purified factors (Manthorpe et al., 1985c). An identical molecular weight of 24,000 characterizes the CNTF extracted from rat astroglial cell microcultures (Rudge et al., 1986). It appears, therefore, that the rat brain CNTF may well be a product of CNS glia.

c. Low Molecular Weight (LMW) CNS-addressing neuronotrophic agents. CNS neurons, seeded at low densities to avoid cooperative support within the culture itself, can be made to survive either with LMW trophic agents found in cell conditioned media or with trophic proteins present in CNS wound fluid and tissue extracts (Manthorpe et al., 1982; Varon et al., 1984b). The CNS-addressing trophic activity which accumulates in astroglia-conditioned medium (Fig. 3, CNS-TA panel) is not destroyed by proteolytic, heat or extreme pH treatments and filters though ultramembranes with a 1,000 dalton cut-off -- thereby readily separating from the other, protein-associated neuronotrophic activities of the astroglial medium. The main LMW trophic agent has been found to be pyruvate (Selak et al., 1985), which can be replaced by α-ketoglutarate or oxaloacetate (but not other Krebs cycle intermediates) and also by amino acids capable of generating the three competent metabolites by transamination with a concurrently supplied α-keto acid (Facci et al., 1985). Many PNS neurons also require exogenous pyruvate (or its equivalent) for their survival in vitro, in addition to their own neuronotrophic factor. The inability of these CNS and PNS neurons to survive with the glucose present in the culture medium suggests that their glucose utilization is faulty in some yet unidentified manner, and that one contribution by astroglial cells may be directed to compensate for such a deficiency. It is tempting to speculate (cf. Varon, 1985) that, even in vivo, adult neurons may express this deficiency under functional overload or pathological stress, and call upon the neighboring astroglia for mobilization and delivery of pyruvate-like molecules -- very much in line with Hyden's (1967) concept of a glia-neuron metabolic unit. These concepts gain further strength from the ability of some neuron-released molecules -- such as K^+, neurotransmitters and certain peptides -- to stimulate glycogen breakdown and mobilization in astroglial cells, thereby closing a trophic/metabolic feedback loop between neurons and glia (Magistretti et al., 1983).

A clue in understanding the neuronal needs met by exogenous pyruvate has been provided by very recent studies on a red blood cell-derived trophic protein for CNS neurons (Longo et al., 1984; Williams et al., 1985). The trophic protein from human erythocytes has been purified, compared with identified red blood cell cytosolic proteins, and found to be a catalase (Walicke et al., 1985). Catalases, other peroxidases and anti-oxidant substances (e.g. vitamin E) are all able to maintain neurons in these test cultures even in the absence of pyruvate. It appears, therefore, that both catalase and pyruvate may provide the same ultimate benefits to the neurons: catalase by "scavenging" hydrogen peroxide and oxygen radicals outside the neuron, and pyruvate by reducing endogenously produced oxygen radicals and organic peroxides. Free radicals have long been suspected to be a main source of cell damage and death, particularly in the aging process (cf. Armstrong, 1984), and the observations just reviewed propose yet another role for glial cells, i.e. anti-peroxidation protection to the neighboring neurons.

d. Anchorage-dependent Neurite-Promoting Factors. These are large macromolecules which bind to the culture substratum and confer to the substratum the ability to promote neuritic growth from peripheral and central neurons. Because of their particular affinity for polycationic surfaces (e.g. polyornithine-coated plastic), these factors were initially designated as polyornithine-binding neurite promoting factors, or PNPFs (Adler and Varon, 1980). These neurite-promoting factors are found in many conditioned media (Collins, 1978; Adler et al., 1981a), including those from Schwann and astroglial cells, are transferred to the surrounding substratum by ganglionic explants (Adler and Varon, 1981a) and appear to occur also on the surface of glial cells in culture (Adler and Varon, 1981b). Because of these and other properties, it was speculated that these NPFs might normally occur both on cell surfaces and in the extracellular matrix which neurites contact during their growth or regrowth in vivo.

Of several purified extracellular matrix constituents laminin was found to behave as an extremely potent NPF (Manthorpe et al., 1983b). A PNPF from rat schwannoma cell conditioned medium and purified laminin are indistinguishable in their effects on a number of neuritic growth parameters (Davis et al., 1985a). However, antibodies against rat laminin blocked the neurite promoting activity of rat laminin but not the PNPF from rat schwannoma conditioned medium -- a feature that initially discouraged an identification of the schwannoma PNPF with laminin. The rat schwannoma PNPF has now been purified and found to be closely related to laminin (Davis et al., 1985b,c). Five properties distinguish, thus far, the schwannoma factor from laminin: the apparent absence of a 400 kD subunit, the presence of additional 135 kD and 35 kD subunits, the resistance to anti-laminin antibodies, an altered short arm structure in its electron microscopy profile, and the association with proteoglycans. It seems reasonable, from these and other reports (e.g. Lander et al., 1985), to presume that laminins and PNPFs from many different cell conditioned media are members of a common family of anchorage-dependent neurite promoting factors.

The astroglial microcultures produce and release a PNPF (Fig. 3), which appears to contain laminin, since it is immunosequestered by immobilized anti-laminin antibodies (unpublished observations). Here, too, the critical questions are to what extent, under what circumstances and in what physical arrangements (soluble, bound to cell surfaces, or incorporated into extracellular matrices) do astrocytes display laminin-like products in vivo. Antibodies against laminin have only recently become available, and investigation by immunochemical techniques of laminin distribution in the CNS has just begun. Laminin antigens are detected in fetal rat brain between capillaries and surrounding astroglial endfeet at E13, shortly after capillary formation (Bignami et al., 1984). In the adult rat brain, laminin is normally only seen in such locations but injury appears to induce its appearance in astrocytes at the lesion site (Liesi et al., 1984). It is possible, therefore, that laminin-coated surfaces are made available for axonal regrowth in the injured CNS, as they are in peripheral nerve (e.g. Keynes et al., 1984; Ide et al., 1983). If so, the abortive outcome of regeneration in the adult mammalian CNS may implicate an inadequately oriented distribution of neurite promoting factors, beside any other failings (cf. Varon, 1985).

4. A neurobiological view of glia-neuron interactions

Traditional views (cf. Varon and Somjen, 1979) have assigned to the glia housekeeping tasks such as physical support, electrical insulation, selective monitoring of blood-borne inputs, nutrient supply, and waste removal. Attention has been correspondingly given to histological, bioelectrical and biochemical properties of the astroglia, as well as to the production and maintenance of myelin by oligodendroglia and Schwann cells. Information and concepts acquired through these approaches will provide an essential basis for future work on their participation in the functional community of cells which constitutes the nervous system. The recent strides of cellular neurobiology have now expanded our perceptions of glia-neuron interactions. Figure 4 summarizes features discussed in this paper. It is already clear, however, that even those represent but a beginning in our understanding of how much richer and pervasive must be the true, in vivo partnership between the glial and neuronal elements in the normal and the pathological neural system.

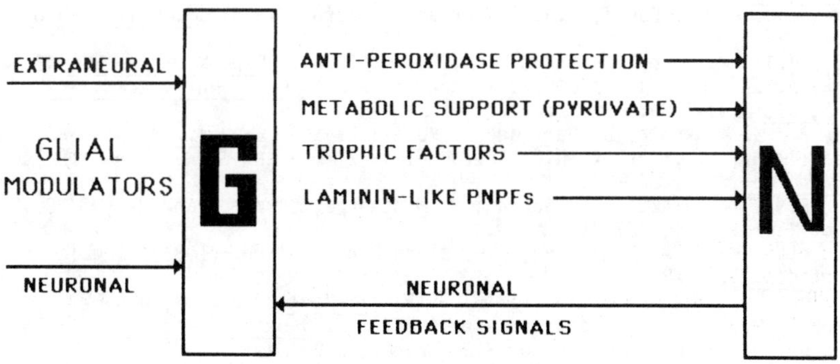

Fig. 4. An overview of glia-neuron interactions in the present perspectives.

ACKNOWLEDGEMENTS

Work reviewed here has been supported by grants from NINCDS: NS-16349 and NSF: BNS-18366.

REFERENCES

Adler, R. and S. Varon (1980). Brain Res., 188, 437-448.
Adler, R. and S. Varon (1981a). Dev. Biol., 86, 69-80.
Adler, R. and S. Varon (1981b). Dev. Biol., 81, 1-11.
Adler, R., M. Manthorpe, S.D. Skaper and S. Varon (1981). Brain Res., 206, 129-144.
Armstrong, D. (1984). Free Radicals in Molecular Biology, Raven Press, New York.
Barbin, G., M. Manthorpe, and S. Varon (1984). J. Neurochem., 43, 1468-1478.
Barde, Y.A., S. Edgar and H. Thoenen (1982). EMBO J., 1, 549-553.
Bignami, A., N.H. Chi and D. Dahl (1984). Int. J. Dev. Neurosci., 2, 367-376.
Carnow, T.B., M. Manthorpe, G.E. Davis and S. Varon (1985). J. Neurosci., 5, 1965-1971.
Collins, F. (1978). Dev. Biol., 65, 50-57.
Crutcher, K.A. and J.P. Chandler (1984). Brain Res., 308, 177-181.
Davis, G.E., M. Manthorpe, E. Engvall and S. Varon (1985a). J. Neurosci., in press.
Davis, G.E., M. Manthorpe and S. Varon (1985b). Dev. Brain Res., 17, 75-84.
Davis, G.E., E. Engvall, S. Varon and M. Manthorpe (1985c). Trends Neurosci., in press.
Facci, L., S.D. Skaper and S. Varon (1985). J. Neurochem., 45, 926-934.
Gage, F.H., A. Bjorklund and U. Stenevi (1984). Nature, 308, 637-639.
Ghahn, H., F. Hefti, R. Heumann, M.E. Schwab and H. Thoenen (1983). Dev. Brain Res., 9, 45-52.
Gundersen, R.W. (1985). J. Neurosci. Res., 13, 199-212.
Honegger, P. and D. Lenoir (1982). Dev. Brain Res., 3, 229-238.
Hyden, H. (1967). IN The Neurosciences: A Study Program, G.C. Quarton, T. Melnechuck, F.O. Schmitt (eds.), Rockefellar Univ. Press. pp. 248-266.
Ide, C., K. Tohyama, R. Yokoto, T. Nitatori and S. Onodera (1983). Brain Res., 288, 61-75.
Katoh-Semba, R., L. Facci, S.D. Skaper and S. Varon (1985). J. Cell Physiol., in press.
Keynes, R.J., W.G. Hopkins and C.L.-H. Huang (1984). Exp. Neurol., 85, 426-436.
Lander, A.D., D.K. Fujii and L.F. Reichardt (1985). Proc. Natl. Acad. Sci. USA, 82, 2183-2189.
Liesi, P., S. Kaakkola, D. Dahl and A. Vaheri (1984). EMBO J., 3, 683-686.
Lindsay, R.M., R.C. Barber, R.C. Sherwood, J. Zimmer and G. Raisman (1982). Brain Res., 243, 329-343.
Longo, F.M., I. Selak, J. Zovickian, M. Manthorpe, S. Varon and H-S U (1984). Exp. Neurol., 84, 207-218.
Magistretti, P.J., M. Manthorpe, F.E. Bloom, and S. Varon (1983). Regulatory Peptides, 6, 71-80.
Manthorpe, M., R. Adler and S. Varon (1979). J. Neurocytol., 8, 605-621.
Manthorpe, M., F.M. Longo and S. Varon (1982). J. Neurosci. Res., 8, 241-250.
Manthorpe, M., M. Nieto-Sampedro, S.D. Skaper, E.R. Lewis, G. Barbin, F.M. Longo, C.W. Cotman and S. Varon (1983a). Brain Res., 267, 47-56.
Manthorpe, M., E. Engvall, E. Ruoslahti, F.M. Longo, G.E. Davis and S. Varon (1983b). J. Cell Biol., 97, 1882-1890.
Manthorpe, M., J. Rudge and S. Varon (1985a). Adv. in Cellular Neurobiol., in press.
Manthorpe, M., S.D. Skaper, L.R. Williams and S. Varon (1985b). Brain Res., in press.
Manthorpe, M., G.E. Davis and S. Varon (1985c). Fed. Proc., in press.
Nieto-Sampedro, M., M. Manthorpe, G. Barbin, S. Varon and C.W. Cotman (1983). J. Neurosci., 3, 2219-2229.
Perez-Polo, J.R., K. Hall, K. Livingston and K. Westlund (1977). Life Sci., 21, 1535-1544.
Rudge, J.S., M. Manthorpe and S. Varon (1985). Dev. Brain Res., 19, 161-172.
Rudge, J., G. Davis, M. Manthorpe and S. Varon (1986). Submitted.
Schwartz, J.P., D.N. Chuang and E. Costa (1977). Brain Res., 137, 369-375.

Selak, I., S.D. Skaper and S. Varon (1985). J. Neurosci., 5, 23-28.
Skaper, S.D. and S. Varon (1985). Dev. Brain Res., in press.
Skaper, S.D., L. Facci, J.S. Rudge, R. Katoh-Semba, M. Manthorpe and S. Varon (1986). Dev. Brain Res., in press.
Varon, S. (1985). Discussions in Neuroscience, FESN Publisher, Geneva, Switzerland, in press.
Varon, S. and M. Manthorpe (1982). Adv. Cell. Neurobiol., 3, 35-95.
Varon, S. and G. Somjen (1979). Neurosci. Res. Prog. Bull., 17, 1-239.
Varon, S., R. Adler, M. Manthorpe and S.D. Skaper (1983). IN Neuroscience Approached through Cell Culture, Vol. 2, S.E. Pfeiffer (ed.), CRC Press, Boca Raton, Florida. pp. 53-77.
Varon, S., M. Manthorpe and L.R. Williams (1984a). Dev. Neuroscience, 6, 73-100.
Varon, S., S.D. Skaper, G. Barbin, I. Selak and M. Manthorpe (1984b). J. Neurosci., 4, 654-658.
Varon, S., C. Raiborn and S. Norr (1974). Exp. Cell Res., 88, 247-256.
Walicke, P., S. Varon and M. Manthorpe (1985). J.. Neurosci., in press.
Williams, L.R., I. Selak, S.D. Skaper, M. Manthorpe and S. Varon (1985). Brain Res., 336, 99-105.

POSTER

Influences of Neurons on Glial Cells Development in Culture

D. Langui, J. P. Delaunoy and M. Sensenbrenner

Centre de Neurochimie di CNRS, 5 rue Blaise Pascal, 67084
Strasbourg CEDEX, France

KEYWORDS

Culture ; glial cells ; neuronal extract ; glial fibrillary acidic protein ; carbonic anhydrase ; glutamine synthetase ; radioimmunoassay ; immunocytochemistry.

Previous studies have shown that brain extracts from chick embryos, newborn and adult rats stimulate the biochemical maturation of astrocytes and oligodendrocytes as seen by the increase of the content of glial fibrillary acidic protein (GFAP), S100 protein and isoenzyme II of carbonic anhydrase (CAII) (Sensenbrenner, et al., 1980). Then, the question arise of a possible neuronal origin of the active factor(s) contained in the brain extract. This let us to investigate the influence of cultured neurons on the development of glial cells in primary culture.
Glial cell cultures were obtained from newborn rat cerebral hemispheres (Booher and Sensenbrenner, 1972) and pure neuronal cultures from 8 day-old chick embryo cerebral hemispheres (Pettmann, et al., 1979). Both cultures were maintained in a serum-free defined medium. To prepare neuronal extracts culture of neurons, at days 5, 7, 9 and 12 were homogenized in phosphate saline buffer. The homogenates were centrifuged and the supernatants were then tested on glial cells at least 2 days before harvesting the cultures. Conditioned media from the neuronal cultures and brain extract from 8 day-old chick embryos were also tested.
Immunocytochemical data using rabbit anti-GFAP and monoclonal anti-CAII antibodies confirmed the presence of these proteins in two different cell types, GFAP in astrocytes and CAII in oligodendrocytes (Langui, et al., 1985).
We have developed polyclonal antibodies against glutamine synthetase (GS), an astroglial marker and against CAII, and used them for measuring the amount of these two proteins by radioimmunoassay. Quantitative results showed that the amount of GS and CAII increased when chick brain or neuronal extract was added to glial cultures (Table 1). However, this effect was seen only when neurons used to prepare the extract have been grown for at least 7 days, that is to say, when they have reached a certain degree of maturity (Pettmann, et al., 1979). Conditioned media affected only the level of GS.

In conclusion, soluble factors from mature neurons influence the maturation of glial cells in culture. Investigations will be undertaken to study the type of neurons which exert this effect.

Table 1. Effects of chick neuronal extract (CNE), chick brain extract (CBE) and conditioned media from chick neurons in culture (NCM) on the content of glutamine synthetase (GS) and carbonic anhydrase (CAII) in primary glial cell cultures (20 days) measured by radioimmunoassay.

Treatment	GS ng/mg protein	CAII ng/mg protein
Control	662 ± 60 (n = 7)	346 ± 67 (n = 7)
NCM (20 %)	848 ± 276 (n = 8)	363 ± 47 (n = 7)
CNE (20 %)	786 ± 137 (n = 9)	600 ± 115 (n = 9)
CBE (10 %)	964 ± 197 (n = 8)	653 ± 152 (n = 6)

REFERENCES

Booher, J., and M. Sensenbrenner (1972). Neurobiology, 2, 97-105.

Langui, D., J.P. Delaunoy, M.S. Ghandour, and M. Sensenbrenner (1985). Neurosci. Letters, in press.

Pettmann, B., J.C. Louis, and M. Sensenbrenner (1979). Nature, 281, 378-380.

Sensenbrenner, M., G. Labourdette, J.P. Delaunoy, B. Pettmann, G. Devilliers, G. Moonen, and E. Bock (1980). Tissue Culture in Neurobiology. Eds. E. Giacobini, A. Vernadakis, A. Shahar, Raven Press, New-York, pp. 385-395.

Potassium as a Signal in Metabolic Interactions Between Neurons and Astrocytes

L. Hertz

Department of Pharmacology, University of Saskatchewan, Saskatoon, Saskatchewan, S7N 0W0, Canada

ABSTRACT

Effects of excess extracellular potasssium concentrations on different aspects of astrocytic energy metabolism are reviewed. Evidence is presented that two different energy-requiring processes are involved, each of them associated with a different mechanism for active uptake of potassium.

KEYWORDS

Astrocytes; energy metabolism; glucose utilization; oxygen uptake; potassium.

INTRODUCTION

It is now overwhelmingly well demonstrated that the potassium concentration in the extracellular space of the central nervous system *in vivo* increases measurably (from 3 to about 4 mM) during physiological stimulation, to a larger extent (from 3 up to 12 mM) during seizures or direct stimulation of afferent pathways, and to exceedingly high values (up to 80-100 mM) during anoxia and spreading depression (for references, see e.g., Somjen, 1979; Walz and Hertz, 1983). It is also well established that the potassium concentration in the extracellular space after stimulation, seizures, spreading depression or short-lasting anoxia rapidly returns to its resting level of 3 mM. Already in the mid 1960's it was suggested that glial cells might be involved in this clearance of extracellular potassium in the central nervous system. Since excess extracellular potassium, at least after neuronal stimulation (including seizures), is likely to originate from neurons this represents a neuronal-astrocytic interaction utilizing the elevated potassium concentration as a signal. Two different hypotheses were suggested, i.e., 1) that extracellular potassium is accumulated *into* glial cells by active uptake processes (Hertz, 1965), or 2) that extracellular potassium is redistributed by a current-carried transport mechanism *through* electrically coupled glial cells which function as perfect potassium electrodes (Orkand *et al*., 1966). The latter of these mechanisms is not directly dependent upon energy metabolism (although it does require membrane polarization which ultimately has been brought about by energy requiring processes). It will therefore not be discussed in the present context. In contrast, the former of these mechanisms is directly dependent upon energy metabolism. A considerable amount of information is now available about increases in energy metabolism in astrocytes and in non-mammalian glial cells during exposure to elevated potassium concentrations. These increases, which at least partly may be metabolic manifestations of active potassium uptake will be reviewed in the present paper. However, in order to demonstrate that these events are of importance also for the function of the normal intact brain, metabolic effects of potassium in the brain *in vivo* and in cellularly more composite structures, e.g., brain slices will also be briefly discussed.

POTASSIUM EFFECTS ON ENERGY METABOLISM *IN VIVO*

Fluorometric measurements *in vivo* of NADH/NAD+ and reduced/oxidized cytochrome ratios have shown that the increased extracellular potassium concentration during excitation, seizures or spreading depression is accompanied by an increase in energy metabolism (relative decrease of NADH level) both in mammalian (Rosenthal and Somjen, 1973; Mayevski and Chance, 1974, 1975; Mayevski *et al.*, 1974; LaManna and Rosenthal, 1975) and amphibian (Sick and Kreisman, 1981) brain. During normal excitation there is a very good quantitative correlation (Fig. 1) between the external potassium concentration and the magnitude of the metabolic increase (e.g., Lewis and Schuette, 1975;

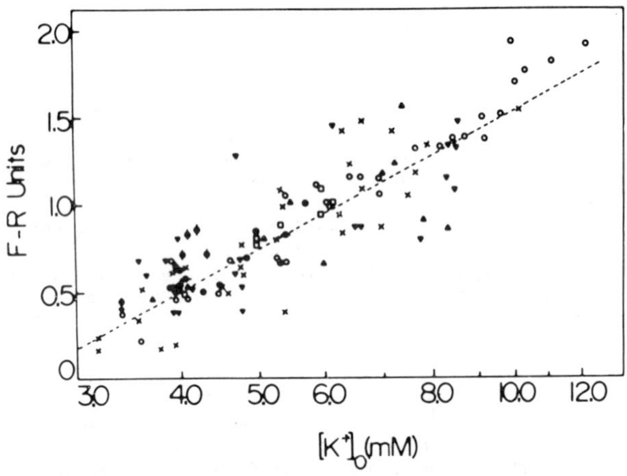

Fig. 1. Correlation between corrected fluorescence (a measurement of energy metabolism) responses and extracellular potassium activity in the cerebral neocortex of the cat exposed to stimulus trains up to 8 sec duration. From Lothman *et al.*, 1975.

Lothman *et al.*, 1975; Rosenthal *et al.*, 1979) suggesting that the increased potassium concentration *per se* triggers a process which requires energy, e.g., cellular uptake of potassium. However, during paroxysmal after-discharges, seizures induced by pentylenetetrazol or spreading depression the potassium concentration rises to at least 8-10 mM and the metabolic activity becomes even higher (Fig. 2) than could be deduced from extrapolation of the good quantitative correlation between metabolic activity and extracellular potassium concentration observed during normal excitation (Lothman *et al.*, 1975). For this reason it has been suggested that there may exist a secondary source of ATP, i.e., another energy requiring process leading to hydrolysis of ATP which becomes evident when the extracellular potassium concentration rises above 10-12 mM (Lothman *et al.*, 1975). It is another indication of an increase in energy metabolism that the uptake of ^{14}C-deoxyglucose in the brain *in vivo* also is considerably enhanced by spreading depression (Shinohara *et al.*, 1979). Thus there is good evidence for an association between increased extracellular potassium concentrations and stimulation of energy metabolism in the central nervous system *in vivo*, but it is not known whether the increased energy metabolism occurs in neurons or astrocytes (or both).

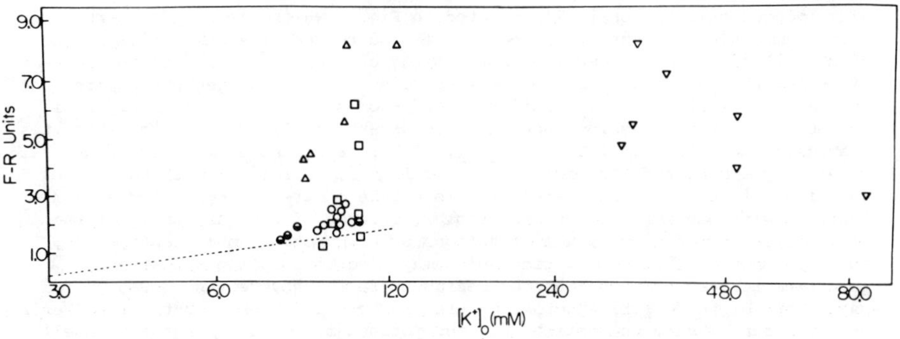

Fig. 2. Correlation between corrected fluorescence responses (a measurement of energy metabolism) and extracellular potassium activity in the cerebral neocortex of the cat exposed to stimulus trains of 10 sec or more (○ and ◒), paroxysmal after-discharges (□), seizures induced by pentylenetetrazol (△) and spreading depression (▽). Dotted line is regression line from Fig. 1. From Lothman *et al.*, 1975.

POTASSIUM EFFECTS ON ENERGY METABOLISM IN MIXED NEURONAL-GLIAL *IN VITRO* PREPARATIONS

Brain slices: It has now been known for 50 years that the rate of oxygen consumption in brain slices is substantially increased during exposure to highly elevated potassium concentrations (Ashford and Dixon, 1935; Dickens and Greville, 1935), and at the same time there is an increase in CO_2 production from labeled glucose (Sims *et al.*, 1983). The potassium-induced stimulation of oxygen uptake is not specific to potassium but can also be seen after exposure to ammonia, cesium, rubidium or lithium whereas an increase in the

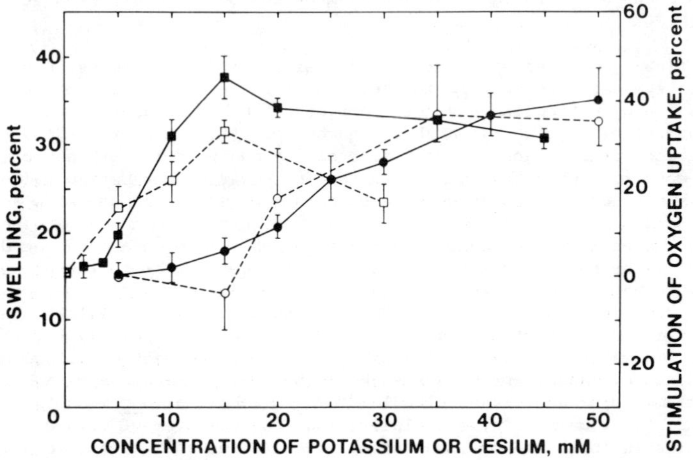

Fig. 3 Swelling (■ and ●) and stimulation of oxygen consumption (□ and ○) in brain slices as a function of the concentration of cesium (□ and ■) or potassium (○ and ●) in the medium. Swelling is expressed as percentage weight increase during the incubation, and stimulation of oxygen consumption as percentage increase over the oxygen consumption during incubation in a "balanced" medium. SEM are shown by vertical bars. From Hertz and Kjeldsen, 1985.

sodium concentration has no such effect (Hertz and Schou, 1962). The potassium induced stimulation is illustrated in Fig. 3 which shows that it requires potassium concentrations of at least 20 mM and reaches its maximum at around 35 mM. It can also be seen that both in the case of potassium and in the case of cesium there is a correlation between stimulation of oxygen consumption and increase in swelling, i.e., fluid uptake. From histological examinations it is known that the potassium-induced fluid uptake mainly occurs into astrocytes (Zadunaisky et al., 1963; Bourke *et al.*, 1970; Moller et al., 1974). There is general agreement that the astrocytic swelling is correlated with an uptake of chloride but it is disputed whether this uptake occurs together with sodium or together with potassium. This is probably because of the complexity of the brain slice and of a multitude of actions by high potassium concentrations, including neuronal depolarization which may lead to a sodium uptake and potassium release. In the case of cesium there is, however, no doubt that a major part of the chloride uptake occurs with cesium as the counter ion (Hertz and Kjeldsen, 1985), suggesting that the potassium induced increase in swelling and in oxygen uptake analogously represent an uptake of potassium chloride which for osmotical reasons leads to an uptake of water, and that these phenomena occur in astrocytes. One half mM of ethacrynic acid, an inhibitor of chloride transport, abolishes the potassium induced stimulation of oxygen uptake and reduces the potassium-induced increase in swelling; barbiturates (0.25-0.5 mM) have a similar, but less pronounced effect whereas the potassium induced stimulation of oxygen uptake is relatively resistant to 100 μM ouabain (C.S. Kjeldsen and L. Hertz, unpublished experiments).

The potassium concentrations required to increase oxygen consumption and swelling in brain slices correspond to those causing the very high rate of energy metabolism *in vivo* which is associated with potassium concentrations $>$ 10 mM and there is no increase in oxygen uptake (Fig. 3) corresponding to the more moderate increase in *in vivo* energy metabolism which occurs in the range 3-12 mM potassium (Figs. 1-2). Nevertheless, the potassium content in brain slices reaches a peak around 15-20 mM potassium and the sodium concentration is slightly but significantly decreased at this level (Lund-Andersen and Hertz, 1970; Hertz, 1973; Grisar and Franck, 1981). This suggests the operation of a Na^+,K^+-ATPase, and maximum activity of this enzyme in brain has been found to occur at 10-20 mM potassium (Samson and Quinn, 1967; Henn *et al.*, 1972).

Drone retina: A series of important studies on potassium homeostasis at the cellular level and the correlation between an increased potassium concentration and energy metabolism has been carried out by Coles and coworkers using the drone retina, a tissue that has a number of practical advantages, notably that intracellular potassium activities can be measured in both neurons (photoreceptors) and glial cells. In this preparation both stimulation with light (which causes a pronounced increase in extracellular potassium activity) (Fig. 4) and an elevation of the potassium concentration in the bathing solution (Coles and Tsacopoulos, 1979; Coles 1981) lead to a pronounced increase of the potassium activity of the glial cells whereas the potassium activity in the neurons either decreases (during light stimulation) or shows only a modest increase (during exposure to increased extracellular potassium). The potassium uptake into the glial cells occurs at least partly together with chloride (Coles and Orkand, 1984) and the increased content of potassium *might* be explained by passive processes such as the re-establishment of a Donnan equilibrium (Coles and Orkand, 1983). This does however *not* mean that the uptake *might* not be active, especially since the turnover of glycogen in the glial cells is increased by light stimulation (Coles and Tsacopoulos, 1981) and since the glial uptake of potassium is inhibited by strophantidine (Tsacopoulos and Coles, 1978).

The rate of oxygen uptake in the drone retina is also increased by light stimulation (Fig. 4). It has been reported that most of the mitochondria in this preparation are located in the photoreceptors so that the enhanced energy metabolism should be a neuronal phenomenon (Coles and Tsacopoulos, 1981). A likely candidate for energy requiring processes in neurons resulting from

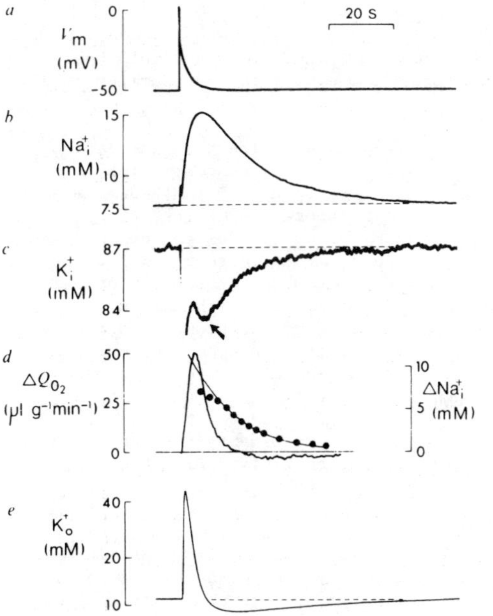

Fig. 4. Comparison of electrical, ionic and oxygen responses to single light flashes. Slices of drone retina 300-800 μm thick were stimulated with a light flash 20-40 ms long. a, Chart recording of an intracellular receptor potential. b, Sodium signal recorded as the difference in potential between barrels of a double-barrelled intraneuronal electrode. c, Potassium signal recorded with a double-barrelled intraneuronal electrode. d, Noisy line is increase in oxygen consumption. Filled circles show changes in intracellular sodium measured from the record in b. e, change in extracellular potassium measured with a double-barrelled electrode. From Tsacopoulos et al., 1983.

excitation would be sodium extrusion. However, as can be seen in Fig. 4 the increase in oxygen consumption after light stimulation is over long before the neuronal intracellular sodium activity has returned to normal, whereas there is a very good temporal correlation between stimulation of oxygen consumption and removal of excess extracellular potassium (Tsacopoulos et al., 1983). Since the latter mainly occurs into glial cells, one might wonder whether an increased metabolism in spite of the reported scarcity of mitochondria, occurs in these cells or whether there might be a transfer of energy-rich compounds from neurons to glial cells. In this context it seems noteworthy that the mitochondria in the neurons are located very close to the cell membranes (Coles and Tsacopoulos, 1981).

Leech and snail ganglia. In leech and snail ganglia stimulation and an increase of the potassium concentration have both been found to increase the incorporation of labeled 2-deoxyglucose into glycogen in glial cells. A maximum effect was observed when the potassium concentration was increased from 4 to 7-8 mM (Pentreath and Kai-Kai, 1982; Pentreath, 1982). No potassium-induced stimulation of oxygen uptake is found in leech ganglia (Hertz and Nissen, 1976).

POTASSIUM EFFECTS ON ENERGY METABOLISM IN ALL GLIAL PREPARATIONS

Glucose utilization: The utilization of the glucose analogue 2-deoxyglucose in primary cultures of astrocytes amounts to 7-13 nmol/min per mg protein at 5.4 mM potassium (Cummins et al., 1979; Brookes and Yarowsky, 1985). A substantial part of utilized glucose in released as lactate (which itself may be further oxidized) and the formation of $^{14}CO_2$ from labeled glucose only corresponds to the degradation of ≃1 nmol/min glucose per mg protein (Hertz and Schousboe, 1986). In addition to glucose, astrocytes are, however, capable of oxidizing several other substrates, including glutamate and glutamine (Yu and Hertz, 1983).

As seen in Fig. 5 an increase of the potassium concentration within the interval 2.3-20 mM causes an increase of glucose utilization (Cummins et al., 1979; Brookes and Yarowsky, 1985). This potassium-induced increase in deoxyglucose utilization is only partly inhibitable by 1 mM ouabain (Brookes and Yarowsky, 1985), a concentration of ouabain that is high enough to cause a large inhibition of potassium uptake into corresponding cultures, even at an elevated potassium concentration (Walz and Hertz, 1982). The lack of a complete inhibition of the potassium-induced stimulation of deoxyglucose utilization by ouabain may therefore suggest that active exchange between sodium and potassium is not the only process responsible for the increase in energy metabolism. At a very high potassium concentration the utilization of deoxyglucose is, in contrast, decreased (Fig. 5) but this could be due to the fact that potassium was substituted for sodium so that the sodium concentration was drastically decreased.

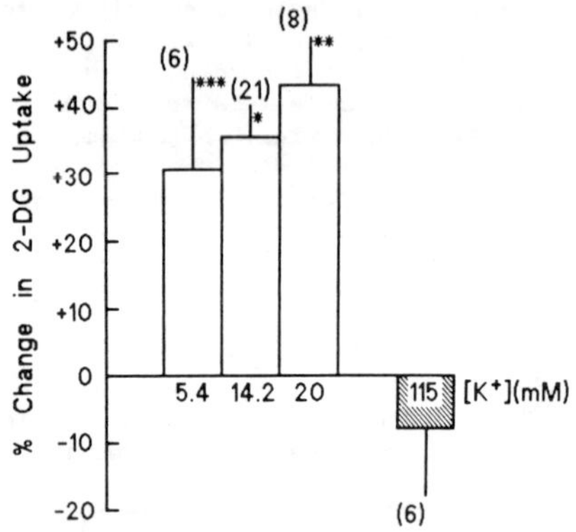

Fig. 5 Effect of increasing potassium concentration on the uptake of [^{3}H]2-deoxyglucose (2-DG) in astrocyte cultures. Percentage changes are related to uptake at the control potassium concentration of 2.3 mM, with the exception of the highest concentration (hatched column) which relates to a 5.4 mM control. Each value is the mean ± SEM for the number of experiments shown in parenthesis. From Brookes and Yarowsky, 1985.

A reduction of the potassium concentration from 2.3 to 0.44 mM causes a distinct decrease of the deoxyglucose uptake and the difference between deoxyglucose utilization at 0.44 and 2.3 mM is totally blocked by ouabain (Brookes and Yarowsky, 1985). Thus, in this interval the increased energy requirement seems entirely to reflect an active exchange between sodium and potassium. It is in agreement with this concept that exposure to potassium free medium at 0°C causes a substantial increase in the sodium content in astrocytes in primary cultures. After re-establishment of 37°C the sodium efflux is enhanced provided there is a relatively high (12 mM) extracellular potassium concentration (Walz and Hertz, 1984b). Previous exposure to cold potassium-free medium also causes an enhancement of deoxyglucose utilization, probably as a metabolic manifestation of the increased sodium transport (Yarowsky et al., 1985). An even more pronounced increase in deoxyglucose utilization is evoked when the intracellular sodium concentration has been raised by exposure to the sodium ionophore monensin (Yarowsky et al., 1985). Utilization of labeled glucose has also been measured in "all glia" optic nerves from necturus obtained two months following surgical removal of the eye (Salem et al., 1975). Exposure to 15 mM potassium increased the glucose uptake by 70%.

Oxygen uptake and CO_2 formation. A substantial (about 50%) stimulation of oxygen consumption by highly elevated potassium concentrations (50-100 mM) has repeatedly been shown in microdissected glial cells (Hertz, 1966; Aleksidze and Blomstrand, 1969), glial cells obtained by gradient centrifugation (Haljamae and Hamberger, 1971) and astrocytes in primary cultures (Hertz *et al*., 1973; E. Hertz and Hertz, 1979; E. Hertz *et al*., 1986). This phenomenon is illustrated in Fig. 6. The CO_2 production from glucose in astrocytes in

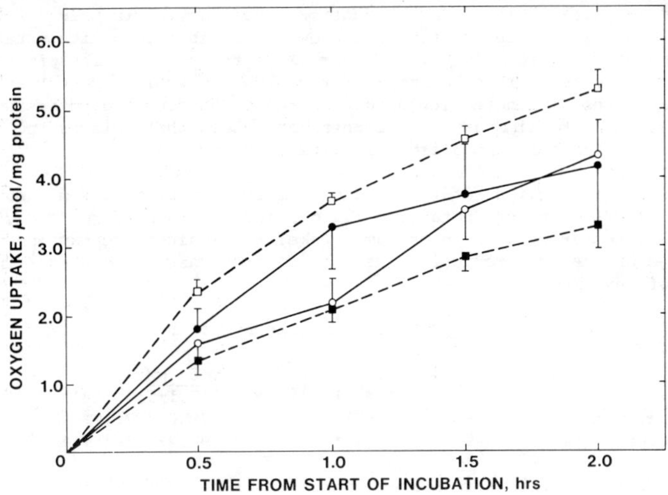

Fig. 6. Cumulative oxygen uptake by astrocytes in primary cultures incubated at a potassium concentration of 5.4 (○ and ●; solid lines) or 54 mM (□ and ■; stippled lines) in the absence (○ and □) or presence (● and ■) of 1 mM pentobarbital. From E. Hertz *et al*., 1986.

primary cultures is also enhanced during exposure to about 50 mM potassium although the effect is somewhat less pronounced (Table 1). According to most, but not all (Hultborn and Hyden, 1974) authors, neurons do not display a corresponding phenomenon (Hertz, 1966; Aleksidze and Blomstrand, 1969; Haljamae and Hamberger, 1971; Hertz *et al*., 1973; E. Hertz *et al*., 1986). Dinitrophenol, which also causes an increase in brain metabolism has similarly a much larger effect on astrocytes (105% stimulation) than on neurons (46% stimulation) (Olson and Holtzman, 1980; Jameson *et al*., 1984).

The high concentrations of potassium required to cause an increase of oxygen uptake suggest that this phenomenon is not evoked by stimulation of a potassium-sodium exchange catalyzed by the Na^+,K^+-ATPase since it was the effect of relatively low potassium concentrations on 2-deoxyglucose utilization in cultured astrocytes which was completely blocked by ouabain and since a ouabain inhibited potassium uptake in astrocytes in primary cultures reaches its maximum around ≈ 12 mM potassium (Walz and Hertz, 1982; Hertz, 1986). However, there is also another uptake mechanism in astrocytes in primary cultures which is found only at elevated potassium concentrations (Walz and Hertz, 1984a). This uptake which is accompanied by cell swelling (Walz and Hinks, 1985) is inhibited by furosemide, an inhibitor of coupled uptake of chloride together with cations (sodium and/or potassium). That the stimulation of oxygen uptake in primary cultures is a metabolic manifestation of this potassium uptake is suggested by the requirement for very high potassium concentrations by both of these phenomenona and by the good correlation between stimulation of oxygen uptake and increase of swelling in brain slices (Fig. 3). This concept is further supported by the fact that the potassium induced stimulation of CO_2 production is inhibited by furosemide (Table 1). Since furosemide and ethacrynic acid have similar mechanisms of action, this is an analogy to the inhibition of the potassium induced stimulation of oxygen uptake in brain slices by ethacrynic acid. Barbiturates which likewise inhibited the potassium induced stimulation of oxygen uptake in brain slices (see above) also inhibit the potassium-induced stimulation of oxygen uptake (E. Hertz _et al._, 1986) and CO_2 production in astrocytes in primary cultures (Fig. 6; Table 1) and these drugs are known to inhibit potassium uptake in the same preparation (Hertz, 1979). If they exert their inhibition of energy metabolism by a similar mechanism as furosemide one would expect that the barbiturate sensitive potassium uptake increases at elevated potassium concentrations. Preliminary experiments have shown that this is the case (L. Hertz and M. Shargool, unpublished results).

Table 1. CO_2 production rates in astrocytes in primary cultures exposed to normal (5.4 mM) or excess potassium (54 mM) and/or drugs (furosemide, pentobarbital) interfering with potassium uptake. All values are expressed as a percentage of the CO_2 production rate at 5.4 mM potassium (control) in the absence of any drugs.

	Potassium concentration	
	5.4 mM	54 mM
control	100.0±3.4 (14)*	117.1±5.4 (14)
furosemide, 2 mM	84.1±5.2 (5)	79.6±7.7 (9)
pentobarbital, 1 mM	82.2±3.2 (4)**	97.2±7.2 (5)**

* Number of experiments are shown in parenthesis
** Results from E. Hertz et al., 1986

Although quite substantial evidence thus is found for a correlation between furosemide sensitive uptake of potassium and the potassium induced stimulation of oxygen uptake it should also be mentioned that a dissociation between these two effects seems to exist in jimpy mice which do show a normal potassium uptake (Hertz and Chaban, 1982; L. Hertz, M. Shargool and E. Hertz, unpublished experiments) in spite of an absence of the potassium induced stimulation of oxygen uptake (Hertz _et al._, 1979). One possible explanation for this may be that the oxygen uptake is measured during a much longer period than the potassium uptake which is measured over 10 sec-1 min and that a fatiguing with time of the uptake mechanism would not be obvious immediately after the exposure to excess potassium.

NADH levels: Experiments in principle similar to the _in vivo_ fluorometric measurements of NADH levels (see above) have been carried out in "all-glia" _necturus_ optic nerves where increases in extracellular potassium concentration to between 6 and 12 mM in 21 out of 37 tests showed a decreased level of NADH (Fig. 7) which is consistent with a stimulation of oxidative phosphorylation (Orkand _et al._, 1973).

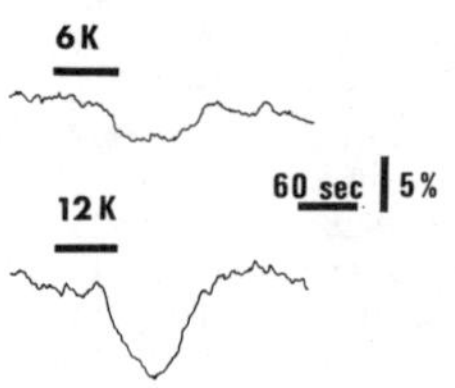

Fig. 7 Fluorescence decrease in "all-glia" _necturus_ optic nerve when the external potassium concentration was raised from normal (3) to 6 mM (top) and 12 mM (bottom). From P.Orkand _et al._, 1973.

ATP levels: High concentrations of potassium (54 mM) cause a decrease of the level of ATP in cultured astrocytes (from the spinal cord) but not in cultured neurons (from dorsal root ganglia) (Schousboe et al., 1970). No information is available about the effects of lower potassium concentrations.

CONCLUDING REMARKS

The present review has focused on effects by potassium on metabolism in astrocytic preparations and on the similarities between these effects and those on more complex preparations. Quite consistently, two different effects have been observed, i.e., one by low concentrations of potassium (generally below 10-20 mM) which seems to be a manifestation of an increased Na^+,K^+-ATPase activity and one by higher concentrations of potassium (generally above 10-20 mM). Evidence was presented that the latter effect both in brain slices and in isolated astrocytes may be a metabolic manifestation of an uptake of chloride which at least partly is coupled with an uptake of potassium. No information is available about the detailed mechanisms of this process. The facts that it is inhibited by barbiturates and that barbiturate-induced coma *in vivo* is associated with a decrease in deoxyglucose utilization (Sokoloff et al., 1977) and oxygen consumption (Kety, 1957) of the brain do suggest that it may be of considerable importance for normal brain function. It appears to be against this concept that the concentrations of potassium which are required for stimulation of energy metabolism by this mechanism have been demonstrated only after rather massive stimulation (Fig.2). However, it should be kept in mind that extracellular potassium concentrations generally have been measured in "pockets" of extracellular fluid which are large enough to accomodate the tip of an ion-selective electrode (about 1 μm) and that the potassium concentrations in the narrow clefts between adjacent cells (0.015 μm) conceivably may be considerably higher.

REFERENCES

Aleksidze, N.G. and C. Blomstrand (1969). Proc. Acad. Sci. USSR, Biochem. Series, 186, 140-141.

Ashford, C.A. and K.C. Dixon (1935). Biochem, J., 29, 157-168.

Bourke, R.S., K.M. Nelson, A. Naumann and O.M. Young (1970). Exp. Brain Res. 10, 427-466.

Brookes, N. and P.J. Yarowsky (1985). J. Neurochem., 44, 473-479.

Coles, J.A. (1981). J. Physiol., 318, 16-17P.

Coles, J.A. and R.K. Orkand (1983). J. Physiol, 340, 157-174.

Coles, J.A. and R.K. Orkand (1984). J. Physiol., 357, 11P.

Coles, J.A. and M. Tsacopoulos (1979). J. Physiol., 290, 525-549.

Coles, J.A. and M. Tsacopoulos (1981). J. Exp. Biol., 95, 75-92.

Cummins, C.J., R.A. Glover and O.Z. Sellinger (1979). J. Neurochem., 33, 779-785.

Dickens, F. and D.G. Greville (1935). Biochem. J., 29, 1468-1483.

Grisar, T. and G. Franck (1981). J. Neurochem., 36, 1853-1857.

Haljamae, H. and A. Hamberger (1971). J. Neurochem., 18, 1903-1912.

Henn, F.A., H. Haljamae and A. Hamberger (1972). Brain Res., 43, 437-443.

Hertz, E. and L. Hertz (1979). In Vitro, 15, 429-436.

Hertz, E., M. Shargool and L. Hertz (1986). Neuropharmacol., in press.

Hertz, L. (1965). Nature, 206, 1091-1094.

Hertz, L. (1966). J. Neurochem., 13, 1373-1387.

Hertz, L. (1973). Trans. Biochem. Soc., 1, 115-118.

Hertz, L. (1979). Neuropharmacol., 18, 629-633.

Hertz, L. (1986). Ann. N.Y. Acad. Sci., in press.

Hertz, L. and G. Chaban (1982) In "Neuroscience Approached through Cell Culture". Vol.1, S.E. Pfeiffer ed., CRC Press, Boca Raton, pp. 157-174.

Hertz, L. and C.S. Kjeldsen (1985). J. Neurosci. Res., 14, 83-93.

Hertz, L. and C. Nissen (1976). Brain Res., 110, 182-188.

Hertz, L. and M. Schou (1962). Biochem. J., 85, 93-104.

Hertz, L. and A. Schousboe (1986). In "Astrocytes", S. Fedoroff and A. Vernadakis, eds., Academic Press, N.Y., in press.
Hertz, L., L. Dittmann and P. Mandel (1973). Brain Res., 60, 517-520.
Hertz, L., G. Chaban and E. Hertz (1979). Brain Res., 181, 482-487.
Hultborn, R. and H. Hyden (1974). Exp. Cell Res., 87, 346-350.
Jameson, N., J. Olson, H. Nguyen and D. Holtzman (1984). J. Neurochem., 42, 470-474.
Kety, S.S. (1957). In "Metabolism of the Nervous System", D. Richter, ed., Pergamon Press, London, pp. 221-237.
LaManna, J.C. and M. Rosenthal (1975). Brain Res., 88, 145-149.
Lewis, D.V. and W.H. Schuette (1975). J. Neurophysiol., 38, 405-417.
Lothman, E., J. LaManna, G. Cordingley, M. Rosenthal and G. Somjen (1975). Brain Res., 88, 15-36.
Lund-Andersen, H. and L. Hertz (1970). Exp. Brain Res., 11, 199-212.
Mayevski, A. and B. Chance (1974). Brain Res., 65, 529-533.
Mayevski, A. and B. Chance (1975). Brain Res., 98, 149-165.
Mayevski, A., T. Zeuthen and B. Chance (1974). Brain Res., 76, 347-349.
Moller, M., K. Mollgard, H. Lund-Andersen and L. Hertz (1974). Exp. Brain Res., 22, 299-314.
Olson, J.E. and D. Holtzman (1980). J. Neurosci. Res., 5, 497-506.
Orkand, P., H. Bracho and R.K. Orkand (1973). Brain Res., 55, 467-471.
Orkand, R.K., J.G. Nicholls and S.W. Kuffler (1966). J. Neurophysiol., 29, 788-806.
Pentreath, V.W. (1982). Trends Neurosci., 5, 339-345.
Pentreath, V.W. and M.A. Kai-Kai (1982). Nature, 295, 59-61.
Rosenthal, M. and G. Somjen (1973). J. Neurophysiol., 36, 739-749.
Rosenthal, M., J. LaManna, Y. Yamada, W. Younts and G. Somjen (1979). Brain Res., 162, 113-127.
Salem, R.D., R. Hammerschlag, H. Bracho and R.K. Orkand (1975). Brain Res., 86, 499-503.
Samson, F.E. and D.J. Quinn (1967). J. Neurochem., 14, 421-427.
Schousboe, A., J. Booher and L. Hertz (1970). J. Neurochem, 17, 1501-1504.
Shinohara, M., B. Dollinger, G. Brown, S. Rapoport and L. Sokoloff (1979). Science, 20, 188-190.
Sick, T.J. and N.R. Kreisman (1981). J. Neurophysiol., 45 998-1012.
Sims, N.R., D.M. Bowen, D. Neary and A.N. Davison (1983). J. Neurochem., 41, 1329-1334.
Sokoloff, L., M. Reivich, C. Kennedy, M.H. Des Rosiers, C. Patlak, K.D. Pettigrew, O. Sakurada and M. Shinohara (1977). J. Neurochem., 28, 897-916.
Somjen, G.G. (1979). Ann. Rev. Physiol., 41, 159-177.
Tsacopoulos, M. and J.A. Coles (1978). Experientia, 34, 903-304.
Tsacopoulos, M., R.K. Orkand, J.A. Coles, S. Levy and S. Poitry (1983). Nature, 301, 604-606.
Walz, W. and L. Hertz (1982). J. Neurochem., 39, 70-77.
Walz, W. and L. Hertz (1983). Progr. Neurobiol., 20, 133-183.
Walz, W. and L. Hertz (1984a). J. Cerebr. Blood Flow & Metabl., 4, 301-304.
Walz, W. and L. Hertz (1984b). J. Neurosci. Res., 11, 231-239.
Walz, W. and E.C. Hinks (1985). Brain Res., 343, 44-51.
Yarowsky, P., A.F. Boyne, R. Wierville and N. Brookes (1985). J. Neurosci., in press.
Yu, A.C.H. and L. Hertz (1983). In "Glutamine, Glutamate and GABA in the Central Nervous System, L. Hertz, E. Kvamme, E.G. McGeer and A. Schousboe, eds, Alan R. Liss; N.Y. pp. 431-439.
Zadunaisky, J.A., F. Wald and E.D.P. de Robertis (1963). Exp. Neurol., 8, 290-309.

Regional Heterogeneity of Astrocytic Transport Processes for Glutamate and GABA

A. Schousboe*, J. Drejer[†], E. Meier[‡], O. M. Larsson[§] and I. Schousboe[‖]

Departments of Biochemistry A* and C[‖], Panum Institute, University of Copenhagen, DK-2200 Copenhagen N; [†]A/S Ferrosan, Research Division, DK-2860 Søborg; [‡]H. Lundbeck & Co.; [§]Department of Pharmacology, DK-2500 Valby and Department of Biology, Royal Danish School of Pharmacy, DK-2100 Copenhagen Ø, Denmark

ABSTRACT

Inactivation of the two neurotransmitters glutamate and GABA takes place by high affinity transport into presynaptic nerve endings and into astrocytic processes enwrapping the synapses. Studies of these high affinity transport processes in astrocytes originating from different brain areas have suggested that regional specialization of astrocytes may be found in the brain. A discussion of possible means by which neurons may influence and regulate this specialization of astrocytic properties will be given.

KEYWORDS

Glutamate; GABA; high-affinity transport; astrocytes; regional specialization; neurotransmission.

I. INTRODUCTION

The amino acids L-glutamate and GABA are the quantitatively most abundant neurotransmitters in the central nervous system acting as respectively an excitatory and inhibitory transmitter (Hertz et al, 1983). During recent years it has been established that astrocytes play an important role for the termination of these neurotransmission processes (Henn & Hamberger, 1971; Schousboe et al, 1977a,b; Hertz et al, 1978; Hertz, 1979; Schousboe, 1981). This termination appears to take place via high affinity transport processes for the amino acids (Kuffler & Edwards, 1958; Elliott & van Gelder, 1958; Curtis & Johnston, 1974; Krnjevic, 1974) and these transport processes are present both in nerve endings and astrocytes (Hertz, 1979; Schousboe, 1981;1982a). Functional aspects of astrocytic transport processes for glutamate and GABA as well as possible neuronal influence on the development of these processes will be discussed in this review.

II. ASTROCYTIC TRANSPORT OF GLUTAMATE AND GABA

a. Glutamate

The kinetic characteristics of astrocytic glutamate uptake as well as the corresponding uptake into neurons have recently been extensively reviewed by Hertz (1979) and Schousboe (1981;1982a). It is quite clear that the capacity for glutamate uptake is higher in astrocytes than in neurons (Schousboe & Hertz, 1983). The transport site is strictly stereospecific with regard to glutamate uptake since only the L-form of glutamate is accepted by the carrier (Schousboe et al, 1977b; Drejer et al, 1982). On the contrary, both enantiomers of aspartate (L- and D-aspartate) are accepted and transported by the carrier as shown by Drejer et al (1983a). The inhibition by D-aspartate of L-glutamate uptake has, however, been shown to exhibit some degree of complexity (Drejer et al, 1982) indicating that the carrier to some extent differentiates between L-glutamate and L-aspartate. These findings are summarized in Table 1 which shows the mutual inhibition pattern of the three amino acids L-glutamate, L-aspartate and D-aspartate.

Table 1. Mutual inhibition pattern for L-glutamate and L- and D-aspartate in astrocytes.

Transported amino acid	Inhibiting amino acid L-glu	L-asp	D-asp
L-glu	67[a]	68	106*
L-asp	73	77[a]	70
D-asp	75	82	83[a]

The table gives K_m values (μM, identified by 'a') for uptake of the amino acids and K_i values (μM) for their mutual inhibition. The asterisk indicates mixed comp./non-comp. inhibition. All other K_i values represent competitive inhibition. From Drejer et al (1983a).

b. GABA

Opposite to what is the case for glutamate, astrocytic uptake of GABA appears to be less intense than the uptake of GABA into nerve endings (Schousboe et al, 1983a; Hertz and Schousboe, 1986). That the uptake of GABA into the astrocytic compartment is nevertheless of physiological importance is shown by the finding that selective inhibition of the glial GABA uptake by GABA analogues like THPO (4,5,6,7-tetrahydroisoxazolo[4,5-c]pyridin-3-ol) is capable of preventing spontaneous or chemically induced seizure activity (Krogsgaard-Larsen et al, 1981; Meldrum et al, 1982; Croucher et al, 1983; Wood et al, 1980; 1983; Schousboe et al, 1983b; Seiler et al, 1985). This anticonvulsive action of THPO seems to be correlated to its ability to increase the GABA content of the neurotransmitter (nerve ending) pool of GABA (Wood et al, 1982). The substrate specificity of the GABA uptake site has been extensively studied

Table 2. Mutual inhibition pattern for GABA and selected GABA analogues in astrocytes.

Transported amino acid	Inhibiting amino acid GABA	Nip.	4-OH-Nip	ACHC
GABA	24[a]	15	148*	1041*
Nip.	75	29[a]	-	-
4-OH-Nip.	35*	-	25[a]	-
ACHC	15	-	-	211[a]

The table gives K_m values (μM, identified by 'a') for uptake of the amino acids and K_i values (μM) for their mutual inhibition. The asterisk indicates non-comp. inhibition. All other K_i values represent competitive inhibition. From Larsson <u>et al</u> (1980; 1983a; 1985b).
Nip: nipecotic acid; 4-OH-Nip: 4-hydroxy-nipecotic acid; ACHC: <u>cis</u>-3-aminocyclohexane carboxylic acid.

(Schousboe <u>et al</u>, 1978; 1979; 1981; 1985; Larsson <u>et al</u>, 1981; 1983a; 1985b) and it is evident that the astrocytic GABA carrier is distinctly different from the neuronal carrier. In addition, the substrate specificity of the astrocytic carrier appears to be so complex (cf. Table 2 and 6) that it has been suggested that a microheterogeneity of this carrier exists (Larsson <u>et al</u>, 1983a; 1985b). This aspect will be discussed further below (Chapter III.b).

Table 3. Kinetic parameters for L-glutamate uptake into astrocytes cultured from different brain regions.

Brain region	K_m (μM)	V_{max} (nmol x min^{-1} x mg^{-1})
Prefrontal cortex	58	13.9
Occipital cortex	33	11.4
Neostriatum	82	27.3
Cerebellum 0-days	34	5.8
Cerebellum 7-days	54	2.2

From Drejer <u>et al</u> (1982).

III. REGIONAL AND CELLULAR HETEROGENEITY

a. Glutamate uptake

Culturing astrocytes originating from different brain regions such as prefrontal and occipital cortex, neostriatum and cerebellum, Schousboe and Divac (1979) observed large differences in the high affinity glutamate uptake between these astrocytes. On the basis of this it was suggested that the astrocytic capacity for glutamate uptake might reflect the glutamatergic activity in a given brain region. This notion is in line with that proposed by Henn (1976) who observed a more intense uptake of the inhibitory neurotransmitter glycine in astrocytes isolated from spinal cord that in astrocytes of central origin, reflecting the fact that glycine mediated neurotransmission is more pronounced at the spinal cord level than in the brain (for references, see Schousboe, 1982a). The kinetics of the glutamate uptake into astrocytes of different regional origin was later studied in detail by Drejer et al (1982) and it was shown that it is primarily the number of transport sites (V_{max}) and not the affinity of the carrier (K_m) which differs in these different astrocytes (Table 3). These findings of differences in the number of glutamate transport sites in different brain areas have recently been confirmed by Hansson (1983) using autoradiographic analysis of glutamate uptake into astrocytes cultured from different brain regions. In addition, it has been shown that a similar regional heterogeneity exists when the concentration of the astrocytic specific glutamate metabolizing enzyme glutamine synthetase is measured immunohistochemically (Norenberg, 1979). Also when glutamine synthetase activity is measured directly in astrocytes cultured from different brain areas large regional variations are observed (Table 4). In analogy with what was found for high affinity uptake (Table 3) astrocytes originating from neostriatum and prefrontal cortex had the highest glutamine synthetase activity.

Table 4. Activity of glutamine synthetase in astrocytes cultured from different brain regions.

Brain area	Enzyme activity ($nmol \times min^{-1} \times mg^{-1}$)
Prefrontal cortex	70.0 ± 11.5 (12)
Occipital cortex	49.7 ± 5.6 (12)
Neostriatum	62.3 ± 5.8 (11)
Cerebellum	31.6 ± 2.6 (13)

Glutamine synthetase activity was assayed as described by Schousboe (1982b) in astrocytes cultured 3 weeks from the respective brain regions of newborn rats. Results are averages ± SEM with number of experiments given in parentheses (A.Schousboe, unpublished exp.).

Table 5. High affinity uptake of GABA into astrocytes cultured from different brain regions.

Brain region	GABA uptake ($pmol \times min^{-1} \times mg^{-1}$)
Cerebral cortex	34.4
Cerebellum	15.1

Astrocytes were kept in culture for 3 weeks as described by Hertz et al (1982) and GABA uptake measured at an external GABA concentration of 1 µM. From Drejer et al (1983b) and Larsson et al (1985b).

b. GABA

As shown in Table 5 the high affinity GABA uptake into astrocytes cultured from brain cortex appears to be quantitatively higher than the corresponding uptake into astrocytes cultured from cerebellum, indicating a regional heterogeneity of astrocytic GABA uptake analogous to what was found for glutamate (III.a.). A detailed kinetic analysis of the GABA uptake into astrocytes of different regional origin is, however, lacking, but on the basis of autoradiographic analysis of GABA uptake into astrocytes cultured from cerebellum it has been demonstrated that different types of astrocytes in these cultures take up GABA with different efficiency (Wilkin et al, 1983).

Another indication that astrocytic GABA transport sites are highly heterogeneous comes from detailed kinetic analyses (Larsson et al, 1983a; 1985b) of the uptake of GABA and some of its structural analogues such as 3-aminocyclohexane carboxylic acid (ACHC), nipecotic acid (Nip) and cis-4-hydroxy-nipecotic acid (4-OH-Nip). As shown in Table 6 both the affinity of the carrier for the different GABA analogues as well as the number of transport sites vary considerably. This indicates at least that a number of distinctly different astrocytic GABA carriers exist and these may to some extent even be present on the same astrocytes. The latter conclusion is somewhat ambiguous since it is not unlikely that a certain degree of cellular heterogeneity may be found within a culture of astrocytes from cerebral cortex (Hansson, 1984; Hallermayer & Hamprecht, 1984).

Table 6. Kinetic characteristics of uptake of GABA and some of its structural analogues into astrocytes cultured from cerebral cortex.

Substrate	K_m (μM)	V_{max} (% of GABA uptake)
GABA	24	100
Nip.	29	95
4-OH-Nip.	25	47
ACHC	211	151

The V_{max} for GABA uptake was 0.870 nmol x min^{-1} x mg^{-1} cell protein. From Larsson et al (1980; 1983a; 1985b).
Nip: nipecotic acid; 4-OH-Nip: 4-hydroxy-nipecotic acid; ACHC: cis-3-aminocyclohexane carboxylic acid.

IV. NEURONAL INFLUENCE ON ASTROCYTIC AMINO ACID UPTAKE

The finding of a characteristic regional heterogeneity in astrocytic glutamate uptake prompted Schousboe et al (1980) to suggest that regional differences in astrocytic function might be a consequence of an interplay between internal astrocytic and external neuronal factors. That such regulatory mechanisms indeed exist was recently shown experimentally by Drejer et al (1983b). It was demonstrated that if astrocytes originating from cerebellum of 7-day-old rats were exposed to conditioned media from different types of neurons, the capacity for uptake of glutamate or GABA could be regulated in a characteristic manner. As illustrated in Fig.1A and B treatment of the astrocytes with conditioned media from cerebral cortex interneurons selectively induced glutamate uptake sites whereas treatment with conditioned media from cerebellar granule neurons selectively induced GABA uptake sites in the same cultures of astrocytes. Analogous experiments were performed using astrocytes derived either from cerebellum or cerebral cortex of newborn rats but in none of these cases was any induction of transport sites observed. These results strongly indicate that a combination of intrinsic and extrinsic stimuli is needed in order to regulate the phenotype of the astrocytes (Drejer et al, 1983b). The observation that the different neuronal cultures selectively affected different amino acid transport mechanisms is thought-provoking. The fact that cerebral cortex interneurons are GABAergic (Larsson et al, 1983b; 1985a; Yu et al, 1984) and the cerebellar granule cells glutamatergic (Drejer et al, 1982; 1983a; 1985a) shows that it is not the type of transmitter released by the neurons that determines the specificity of the factors which are responsible for the induction of amino acid transport sites on the astrocytes. One other characteristic difference between the two types of neurons seems to be the prevailing types of receptors for glutamate and GABA. The cerebellar granule neurons are particularly rich in GABA receptors (Meier & Schous-

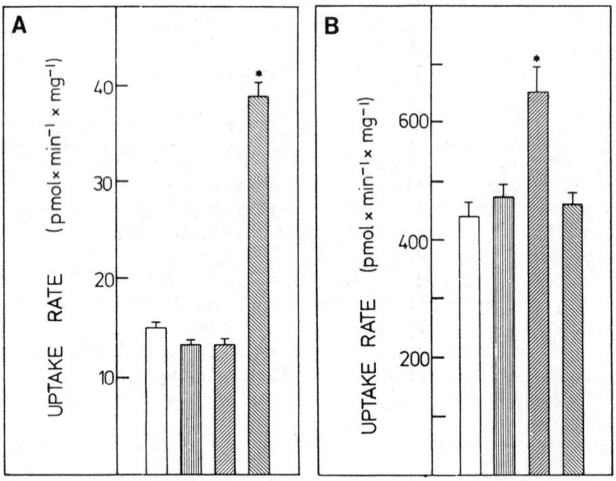

Fig.1.Velocity of [^{14}C]GABA (A) or [^{3}H]glutamate (B) uptake into astrocytes cultured from cerebellum of 7-day-old rats. Astrocytes were during the last two weeks of culture exposed to neuronal medium (□); neuronal medium containing 50 µM GABA (▥), conditioned medium from cultures of cerebral cortical neurons (▨) or conditioned medium from cultured cerebellar granule cells (▧). Uptake experiments were performed at 1 µM GABA or 5 µM glutamate. Results are averages ± S.E.M. of 12 experiments, i.e. 12 individual wells. Asterisks indicate statistically significant differences from the control uptake rates shown in the first column (A:$P<0.001$; B:$P<0.01$). From Drejer et al (1983b).

boe, 1982; Meier et al, 1984) but do not express glutamate receptors of the N-methyl-D-aspartate, quisqualate or kainate preferring types (Drejer et al, 1985b). On the other hand, the cerebral cortex interneurons seem to be relatively rich in glutamate receptors belonging to the N-methyl-D-aspartate as well as the quisqualate and kainate types of glutamate receptors (Drejer et al, 1986) and to have fewer GABA receptors (DeFeudis et al, 1979; Ticku et al, 1980; Schousboe & Hertz, 1986). On the basis of this it may be hypothesized that a determining factor for the specificity of the signal substance produced by the neurons could be the prevailing receptor population of the neurons which, in turn, must reflect the nature of the transmitter input to the neuron.

V. CHARACTERIZATION OF NEURONAL FACTORS FOR ASTROCYTIC AMINO ACID UPTAKE

Recently we have initiated studies by which we hope to at least partially characterize the neuronal factors which are responsible for the induction of amino acid transport sites on astrocytes. For the sake of simplicity we have chosen to concentrate on the factor from cerebellar granule neurons which induces GABA uptake sites on astrocytes. By the use of ultrafiltration on Amicon filters and by heat inactivation it has been established that the factor is heat sensitive and has a molecular weight equal to or larger than 5-10 KD. Assuming that the factor is a protein we have started to label releasable proteins by [^{35}S]methionine. Cultured granule cells were incubated for 8 h in a serum-free culture medium devoid of cold methionine but containing [^{35}S]methionine (10 µCi/ml). Subsequently, the culture medium was collected and chromatographed on Sepharose conjugated with an antibody produced against serum-free culture medium in order to remove proteins endogenous to the medium. The eluate from the Sepharose column was precipitated with ethanol and the precipitate subjected to reducing SDS-PAGE. After electrophoresis gels were stained for proteins and radioactive bands were visualized by fluorography. As expected (cf. Annunziata et al, 1983) several protein bands were found to be labelled with [^{35}S]methionine, one band with a M_r around 110 KD and several bands with M_r of 5-10 KD being particularly heavily labelled. Whether or not any of these labelled proteins are pertinent for the effect on astrocytic GABA uptake remains to be established e.g. by adsorption to astrocytic membranes.

VI. CONCLUDING REMARKS

At the time of the first conference on the role of astrocytes in the CNS (Schoffeniels et al, 1978) it was firmly established that astrocytes, among other functions, are involved in the inactivation of amino acid neurotransmitters (Schousboe, 1978). Since this time a new concept of regional diversity and specialization of astrocytic functions has evolved. If indeed this concept is correct it will be an immense task to characterize astrocytic functions in different brain areas. However, since astrocytes via different populations of receptors for neurotransmitters, peptides and drugs (Hamprecht & van Calker, 1980; Henn & Henn, 1980; Hertz, 1982; Hertz et al, 1984; McCarthy, this volume; Bender & Hertz, this volume) are likely to be involved in essentially all functions of the control of nervous system activity it would seem logical to have a certain degree of specialization of astrocytes analogous to what is known for the neurons. Future research will establish to what extent astrocytes exhibit regional specialization and how such specialization is regulated, e.g. on the basis of neuronal signaling.

ACKNOWLEDGEMENTS

The expert technical assistance of Hanne Fosmark, Ulla Bjørnsen and Margit Søe Rasmussen is gratefully acknowledged. The work has been supported financially by grants from The Danish State Medical and Natural Science Research Councils, The Novo Foundation, Gerda & Aage Haensch's Foundation and P.Carl Petersen's Foundation.

VII. REFERENCES

Annunziata, P., C. Regan and R. Balazs (1983). Dev. Brain Res. 8, 261-273.

Croucher, M.J., B.S. Meldrum and P. Krogsgaard-Larsen (1983). Eur. J. Pharmacol. 89, 217-228.

Curtis, D.R. and G.A.R. Johnston (1974). Ergb. Physiol. 69, 97-188.

DeFeudis, F.V., L. Ossola, G. Schmitt and P. Mandel (1979). Neurosci. Lett. 14, 195-199.

Drejer, J., O.M. Larsson and A. Schousboe (1982). Exp. Brain. Res. 47, 259-269.

Drejer, J., O.M. Larson and A. Schousboe (1983a). Neurochem. Res. 8, 231-243.

Drejer, J., E. Meier and A. Schousboe (1983b) Neurosci. Lett. 37, 301-306.

Drejer, J., O.M. Larsson, E. Kvamme, G. Svenneby, L. Hertz and A. Schousboe (1985a). Neurochem. Res. 10, 49-62.

Drejer, J., T. Honoré, E. Meier and A. Schousboe (1985b). Life Sci. Submitted.

Drejer, J., T. Honoré and A. Schousboe (1986) In: Excitatory Amino Acids (P.J.Roberts,Ed.) McMillan Press, Oxford, in press.

Elliott, K.A.C. and N.M. van Gelder (1958) J. Neurochem. 3, 28-40.

Hallermayer, K. and B. Hamprecht (1984). Brain Res. 295, 1-11.

Hansson, E. (1983) Brain Res. 289, 189-196.

Hansson, E. (1984) Neurochem. Res. 9, 153-172.

Henn, F.A. (1976) J. Neurosci. Res. 2, 271-282.

Henn, F.A. and A. Hamberger (1971) Proc. Natl. Acad. Sci. USA 68, 2686-2690.

Henn, F.A. and S.W. Henn (1980) Prog. Neurobiol. 15, 1-17.

Hertz, L. (1979). Prog. Neurobiol. 13, 277-323.

Hertz, L. (1982) In: Handbook of Neurochemistry, 2.ed. Vol.1 (Lajtha,A.,ed.) pp. 319-355. Plenum Press, New York.

Hertz, L. and A. Schousboe (1986) In: Model Systems of Development and Aging of the Nervous System (A.Vernadakis and E.Giacobini, Eds.) M.Nijhoff Publ.Comp. Boston, in press.

Hertz, L., A. Schousboe, N. Boechler, S. Mukerji and S. Fedoroff (1978). Neurochem. Res. 3, 1-14.

Hertz, L., B.H.J. Juurlink, H. Fosmark and A. Schousboe (1982). In: S.E.Pfeiffer(Ed.): Neuroscience Approached Through Cell Culture, Vol. 1. Boca Raton, Fl: CRC Press, pp. 175-186.

Hertz, L., E. Kvamme, E. McGeer and A. Schousboe (Eds.) (1983). Glutamine, Glutamate and GABA in the Central Nervous System, Alan Liss Inc., New York.

Hertz, L., I. Schousboe, L. Hertz and A. Schousboe (1984). Prog. Neuro-Psychopharmacol. Biol. Psychiat. 8, 521-527.

Krnjevic, K. (1974) Phys. Rev. 54, 418-540.

Krogsgaard-Larsen, P., I. Labouta, B. Meldrum, M. Croucher and A. Schousboe (1981). In: Neurotransmitters, Seizures and Epilepsy (Morselli,P.L., Reynolds,E.M., Lloyd,K.G., Löscher,W. and Meldrum,B.S.,Eds.) Raven Press, New York, pp. 23-33.

Kuffler, S.V. and C. Edwards (1958). J. Neurophysiol 21, 589-610.

Larsson, O.M., P. Krogsgaard-Larsen and A. Schousboe (1980). J. Neurochem. 34, 970-977.

Larsson, O.M., P. Thorbek, P. Krogsgaard-Larsen and A. Schousboe (1981). J. Neurochem. 37, 1509-1516.

Larsson, O.M., G.A.R. Johnston and A. Schousboe (1983a). Brain Res. 260, 279-285.

Larsson, O.M., J. Drejer, L. Hertz and A. Schousboe (1983b) J. Neurosci. Res. 9, 291-302.

Larsson, O.M., J. Drejer, E. Kvamme, G. Svenneby, L. Hertz and A. Schousboe (1985a). Int. J. Dev. Neurosci. 3, 177-185.

Larsson, O.M., P. Krogsgaard-Larsen and A. Schousboe (1985b). Neurochem. Int. 7, 853-860.

Meier, E. and A. Schousboe (1982). Dev. Neurosci. 5, 546-553.

Meier, E., J. Drejer and A. Schousboe (1984). J. Neurochem. 43, 1737-1744.

Meldrum, B.S., M.J. Croucher, M.J. and P. Krogsgaard-Larsen (1982). In: Problems in GABA Research from Brain to Bacteria (Y.Okada & E.Roberts, Eds.). Excerpta Medica, Amsterdam. pp. 182-191.

Norenberg, M.D. (1979). J. Histochem. Cytochem. 27, 756-762.

Schoffeniels, E., G. Franck, D.B. Tower and L. Hertz (Eds.)(1978). Pergamon Press, Oxford pp. 3-460.

Schousboe, A. (1978). In: Dynamic Properties of Glia Cells (E.Schoffeniels, G.Franck, D.B.Tower and L.Hertz, Eds.). pp. 173-182, Pergamon Press, Oxford.

Schousboe, A. (1981). Int. Rev. Neurobiol. 22, 1-45.

Schousboe, A. (1982a). In: Pfeiffer,S.E.(Ed.) Neuroscience Approached Through Cell Culture. Vol.I. CRC Press, Boca Raton, Florida, pp. 107-141.

Schousboe, A. (1982b). Scand. J. Immunol. 15 supl.9, 339-356.

Schousboe, A. and I. Divac (1979). Brain Res. 177, 407-409.

Schousboe, A. and L. Hertz (1983). In: Osborne NN (Ed.): Dale's Principle and Communication Between Neurones. Pergamon Press, Oxford, pp. 113-141.

Schousboe, A. and L. Hertz (1986). In: Model Systems of Development and Aging of the Nervous System. (A.Vernadakis & E.Giacobini, Eds.). M.Nijhoff Publ.Comp. Boston. In press.

Schousboe, A., L. Hertz and G. Svenneby (1977a). Neurochem. Res. 2, 217-229.

Schousboe, A., G. Svenneby and L. Hertz (1977b). J.Neurochem. 29, 999-1005.

Schousboe, A., P. Krogsgaard-Larsen, G. Svenneby and L. Hertz (1978). Brain Res. 153, 623-626.

Schousboe, A., P. Thorbek, L. Hertz and P. Krogsgaard-Larsen (1979). J. Neurochem. 33, 181-189.

Schousboe, A., J. Drejer and I. Divac (1980). Trends. Neurosci. 3, 11, XIII-XIV.

Schousboe, A., O.M. Larsson, L. Hertz and P. Krogsgaard-Larsen (1981). Drug Dev. Res. 1, 115-127.

Schousboe, A., O.M. Larssson, J. Drejer, P. Krogsgaard-Larsen and L. Hertz (1983a). In: Glutamine, Glutamate and GABA in the Central Nervous System. (L.Hertz, E.Kvamme, E.G.McGeer & A.Schousboe,Eds.) pp. 297-315. Alan R. Liss, Inc. N.Y.

Schousboe, A., O.M. Larsson, J.D. Wood and P. Krogsgaard-Larsen (1983b). Epilepsia 24, 531-538.

Schousboe, A., O.M. Larsson and P. Krogsgaard-Larsen (1985). Neurochem.Int. 7, 505-508.

Seiler, N., S. Sarhan, P. Krogsgaard-Larsen, H. Hjeds and A. Schousboe (1985). Gen. Pharmacol. 16, 509-511.
Ticku, M.K., A. Huang and J.L. Barker (1980). Mol.Pharmacol. 17, 285-289.
Wilkin, G.P., G. Levi, S.R. Johnstone and P.N. Riddle (1983). Dev. Brain Res. 10, 265-277.
Wood, J.D., A. Schousboe and P. Krogsgaard-Larsen (1980). Neuropharmacol. 19, 1149-1152.
Wood, J.D., D.D. Johnson, P. Krogsgaard-Larsen and A. Schousboe (1983). Neuropharmacol. 22, 139-142.
Yu, A.C.H., E. Hertz and L. Hertz (1984). J. Neurochem. 42, 951-960.

Heterogeneity Among Astrocytes, Evaluated by Biochemical Parameters

E. Hansson, Å. Nilsson[+], P. Eriksson*, M. Nilsson* and Å. Sellström

*Institute of Neurobiology, University of Göteborg, Göteborg, Sweden
[+]Division of Experimental Medicine, National Defence Research Institute, Umeå, Sweden

INTRODUCTION

Cell and tissue cultures constitute sources of metabolically active glial cells. The different types of cultures and their usefulness in studying specific research problems have previously been described (Murray, 1971; Herschman, 1974; Varon, 1975; Pfeiffer et al., 1978). The present review is mainly concerned with studies performed in astroglial-enriched, primary cultures from the central nervous system of rats.

Primary cultures originate from cells, tissues or organs taken directly from the organism and are regarded as primary until subcultivated. Such cultures provide the option of choosing the brain region and animal age at seeding. The newborn rat brain contains immature cells of various kinds which could survive in the tissue culture. The first subsection of this paper reviews the cellular composition of the primary astroglial culture. The second subsection reviews some results obtained with this culture regarding functional parameters expressed by astroglial cells in situ, i.e. transmitter uptake and catabolism, and the presence of receptors for some neurotransmitters (Schousboe et al., 1977a,b; Borg et al., 1980; Van Calker and Hamprecht, 1980; Hallermayer et al., 1981; Hansson, 1985b, Hansson et al., 1985a). The response of receptors was measured either as an increased adenyl-cyclase activity or as a depolarisation of the resting membrane potential. The functional differentiation of the culture is compared with that of the developing brain.

Specialization of astroglial cells in various brain regions has been suggested. Such specialization is most likely secondary to the functioning neuronal pathways (Henn, 1976; Drejer et al., 1982; Norenberg, 1979; Hansson, 1983, 1984a; Hansson et al., 1984c, 1985a) and the third subsection reviews some of the evidence for this functional specialization of astroglial cells.

The use of primary cultures as a tool to study the influence of environmental factors on the process of differentiation is also evaluated.

MORPHOLOGICAL AND BIOCHEMICAL CHARACTERIZATION OF PRIMARY CULTURES

Primary cultures of neonatal mammalian brain cells have been regarded as "pure" astroglial cultures. The majority of the cells assume the morphology of mature astrocytes, sending out long processes when treated with dB-cAMP or brain extract (Hertz et al., 1978a; Kimelberg et al., 1978; Moonen and Sensenbrenner, 1975; Shapiro, 1973; Lim et al., 1973; Sensenbrenner et al., 1980). Moreover, the astrocyte specific glial fibrillary acidic protein (GFAp) could be immunocytochemically detected in most cells (Bignami and Dahl, 1977; Eng et al., 1971).

A detailed analysis using specific markers for various cell types surviving in the culture is important as cells other than astrocytes have been detected (Hansson, 1984b; Hansson et al., 1984b). Additional astroglial markers might be the S-100 protein and glutamine synthetase activity, while markers for other cells might be the 14-3-2 protein or neuron-specific enolase (NSE) (neurons), macrophage acidic protein (MAP) (macrophages), alkaline phosphatase (endothelial cells), myelin basic protein (MBP) and 2',3'-cyclic nucleotide 3'-phosphohydrolase (CNP) (oligodendroglia). Using the above mentioned markers, astroglial-like cells were found to represent 60-70% of the cells present in cultures from cerebral hemispheres, cerebral cortex, striatum, hippocampus, brain stem and cerebellum. Also, phagocytic cells (macrophages), endothelial-like cells, mesenchymal-like cells, ependymal-like cells and oligoblasts were found. Neither mature neurons, nor mature oligodendroglial cells were observed (Hansson, 1984b; Hansson et al., 1984b).

FUNCTIONAL CHARACTERIZATION OF PRIMARY CULTURES

A characterization of the "functional" properties of primary cultures was used to evaluate the degree of cell differentiation. Astroglial cells are equipped with transport systems capable of translocating various amino acid neurotransmitters. A high-affinity, high-capacity uptake for glutamate, aspartate and GABA has been observed (Schousboe et al., 1977a, b; Hertz et al., 1978b; Borg et al., 1980; Drejer et al., 1983; Hansson, 1985b; Hansson et al., 1984a; 1985a). High-affinity uptake of norepinephrine and 5-hydroxytryptamine (5-HT) into astroglial cells has also been reported to be present (Kimelberg and Pelton, 1983; Suddith et al., 1978; Tardy et al., 1982). In our hands, however, this latter transport represents diffusion (Hansson et al., 1985a; Hansson, 1985b), the nature of which might be of physiological significance. The termination of neurotransmitter function is followed by intracellular enzymatic catabolism. The enzymes monoamine oxidase (MAO), catechol-O-methyltransferase (COMT), γ-aminobutyric acid transaminase (GABA-T) and glutamine synthetase (GS), which catabolize monoamines, GABA and glutamate, have all been determined in astroglial cell cultures (Schousboe et al., 1977a, b; Bardakdjian et al., 1979; Hallermayer et al., 1981; Hertz et al., 1978a; Juurlink, 1982; Hansson, 1984a; 1985c; Hansson and Sellström, 1983). The presence of receptors for different neurotransmitters and neuromodulators has also been demonstrated on astroglial cells (Van Calker and Hamprecht, 1980; Van Calker et al., 1980; 1983; Hansson et al., 1984c; Magistretti et al., 1983; Hertz et al., 1979; Hösli et al., 1984).

Tissue culture provides a unique opportunity to perform intracellular recordings in glial cells under direct visual inspection, allowing prior identification of the different cell types by morphological (Hösli and Hösli, 1978) and immunohistochemical criteria. A stable resting membrane potential, with the characteristics of a potassium diffusion potential, was obtained from astrocytes (Hösli et al., 1974; 1976; Kimelberg et al., 1979; Moonen et al., 1980; Bowman et al., 1984; Nilsson et al., 1985b). Recently, astroglial cells were shown to depolarise on the application of L-glutamate (Bowman and Kimelberg, 1984; Kettenmann et al., 1984), GABA and aspartate (Kettenmann et al., 1984) in a dose-dependent manner.

HETEROGENEITY AMONG ASTROCYTES

Regional variations in biochemical parameters are found in the brain and these may be related to the presence of neuronal pathways utilizing the different neurotransmitters. It is less well recognized that astroglial cells, which contribute to the control of neuronal activity, may also exhibit a regional heterogeneity. Morphological as well as biochemical studies have recently indicated that such heterogeneities may exist (Henn, 1976; Norenberg, 1979; Schousboe and Divac, 1979; Fedoroff, 1978; Hansson, 1983; 1984a; Hansson et al., 1984c; 1985a).

The presence of norepinephrine- and dopamine-sensitive adenylate cyclase was reported in bulk-prepared glial fractions with some variation in activity in the various brain regions examined (Palmer, 1973; Palmer and Manian, 1976; Henn et al., 1977). Although dopamine interacts to some extent with the β-adrenoceptors, the data from experiments with primary astrocyte cultures indicate the presence of dopamine receptors on groups of astroglial cells from striatum, further suggesting a specialization of these cells (Hansson et al., 1984c) (Fig. 1). Studies of primary astroglial cultures from various brain regions have revealed that the number of β-adrenoceptors per astroglial cell and/or their sensitivity vary from one brain region to another (Hansson et al., 1984c).

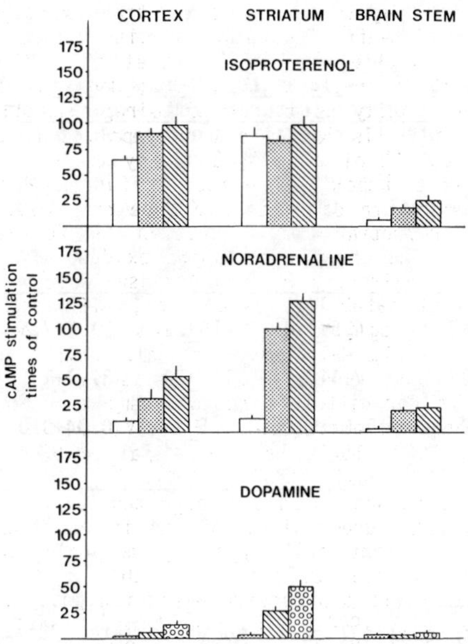

Fig. 1
cAMP accumulation by 10^{-7}-10^{-5}M isoproterenol, 10^{-7}-10^{-5}M noradrenaline and 10^{-6}-10^{-4}M dopamine in primary astroglial cultures from cerebral cortex, striatum and brain stem. The cultures were seeded from newborn rats and cultured for 2 weeks. For details, see Hansson (1985a).
□ 10^{-7}M; ▦ 10^{-6}M; ▧ 10^{-5}M; ▩ 10^{-4}M.

Astroglial cells participate in the inactivation and metabolic homeostasis of amino acid neurotransmitters (Hertz and Schousboe, 1980). The distribution of the glial transport system for glycine was shown to parallel the presumed distribution of glycine as an inhibitory neurotransmitter in the spinal cord but not in the cortex (Henn, 1976). Moreover, astroglial cells cultured from different brain regions exhibit different capacities for high-affinity glutamate, aspartate and GABA transport (Drejer et al., 1982; Hansson et al., 1985a). It was shown that the differences in their rates of the uptake were based on differences in V_{max} values (Table I). The highest uptake of glutamate, aspartate and GABA seems to be in astrocytes that originate from brain regions where the transmission of the respective amino

TABLE 1

Kinetic constants of ^{3}H-glutamate, ^{3}H-aspartate and ^{3}H-GABA uptake into primary astroglial cultures from cerebral cortex, striatum and brain stem. For details see Hansson et al. (1985a).

	Cerebral cortex		Striatum		Brain stem	
	K_m	V_{max}	K_m	V_{max}	K_m	V_{max}
Glu	34±10	4.5±1.4	110±22	39.1±7.5	11±3	0.4±0.1
Asp	97±32	7.4±2.1	187±64	37.1±11.6	88±30	3.1±0.9
GABA	6±2	0.01±0.001	5±1	0.04±0.01	17±6	0.1±0.04

acid is most pronounced (Fonnum et al., 1981). This suggests a relationship between the capacity of glial cells to accumulate glutamate, aspartate and GABA and the number of the respective terminals in a given brain region (Drejer et al., 1982). The highest density of high-affinity glial glutamate and aspartate transport sites were found on astrocytes cultured from striatum, and the highest density of glial GABA transport sites were on brain stem cultures. The uptake of the respective amino acids found in the transport studies is well in line with the accumulation of these substances by autoradiography (Hansson, 1983). Glutamine synthetase (GS), the enzyme that converts glutamate to glutamine, has been localized to astroglial cells by immunohistochemical methods (Norenberg, 1979). Consequently, the glutamate accumulated in the astroglial cell is inactivated by this enzyme, and the capacity for glutamate uptake has been shown to correlate with the activity of GS (Hertz et al., 1978a; Hansson, 1985c) (Fig. 2).

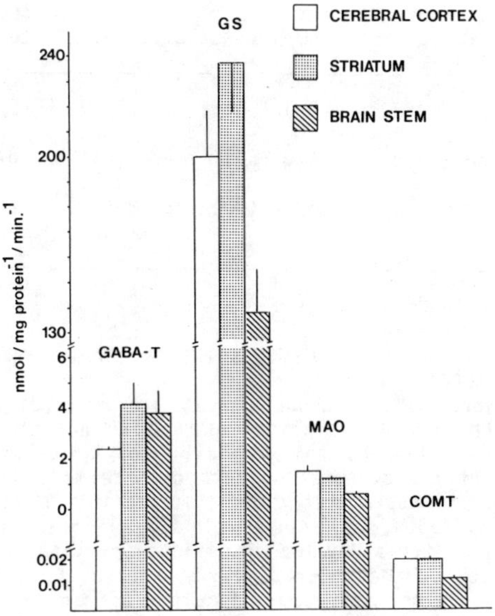

Fig. 2
GABA-T, GS, MAO and COMT activities in primary astroglial cultures seeded from newborn rats and grown 2 weeks in culture.
For details, see Hansson (1984a, 1985c).

The technique of intracellular recording has also been used to illustrate the heterogeneity of the primary astroglial culture (Nilsson et al., 1985a). Recently the response of astroglial-enriched cultures from the cerebral cortex, striatum and brain stem to L-glutamate, dopamine, isoproterenol and K^+ was compared (Table II) (Nilsson et al., 1985a). In all three cultures most cells responded distinctly by depolarisation when K^+ and L-glutamate was applied. Depolarising responses to applied dopamine and isoproterenol were less common. The variations in the sensitivity of the three regions to the three compounds are illustrated when the relative frequencies of positive responses are compared. Examples were observed of cells responding to more than one of the drugs, to one, two or all three of them.

TABLE II
Intracellular recording in astroglial cells from cerebral cortex, striatum and brain stem seeded from newborn rats and cultivated for 2 weeks

Glial cell origin	Membrane potential mV	Number of cells reacting positive (percentage)			
		Dopamine	Glutamate	Iso-pro-terenol	K^+
Cerebral cortex	-57 ± 11 (11)	5/9 (56)	10/12 (83)	5/7 (71)	6/6 (100)
Striatum	-69 ± 12 (9)	4/6 (67)	5/8 (63)	0/2 (0)	3/4 (75)
Brain stem	-62 ± 8 (12)	4/12 (33)	8/12 (67)	2/8 (25)	3/4 (75)

ASPECTS OF ASTROGLIAL CELL DIFFERENTIATION IN PRIMARY CULTURES FROM DIFFERENT BRAIN REGIONS.

Some brain regions of the neonatal rat are more differentiated than others. Therefore, the age of the animal at seeding and the time in culture might have variable influences on the cell growth, cellular content and maturational degree of the cultures. The importance of this aspect for the expression of different astroglial cell characteristics has recently been enlightened (Hansson, 1985a, c; Hansson et al., 1985b). Embryonic (17-day old fetus rats) cell cultures develop specific properties generally one week later than neonatal or postnatal ones (7-day old rats). Cultures from newborn rat cerebral cortex and striatum and cultures from newborn or embryonic rat brain stem show many similarities in the expression and development of some astroglial properties during cultivation although the extent of some functional parameters vary. It thus seems that the regional specialization of astroglial cell characteristics is not only a product of the degree of cell maturation in culture.

The differentiation of astroglial cells in primary cultures is relatively low (Hansson, 1982). Ways of increasing cell differentiation include the addition to the medium of partially isolated protein or peptide fractions from embryonic brain (Sensenbrenner et al., 1980), treatment of the cultures with conditioned media, i.e. serum-free media from neuronal containing cultures (Drejer et al., 1983; Peterson and Walum, 1983), or co-cultivation with neuronal containing cultures (Hansson, 1985d). These models proved interesting as they have led to some morphological and biochemical differentiation of astroglial cells in culture. However, the tissue culture technique has been limited by the necessity of using fetal calf-serum in the medium. Serum contains growth factors and hormones that could alter or mask the action of other compounds being examined. To eliminate the influence of unidentified factors, chemically defined media have been introduced (Morrison and de Vellis, 1983; Honegger et al., 1979; Romijn et al., 1984; Pettmann et al., 1982). The advantages are to create a defined model system, to facilitate multiplication and differentiation of desired cells and to suppress survival and growth of other cells. With these improvements, the regional heterogeneity of astroglial cells in brain can be examined in a defined culture system with more differentiated astroglial cells having at least some properties similar to those *in vivo*.

CONCLUSION

This review deals with the importance of characterization of the cellular composition of primary brain cultures by using different markers for the cells expected to be present. Although the cultures are not pure in astroglia, nor are the cells fully differentiated, they represent a valuable tool to study at least some astroglial properties, such as uptake and catabolism of neurotransmitters and receptor set-ups. Evidences for "functional" specializations of astroglial cells in various brain regions are discussed. The degree of cell differentiation might be increased by using e.g. defined growth media or by co-cultivation with neuronal-containing cultures.

ACKNOWLEDGEMENTS

The work presented in this paper was supported by the Swedish Medical Research Council (project Nos. 12X-06812 and 12P-7308).

REFERENCES

Bardakdjian, J., Tardy, M., Pimoule, C., and Gonnard P. (1979). Neurochem. Res. 4, 517-527.

Bignami, A. and Dahl, D. (1977). J. Histochem. Cytochem. 25, 466-469.

Borg, J., Ramaharobandro, N., Mark, J., and Mandel, P. (1980). J. Neurochem. 34, 1113-1122.

Bowman, C.L., and Kimelberg, H.K. (1984). Nature 311, 656-659.

Bowman, Ch.L., Kimelberg, H.K., Frangakis, M.V., Berwald-Netter, Y., and Edwards, Ch. (1984). J. Neurosci. 4, 1527-1534.

Drejer, J., Larsson, O.M., and Schousboe, A. (1982). Exp. Brain Res. 47, 259-269.

Drejer, J., Meier, E. and Schousboe, A. (1983). Neuroscience Letters 37, 301-306.

Eng, L.F. , Vanderhaeghen, J.J., Bignami, A. and Gerstl, B. (1971). Brain Research 28, 351-354.

Fedoroff, S. (1978). In "Dynamic Properties of Glial Cells" (E. Schoffeniels, G. Frank, L. Hertz, and D.B. Tower, eds.), pp. 83-92. Pergamon Press, New York.

Fonnum, F., Storm-Mathisen, J., and Divac, I. (1981). Neuroscience 6, 863-875.

Hallermayer, K., Harmening, C. and Hamprecht, B. (1981). J. Neurochem. 37, 43-52.

Hansson, E. (1982). Primary astroglial cultures. Aspects of morphology, biochemistry and transmitter metabolism. Thesis, Göteborg.

Hansson, E.(1983). Brain Research 289, 189-196.

Hansson, E. (1984a). Neurochem. Res. 9, 45-47.

Hansson, E. (1984b). Neurochem. Res. 9, 153-172.

Hansson, E. (1985a). Dev. Brain Res. 21, 187-192.

Hansson, E. (1985b). Neurochem. Res. 10, 667-675.

Hansson, E. (1985c). Dev. Brain Res. (in press).

Hansson, E. (1985d). Brain Res. (in press).

Hansson, E., Eriksson, P. and Nilsson, M. (1985a). Neurochem. Res. 10, 1335-1341.

Hansson, E., Isacsson, H. and Sellström, Å. (1984a). Acta Physiol. Scand. 121, 333-341.

Hansson, E., Rönnbäck, L., Lowenthal, A. and Noppe, M. (1985b). Dev. Brain Res. 21, 175-185.

Hansson, E., Rönnbäck, L., Persson, L.I., Lowenthal, A., Noppe, M., Alling, C., and Karlsson, B. (1984b). Brain Research 300, 9-18.

Hansson, E., Rönnbäck, L., and Sellström, Å. (1984c). Neurochem. Res. 9, 679-689.

Hansson, E. and Sellström, Å. (1983). J. Neurochem. 40, 220-225.
Henn, F.A. (1976). J. Neurosci. Res. 2, 271-282.
Henn, F.A., Anderson, D.J. and Sellström, Å. (1977). Nature 266, 637-638.
Herschman, H.R. (1974). In: Research Methods in Neurochemistry. Marks, N. Rodnight, R. (eds.) New York, Plenum Press Vol. II, pp. 101-160.
Hertz, L., Baldwin, F., and Schousboe, A. (1979). Can. J. Physiol. Pharmacol. 57, 223-226.
Hertz, L. Bock, E. and Schousboe, A. (1978a). Dev. Neurosci. 1, 226-238.
Hertz, L. and Schousboe, A. (1980). Brain Research Bulletin 5, Suppl. 12, 389-395.
Hertz, L., Schousboe, A. Boechler, N., Mukerji, S., and Fedoroff, S. (1978b). Neurochem. Res. 3, 1-14.
Honegger, P., Lenoir, D. and Favrod, P. (1979). Nature 282, 305-308.
Hösli, L. and Hösli, E. (1978). Rev. Physiol. Biochem. Pharmacol. 81, 115-188.
Hösli, L., Andrès, P.F. and Hösli, E. (1976). Neurosci. Letters 2, 223-227.
Hösli, L., Hösli, E and Andrès, P.F. (1974). In: Dynamics of Degeneration and Growth in Neurons. Fuxe, K., Olsson, L. and Zottermand, Y. (eds.) Oxford and New York, Pergamon Press, pp. 521-532.
Hösli, L., Hösli, E., Schneider, U., and Wiget, W. (1984). Neurosci Lett 48, 287-291.
Juurlink, B.H.J. (1982). Neurochem. Res. 7, 905-910.
Kettenmann, H., Backus, K.H., and Schachner, M. (1984). Neurosci. Lett. 52, 25-29.
Kimelberg, H.K. and Pelton, E.W. (1983). J. Neurochem. 40, 1265-1270.
Kimelberg, H.K., Narumi, S. and Bourke, R.S. (1978). Brain Research 153, 55-77.
Kimelberg, H.K.,Bowman, C.L., Biddlecome, S., and Bourke, R.S. (1979). Brain Research 177, 533-550.
Lim, R., Mitsunobu, K. and Li, W.K.P. (1973). Exp. Cell Res. 79, 243-246.
Magistretti, P.J., Manthorpe, M. Bloom, F.E. and Varon, S. (1983). Regulatory Peptides 6, 71-80.
Moonen, G. and Sensenbrenner, M. (1975). Experientia 32, 40-42.
Moonen, G., Franck, G., and Schoffeniels, E. (1980). Neurochem. Intl. 2, 299-310.
Morrison, R.S. and de Vellis, J. (1983). Dev. Brain Res. 9, 337-345.
Murray, M.R. (1971). In: Handbook of Neurochemistry. Lajtha, A. (ed.) New York-London, Plenum Press, Vol. 5, pp. 373-438.
Nilsson, Å., Hansson, E. and Sellström, Å. (1985a). (Submitted).
Nilsson, Å., Lidén, E., and Sellström, Å. (1985b). Submitted for publication.
Norenberg, M.D. (1979). J. Histochem. Cytochem. 27, 756-762.
Palmer, G.C. (1973). Res. Commun. Chem. Pathol. Pharmacol. 5, 603-613.
Palmer, G.C. and Manian, A.A. (1976). Biochem. Pharmacol. 25, 63-71.
Peterson, A. and Walum, E. (1983). In Vitro 19, 875-880.
Pettmann, B. Weibel, M., Daune, G., Sensenbrenner, M. and Labourdette, G. (1982) J. Neurosci. Res. 8, 463-476.
Pfeiffer, S.E., Betschart, B. Cook, J. Mancini, P., and Morris, R. (1978). In: "Cell, Tissue and Organ Cultures in Neurobiology" (S. Fedoroff and L. Hertz, Eds.) pp. 287-346. Academic Press, New York.
Romijn, H.M., van Huizen, F. and Walters, P.S. (1984)Neuroscience & Biobehavioral Reviews 8, 301-334.
Schousboe, A. and Divac, I. (1979). Brain Research 177, 407-409.
Schousboe, A. Hertz, L. and Svenneby, G. (1977a). Neurochem. Res. 2, 217-229.
Schousboe, A., Svenneby, G. and Hertz, L. (1977b). J. Neurochem. 29, 999-1005.
Sensenbrenner, M. , Devilliers, G., Bock, E. and Porte, A. (1980) Differentiation 17, 51-61.
Shapiro, D.L. (1973). Nature 241, 203-204.
Suddith, R.L., Hutchison, H.T., and Haber, B. (1978). Life Sci. 22, 2179-2188.

Tardy, M., Costa, M.F.D., Fages, C., Bardakdjian, J., and Gonnard, P. (1982). Dev. Neurosci. 5, 19-26.
Van Calker, D. and Hamprecht, B. (1980). In: "Advances in Cellular Neurobiology" (S. Fedoroff and L. Hertz, eds.) Vol. 1, pp. 31-67. Academic Press, New York.
Van Calker, D., Löffler, F., and Hamprecht, B. (1983). J. Neurochem. 40, 418-427.
Van Calker, D., Müller, M., and Hamprecht, B. (1980). Proc. Nat. Acad. Sci (USA) 77, 6907-6911.
Varon, S. (1975). Exp Neurol. 48, 93-134.

The Role of Neurosecretory Cells in Learning and Memory

V. E. Shashoua

Ralph Lowell Laboratories, Harvard Medical School, McLean Hospital, Belmont, MA 02178, USA

INTRODUCTION

The cellular and molecular machinery of the nervous system has the capacity to code learned information in a suitable form for storage as a long-term memory. In man, such memories are stable for upwards of 50 years. If we search the components of the CNS looking for molecules that can fulfill the criterion of long-term stability, we find that everything, except the DNA, is in a dynamic state. Proteins have an average half-life of less than 14 days (Lajtha and Toth, 1966), RNA turnover rates vary between one-half to 24 hr (Appel, 1967), and lipids and carbohydrates are in a rapid state of flux (Bourre et al., 1977). The possibility that DNA itself could somehow be modified to encode learning has, as yet, no supporting evidence. It would seem, therefore, that only the patterns of structure and connectivity of the central nervous system have features with sufficient stability for such a purpose. This idea was initially proposed in 1893 by Tanzi, soon after Ramon y Cajál demonstrated that the neuron is the basic unit of the nervous system. If patterns of structure and connectivity are indeed altered during the formation of a long-term memory, then the search for biochemical mechanisms essential for the process reduces to the identification of molecular events that can ultimately lead to alterations of neural circuits. In this respect, the process of learning and memory can be regarded as a special adaptation of the neural plasticity that occurs during development (Shashoua, 1968, 1972). Thus the nervous system, in response to learning, must be capable of forming new synaptic connections via molecular mechanisms that can be turned on and off at highly localized microscopic sites.

In our laboratory we have used goldfish and mice as experimental animals and double-labeling techniques to identify a specific class of glycoproteins whose turnover rate is enhanced after training (Shashoua, 1976, 1982). These macromolecules were found subsequently to be components of the brain extracellular fluid (Shashoua, 1979). Immunohistochemical methods, using polyclonal and monoclonal antisera, showed that the proteins were released from a class of cells which are highly concentrated in the "zona ependyma" of goldfish brain. This zone has a preponderance of glial cells. We have proposed the name "Ependymins" for this class of protein. Investigations of the chemical properties of the ependymins suggest that they may have a functional role in the modification of neural circuits (Shashoua, 1985).

Advances in the Biosciences Vol. 61

Experiments Identifying the Role of Ependymins in Learning

Figure 1 illustrates the behavioral paradigm used for studying the effect of training on the pattern of protein synthesis in goldfish brain. In this "vestibular-motor" conditioning experiment, goldfish initially suspended in an upside-down position by a float sutured 1 mm caudal to their first pair of pectoral fins adapt to the float through a series of reproducible stages until they can swim upright in a horizontal posture (Shashoua, 1968). The task involves the conditioning of vestibular, cerebellar, and lateral line systems in association with motor circuits that govern the swimming behavior of the animals. The goldfish master the training in 4-5 hr. The procedure was devised to provide a large challenge to each animal in an attempt to accentuate any brain biochemical changes that may occur and thus make them detectable. Experimentally the performance of each group of 7 goldfish is evaluated according to a fixed set of criteria as outlined in Figure 1. When trained goldfish are retested 3 or even 14 days later, they are able to swim in a horizontal posture within 5-10 min after the floats are reattached, demonstrating that there is retention of the previous experience. We have used this behavioral paradigm to look for macromolecular changes in the central nervous system of the goldfish. Using double-labeling methods, the trained goldfish, at 1 hr after completion of training, are injected with tritiated valine while 7 untrained controls in a separate tank receive [^{14}C]valine into their 4th brain ventricles (Shashoua, 1976). Following a labeling time of 1 hr, the brains of the two groups are homogenized together and separated into various subcellular fractions (Whittaker, 1959). Each of these fractions contains [^{3}H]-labeled proteins derived from the trained fish and [^{14}C] from the controls (in some experiments the labels were reversed). After separation of the proteins by SDS gel electrophoresis, the [^{3}H]/[^{14}C] ratio was found to show changes in only the cytoplasmic fraction at three specific protein bands corresponding to molecular weights of 43,000 (α), 37,000 (β) and 31,000 (γ). The average increased label incorporated was found to be 46%, 59% and 72%, respectively. A variety of control experiments, examining the effects of physiological activity, stress, and the repeat performance of the training by previously trained groups (i.e., without any significant learning), showed no such protein changes. These results suggested that the effect of the training procedure itself was a major factor in producing the protein changes. The data indicate that the turnover rate of the α, β and γ proteins is enhanced after training and that the metabolism of these proteins may have a relationship to the process of acquisition of a new pattern of behavior.

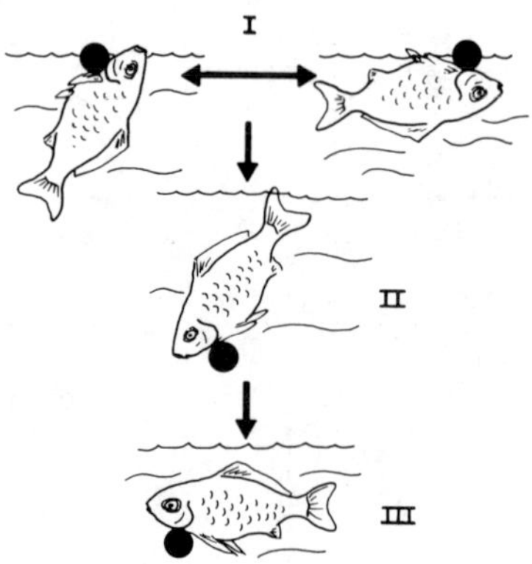

Fig. 1. Diagram showing principal stages in the adaptation of a goldfish to swimming with a float in the vestibular-motor conditioning experiment. I, II and III represent swimming postures at 0%, 50% and 100% training stages.

Isolation and Partial Characterization of the Ependymins

The β and γ proteins were isolated and purified from extracts of goldfish brain by a combination of affinity chromatography, gel electrophoresis and sephadex gel chromatography (Shashoua, 1977a). Both were found to be acidic proteins (see Table 1). Additional studies (Schmidt and Shashoua, 1983) indicated that β and γ exist as disulfide-linked β_2, βγ and γ_2 dimers in the native state. All three forms of the dimer are convertible by heating in 1% SDS or by the addition of mercaptoethanol to the monomeric β and γ chains. β and γ contain at least 5% and in some preparations as much as 20% covalently bound carbohydrate to each chain. They can bind to concanavalin A (Shashoua, 1982). Recent studies indicate that the α component (M.W.43K) is also derived by proteolysis from the β_2 dimer (Shashoua, 1986). Polyclonal antisera raised against the purified β or γ chains were found to cross-react with the α, β and γ forms of the molecules. Peptide maps of proteolytic digests of β and γ were found to be identical, with the exception of one extra β-derived peptide fragment. $[^{125}I]$-labeled β gives rise to $[^{125}I]$-labeled γ polypeptide among its protease digestion products, whereas $[^{125}I]$-labeled γ gave no component corresponding to β. These observations suggest that β_2 might be the precursor which gives rise to the other forms of the molecule (Schmidt and Shashoua, 1983).

Table 1. Amino Acid Compositions of Ependymins[a]

Amino Acid	Ependymin β (mol %)	Ependymin γ (mol %)
Asx	8.9	9.2
Thr	5.8	6.3
Ser	8.9	9.2
Glx	13.1	12.5
Pro	5.5	6.3
Gly	10.0	10.8
Ala	6.2	6.3
Val	5.8	5.8
Cys	1.0	0.8
Met	0.7	0.4
Ile	3.4	3.3
Leu	7.2	7.5
Tyr	2.4	2.5
Phe	5.8	5.4
Orn	0.7	0.8
Lys	6.2	5.8
His	4.8	3.8
Arg	3.4	3.3
% Acidic AA	22	21.7
% Basic AA	15.1	13.7

[a]The data represent the mean of 10 independent preparations. Tryptophan was not analyzed.

Immunohistochemical studies of the neuroanatomical distribution of the β and γ proteins in the brains of untrained and trained goldfish (Shashoua, 1977b) showed them to be present in the cytoplasm of specific cells as well as in fibrous processes (see Fig. 2, 3). A goldfish brain has a system of ependymin-containing cells distributed along the periventricular gray throughout its neuraxis (Benowitz and Shashoua, 1977). The most prominent locus for such cells (E-cells) was found to be in the "zona ependyma" beneath the optic tectum of goldfish brain. This periventricular localization of E-cells raises the possibility that ependymins may be a class of glycoproteins which are released from the site of their synthesis into the extracellular fluid for a function at some distant target.

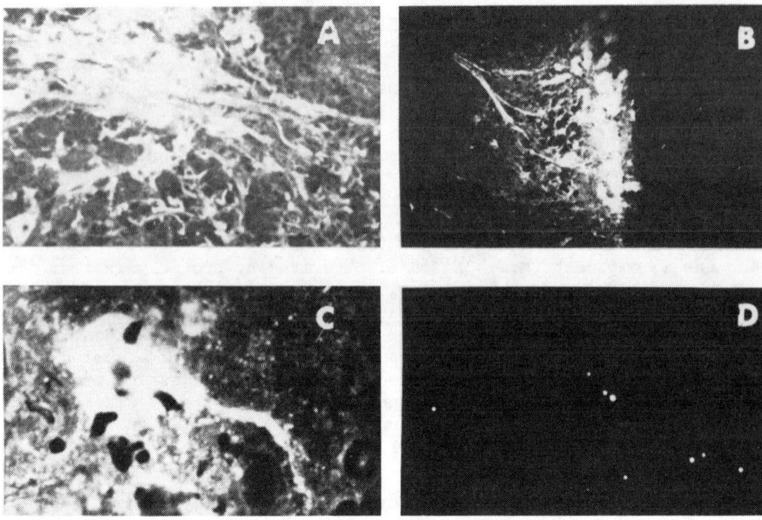

Fig. 2. The immunohistofluorescence localization of cells and fibers stained with polyclonal antisera to ependymin β and γ. A = zona ependyma beneath optic tectum (150 X); B = basal forebrain (150 X); C = single cell in dorsal tegmentum (1000 X); D = zona ependyma stained with preimmune serum: Note that nuclei do not stain.

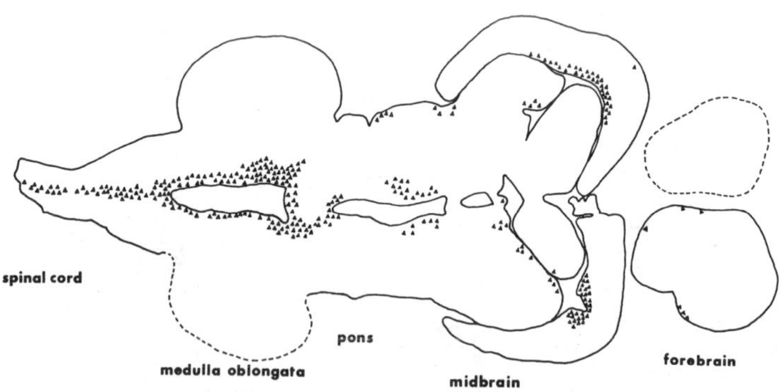

Fig. 3. Horizontal section of goldfish brain showing the immunohistochemical localization of ependymin β cells.

Evidence for Release of Ependymins in the Brain Extracellular Fluid

Several types of experiments indicate that the ependymins are released from their site of synthesis into the extracellular space to become highly localized in the brain extracellular fluid (ECF). (1) Goldfish brain cerebrospinal fluid (CSF) and ECF, when plated against ependymin antisera, were found to give precipitin bands identical to those of the pure antigen in an Ouchterlony assay (Shashoua, 1979). (2) Quantitative measurements, using [^{125}I]-labeled ependymin in a radioimmunoassay, showed that about 14% of the total proteins present in goldfish brain ECF fractions consist of ependymins (Schmidt and Shashoua, 1981).

(3) Double-labeling studies (Shashoua, 1979) of the pattern of protein synthesis after training followed by isolation of an enriched ECF fraction indicated that most of the enhanced labeling obtained after training was localized at the molecular weight position of the β and γ ependymins. This suggests that the ependymins are synthesized in specific cells and rapidly released into the extracellular space after learning.

Additional evidence (Majocha et al., 1982) for the release of these proteins into the ECF was obtained from tissue culture experiments. The "zona ependyma" cells of goldfish brain were grown _in vitro_ on polylysine-coated plates. These contained a variety of glial as well as some neural cells. Immunohistochemical studies of the cultured cells indicated that a large number of cells resembling glial cells were stained with polyclonal antisera to the ependymins. Analysis of the growth medium by radioimmunoassay showed that a progressive increase in the concentration of the ependymins occurred in the media without any significant change in the amount present in cells. In addition, labeling studies with [^{3}H]-valine showed that newly synthesized ependymins were rapidly released into the medium. These data support the hypothesis that certain types of cells present within the "zona ependyma" (E-cells) release the ependymins into the brain ECF and suggest that the role of such cells may be to synthesize the ependymins and maintain homeostatic levels of proteins within the extracellular space, ready for use at targets distant from the place of synthesis. The generality of this observation is supported by the finding that mammalian brain also contains ependymin-like proteins which are released into ECF and CSF and whose turnover rate is enhanced by the acquisition of a new pattern of behavior (Shashoua and Moore, 1980).

Table 2. Studies of the Effect of Ependymin Antisera on Behavior

		Trained + Antisera[b]		Trained + Preimmune Serum[c]		
Antiserum	Time (hr)[a]	No. of Animals	Retention Score (%)	No. of Animals	Retention Score (%)	t-Test P Value
β + γ	+8	21	36 ± 19	21	94 ± 7	0.0025*
β	+8, +20	21	25 ± 11	21	94 ± 16	0.0005*
β + γ	+8, +20	98	49 ± 21	97	92 ± 27	0.0005*
β + γ	+48	15	28 ± 33	14	81 ± 27	ns[d]
β + γ	+72	13	70 ± 32	12	77 ± 37	ns
			Control experiments			
β + γ	-0.5	21	90 ± 13	21	92 ± 6	ns
β + γ	-24	21	94 ± 11	21	120 ± 35	ns
NS-6[e]	+8	20	89 ± 39	20	120 ± 38	ns
AE[f]	+8, +20	12	82 ± 32	11	71 ± 24	ns

[a]Injection time before or after initiation of training. [b]Trained goldfish injected with antisera to ependymins. [c]Trained goldfish injected with preimmune sera. [d]Not significant. [e]NS-6 antiserum to another brain antigen. [f]AE antiserum to ependymin presorbed with antigen to remove the ependymin-specific IgG. *Highly statistically significant.

Amnestic Effects of Antisera to the Ependymins

The observation that goldfish brain contains a class of macromolecules whose metabolism is enhanced after training suggests a number of explanations for the role of such molecules in the CNS. One possibility is that the whole system of the β and γ cells is activated by training. In this case, the ependymins would be merely markers for specific neural circuits that are activated by the training procedure and thus would have no direct functional role in the mechanism of long-term memory formation. Another possibility is that the ependymins might be participating directly in some aspect of the neural plasticity process. To test this hypothesis, we investigated the effects of antisera on the retention of training (Shashoua and Moore, 1978). The procedure was to train animals and then inject antisera to the ependymins into the 4th ventricle of goldfish brain. Control animals from the same groups, trained at the same time, were injected with normal or preimmune rabbit serum. The experiments were run blind. After 3 days the goldfish were tested for recall. Table 2 shows the results of such experiments. Animals that received injections of antisera to the ependymins were not able to recall the behavior, whereas animals receiving preimmune sera had good retention. Thus far, 500 animals have been tested by this procedure, yielding results that are statistically very highly significant.

There are several explanations for these observations. One is that the antisera might generate a toxic reaction and thus indirectly cause amnestic behavior. As a control for this, the effects of injections of the anti-ependymin sera prior to training were tested. Such injections at 30 min and 24 hr before training were found to have no influence on the rate of learning or the recall of the behavior. This suggests that neither acute nor delayed toxicity factors can be important in the phenomenon. Another possibility is that any antiserum to any brain-specific protein might produce amnesia. To test for this, an antiserum to brain-specific protein NS-6 (Chaffee and Schachner, 1978) was tested and found to have no effect. In addition, antisera to ependymins adsorbed with the antigen were found to have no effect on either the rate of acquisition or the recall of behavior, suggesting that the specific IgG molecules directed against the ependymins may be important for the observed biological effects. A study of the time course of the amnestic effects of the ependymin antisera indicated that there exists a window of time between 3 and 24 hr after training during which injections of antisera are effective in blocking the recall of the behavior. Injections between 48 and 72 hr after training were found to produce no statistically significant amnestic effects. These observations suggest that the learning process may actually initiate a series of sequential biochemical events which at some stage require the participation of extracellular proteins such as the ependymins. Once this step is completed, i.e., about 24 hr later, the process seems to be no longer vulnerable to attack by the antisera to the ependymins.

Isolation and Characterization of Goldfish Brain Extracellular Fluid Proteins

Goldfish brain is surrounded by a large external fluid that is commonly referred to as the extradural fluid (EDF). This occupies most of the space in the cranial cavity and can be removed from perfused goldfish brains by direct aspiration without disturbing the nervous system or causing the release of any other brain fluids (Shashoua, 1981). In addition, the brain contains the CSF which is localized within the ventricular spaces, and the ECF, which is present throughout the brain extracellular space. We have developed methods for isolating enriched fractions of each of these fluids and have analyzed some of their protein composition and properties.

Goldfish (7-9 g weight) anesthetized in a solution of 0.01% tricanemethanesulfonate are perfused with 20 cc of normal saline through the heart. This procedure displaces most of the blood and serum proteins from the brain capillaries. After perfusion, the heart-shaped skull cap can be carefully cut out to expose the brain with its surrounding EDF, which can then be aspirated with a Teflon capillary 0.5 mm I.D. without disturbing the brain. An average of 2-3 µl of fluid can be easily removed from each animal and placed into 0.32 N sucrose containing 2.5 mM calcium acetate. Such fluids contain an average of 0.5% protein. For isolation of CSF, the intact brain tissue exposed in the cranial cavity is irrigated with isotonic sucrose containing 2.5 mM calcium at 0°C. This allows the leakage of the CSF from the ventricular fluid into the extraction medium to yield an enriched CSF protein fraction in 0.32 N sucrose. The ECF fraction is obtained by carefully dissecting out the brain from the cranial cavity and extracting it at 0°C in 0.32 N sucrose containing 2.5 mM calcium for 2 hr. It is essential to maintain the 0°C temperature to allow the diffusion of the ECF into the extraction medium in order to minimize protein leakage from intact cells. Enriched protein fractions of ECF obtained by this method yield approximately 1 mg protein from each goldfish brain. Studies of kinetics of the extraction process and measurements of the volume of the extracellular space using [^{14}C]-inulin (Shashoua, 1981) indicate that the ECF contains approximately 1% protein. Similar methods were used to isolate enriched ECF protein fractions from mouse brain (Hofstein et al., 1983). The gel electrophoretic patterns of ECF and CSF proteins of goldfish brain are essentially identical to the pattern of the EDF fraction, which is isolated without disturbing the brain. This suggests that there is no major contamination of the ECF by leakage of cytoplasmic proteins. Additional evidence for the validity of the extraction method was obtained by enzyme assays. Measurement of lactate dehydrogenase (LDH) and tyrosine hydroxylase (characteristic markers for cytoplasmic fraction) showed that in the enriched mammalian ECF extracts there could be no more than 2% contamination by specific leakage of these markers. In addition, two-dimensional gel electrophoretic patterns of the enriched ECF extract were completely different from those of the cytoplasmic protein fraction (Hofstein et al., 1983).

Additional evidence for the release of extracellular proteins by brain tissue was obtained by studying the release of proteins from rat brain hippocampal slices (Hesse et al., 1984). Such slices can be maintained in artificial CSF with their physiological and biochemical characteristics intact (Dingledine et al., 1980; Duffy et al., 1981). Physiologically viable slices of the rat hippocampus in vitro were found to continuously release proteins into a superfusion medium at a rate of about 2 μg/mg tissue per hr. The level of LDH in the released fraction was equal to that present in CSF, indicating that the released material was not the result of cell lysis. Pulse-chase experiments using [^{3}H]-valine indicate that the releasable protein pool has an apparent half-life of about 4 hr. Simultaneous labeling of the newly synthesized protein with [^{3}H]-fucose and [^{14}C]valine showed a 3-fold higher ratio of the fucose:valine in the released protein products as compared to the soluble cytoplasmic protein and the crude membrane protein fraction. This suggests that the soluble released proteins are more highly glycosylated than the proteins retained in the tissue. The released proteins had molecular weights and immunostaining properties (Western blots) identical to the monomeric and dimeric forms of the ependymins. Similar results were obtained using organ culture of goldfish brain optic tecta.

Evidence for the Presence of Ependymins in Mammalian Brain

Ependymin-like proteins are present in mammalian brain (Shashoua and Moore, 1980). Double-labeling studies using a specific strain of mice (Balb/c) as the experimental animal and a self-return T-maze indicated that the only proteins whose turnover rate was increased after training had similar molecular weights to the goldfish brain ependymins (Shashoua, 1982). Such proteins were released into the ECF after training and consisted of a variety of molecular weight types. Dimeric products, molecular weights 68K, 54K, and 59K, were present, and these could be dissociated into 38K, 30K, and 19K proteins by incubation with ECF at room temperature for several hours. Polyclonal antisera raised against the 30K protein were found to cross-react with the monomeric 38K as well as with the dimeric forms of the molecule (Western blots). Immunohistochemical studies localize the cells that contain ependymin-like proteins in the brains of both trained and untrained mice. These cells were found to be highly concentrated in the hippocampus, caudate, putamen and thalamus, as well as in some deep cortical layers (Shashoua, 1985). More recent studies indicate that certain monoclonal antibody preparations were equally active against mouse and goldfish brain ependymins by Western blot analysis, suggesting that common epitopes may be present in the molecules derived from the two species.

Aggregation Properties of Ependymins and Extracellular Fluid Proteins

The brain ECF is an important communication medium in the CNS. It contains proteins and peptide factors which can initiate various aspects of growth

(Patterson and Chun, 1977), the differentiation of neural cell types (Arenander and De Vellis, 1981; Williams and Varon, 1985), and the formation of long-term memory. The ependymins are the principal components of goldfish brain ECF. They are highly soluble glycoproteins in the presence of aqueous media containing physiological concentrations of electrolytes and calcium (2.5 mM in goldfish CSF). If calcium is removed from such solutions by dialysis at 4°C or by the addition of a calcium chelating agent (EGTA), the proteins spontaneously form insoluble aggregates (Shashoua, 1985). Similar changes are obtained if pure ependymins are used for the polymerization reaction. The initiation and propagation of the polymerizations were found to require critical concentrations of EGTA. Figure 4 shows the light- and electron-microscopic data for the type of aggregated products obtained. The extracellular matrix once formed could not be redissolved in high-calcium media or in most protein solvents including boiling 1% SDS in 6 M urea and even trifluoroacetic acid. Monoclonal and polyclonal antisera raised against the ependymins were found to stain the fibrous aggregates, whereas preimmune sera did not. These results suggest that the capacity to aggregate may be built into the molecular structure of the ependymins and that the aggregated products may perhaps be related to the formation of cross-linked structures.

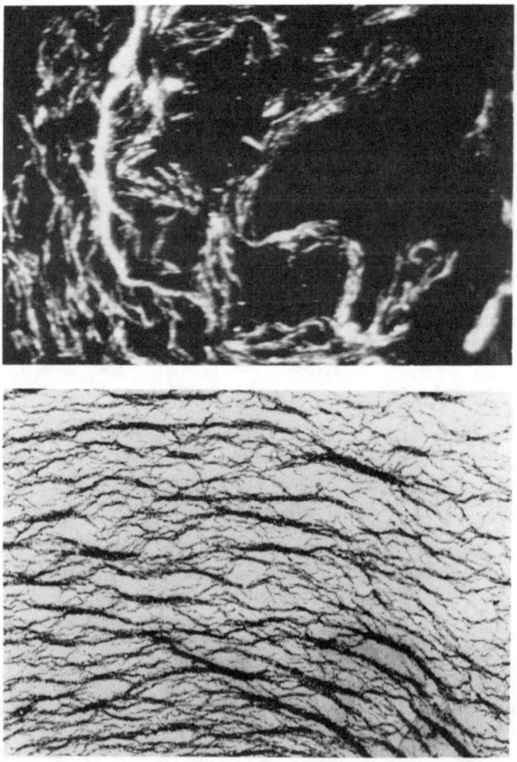

Fig. 4. Optical and electron-microscopic photographs of polymerized goldfish brain extracellular proteins. (A) Photographs (dark-field optical microscopy) of fibrous ependymin aggregates obtained 10 min following the initiation by EGTA. (B) An electron micrograph, negatively stained, of the fibrous products obtained at 45 sec after the addition of EGTA. (A) 100X, (B) 2000 X.

A Hypothesis for the Role of ECF Proteins and Ependymins in Learning and Memory

Figure 5 illustrates a working hypothesis for the role of extracellular matrix-forming proteins in neural plasticity and learning. The ependymins are normally present as soluble molecules throughout the brain extracellular space. During learning, at specific sites where an association of input occurs, a transient depletion of ECF calcium could occur to convert the soluble ependymins into an insoluble fibrous matrix. Such a matrix essentially defines the locus at which changes in pre-existing pre- and post-synaptic elements or the growth of new synapses will eventually take place. The notion that an extracellular matrix can provide a pathway for the development of neural connections has been proposed previously in studies of neural development (Bonner-Frazer and Cohen, 1980) and the regeneration of neuromuscular junctions (McMahan et al., 1980; Sanes, 1983). Thus ependymins are always present in the ECF as soluble proteins ready to be converted into an insoluble matrix at specific sites where a signal for their transformation is initiated by Ca^{2+} depletion.

The hypothesis essentially has three requirements: (1) The ECF must have a steady supply of soluble ependymins. (2) A transient signal-generating mechanism must exist in the CNS which can distinguish normal use during general activity from the type of novel use that occurs during training, i.e., a process for removal of calcium from the extracellular environment which is much larger during stimulation of multiple converging inputs at a locus than during the functional use of a single pathway. (3) The initiation signal for forming an extracellular matrix (i.e., the calcium-depletion mechanism) should be rapid, transient and localized at specific sites. All these aspects of the hypothesis have some supporting evidence. In the goldfish the supply of ependymins in ECF seems to be well regulated. They are among the rapidly turning-over proteins in the brain with an approximate turnover half-life of 4 hr. Moreover, the ependymin neurosecretory cells with their multiple processes are capable of maintaining a homeostatic level of the proteins in the ECF. A number of electrophysiological studies have established the concept that calcium fluxes are the first chemical changes that occur during associative learning in the marine snail *Hermissenda* (Alkon, 1980). A transient accumulation of calcium occurs within specific cells as a result of a pairing of inputs. Similar changes have been postulated in another marine snail, *Aplysia* (Carew et al., 1981). In both experiments the opening of

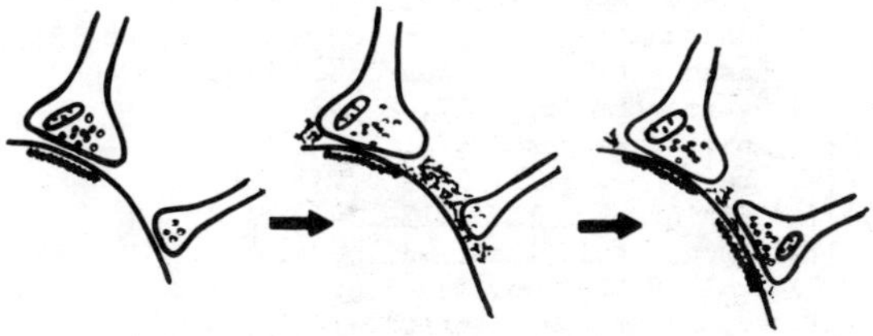

Fig. 5. Diagram illustrating a working hypothesis for the role of ependymins in learning. The figure depicts two pathways (a "strong" and a "weak") converging onto a single dendrite. During associative learning a sequential or simultaneous firing of the strong (large synaptic area) together with the weak pathway (synapse with small contact area) results in a depletion of extracellular Ca^{2+} at a locus between the two inputs. An extracellular fibrous matrix forms (see middle drawing) by the polymerization of the ependymins at the region of greatest depletion of Ca^{2+} within the extracellular space. This is then followed by a growth of the weak (developmentally immature) synapse along the path defined by the extracellular matrix to form a strong synapse. As a consequence of this event, an originally weak signaling point in a neural circuit is transformed into a powerful one which can independently signal the dendrite.

calcium channels presumably allows the influx of extracellular calcium to deplete Ca^{2+} at a local site in the extracellular space. It is conceivable that during associative learning the depleted calcium will initiate polymerization of the ependymins. It is also possible that local Ca^{2+} depletion in ECF may also be caused by the release of a Ca^{2+} chelating factor. *In vivo* and *in vitro* experiments in mammalian brain (Krnjević et al., 1982a,b) have shown that the extracellular environment in hippocampal pyramidal cells during the stimulation of multiple pathways shows a transient Ca^{2+} decrease by about 70%. Such changes are certainly within the concentration levels necessary to initiate the rapid polymerization of ependymins into a localized extracellular matrix. These findings support the ependymin polymerization hypothesis. We do not yet know if it is valid. It is, however, a testable one and suggests a role for brain extracellular proteins: one step in a complex sequence of cell biological events that ultimately lead to the formation of long-term memory.

References

Alkon, D.L. (1980) Biol. Bull., 159, 505-560.

Appel, S.H. (1967) Nature (Lond.), 213, 1253-1254.

Arenander, A.T. and J. De Vellis (1981) Brain Res., 224, 117-127.

Benowitz, L.I. and V.E. Shashoua (1977) Brain Res., 136, 227:242.

Bonner-Frazer, M. and A.M. Cohen (1980) Dev. Biol., 77, 130-141.

Bourre, J.M., S. Pollet, M. Paturneau-Jovas, and N. Baumann (1977) Adv. Exp. Med. Biol., 83, 103-109.

Chaffee, J. and M. Schachner (1978) Dev. Biol., 62, 173-184.

Carew, T.L., E.T. Walters, and E.R. Kandel (1981) Science, 211:501-504.

Dingledine, R., J. Dodd, and J.S. Kelly (1980) J. Neurosci. Meth., 2, 323-362.

Duffy, C., T.J. Teyler, and V.E. Shashoua (1981) Science, 212, 1148-1151.

Hesse, G., R. Hofstein, and V.E. Shashoua (1984) Brain Res., 305, 61-66.

Hofstein, R., G. Hesse, and V.E. Shashoua (1983) J. Neurochem., 40, 1448-1455.

Krnjević, K., M.E. Morris, and R.J. Reiffenstein (1982a) Can. J. Physiol. Pharmacol., 60, 1643-1657.

Krnjević, K., M.E. Morris, R.J. Reiffenstein, and N. Ropert (1982b) Can. J. Physiol. Pharmacol., 60, 1658-1671.

Lajtha, A. and J. Toth (1966) J. Biochem. Biophys. Res. Comm., 23, 249-299.

Majocha, R.E., R. Schmidt, and V.E. Shashoua (1982) J. Neurosci. Res., 8, 331-342.

McMahan, V.J., D.R. Edgington, and D.P. Kuffler (1980) J. Exp. Biol., 89, 31-38.

Patterson, P.H. and L.L.Y. Chun (1977) Devel. Biol., 56, 263-280.

Sanes, J.R. (1983) Annu. Rev. Physiol., 45, 581-600.

Schmidt, R. and V.E. Shashoua (1981) J. Neurochem., 36, 1368-1377.

Schmidt, R. and V.E. Shashoua (1983) J. Neurochem., 40, 652-660.

Shashoua, V.E. (1968) Nature (Lond.), 217, 238-240.

Shashoua, V.E. (1972) Int. J. Neurosci., 3, 299-304.

Shashoua, V.E. (1976) Brain Res., 111, 347-367.

Shashoua, V.E. (1977a) Brain Res., 122, 113-124.

Shashoua, V.E. (1977b) Proc. Natl. Acad. Sci. USA, 74, 1743-1747.

Shashoua, V.E. (1979) Brain Res., 166, 349-358.

Shashoua, V.E. (1981) Neurochem. Res., 6(10), 1129-1147.

Shashoua, V.E. (1982) Adv. Cell. Neurobiol., 3, 97-141.

Shashoua, V.E. (1985) Cell. Molec. Neurobiol., 5, 183-207.

Shashoua, V.E. and M.E. Moore (1978) Brain Res., 148, 441-449.

Shashoua, V.E. and M.E. Moore (1980) Neurosci. Abstr., 6, 290.4.

Whittaker, V.P. (1959) Biochem. J., 72, 694-706.

Williams, L.R. and S. Varon (1985) J. Comp. Neurol., 231, 209-220.

Involvement of Oligodenrocytes in Amino Acid Neurotransmitter Metabolism

R. Reynolds and N. Herschkowitz

Neurochemistry, Department of Pediatrics, University of Berne, Inselspital, CH-3010 Berne, Switzerland

ABSTRACT

Oligodendrocytes in dissociated neonatal mouse CNS cultures, recognised by their characteristic anti-galactocerebroside membrane staining, rapidly accumulated ^{3}H-GABA and ^{3}H-D-aspartate by sodium dependent mechanisms, becoming labelled over both cell bodies and processes after short incubations. Both GC-positive oligodendrocytes and GC-negative oligodendrocyte precursors accumulated both GABA and D-aspartate, suggesting that this may be a function expressed early in the differentiation of these cells. A small number of oligodendrocytes were lightly labelled with ^{3}H-taurine after long incubations. Net uptake of ^{3}H-ß-alanine and ^{3}H-glycine by oligodendrocytes was not observed under any conditions tested. The uptake of certain neuroactive amino acids is thus a property shared by astrocytes and oligodendrocytes, both of which may participate in the control of the extracellular concentration of these amino acids.

KEYWORDS

^{3}H-GABA - ^{3}H-D-aspartate - amino acid uptake - oligodendrocytes - astrocytes - tissue culture - autoradiography - immunocytochemistry

INTRODUCTION

It has been suggested that CNS glial cells may play an important role in the control of overall neuronal excitability via control of the levels of neuroactive amino acids in the extracellular space around neurons, both at the synapse and at regions remote from the synapse (8,12). Certain types of glial cells have been shown to share with neurons the capacity to accumulate GABA (11,13,17,28,30), and also the putative excitatory amino acid transmitters, L-glutamate and L-aspartate (3,9,29,32). The use of primary dissociated CNS tissue culture offers significant advantages for such studies on the autoradiographical localization of accumulated neurotransmitters in different cell types (17,18,24). Specific antisera produced against established cell-type specific antigenic markers now provide reliable identification and therefore qualitative and quantitative knowledge of the cellular composition of dissociated cultures (23).

Previous studies on the glial involvement in amino acid neurotransmitter metabolism have investigated the role of astrocytes (3,10,11,33), and oligodendrocytes have been envisaged as having no interactions with either neurons or astrocytes in the turnover of these transmitters (9). However, in recent studies, we have shown that oligodendrocytes, in primary dissociated mouse CNS cultures, accumulated ^{3}H-GABA via a sodium dependent, ouabain sensitive transport mechanism (24), with an inhibitor sensitivity similar to the neuronal GABA transport system. Uptake was cumulative and not due to homoexchange. In a further study ^{3}H-D-aspartate was shown to be avidly accumulated by these cells (25,26). Thus it is suggested that oligodendrocytes, in addition to astrocytes, may be involved in the modification of neuronal function by the uptake and inactivation of neuroactive amino acids.

In the present paper we demonstrate autoradiographically the uptake of certain neuroactive amino acids by oligodendrocytes in dissociated CNS cultures, prepared from neonatal mouse brain. Cell populations present in the cultures were defined by immunocytochemistry.

METHODS

Cell Cultures

Cells were dissociated from cerebral hemispheres of Swiss albino mice less than 12 hours after birth, according to previously described procedures (24,26), and suspended in Dulbecco's Modified Eagle's Medium containing 10% fetal calf serum, 30mM D(+)-glucose, 7.4mM L-glutamine and 100 units/ml penicillin. The suspended cells were plated at 2.5 x 10^5 cells/cm^2 into 35 mm plastic dishes containing glass coverslips.

[^{3}H]amino acid uptake

Uptake experiments were performed as previously described (24,26). Cultures were rinsed once in a Tris-buffered Krebs salt solution (TK; pH 7.4; 37°C), preincubated for 3 min, and incubated for 1-30 min at 37°C with the ^{3}H-amino acid. Final concentrations of the amino acids used were: 5-6x10^{-8}M for GABA, L-glutamate, ß-alanine and glycine, 1.1x10^{-7}M for taurine; 1.4x10^{-7}M for L and D-aspartate. The cultures were then washed five times with sodium free TK (4°C), containing 1mM unlabelled amino acid to minimize both specific and non-specific binding to membrane receptors, and fixed for 20 min in 2% glutaraldehyde in isotonic phosphate buffered saline (PBS). Following washing in PBS and air drying the coverslips were covered with Ilford K2 emulsion, exposed and developed as described previously (24).

Immunocytochemistry

Cultures were stained using antisera against the following markers: (a) galactocerebroside (GC; 23); myelin basic protein (MBP; 2); glial fibrillary acidic protein (GFAP; 1); neuron-specific enolase (NSE; 20); and fibronectin (23). The rabbit anti-GC and anti-MBP sera were produced as described previously (26). The anti-GFAP, anti-NSE and anti-fibronectin antisera were purchased from Dakopatts, Polysciences and Cappel Labs. respectively. For the cell membrane markers, GC and fibronectin, living cultures were incubated sequentially with primary antiserum and FITC-labelled anti-rabbit, or anti-goat, IgG, and then fixed with 3.7% formaldehyde in PBS. For the intracellular markers GFAP and MBP, cultures were first fixed in 3.7% formaldehyde in PBS followed by 3.7% formaldehyde containing 0.2% Triton X-100, prior to incubation with primary antiserum and FITC-labelled anti-rabbit IgG. NSE localization was investigated using the PAP method as described previously (24). Slides were examined with a Leitz Orthoplan microscope equipped with fluorescence optics (filter block I2).

Simultaneous autoradiography and immunocytochemistry

Living cultures were incubated sequentially with anti-GC serum (20 min), TRITC-labelled goat anti-rabbit IgG (20 min), and ^{3}H-GABA (10 min, 37°C, as described above). Cultures were then fixed for 20 min in 4% paraformaldehyde in

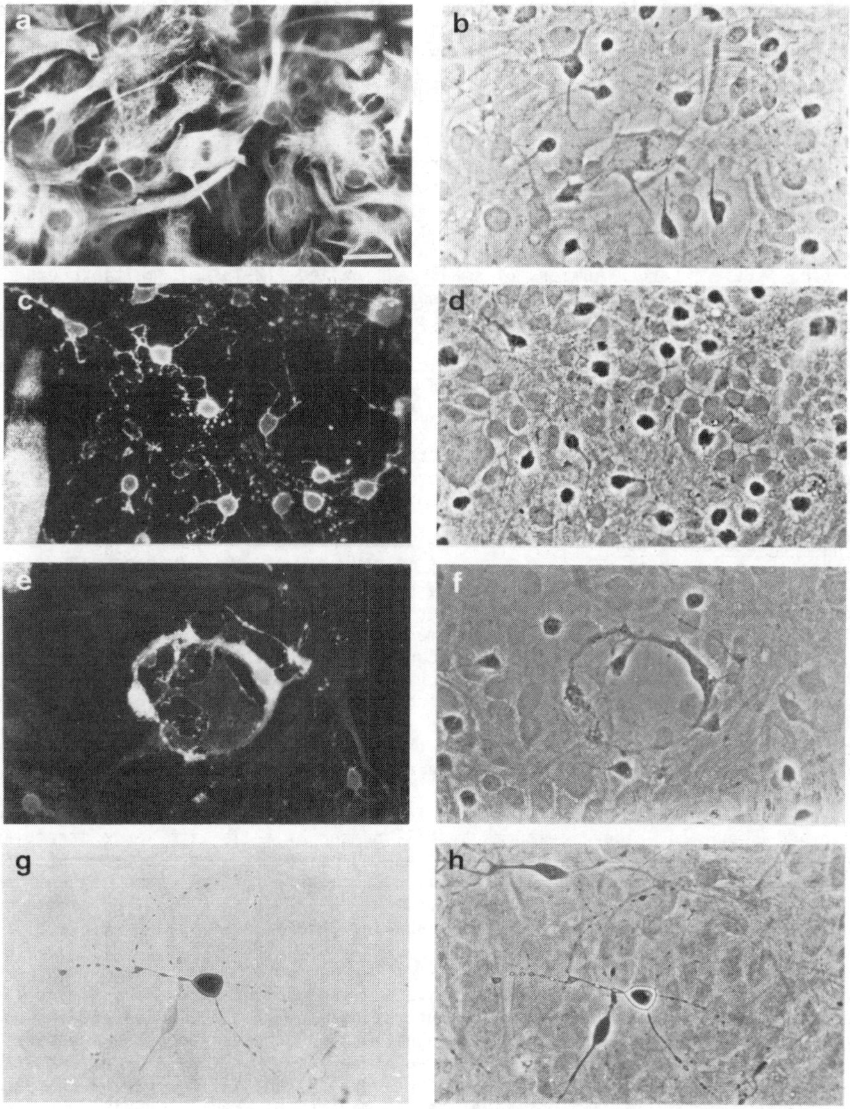

Fig. 1. Immunocytochemical staining of primary dissociated neonatal mouse CNS cell cultures. a and b: Anti-GFAP immunofluorescence of a 10 DIV culture and corresponding phase contrast picture. The cells on the bottom carpet layer of the culture are positively stained whilst the small phase contrast dark cells remain unlabelled. c and d: Anti-GC immunofluorescence of a 10 DIV culture, showing the characteristic surface labelling of the cell body and processes of oligodendrocytes. e and f: Anti-MBP staining of a large oligodendrocyte, 10 DIV. Note the extensive areas of stained flat membranous processes. g and h: Anti-NSE PAP staining of an isolated neuron, 10 DIV. Bar represents 30 μm.

0.1M sodium phosphate buffer (pH 7.2) containing 0.1% glutaraldehyde. This fixation results in a reduced retention of ^{3}H-GABA but an absence of autofluorescence (Reynolds and Herschkowitz, in preparation).

RESULTS

Immunocytochemistry

Dissociated mixed cell cultures prepared from neonatal mouse brain displayed a multilayered structure as described previously (2,26). The bottom layer of the cultures consisted of a confluent layer of large flattened GFAP-positive astrocytes (Fig.1a,b) and only few small patches of fibronectin-positive fibroblasts. Occasionally ependymal cells were seen, identified by their beating cilia. Large numbers of small cell bodied (12-15 µm), multiprocessed, cells were evenly distributed on top of the astroglial layer (Fig.1d). Approximately 50% of these cells (26) were identified as oligodendrocytes by their characteristic anti-GC membrane immunofluorescence (Fig.1c,d), and 2-6% expressed the oligodendrocyte specific protein MBP (Fig.1e,f). A small number of surviving neurons were identified by their intracellular staining with anti-NSE serum (Fig.1g,h) representing approx. 2% of the small process bearing cells at both 10 and 21 days in vitro (DIV). The remaining small process bearing cells (approx. 50%) did not express any of the markers employed in this study and have been suggested in other studies (21,27) to be immature oligodendrocytes which differentiate into GC-positive cells.

[^{3}H]GABA autoradiography

After short incubations (3-5 min) of 10 DIV cultures with ^{3}H-GABA 81% of the small process bearing cells had become heavily labelled over the cell body and some processes (Fig.2a). After longer incubations (10-20 min; Fig.2b) the distribution of silver grains was more uniform with all the labelled cells having lightly or heavily labelled processes. In 21 DIV cultures (Fig.2c,d) the accumulation of ^{3}H-GABA by oligodendrocytes was more pronounced with 99% of the small process bearing cells heavily labelled. Heavily labelled cell processes often formed a network on the surface of the astroglial layer (Fig.2d). Whereas only 50% of these cells expressed the myelin lipid GC, they represented a homogenous population with regard to their ability to accumulate ^{3}H-GABA (26).

In cultures of all ages the astroglial cell layer was only lightly labelled with ^{3}H-GABA. In older cultures a small number of cells with large irregular shaped cell bodies and broad processes, growing on the surface of the astroglial cell layer, became heavily labelled, to the same extent as oligodendrocytes (Fig.2e,f). Accumulation of ^{3}H-GABA either, in the absence of sodium, or in the presence of 1mM unlabelled GABA, was reduced by 98% and 99% respectively.

The uptake of ^{3}H-GABA by GC-positive oligodendrocytes was confirmed by simultaneous immunocytochemistry and autoradiography (Fig.3). Silver grains and anti-GC immunofluorescence could be localized to both cell body and processes. This method, however, leads to a reduction in the intensity of both the immunofluorescence and GABA uptake autoradiography, and it is apparent that the fluorescence from weakly positive cells may be obscured by overlying silver grains or reduced to background levels due to fading.

[^{3}H]Glutamate and aspartate uptake

In cultures of all ages both ^{3}H-L-glutamate and ^{3}H-L-aspartate labelled predominantly astroglial cells in the carpet layer of the culture. A small number of small process bearing cells were lightly labelled over the cell body but the majority were completely unlabelled (26). Incubation of 10 DIV cultures with ^{3}H-D-aspartate, a non-metabolised analogue of L-glutamate, for times as short as 1-3 min, led to heavy labelling over the cell body and some processes of these cells (Fig.4a). The intensity of the labelling increased after longer incubation times (Fig.4b), and in older cultures (21 DIV; Fig.4c,d).

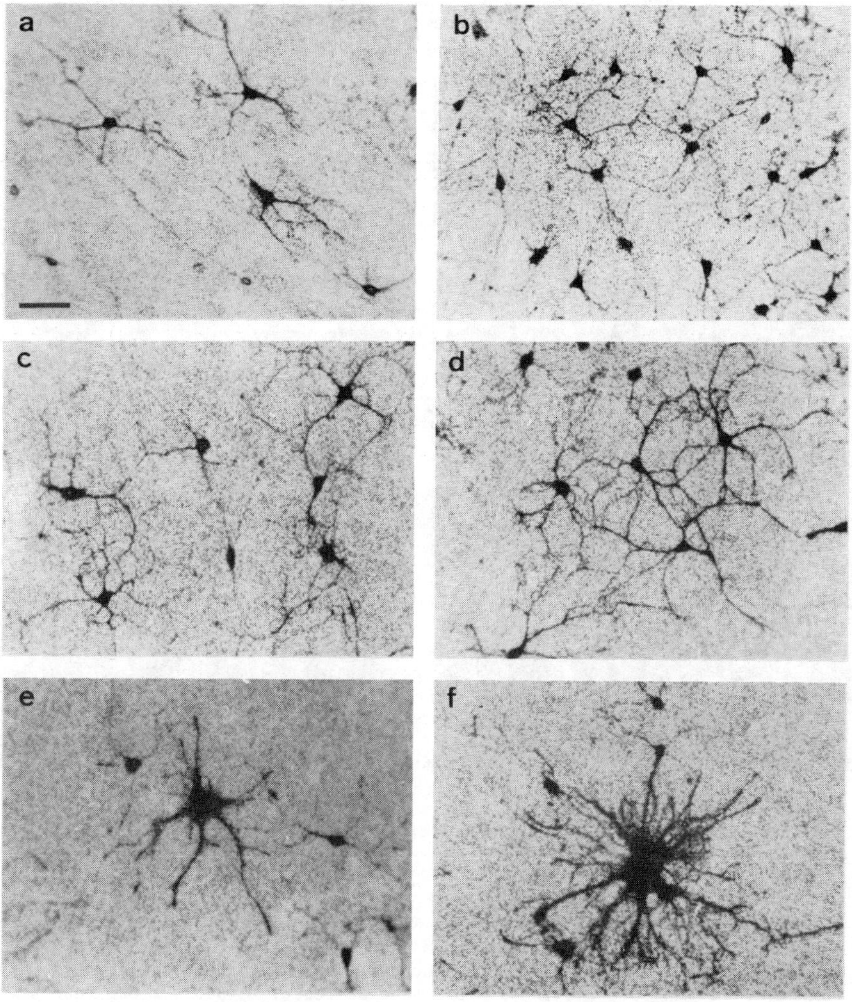

Fig. 2. Autoradiographs of ^{3}H-GABA uptake by oligodendrocytes and astrocytes in 10 and 21 DIV cultures. a and b: 10 DIV cultures, labelling of oligodendrocytes, both perikarya and processes, after 5 min (a) and 10 min (b) incubations. c and d: 21 DIV cultures, oligodendrocyte labelling after 10 min (c) and 20 min (d) incubations. Note the intensity of the labelling occuring over oligodendrocytes compared to the astroglial cell layer. e and f: Heavy labelling of large irregular shaped astrocytes growing on the surface of the astroglial carpet cell layer in 21 DIV cultures, 10 and 20 min incubations respectively. Bar represents 30 µm.

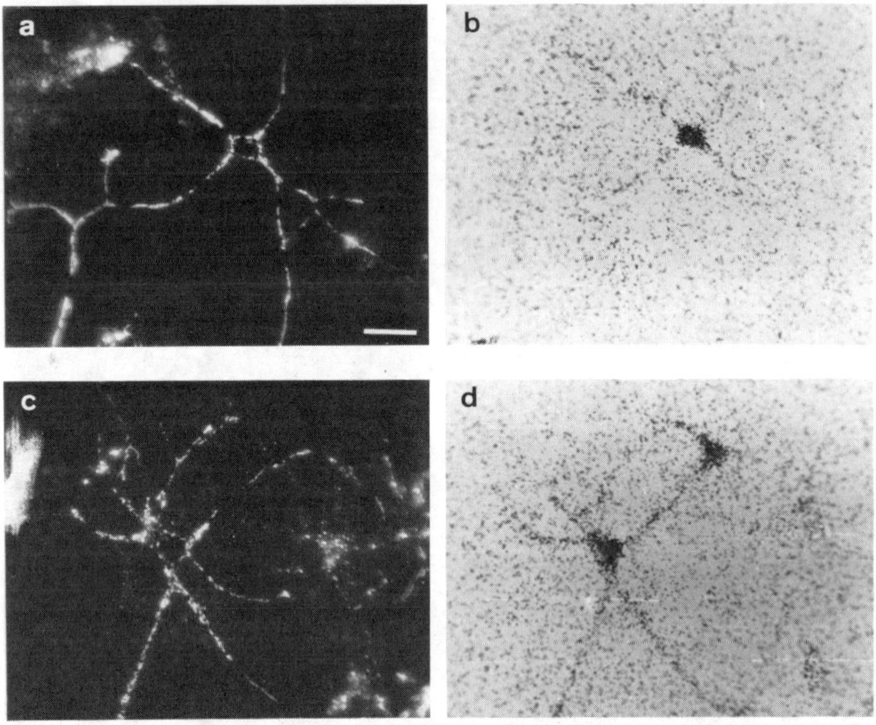

Fig. 3. Simultaneous anti-GC immunofluorescence and ^{3}H-GABA autoradiography. 16 DIV cultures were labelled with anti-GC antibodies followed by a) TRITC or c) Texas Red labelled anti-rabbit IgG conjugates, and then incubated with ^{3}H-GABA for 10 min. Two GC positive oligodendrocytes a) and c) have accumulated GABA, becoming labelled over both cell body and processes b) and d). Bar represents 16 µm.

D-aspartate accumulating cells represented 75% and 82% of these small process bearing cells in 10 and 21 DIV cultures respectively (26). Astroglial cells in the bottom layer of the culture became more heavily labelled with D-aspartate than with GABA (cf. Fig.2b,d and Fig.4b,d). After longer incubation times there was little difference in the labelling intensity of oligodendrocytes and astrocytes. As with GABA, some irregularly shaped astrocytes in the top layer of the cultures became heavily labelled with D-aspartate (Fig.4e,f). Uptake of D-aspartate by the cultures was 84% sodium dependent, and reduced by 96% in the presence of 1mM unlabelled D-aspartate.

Other amino acids

Net uptake of ^{3}H-ß-alanine or ^{3}H-glycine by oligodendrocytes was not apparent

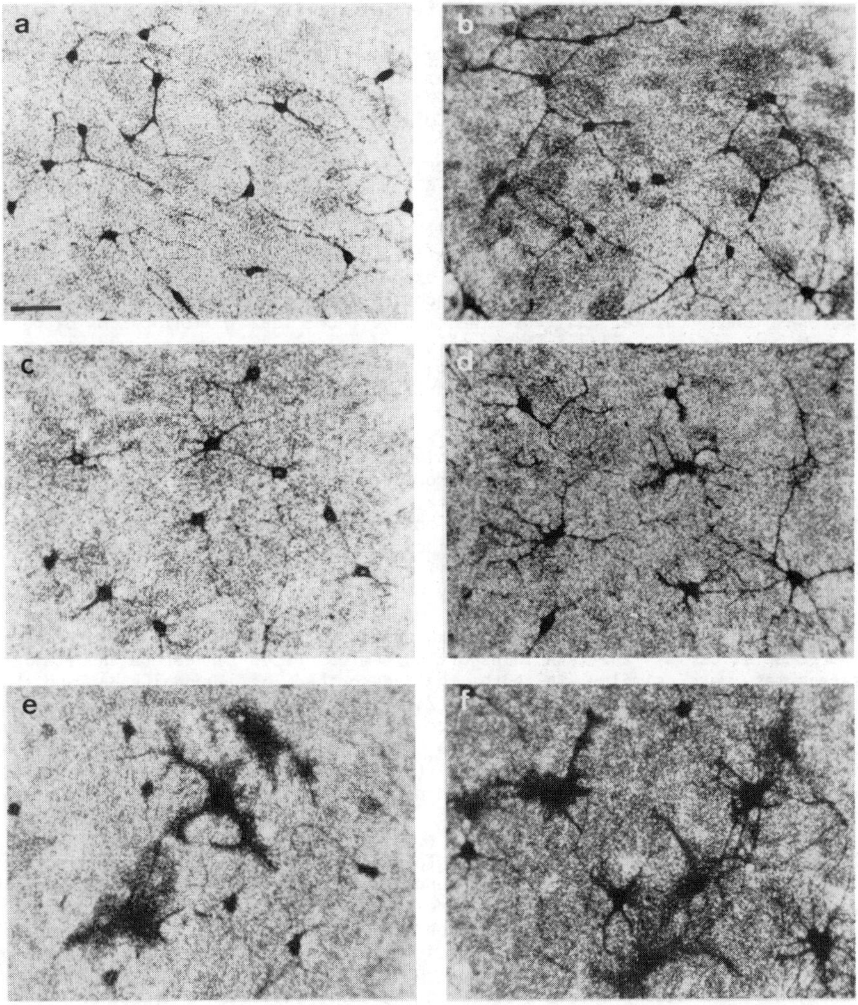

Fig. 4. Autoradiographs of ^{3}H-D-aspartic acid uptake by oligodendrocytes and astrocytes in 10 and 21 DIV cultures. a and b: Labelling of oligodendrocytes, both perikarya and processes, after 3 min (a) and 20 min (b) incubations in 10 DIV cultures. c and d: heavy labelling of oligodendrocytes after 3 min (c) and 20 min (d) incubations in 21 DIV cultures. e and f: Heavy labelling of large irregular shaped astrocytes growing on the surface of 21 DIV cultures after 3 min (e) and 10 min (f) incubations. Bar represents 30 μm.

under any conditions tested, these two amino acids labelling only astroglial cells in the carpet layer of the cultures. A small number of oligodendroglial-like cells in 21 DIV cultures became lightly labelled over the cell body with ^{3}H-taurine after longer incubation times (10-30 min; 26).

DISCUSSION

The present study confirms our previous results, obtained with cultures prepared from fetal mouse brain (24), that oligodendrocytes possess a rapid sodium dependent uptake system for the inhibitory neurotransmitter GABA. Oligodendrocytes became heavily labelled after short incubation times, similar to those used in biochemical estimations of neuronal GABA uptake (22,26,35), indicating that caution should be exercised when using uptake after short incubation times and at low GABA concentrations as an estimate of exclusively neuronal GABA uptake. Without the immunocytochemical identification of cells present in primary glial cultures oligodendroglial GABA uptake could also be a source of error in studies of astroglial GABA uptake (7). This error could be quite substantial considering the greater uptake of GABA by oligodendrocytes compared to astrocytes in our cultures.

It is apparent from our study that two populations of astrocytes present in our cultures exhibited a degree of heterogeneity with respect to their ability to accumulate GABA and D-aspartate. The large astrocytes growing on top of the astroglial carpet layer accumulated both amino acids to a greater extent than those in the carpet layer. Two distinct types of astrocytes in rat cerebellar cultures, stellate and polygonal, differed not only in morphology but also in their surface properties and ability to accumulate GABA (14,18,33). These two types both accumulated D-aspartate but only the stellate shaped cells became heavily labelled with GABA, in close agreement with our results. The structural characteristics of the GABA transport system described for cerebellar stellate astrocytes were similar to both that described as neuronal (18), and that described by us for oligodendrocytes (24), suggesting little difference between these three CNS cell types, contrary to previous suggestions (12). However GABA uptake by Schwann cells in culture was inhibited by β-alanine rather than inhibitors of neuronal GABA uptake (5), suggesting that the glial GABA transport systems in the CNS and PNS may have different structural characteristics.

It has been suggested that during ontogenesis GABA may play several important functional roles before synaptic functioning is fully established. Certain glial cells in the rat neocortex were shown to have the ability to accumulate GABA as early as day 16 in gestation (34). Our finding that both GC-positive oligodendrocytes and GC-negative undifferentiated oligodendrocyte precursor cells rapidly accumulated GABA and D-aspartate suggests that this may be a function expressed relatively early in the differentiation of these cells, and is not related to membrane changes occuring after interaction with neurons. During early CNS development, glia, both astrocytes and oligodendrocytes, may play an important role in stabilising the extracellular environment for neuronal migration and differentiation. In contrast to rat cerebellar stellate astrocytes, which lost their ability to accumulate GABA with time in culture (33), mouse oligodendrocytes still maintained this ability at 8 weeks in culture.

The inactivation of synaptically released L-glutamate and aspartate, and maintenance of their low extracellular concentrations, is proposed to occur predominantly via uptake into surrounding glial cells (3,9). Previous investigations have shown that L-glutamate and L- and D-aspartate share the same transport mechanism, and that D-aspartate is only slowly metabolised in comparison to the L-amino acids (4). Thus our results show that oligodendrocytes do indeed have a transport system for the putative excitatory amino acid transmitters and also possess the necessary enzymatic machinery for their rapid metabolism. The most important glutamate metabolising enzymes in astrocytes are glutamine synthetase, generally considered to be an astrocyte marker enzyme (29), and glutamate dehydrogenase (36). The former has been shown to be absent from oligodendrocytes in primary cultures (6), which means that oxidative deamination to α-ketoglutarate is likely to be the main route of glutamate metabolism in these cells. However, there is no information available concerning the presence and activity of glutamate dehydrogenase in oligodendrocytes.

The uptake of certain neuroactive amino acids is thus a property shared by astrocytes and oligodendrocytes. Whereas astrocytes are thought to play an important role in removing neuronally released transmitters from the synaptic region, oligodendrocytes probably act in a protective fashion by reducing the concentration of these amino acids in the vicinity of axons, in the case of interfasicular oligodendrocytes, and around neuronal cell bodies, in the case of perineuronal satellite oligodendrocytes. Very little is at present known about the function of this latter cell type in the CNS. These cells have been shown to play a role both, in normal myelination in the rat spinal cord (31), and in remyelination following Cuprizone induced demyelination in the mouse cerebellum (19). However, in view of our evidence in vitro (24,26), and that of an earlier study in vivo (13) showing that the most prominant glial labelling following in vivo injections of ^{3}H-GABA occured over oligodendrocytes in satellite positions adjacent to large neurons in all regions of the rat CNS studied, an additional function for these perineuronal satellite cells should now be considered.

In conclusion, immunocytochemical and autoradiographical investigations in our laboratory (24,26), and electrophysiological studies in another (15), have shown that oligodendrocytes can no longer be considered as being solely myelin producing cells, their recognised role in the CNS, but also have the ability to actively accumulate and respond to certain neuroactive amino acids. Thus glial cells, both astrocytes and oligodendrocytes, are envisaged as playing an important role in brain homeostasis by stabilising and protecting the extracellular environment of neurons (8,9,16).

ACKNOWLEDGEMENTS

This work was supported by Grant 3.493.83 from the Swiss National Research Foundation and a grant from the Swiss Multiple Sclerosis Society. We thank Mrs. C. Steffen for technical assistance.

REFERENCES

1. Bignami, A. and Dahl, D.(1977) J. Histochem. Cytochem., 25 , 466-469.
2. Bologa, L., Siegrist, H.P., Z'Graggen, A., Hofmann, K., Wiesmann, U., Dahl, D. and Herschkowitz, N.(1981) Brain Research, 210 , 217-229.
3. Currie, D.N. and Kelly, J.S.(1981) J. exp. Biol., 95 ,181-193.
4. Davies, L.P. and Johnston, G.A.R.(1976) J. Neurochem., 26 , 1007-1014.
5. Gavrilovic, J., Raff, M.C. and Cohen, J.(1984) Brain Research, 303 , 183-185.
6. Hallermayer, K. and Hamprecht, B.(1984) Brain Research, 295 , 1-11.
7. Hansson, E., Isacsson, H. and Sellstrom, A.(1984) Acta Physiol. Scand., 121 , 333-341.
8. Henn, F.A.,(1976) J. Neurosci. Res., 2 , 271-282.
9. Hertz, L.(1979) Prog. Neurobiol., 13 , 277-323.
10. Hertz, L., Wu, P.H. and Schousboe, A.(1978) Neurochem. Res., 3 , 313-323.
11. Hösli, L. and Hösli, E.(1978) Rev. Physiol. Biochem. Pharmacol., 81 , 135-188.
12. Iversen, L.L. and Kelly, J.S.(1975) Biochem. Pharmacol., 24 , 933-938.
13. Iversen, L.L. and Schon, F.E.(1973) In A. Mandell (Ed.), New Concepts in Neurotransmitter Regulation , Plenum Press, New York, pp. 153-193.

14. Johnstone, S.R., Levi, G., Wilkin, G.P., Schneider, A. and Ciotti M.T.(1985) Develop. Brain Res., in press .
15. Kettenmann, H., Gilbert, P. and Schachner, M.(1984) Neurosci. Lett., 47 , 271-276.
16. Krnjevic, K.(1984) Neurosci. Lett., 47 , 283-287.
17. Lasher, R.S.(1974) Brain Research, 69 , 235-254.
18. Levi, G., Wilkin, G.P., Ciotti, M.T. and Johnstone, S.(1983) Develop. Brain Res., 10 , 227-241.
19. Ludwin, S.K.(1979) Acta Neuropathol., 47 , 49-53.
20. Marangos, P.J., Zomzely-Neurath, C., Luk, D.C.M. and York, C.(1975) J. Biol. Chem., 250 , 1884-1891.
21. McCarthy, K.D. and DeVellis, J.(1980) J. Cell Biol., 85 , 890-902.
22. Messer, A.(1977) Brain Research, 130 , 1-12.
23. Raff, M.C., Fields, K.L., Hakomori, S., Mirsky, R., Pruss, R.M. and Winter, J.(1979) Brain Research, 174 , 283-308.
24. Reynolds, R. and Herschkowitz, N.(1984) Brain Research, 322 , 17-31.
25. Reynolds, R. and Herschkowitz, N.(1985) J. Neurochem., 44 (Suppl) , S142.
26. Reynolds, R. and Herschkowitz, N.(1985) Brain Research, in press .
27. Schachner, M., Kim, S.K. and Zehnle, R.(1981) Develop. Biol., 83 , 328-338.
28. Schon, F. and Kelly, J.S.(1974) Brain Research, 66 , 275-288.
29. Schousboe, A.(1981) Int. Rev. Neurobiol., 22 , 1-45.
30. Schousboe, A., Hertz, L. and Svenneby, G.(1977) Neurochem. Res., 2 , 217-229.
31. Sternberger, N.H.(1984) In W. Norton (Ed.), Oligodendroglia , Plenum Press, New York, pp. 125-173.
32. Wilkin, G.P., Garthwaite, J. and Balazs, R.(1982) Brain Research, 244 , 69-80.
33. Wilkin, G.P., Levi, G., Johnstone, S.R. and Riddle, P.N.(1983) Develop. Brain Res., 10 , 265-277.
34. Wolff, J.R., Rickmann, M. and Chronwall, B.M.(1979) Cell Tissue Res., 201 ,239-248.
35. Yu, A.C.H. and Hertz, L.(1982) J. Neurosci. Res., 7 , 23-35.
36. Yu, A.C., Schousboe, A. and Hertz, L.(1982) J. Neurochem., 39 , 954-960.

Presence of Calcitonin Gene-related Peptide in Glial Cells

C. W. Cooper*, S. A. Borosky*, L. Hertz† and B. Haber*

*Department Pharmacology & Toxicology and Marine Biomedical Institute, University of Texas Medical Branch, Galveston, Texas, USA
†Department Pharmacology, University of Saskatchewan School of Medicine, Saskatoon, Saskatchewan, Canada

ABSTRACT

Calcitonin gene-related peptide (CGRP), an alternative product of the calcitonin gene, may be an important neuromodulator. We report the presence of CGRP in normal mouse astrocytes and in rat and human glioma cell lines.

KEYWORDS

CGRP, glial cells, peptide neuromodulator.

INTRODUCTION

CGRP is a 37 amino acid peptide produced by the calcitonin gene (Rosenfeld et al, 1983). The peptide has been found in high amounts in central and peripheral nervous tissue and may be an important neuromodulator (Gibson et al, 1984, Tschopp et al, 1984). In an effort to more precisely define potential sources of CGRP in the nervous system, we examined glial cells to see whether they too contain CGRP.

METHODS

Cells from cell lines in culture were extracted in RIA phosphosaline buffer, and CGRP was measured by RIA using a guinea pig antibody to synthetic rat CGRP which detects as little as 50 pg CGRP and does not crossreact with calcitonin or other peptides. Synthetic rat CGRP (BACHEM) was used as reference standard and for iodination with ^{125}I. Cell lines tested were a rat astrocytoma (C62B), a rat schwannoma (RN22), and a human glioma (U251-214). Rat cell lines also were tested after they had been cultured in the absence of serum for 24 hr. Normal mouse astrocytes were obtained from newborn cerebral cortex by mechanical disruption and sequential filtration through 80 and 10 micron nylon mesh screens. Filtered cells were cultured 2 weeks, incubated 24 hr in serum-free medium containing dibutyryl cAMP, and extracted for assay. Results are expressed as quantity of immunoreactive CGRP per mg Lowry protein.

RESULTS

We found the rat astrocytoma C62B cell line to contain relative large and easily measured amounts of immunoreactive CGRP (Fig 1). CGRP was detected regardless of whether the cells had been cultured for the previous 24 hr in the absence of serum or not. Likewise, the human glomia line tested also showed relatively large amounts of CGRP, whereas the rat schwannoma contained much smaller quantities of CGRP which approached the RIA detection limits (1-2 ng/mg protein). Fig 1 also shows that the peptide was measurable in normal mouse astrocytes as well as in the glioma cell lines cell lines.

Fig 2 shows an RIA curve for standard rat CGRP and a curve generated using multiple dilutions of an extract of rat C62B astrocytoma cells. The extract yielded a curve parallel to that generated with synthetic rat CGRP.

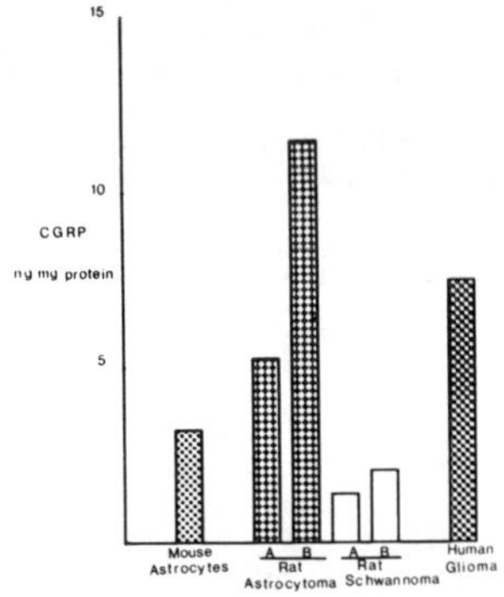

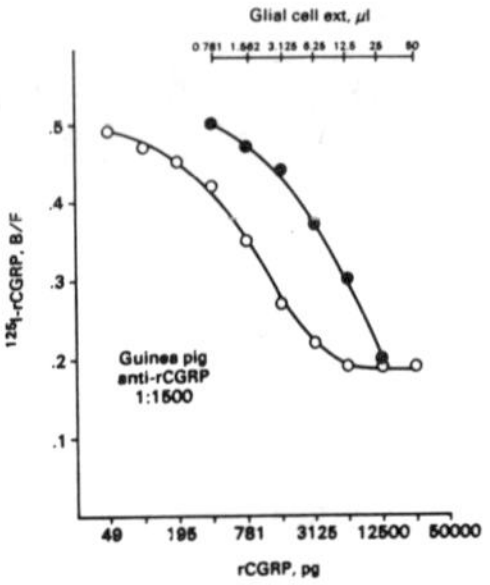

Fig.1. Presence of CGRP in rat and human glial cell lines and in normal mouse astrocytes. (A = cultured with serum; B = cultured 24 hr in the absence of serum)

Fig.2. RIA for rat CGRP. Extract of rat astrocytoma cells gave a curve parallel to that with CGRP standard.

DISCUSSION

Our results show that rat, mouse, and human glial cells contain CGRP. The fact that cell lines cultured in the absence of serum did not have lowered amounts of CGRP indicates the peptide is cell-derived and not serum-derived. The presence of CGRP in normal mouse astrocytes shows that the presence of the peptide in glial cells is not restricted to tumor cells. The findings suggest that glial cells may be a source of CGRP in the nervous system and may provide CGRP to act on neurons as a neuromodulator. Whether glial cells themselves also have receptors for CGRP and whether they actually biosynthesize the peptide are questions that require additional study.

ACKNOWLEDGEMENT

This work was supported by USPHS (NIH) grants AM32060, NS17696, and NS11255, by a grant from the Welch Foundation (H-504), and by a grant from the MRC of Canada (MT-5957).

REFERENCES

Gibson, S.J., J.M. Polak, S.R. Bloom, I.M. Sabate, P.M. Mulderry, M.A. Ghatei, G.P. McGregor, J.F.B. Morrison, J.S. Kelley, R.M. Evans and M.G. Rosenfeld (1984). J. Neurosci., 4, 3101-3111.

Rosenfeld, M.G., J.-J. Mermod, S.G. Amara, L.W. Swanson, P.E. Sawchenko, J. Rivier, W.W. Vale and R.M. Evans (1983). Nature, 304, 129-135.

Tschopp, F.A., P.H. Tobler and J.A. Fischer (1984). Mol. Cell. Endocrinol., 36, 53-57.

Serotonin Uptake in Primary Astrocyte Cultures

H. K. Kimelberg and D. M. Katz

Division of Neurosurgery, Albany Medical College, Albany, New York 12208, USA

INTRODUCTION

Termination of the action of most neurotransmitters in the CNS after release is thought to be due to re-uptake into nerve endings (Fonnum, et al, 1980). However, many synapses are surrounded by astrocytic processes (Peters, et al, 1976) and these could potentially serve as additional sites of uptake. Indeed, it has been shown that astrocytes have very active uptake systems for amino acid transmitters (Henn and Hamberger, 1971; Levi, et al, 1983; Schousboe, et al, 1977). Uptake of monoamine transmitters by mammalian astrocytes in situ has, in general, not been detected (Descarriers and Lapierre, 1973; Hökfelt and Ljungdahl, 1971), except for one report of uptake of serotonin into astrocytic processes in cat medulla (Ruda and Gobel, 1980). This may be due to astrocytes having less ability to concentrate the transmitters in the absence of a neuronal-type vesicular storage system. Uptake of serotonin into glia has been reported in the glial filum terminale of the frog (Ritchie, et al, 1981) and uptake of both catecholamines and serotonin has been observed in transformed or primary glial cell cultures by measuring uptake of radiolabelled transmitters (Hansson, 1983; Suddith, et al, 1978; Tardy, et al, 1982; Whitaker, et al, 1983). We have used primary astrocyte cultures prepared from rat brain to study the uptake of labelled catecholamines (Kimelberg and Pelton, 1983; Pelton, et al, 1981; Semenoff and Kimelberg, 1985) and serotonin (Katz and Kimelberg, 1985; Kimelberg and Katz, 1985). In this paper we present a summary of our results on the uptake of [^{3}H] labelled serotonin.

RESULTS AND DISCUSSION

Rat primary astrocyte cultures prepared from 1 day old rat brains (Frangakis and Kimelberg, 1984) and used after 3-5 weeks growth in vitro showed significant accumulation of [^{3}H] labelled serotonin ([^{3}H]5HT). At concentrations in the range of 0.01 to 0.7μM [^{3}H]5HT, this uptake was 70-80% Na^+ dependent and gave a Km of 0.40 ± 0.11μM and a Vmax of 6.42 ± 0.85 (± S.E.M.) pmoles [^{3}H]5HT/mg protein/ 4 min for the Na^+ dependent component. In the absence of Na^+ the uptake was non-saturable. Uptake of [^{3}H]5HT depended on the presence of Cl^- as well as Na^+ in the medium, and the effect of omission of both ions was non-additive. Varying the concentration of K^+ in the media from 1 to 50mM had a limited effect on [^{3}H] 5HT uptake. Omission of the monoamine oxidase (MAO) inhibitor pargyline markedly reduced the Na^+ dependent component of [^{3}H]5HT uptake but had a negligible effect on the Na^+ independent component, suggesting significant oxidative deamination of serotonin and release of its metabolite, after it has been taken up by

Advances in the Biosciences Vol. 61

the high affinity system. The specific MAO-A inhibitor clorgyline was more effective than pargyline, since 10^{-9}M was sufficient to obtain increased uptake of [^{3}H]5HT, whereas 10^{-5} to 10^{-4}M pargyline was necessary for a comparable effect. Again, only the Na^+ dependent component was affected. This further supports the interpretation that oxidative deamination of serotonin is responsible for the lower level of accumulation in the absence of MAO inhibitors, since serotonin is a preferred substrate for MAO-A. We estimated that the [^{3}H]5HT uptake system enabled the cells to concentrate [^{3}H]5HT up to 50-fold at 10^{-7}M external [^{3}H]5HT.

Inhibition of [^{3}H]5HT uptake by a number of clinically effective antidepressants and specific 5HT uptake inhibitors was also consistent with a specific high affinity uptake mechanism for 5HT. The order of effectiveness of inhibition for a series of these is as follows (the numbers in brackets refer to the respective IC_{50} values in µM for inhibition of uptake of 10^{-7}M5HT measured over 4 minutes): chlorimipramine (0.009) > fluoxetine (0.023) > imipramine = amitriptyline (0.14) > desmethylimipramine (0.62) > iprindole (2.8) > mianserin (4.9).

Autoradiography of cultures after uptake of 3×10^{-7}M [^{3}H]5HT in the presence of Na^+ showed that virtually all the cells had a silver grain density that was above background (Katz and Kimelberg, 1985). Since our cultures are $\geq$ 95% GFAP(+) this strongly suggests localization of such uptake to astrocytes. As further support for this localization we have also been able to simultaneously stain cells for GFAP by immunofluorescence and do autoradiography for [^{3}H]5HT. We Found that GFAP(+) cells had a clearly increased grain density; this was not seen when Na^+ was omitted from the medium or 10^{-7}M chlorimipramine or fluoxetine had been added (Kimelberg and Katz, 1985).

In conclusion, we have shown that astrocytes in primary culture take up [^{3}H]5HT by a process that shows all the characteristics of high affinity uptake; a process which is thought to be specific to nerve-endings (Fonnum, et al, 1980).

REFERENCES

Descarries, L. and Y. Lapierre (1973). Brain Res. 51, 141-160.

Fonnum, F., R.L. Karlsen, D. Malthe-Sorenssen, S. Sterri and I. Walaas (1980). The Cell Surface and Neuronal Function, North-Holland Publish. Co., Amsterdam, pp. 455-504.

Frangakis, M.V. and H.K. Kimelberg (1984). Neurochem. Res. 9, 1689-1698.

Hansson, E. (1983). Brain Res. 289, 189-196.

Henn, F.A. and A. Hamberger (1971). Proc. Nat. Acad. Sci. USA 68, 2686-2690.

Hökfelt, T. and A. Ljungdahl (1971). Progr. Brain Res. 34, 87-102.

Katz, D.M. and H.K. Kimelberg (1985). J. Neurosci. 5, 1901-1908.

Kimelberg, H.K. and D.M. Katz (1985). Science, 228, 889-891.

Kimelberg, H.K. and E.W. Pelton (1983). J. Neurochem. 40, 1265-1270.

Levi, G., G.P. Wilkin, M.T. Ciotti and S. Johnstone (1983). Dev. Brain Res. 10, 227-241.

Pelton, E.W., H.K. Kimelberg, S.V. Shipherd and R.S. Bourke (1981). Life Sci. 28, 1655-1663.

Peters, A., S.L. Palay and H. de F. Webster (1976). The Fine Structure of the Nervous System: The Neurons and Supporting Cells, W.B. Saunders, pp. 231-263.

Ritchie, T., S. Glusman and B. Haber (1981). Neurochem. Res. 6, 441-482.

Ruda, M.A. and S. Gobel (1980). Brain Res. 184, 57-83.

Schousboe, A., G. Svennby and L. Hertz (1977). J. Neurochem. 29, 999-1005.

Semenoff, D. and H.K. Kimelberg (1985). Brain Res. in press.

Suddith, R.L., H.T. Hutchison and B. Haber (1978). Life Sci. 22, 2179-2188.

Tardy, M., M.F.D. Costa, C. Fages, J. Bardakdjian and P. Gonnard (1982). Dev. Neurosci. 5, 19-26.

Whitaker, P.M., C.K. Vint and R. Morin (1983). J. Neurochem. 41, 1319-1323.

Receptor-mediated Inositol Phospholipid Hydrolysis in CNS Astrocytes

B. Pearce, M. Cambray-Deakin, C. Morrow and S. Murphy

Brain Research Group, Biology Department, The Open University, Milton Keynes MK7 6AA, UK

ABSTRACT

Receptor-activated inositol phospholipid hydrolysis in astrocytes was investigated by determining agonist-induced increases in the accumulation of intracellular inositol phosphates.

KEY WORDS

Astrocytes; receptors; inositol phospholipids; inositol phosphates.

INTRODUCTION

In many tissues, including brain, the activation of certain receptors results in the mobilisation of intracellular Ca^{2+} (Berridge, 1984). The initial event in this process is the receptor-mediated breakdown of membrane inositol phospholipids leading to the formation of inositol phosphates (IP). The cellular location of this response in the brain has yet to be fully established, thus our aim was to determine whether CNS astrocytes in primary culture possess receptors coupled to inositol phospholipid metabolism.

METHODS

Astrocyte-enriched cultures of the neonatal rat cortex were preapred and maintained as previously described (Pearce et al, 1985). At 18-22 days in vitro the cultures were incubated for 24 hours with 10μM [^{3}H] myo-inositol to prelabel the membrane inositol lipids. Cultures were then exposed to receptor agonists, sometimes in the presence of antagonists, for 60 minutes in medium containing 5mMLiCl to block the hydrolysis of the [^{3}H] IP formed (Berridge et al, 1982). The [^{3}H] IP accumulated were determined as previously described (Berridge et al, 1982; Pearce et al, 1985).

RESULTS AND DISCUSSION

[^{3}H] IP accumulation in response to either carbachol (CARB) or noradrenaline (NA) was linear over incubation periods up to 80 minutes. Figure 1A shows the concentration dependence of [^{3}H] IP accumulation. NA was the most effective agonist with an EC_{50} of 2μM. Both CARB and acetylcholine (ACh) produced maximal responses which were approximately half that evoked by NA with EC_{50} values of 120μM and 40μM respectively. Figures 1B and C show the inhibition of NA-and CARB-induced [^{3}H] IP accumulation respectively. The α_1-adrenergic receptor antagonist prazosin (PRAZ) was the more effective blocker of NA(1mM)-evoked [^{3}H]

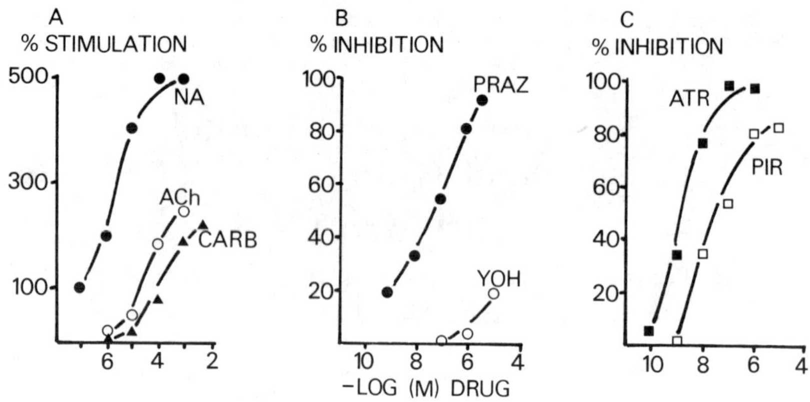

IP formation with an IC_{50} of 70nM. The α_2-adrenergic antagonist yohimbine (YOH) was without effect. The response elicited by 1mM CARB was reversed by both atropine (ATR, IC_{50} 2nM) and pirenzepine (PIR, IC_{50} 40nM). This would suggest that CARB-elicited [^{3}H] IP accumulation is the result of M_1 muscarinic receptor activation in these cells. Astrocyte M_2 receptors appear to be involved in the regulation of adenylate cyclase activity (Murphy et al, 1985b). The simultaneous addition of maximally effective doses of NA and CARB to astrocytes produced an increase in [^{3}H] IP accumulation equal to the sum of their individual effects. This additivity suggests that either the receptors are present on the same cells and linked to separate inositol lipid pools or that they are present on separate subpopulations of astrocytes in these cultures. The result of inositol lipid breakdown in astrocytes is yet to be elucidated, although one possibility is the release of prostaglandins (Berridge, 1981). Indeed, our recent work has shown that astrocytes synthesise and release thromboxane A_2 (Murphy et al, 1985a).

ACKNOWLEDGMENTS
This work is supported by the Science and Engineering Research Council (UK).

REFERENCES

Berridge, M.J. (1981) Mol. Cell. Endocrinol., 24, 115-140.
Berridge, M.J. (1984) Biochem. J. 220, 345-360.
Berridge, M.J., C.P. Downes and M.R. Hanley (1982) Biochem. J., 206 587-595.
Murphy, S., J. Jeremy, B.R. Pearce and P. Dandona (1985a) Neurosci. Lett., in press.
Murphy, S., B.R. Pearce and C. Morrow (1985) Brain Res., in press.
Pearce, B.R, M.A. Cambray-Deakin, C. Morrow, J. Grimble and S. Murphy (1985) J. Neurochem., in press.

Glutamate Stimulates the Respiratory Activity in Cultured Rat Astroglial Cells

A. Peterson

Unit of Neurochemistry and Neurotoxicology, University of Stockholm, Enköpingsvägen 126, S-172 46 Sundyberg, Sweden

ABSTRACT

In the present study the effect of glutamate on the respiratory activity of cultured primary rat astroglial cells was investigated. A respirometer (Internal circulation unit; ICU) which permits measurements on cultures attached to the culture dish was utilized and the results were compared to those obtained for detached cells. It was found that glutamate stimulates astrocytic respiration in attached whereas detached cells remained unaffected.

KEYWORDS

Astrocytes; glutamate; respiratory activity; respirometer.

INTRODUCTION

It has been hypothesized that inactivation of transmitter glutamate in the brain proceeds via uptake into astroglial cells (Kimelberg, 1983). This theory is supported by the presence of a high affinity transport system for glutamate (Schousboe, 1981) and an exclusive localization of the glutamate metabolizing enzyme glutamine synthetase in astrocytes (Norenberg and Martinez-Hernandez, 1979). In order to investigate whether energy metabolic pathways provide an additional mechanism for degradation of glutamate, initially its effect on respiratory activity was studied.

MATERIALS AND METHODS

Rat astroglial cells were prepared according to the method of Booher and Sensenbrenner (1972). Respiratory activities were measured using polarographic oxygen electrodes. Two different respirometer designs were used, a traditional stirring chamber and an ICU unit. Measurements were carried out in a phosphate buffered solution. For measurements in the stirring chamber cell monolayers were removed by a rubber spatula.

Advances in the Biosciences Vol. 61

RESULTS

Table 1. Respiratory activities of rat astroglial cells.

Additives	Stirring chamber	ICU-unit
Glucose (10 mM)	0.094 ± 0.022 (5)	0.106 ± 0.004 (5)
+ L-glutamate (0.1 mM)	n.d.	0.164 ± 0.011 (4)
+ L-glutamate (1.0 mM)	n.d.	0.171 ± 0.010 (5)
+ L-glutamate (10 mM)	0.096 ± 0.023 (5)	0.153 ± 0.005 (5)
+ Ouabain (1 mM) + L-glutamate (1 mM)	n.d.	0.069 ± 0.012 (5)

Results expressed as µl O_2 per minute and milligram protein ± S.E.M., number of experiments in brackets.

From table 1 it can be seen that in the stirring chamber no effect of glutamate addition on the respiratory activity was observed. Contradictory to this were results obtained in the ICU-unit, where the addition of 0.1 mM, 1 mM or 10 mM caused an increase in respiration of approx. 50%. When one millimolar of ouabain was added prior to glutamate, the stimulatory response was abolished.

DISCUSSION

The mechanical detachment of cultured anchorage dependent cells is known to give raise to impaired plasma membranes (Peterson and Walum, 1985). Furthermore, it is well established that the glutamate transport system is extremely labile to mechanical disruption of cells (Fonnum, 1984). It was found that glutamate failed to stimulate the respiratory activity of both detached and ouabain treated cells. Thus, it seems justified to assume that the stimulation of respiration caused by glutamate in attached cells, primarily is the result of an increased energy demand during glutamate uptake. If glutamate in this situation also is utilized as an energy metabolic substrate as proposed by Hertz (1981) remains to be investigated.

REFERENCES

Booher, F. and Sensenbrenner, M. (1972). *Neurobiol.*, *2*, 97-105.
Fonnum, F. (1984). *J. Neurochem.*, *42*, 1-11.
Hertz, L. (1981). *J. Cereb. Blodd flow and Metab.*, *1*, 143-153.
Kimelberg, M.K. (1983). *Cell. Molecul. Neurobiol.*, *3*, 1-16.
Norenberg, M.D. and Martinez-Hernandez, A. (1979). *Brain Res.*, *161*, 303-310.
Peterson, A. and Walum, E. (1985). *In Vitro*, in press.
Schousboe, A. (1977). in *Cell tissue and organ cultures in neurobiology.* (eds) Fedoroff, S. and Hertz, L., Adademic Press, 441-447.

Heterogeneity Among Astrocytes Demonstrated by Protein Synthesis and Soluble Protein Profiles

L. Rönnbäck*,† and E. Hansson*

*Institute of Neurobiology and †Department of Neurology, University of Göteborg, Göteborg, Sweden

ABSTRACT

The serum-free incubation media for released proteins from astroglial-enriched primary cultures were investigated. Electrophoretic profiles of ^{3}H-valine-labelled proteins showed a wide range of isoelectric points and molecular weights which differed from the profiles of intracellular proteins. Medium proteins from astroglial cultures derived from various brain regions exhibit characteristics suggesting specific astroglial properties.

INTRODUCTION

Astroglial cells participate in physiological activities such as regulation of neurotransmitters and of ion exchange (Henn & Hamberger, 1971; Kimelberg, 1983), thereby influencing neuronal survival, proliferation and differentiation. There is experimental evidence for a neuronal-glial interdependency and for glia acting both as target cells reacting to central and peripheral stimuli and as effector cells regulating neuronal activity by releasing various factors. Astroglial cells are, to some degree, functionally specialized for the specific neurons which they serve (see e.g. Hansson, et al., this volume).

METHODS

The primary cultures from newborn rat brain were grown according to Hansson (1984) and Hansson, et al. (1984). Thereafter the cultures were washed three times with Eagle's MEM in the absence of fetal calf serum (fcs) and incubated in serum-free medium for 20-24 h. After incubation in ^{3}Hvaline (5 μ Ci/ Petri dish) for 4 hrs, the media were collected, filtered to remove cells and particulate material and concentrated by dialysis through polyethyleneglycol keeping molecules with m.w. >5,000. Soluble cell proteins were prepared by harvesting the cells from the Petri dishes, followed by homogenisation and centrifugation (105,000xg for 60 min). The supernatants contained the soluble proteins which could be fractionated using isoelectric focusing (IEF) (Vesterberg, 1972) or a discontinuous SDS-PAGE system (Laemmli, 1970).

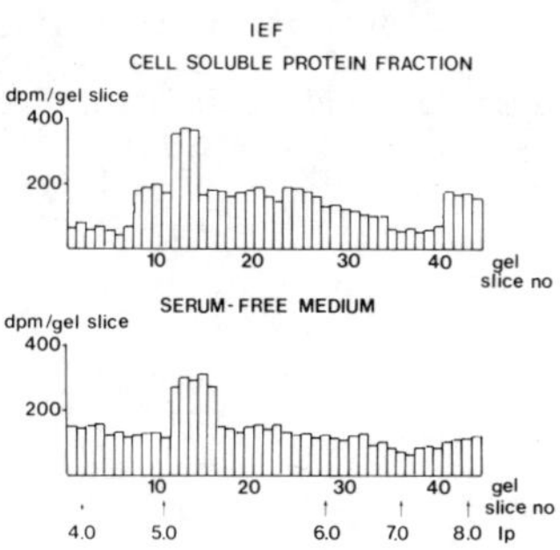

Fig. 1

RESULTS

Electrophoresis revealed proteins in the serum-free incubation medium which had a broad range of Ip:s and molecular weights. These electrophoretic gels were

Printed in Great Britain

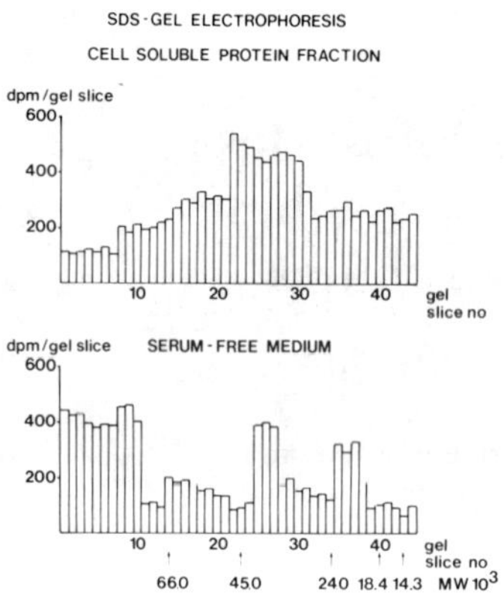

Fig. 2

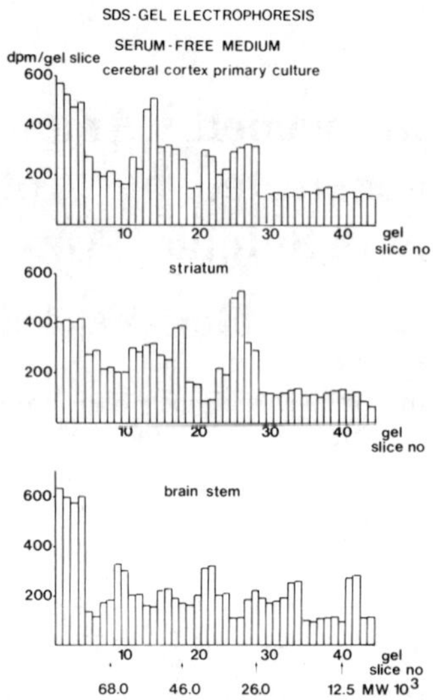

Fig. 3

stained with Coomassie Briliant Blue. The medium proteins from the astroglial-enriched primary cultures showed specific profiles, which differed both from the fcs and from the intra cellular soluble proteins. Fig. 1 shows the ^{3}H-valine distribution in the isoelectric focus gels of cell soluble proteins and serumfree incubation media from cultures of cerebral hemispheres. There are striking differences in ^{3}H-labelling of the gels from the two samples. Most prominent is a high ^{3}H-labelling of acidic proteins (pH 3.5-4.5) in the extracellular fluid proteins. Fig. 2 demonstrates stained SDS-PAGE gels of cell-soluble proteins and serum-free extracellular fluid proteins from cerebral hemisphere cultures, 14 days old. The cultures were treated as in Fig. 1. The most prominent difference between the two samples is the pronounced ^{3}H-labelling of high molecular weight proteins (> 75,000) in the extracellular fluid, while the ^{3}H-valine incorporation into many proteins in the M.W. region of 10,000-50,000 was higher in the cell-soluble protein fraction. Fig. 3 shows the 4 h labelling with ^{3}H-valine of primary cultures from 3 different brain regions. Incubation media material was separated in SDS-PAGE gels. Some specific differences were obtained in the ^{3}H-pattern of the gels.

DISCUSSION

The actual origin of the proteins in the serum-free culture medium remains unknown. The reproducibility of the protein peaks of medium proteins argue against cell damage or lysis as contributing significantly to the profile. Comparison of the protein pattern from serum-free medium with the soluble tritiated components of the cells further suggest a specific, non-random source of labelled medium proteins. Alternative sources of medium proteins include release of substrate attached material, cell surface membrane turnover, or specific cell secretion, the two last possibilities remaining probable.

ACKNOWLEDGEMENT

The work was supported by grants from the Swedish Medical Research Council (project Nos. 12X-06812 and 12P-7308).

REFERENCES

Hansson, E. (1984). Neurochem. Res. 9, 153-172.

Hansson, E., L. Rönnbäck, L.I. Persson, A. Lowenthal, M. Noppe, C. Alling, and B. Karlsson (1984). Brain Research 300, 9-18.

Henn, F.A., and A. Hamberger (1971). Proc. Nat. Acad. Sci. (USA) 68, 2686-2690

Kimelberg, H.K. (1983). Cell. Mol. Neurobiol. 33, 1-16.

Laemmli, U.K. (1970). Nature 227, 680-685.

Vesterberg, O. (1972). Biochem. Biophys. Acta 257, 11-19.

Pituicytes: a Promising Model System for Studying Neuronal–Glial Interactions with a Clear Physiological Background

F. W. van Leeuwen

Netherlands Institute for Brain Research, Meibergdreef 33, 1105 AZ Amsterdam ZO, The Netherlands

ABSTRACT

Enkephalin immunoreactivity was found in synaptoid contacts on pituicytes, which suggests a neuronal-glial interaction and a role for astrocytes in the modulation of oxytocin and vasopressin release into the bloodstream.

KEYWORDS

Enkephalin; vasopressin; oxytocin; pituicytes; astrocytes

Vasopressin (VP) and oxytocin (OXT) are synthesized in the hypothalamus and transported towards the neural lobe, where they are released into the bloodstream. They are involved in a number of functions (anti-diuresis, blood pressure elevation, milk ejection, parturition). Interspersed among the nerve fibers, glial cells of the astrocytic type are present which were termed pituicytes on the basis of a different morphology and unique staining properties. Synaptoid contacts are present on pituicytes (Van Leeuwen et al., 1983; Wittkowski, 1985).

Pituicytes react very strongly to conditions of enhanced VP and OXT release into the bloodstream (e.g., reduced enclosure of nerve fibers by their processes and an increase in the number of synaptoid contacts upon soma and processes). For this reason, speculations have been advanced regarding the role of pituicytes in the modulation of VP and OXT release from the neural lobe (Hatton et al., 1984).

In the course of the last decade it has become apparent that the neural lobe is innervated not only by VP and OXT fibers, but also by fibers containing a large number of other compounds (e.g., amino acids, amines or peptides). One of the most interesting compounds are the opioid peptides, first demonstrated by Rossier et al. (1979). Electrophysiological data from Clarke et al. (1979) indicated that the suckling-induced OXT release in the neural lobe of lactating rats was inhibited by opiates. They hypothesized that the opiates modulate the OXT release via a presynaptic modulation of OXT fibers. We have found that enkephalin-immunoreactive fibers terminate in a synaptoid fashion on pituicytes only (fig. 1), and hypothesized that the modulatory effect of opiates might be effectuated not directly on VP and OXT endings but via the pituicytes (fig. 2). This hypothesis was supported by Lightman et al. (1983), who found evidence for the presence of opiate receptors on pituicytes.

The mechanism by which the pituicytes exert their function remains to be determined. However, recent data suggest that pituicytes are able to regulate the $[Ca^{2+}]$ of the perivascular space, directly influencing OXT and VP release (Bam-

Advances in the Biosciences Vol. 61

Printed in Great Britain

bauer et al., 1984; Shaw and Dyball, 1984).

In conclusion, the pituicytes are most probably involved in the modulation of OXT and VP release. Opioid peptides are only one class of compounds found in synaptoid terminals on pituicytes, while there is morphological evidence for the presence of a number of other synaptoid contacts that contain a variety of releasable compounds (see Van Leeuwen et al., 1983). Indeed, GABAergic synaptoid contacts and binding sites were recently found (Buijs and Van Vulpen, personal communication; Brown and Martin, 1984), while VP and OXT binding sites are also present. These findings indicate that VP and OXT might be able to regulate their own release and/or that of the other compound as well (e.g., via the pituicytes) (Van Leeuwen and Wolters, 1983).

REFERENCES

Bambauer, H.J., S.Ueno, H. Umar and M. Ueck (1984) Cell.Tiss.Res. 237, 491-497.
Brown, C. and I.C. Martin (1984) Eur.J.Pharmacol., 102, 563-564.
Clarke, G., P. Wood, L. Merrick and D.W. Lincoln (1979) Nature, 282, 746-748.
Hatton, G.I., L.S. Perlmutter, A.K. Salm and C.D. Tweedle (1984) Peptides, 5, 121-138.
Lightman, S.L., M. Ninkovic, S.P. Hunt and L.L. Iversen (1983) Nature, 305, 235-237.
Rossier, J., E. Battenberg, Q. Pittman, A. Bayon, L. Koda, R. Miller, R. Guillemin and F. Bloom (1979) Nature 277, 653-655.
Shaw, F.B. and R.E.J. Dyball (1984) Neuroendocrinology, 38, 504-510.
Van Leeuwen, F.W., C.W. Pool and A.A. Sluiter (1983) Neuroscience, 8, 229-241.
Van Leeuwen, F.W. and P. Wolters (1983) Neurosci.Lett., 41, 61-66.
Wittkowski, W. (1985) In: Astrocytes (S. Federoff, ed.), in press.

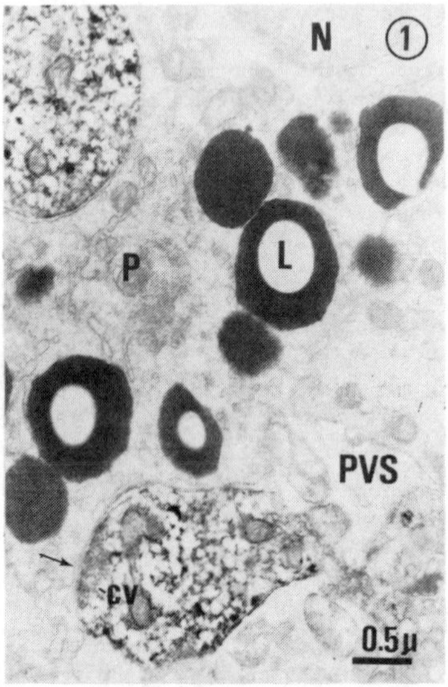

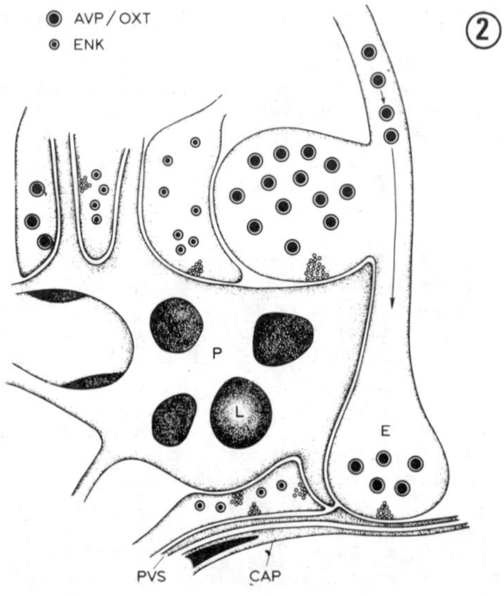

Fig. 1. Enkephalin - immunoreactive fibers surrounding a pituicyte (P), containing partly extracted lipid droplets (L) and a nucleus (N). A synaptoid contact is shown (arrow). cv=clear vesicles, PVS= perivascular space. Fig. 2. Schematic illustration of the relationship between enkephalin (ENK) fibers, a pituicyte (P) and VP or OXT fibers, resulting in modulation of VP or OXT release. E=ending, CAP=capillary. It must be noted that fibers containing other releasable compounds (e.g., GABA) terminate on pituicytes in a synaptoid fashion as well.

Antisera to Glycoproteins Associated with PNS Myelin Immunostain Neurons in Chicken Dorsal Root Ganglia

F. X. Omlin*, E. Philippe*, A. Rochat*, J.-M. Matthieu[+] and B. Droz*

*Institute d'Histologie et d'Embryologie, Ecole de Médicine, Université de Lausanne, CH-1005 Lausanne, Switzerland
[+]Service de Pédiatrie, Ecole de Médicine, Université de Lausanne, CH-1011 Lausanne-CHUV, Switzerland

ABSTRACT

Myelin-associated glycoprotein (MAG) is an intrinsic membrane glycoprotein, which is specifically expressed by myelinating oligodendrocytes and Schwann cells but not by neurons of mammalian nervous tissue. However, here we report the presence of MAG, or at least of a MAG related molecule, which is expressed by small sensory neurons of dorsal root ganglia (DRG) in chickens. These labeled nerve cells of type B were mainly concentrated in the mediodorsal region of the ganglion. In contrast, large A-neurons, which are mostly located in the lateroventral part of the DRG were never immunostained. The same cells were also immunostained for another PNS myelin glycoprotein, called Po. Immunocytochemical investigations of MAG at the EM level have shown that the small sensory neurons exhibit intensely immunostained Golgi apparatus. The cisternae of the rough endoplasmic reticulum were also stained but to a lesser degree, while the cell surface displayed a patchy distribution of immunoprecipitates. Immunoblots of chicken DRG showed the typical pattern of MAG binding, which was identical to that observed in the chick sciatic nerve. To determine whether MAG immunoreaction was expressed by B-ganglion cells at early or late stages of development, DRG of chick embryos were investigated at various embryonic ages. Up to the 6th embryonic day, no MAG immunoreactive material could be detected. At 7 days most immunostained cells were confined to the mediodorsal region of the DRG. By 7-12 days the staining was restricted to one pole of the cell forming a "nuclear cap". This distribution pattern of the immunostaining corresponds to developing Golgi elements in embryonic DRG cells. From 15 days to hatching the immunostaining gradually invaded the perinuclear region and appeared finally at the cell surface. The question was then raised whether neuronal MAG persists in cultured explants of DRG _in vitro_. When such explants taken at 10 days after incubation were immunostained after 6 days in culture the labeled neurons were more or less uniformly distributed throughout the explant. However after 13 days most of the small ganglion cells, which display MAG immunoreactive material moved toward the periphery and even out of the explanted DRG, while unstained large ganglion cells remained in the central part of the ganglion. These results indicate that MAG is already expressed in undifferentiated nerve cells, which are capable of migration. Since MAG appears first in the Golgi apparatus of certain embryonic DRG cells, then at later stages of differentiation at the cell surface, we suggest that MAG, an intrinsic membrane glycoprotein, could play an important role to control migration of small sensory B neurons.

KEYWORDS

Chicken dorsal root ganglia, sensory neurons, immunocytochemistry, immunochemistry, glycoproteins, myelin-associated glycoprotein, PNS myelin glycoprotein, Po-glycoprotein, cell migration.

INTRODUCTION

Myelin-associated glycoprotein (MAG, Mr 100 KD) is an intrinsic membrane glycoprotein, which acts as a specific marker for both myelin and myelinating glia but not for neurons in mammalian nervous tissue (Quarles, 1979). Recently it has been reported, that MAG, or at least a MAG-related molecule, is expressed by small sensory neurons of dorsal root ganglia (DRG) in the chicken (Omlin et al. 1985). This study has shown the presence of MAG immunoprecipitates in the Golgi apparatus, at the cisternal membranes of rough endoplasmic reticulum as well as at the surface of the cell body. It has been suggested that MAG could play an important role to control migration of small B sensory neurons. To define whether MAG immunoreaction is expressed by B neurons at early or late stages of development, DRGs of chicken embryos were investigated immunocytochemically *in situ* and *in vitro*, respectively, at various embryonic ages (Philippe et al. 1985) as well as at different times in culture (Rochat et al. 1985), respectively. Furthermore to determine whether other glycoproteins, which are also specific markers for myelin and myelinating glia in mammals are similarly expressed, we used antibodies against Po.

MATERIAL AND METHOD

Tissue and antibodies: Dorsal root ganglia (DRG), spinal cord and optic nerves from White Leghorn chickens were used at various stages of embryonic and postembryonic development either for immunocytochemical or immunochemical investigations. In parallel, tissue samples from the same regions of 15-day-old mice and rats were tested using the same procedures. In the present study rabbit antisera to both MAG from rat brain and Po from bovine PNS were used. These polyclonal antibodies were generously provided by Drs. R.H. Quarles and B.D. Trapp. Both immunological and immunocytochemical specificity of these antisera have been described (Quarles et al. 1981; Trapp et al. 1981; Trapp and Quarles, 1984).

Light and electron microscopic immunocytochemistry: For light microscopic immunocytochemistry, both chickens and mice were perfused through the heart with a $HgCl_2$/formaldehyde solution (Sternberger et al. 1978). After dissection, the removed pieces of tissue were cut on a vibrating microtome. The 15 μm thick adjacent sections were immunostained either for MAG or for Po according to the PAP method of Sternberger (Sternberger et al. 1970). For electron microscopic immunocytochemistry, chickens were perfused through the heart with a solution containing 0.5% formaldehyde and 0.02% glutaraldehyde in 0.1 M phosphate buffer. The tissue was processed as previously described (Omlin et al. 1982, 1985). The following immunocytochemical controls were carried out: (a) MAG antisera was substituted by normal rabbit serum; no immunoreaction occured in DRG (Fig. 5); (b) vibratome sections were pretreated with chloroform-methanol in order to extract gangliosides and other glycolipids; the immunoreaction with MAG remained unchanged; (c) in DRG of mice, nerve cells did not bind the antisera; (d) optic nerves of both mice and chicken (Figs. 1 and 2) showed immunostained oligodendrocytes.

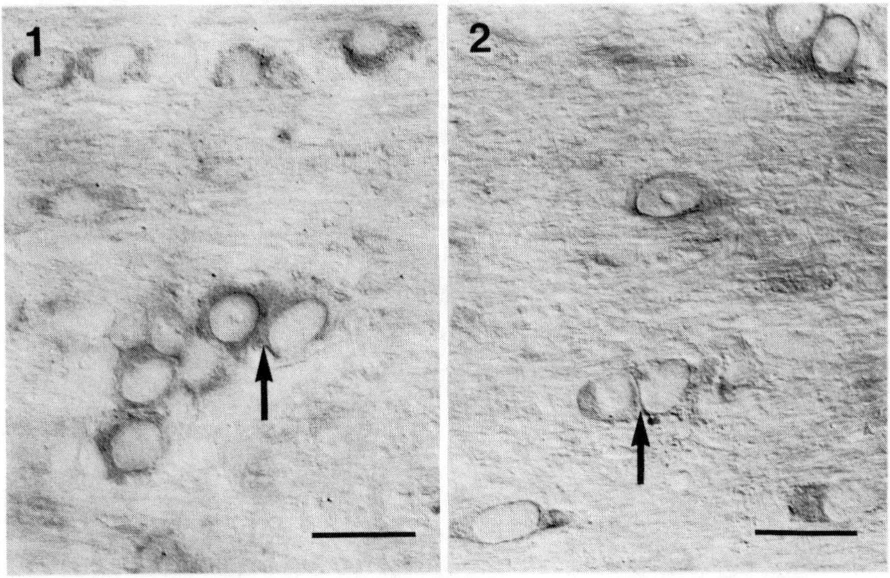

Figs. 1 and 2: Light micrographs (Nomarski optics) of chicken (Fig. 1) and mouse optic nerves (Fig. 2), respectively. These 15 µm thick vibratome sections were immunostained for myelin-associated glycoprotein (MAG). Arrows indicate immunolabeled glial cells, which are myelinating oligodendrocytes. The unstained cells are probably astrocytes. Some of these glial cells appear in clusters or rows, which are intercalated between myelinated and unmyelinated axons. bar 20 µm.

Cell cultures: Explants of DRGs from 10-day-old chick embryos were cultured in collagen coated petri dishes in a medium containing horse serum and NGF over a period of 6 to 42 days (Rochat et al. 1985). The immunocytochemical procedure used was as follows. After several rinses in 0.5 M Tris buffer, the explants were fixed with HgCl2/formaldehyde at 4°C overnight. Crystalline precipitates of HgCl2 were removed from the surface of the explant with 1% Lugol's iodine. Before the cultures were exposed to the antibodies at room temperature (MAG, 1:250 for 2 h; sheep anti-rabbit IgG, 1:50 for 1 h; PAP, 1:100 for 30 min), several rinses in 0.5 M Tris over periods of 2-3 hours each were carried out. These prolonged rinses helped to reduce the nonspecific background immunostaining.

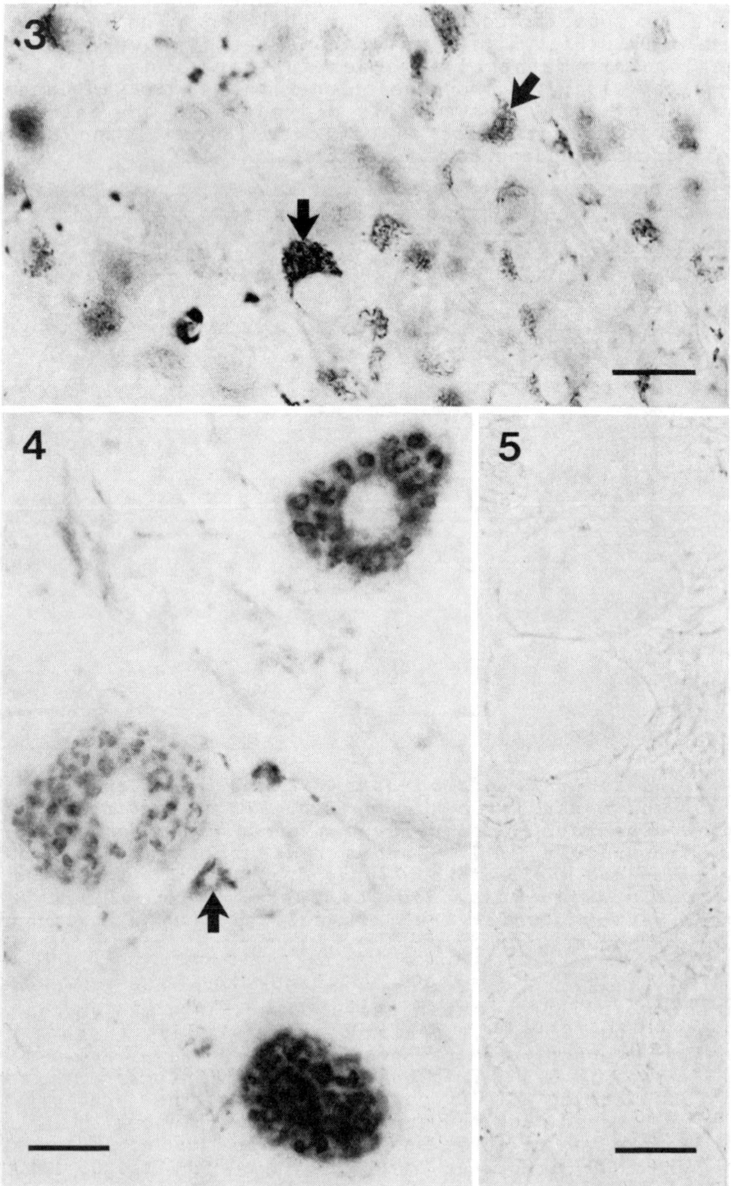
3
4
5

RESULTS

At the embryonic age of 7 days, vibratome sections of chicken DRG showed both stained and unstained nerve cells after incubation in antisera to MAG. The immunolabeled cells were mainly distributed

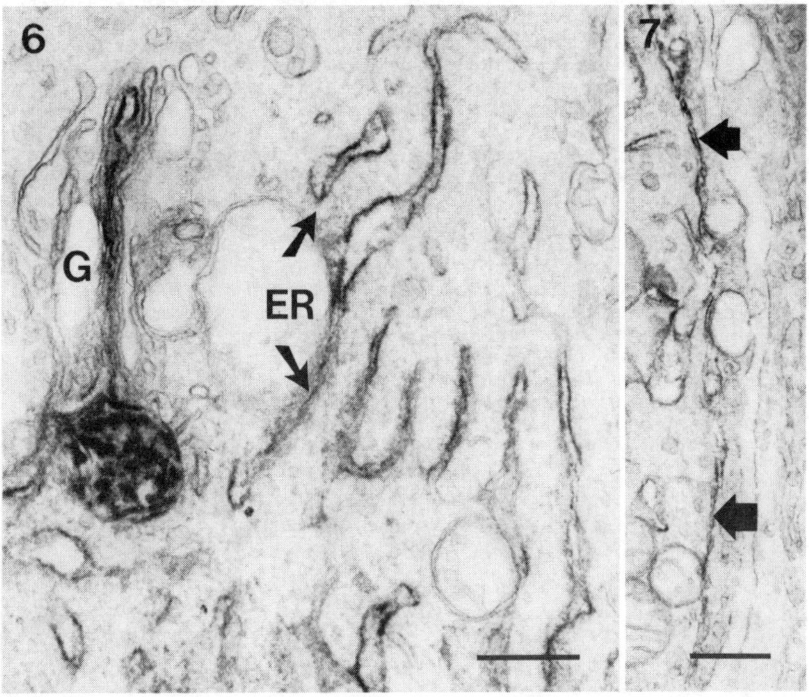

Figs. 3-7.: Figs. 3-5; Light micrographs (bright field) of chicken dorsal root ganglia (DRG) during development. These vibratome sections were immunostained for MAG (Figs. 3 and 4) or treated with normal rabbit serum instead of MAG antiserum, as a control (Fig. 5). At embryonic day 7 (Fig. 3) most of the neuroblasts show a regular shape. The immunoprecipitates (arrows) are located at one pole of the cell, forming a "nuclear cap". At 7 days after hatching (Fig. 4), small sensory neurons of type B show intense immunoprecipitates associated with the Golgi apparatus. A more patchy distribution of reaction products is present at the surface of the cell body. An immunostained Schwann cell is labeled with an arrow. Large neurons remained unstained. bar 20 μm.
Figs. 6 and 7; Electron micrographs of a B-sensory neuron of chicken DRG. This Epon thin section was cut after the treatment with antibodies to MAG (preembedding immunostaining). Intense immunoprecipitates (Fig. 6) are associated with the Golgi apparatus (G) and the endoplasmic reticulum (ER, arrows). A patchy distribution of MAG immunostaining (Fig. 7, arrows) is visible at the surface of the cell body of this small neuron. bar 0.3 μm Fig. 6; bar 0.5 μm Fig. 7.

in the medio-dorsal region of the ganglion. Stained neuroblasts were an identical in shape to the unstained ones. The immunopre-

cipitates of the stained cells were at one pole of the cell-body forming a "nuclear cap" (Fig. 3, arrows).
At 5-7 days after hatching, a similar distribution of labeled cells could be observed within the medio-dorsal region of the ganglion. However, at this stage of development two clearly distinguishable populations of nerve cells are present. The small sensory B-neurons, which were intensely immunostained for both MAG (Fig. 4 and 11) and Po (Fig. 12), and the large neurons of type A (Figs. 4,11 and 12, LN), which remained unstained. Immunoprecipitates in the B neurons were localized in the Golgi region, the cisternae of the endoplasmic reticulum and at the surface of the nerve cell body (Figs. 4,6,7, 11 and 12). Furthermore, some Schwann cells within the DRG were also immunostained (Fig. 4, arrow). Immunochemical investigation using the immunoblot technique (Fig. 8) showed that chicken DRGs display a pattern of MAG binding (130 KD band) which is identical to that observed in the sciatic nerve (Fig. 8 C and D).
In cultured explants of chicken DRG after 7 days in culture (Fig. 9), most of the cells which are immunostained for MAG were located in the central part of the ganglion. After 13 days in culture (Fig. 10) the immunostained cells moved out of the center toward the periphery of the explant. Immunostained cells showing a spindle like form were identified as Schwann cells, whereas the round cells were presumably small sensory neurons. Most of the large unstained cells remained in the center part of the explant.

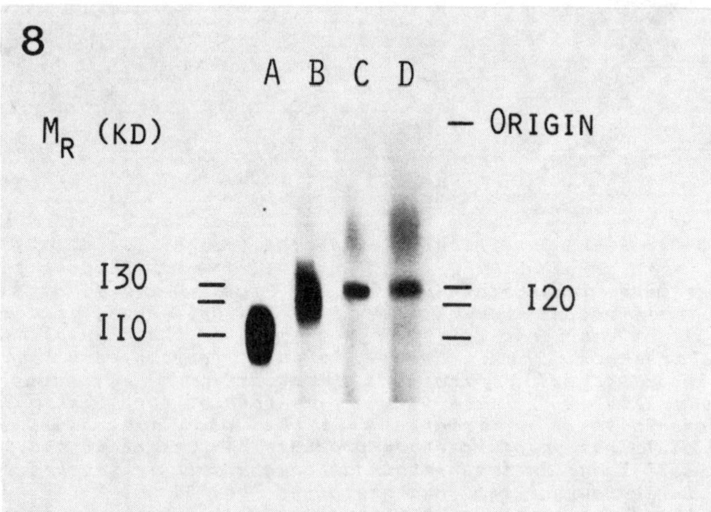

Fig. 8.: Autoradiographs of an immunoblot using an antiserum to MAG. A: rat brain myelin; B: rat sciatic nerve; C: chicken sciatic nerve; D: chicken DRG. Both chicken sciatic nerve and chicken DRG show the same pattern of MAG binding which corresponds to a molecular weight at 120 KD.

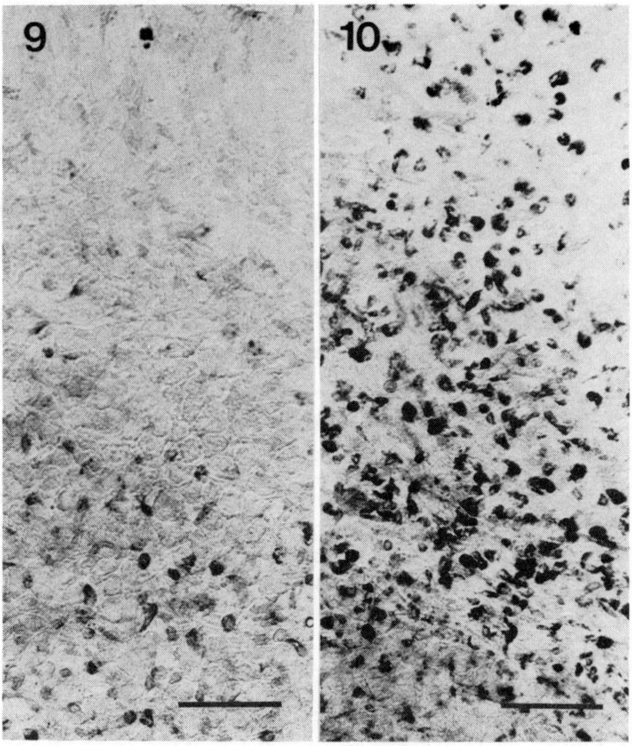

Figs. 9 and 10.: Light micrographs (bright field) of chicken DRG explant cultures. These cultures were immunostained for MAG after 7- (Fig. 9) and 13 days (Fig. 10) in culture, respectively. After 7 days in culture (Fig. 9) most of the labeled neurons are located within the central part of the ganglion, whereas at 13 days many of these immunostained cells have moved out of the center toward the periphery of the explant. Some of the immunostained cells are Schwann cells, which show a spindle like shape. bar 100 µm.

DISCUSSION

Both, MAG and Po are glycoproteins which are specifically associated with myelin and myelinating glia in mammals. They have never been observed in neurons from mammals. Our results show the presence of two glycoproteins, or of polypeptides which share a common epitope with these molecules, in small sensory type B neurons of chicken DRGs. Therefore, these antibodies represent new markers for small sensory neurons in the chicken. These results suggest that some glycoproteins play a more general functional role in chicken than in the nervous tissue of mammals. Possible functions of these glycoproteins in the chicken could be cell recognition, cell adhesion and morphogenesis as suggested by Edelman (1983). Further immunochemical investigations will show whether the polypeptide or the carbohydrate moieties of these two glycoproteins are responsible for the described immunoreaction.

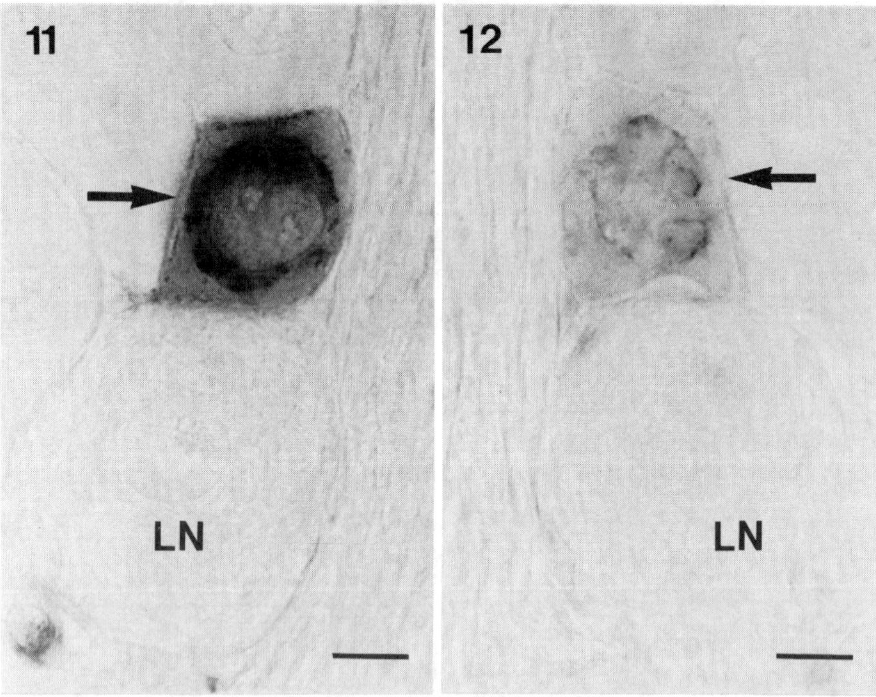

Figs. 11 and 12.: Light micrographs (bright field) of small and large neurons of chicken DRG at the developmental stage of 7 days after hatching. These adjacent 15 µm thick vibratome sections are immunostained for MAG (Fig. 11) and Po (Fig. 12), respectively. These pictures, which are mounted in a mirror configuration, show parts of the same small sensory neuron (arrows), which present immunoprecipitates for MAG and for Po, respectively. In both sections, the large neuron (LN) of type A remained unstained. bar 10 µm.

REFERENCES

Edelman, G.M. (1984). Proc. Natl. Acad. Sci. USA., 81, 1460-1464.

Edelman, G.M. (1983). Science 219, 450-457.

Omlin, F.X., H. deF. Webster, C.G. Palkovits, and S.R. Cohen (1982). J. Cell Biol., 95, 242-248.

Omlin, F.X., J.-M. Matthieu, E. Philippe, J.-M. Roch, and B. Droz (1985). Science 227, 1359-1360.

Philippe, E., F.X. Omlin, and B. Droz (1985). Neurosci. Lett. Suppl., 22, S35.

Quarles, R.H. (1979). Complex Carbohydrates of the Nervous System, Margolis R.U. and R.U. Margolis (Eds). Plenum New York, 209-233.

Quarles, R.H., R.O. Brady and N.H. Sternberger (1981). Neurochem. Res., 6, 1109-1121.

Rochat, A., F.X. Omlin and B. Droz (1985). Neurosci. Lett. Suppl., 22, S331.

Sternberger, L.A., P.H. JR Hardy, J.J. Cuculis and H.G. Meyer (1970). J. Histochem. Cytochem., 18, 315-333.
Sternberger, N.H., Y. Itoyama, M.W. Kies and H. deF. Webster (1978). J. Neurocytol., 7, 251-263.
Trapp, B.D., Y. Itoyama, N.H. Sternberger, R.H. Quarles and H. de F. Webster (1981). J. Cell Biol. 90, 1-6.
Trapp, B.D. and R.H. Quarles (1984). J. Neuroimmunol., 6, 231-249.

ACKNOWLEDGEMENT

The authors like to acknowledge gratefully T. Delapierre, M. Pasquier and Dr. O. Reymond for their excellent assistance. Supported by the Swiss National Science Foundation (3.447.83 and 3.176.82).

Lipid Synthesis in Oligodendroglia and Putative Glioblasts Isolated from Developing Rat Brian

M. J. Brammer

Department of Biochemistry, Institute of Psychiatry, De Crespigny Park, London SE5 8AF, UK

ABSTRACT

Oligodendroglia and probable glial precursors were isolated from developing rat brain and characterised by immunofluorescence and morphology. Lipid synthesis was measured in whole cells and ceramide glycosyltransferase activities determined in cell homogenates.

KEYWORDS

Oligodendroglia, glioblasts, morphology, immunology, lipid biosynthesis

INTRODUCTION

In the fifteen years or so since Bunge (1968) detailed the evidence for the identification of oligodendroglia as the myelin-forming elements of the central nervous system, an increasing amount of effort has been directed towards the elucidation of the properties of these cells during their differentiation and subsequent production of myelin. These studies have normally employed rodents (e.g. rats, mice) in which oligodendroglial differentiation and myelination are largely post-natal and therefore amenable to convenient experimental investigation.

There are now a number of techniques for obtaining oligodendroglial-enriched cell populations from rodent brains (cf. Chao and Rumsby, 1977, McCarthy and deVellis, 1980, Snyder et al, 1980) and one particularly fruitful approach to the study of oligodendroglial differentiation has been the isolation of these cells from neonatal animals followed by studies of their development in culture (see review by Pfeiffer, 1984).

Studies of bulk-isolated cells obtained at different time points during post-natal development has been a less popular approach but we feel that it has considerable merit and can yield valuable information which complements that obtained in the culture studies.

This chapter concerns studies of lipid biosynthesis in cells obtained from post-natal rat brain using the technique developed by Snyder et al (1980) for oligo-dendroglial isolation. In assessing metabolic changes we have also characterised the cells morphologically and by immunofluorescence.

CELL CHARACTERISATION

In their original study, Snyder et al (1980) utilised rats aged 10, 30 and 60 days and reported yields of 3-4 x 10^6 cells/g brain at 30 and 60 days and a somewhat higher yield per g brain (7 x 10^6) at 10 days. Fig. 1 shows that we also obtained relatively constant cell yields per g. brain (2 -3 x 10^6) between 15 and 30 days <u>post partum</u> but a much higher yield in younger animals, peaking at 4-5 days after birth. We suggest that this apparent peak in cell yield may reflect the interplay of two effects. Firstly, a rise in yield due to the increased numbers of cells after glial cell division (Roback and Scherer, 1935; Schonbach et al, 1968; Fulcrand et al, 1975; Skoff et al, 1976; Tennekoon et al, 1977; Privat et al, 1981) and secondly a fall in yield due to the greater difficulty in isolating cells once myelination has reached a significant level.

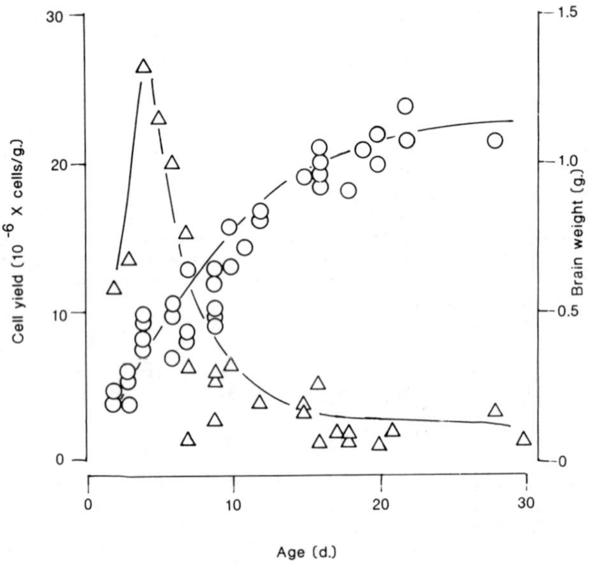

<u>Fig. 1. The dependence of cell yield on age in developing rat brain</u>

Cell yield (triangles) indicates the number of phase-bright cells obtained per g of forebrain. Brain weight (circles) is also shown as an index of development.

It was reported by Snyder et al (1980) that trypsinisation in a balanced salt medium at pH 7.4 selectively destroyed astroglia and neurones, thus facilitating the isolation of oligodendroglia which were primarily identified by morphological criteria. Snyder et al (1980) reported that their preparations were around 90% oligodendroglia. We have used light and electron microscopy and immunofluorescence criteria to characterise the cells which we have isolated, with the following results.

Examination of the isolated cells by phase-contrast microscopy showed that large numbers of small, round, phase bright cells were present. We could detect no cells with apparent neuronal morphology but a small number of red cells ($<$5% of total) and occasional ependyma were seen. The appearance of the cells by light microscopy was broadly similar at all ages.

A subsequent detailed electron microscopic examination of the preparations throughout the age range, using the criteria of Raine et al (1971) for identification of isolated cells, revealed certain interesting facts. Firstly, there was no consistent evidence for serious contamination of the preparations by neurones or astroglia but ependyma and red cells were seen in small numbers, confirming the light microscopic data. Below about five days Post Partum, the bulk of the cells could not be identified with certainty as any of the major differentiated brain cell types. "Dark" cells with many ribosomes and dense cytoplasm were seen as were "light" cells with prominent Golgi apparatus, bunched heterochromatin and some microtubles.

At five days and above the cells appeared to possess many of the features commonly ascribed to oligodendroglia, i.e. dense bunched heterochromatin, large nuclear cytoplasmic ratio, many free ribosomes, densely staining cytoplasm large numbers of microtubules. This is essentially the electron microscopic appearance described by Snyder et al (1980) for their cells.

As identification of undifferentiated cells (likely to be present in our preparations) by morphology alone is difficult, we also attempted independent classification by indirect immunofluorescence. The antibodies chosen were accepted markers for oligodendroglia (Gal-C, monoclonal, Ranscht et al, 1982), astroglia (GFAP, commercial polyclonal), neurones (RT 97 anti-neurofilament, monoclonal, raised by Dr B Anderton) and neurones/developing glia (A2B5, anti ganglioside, Eisenbarth et al, 1979). We utilised these reagents to identify cells using cell suspensions to avoid problems of differential attachment to coverslips.

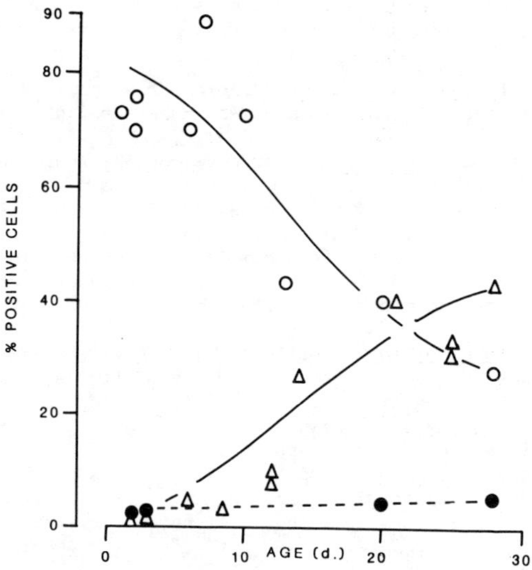

Fig. 2 Age-dependence of the distribution of antigenic markers in isolated cells

The figures indicated are the percentages of total phase-bright cells which were Gal-C(triangles), A2B5 (open circles) or GFAP (filled circles) positive.

The data presented in Fig. 2 concern GFAP, A2B5 and Gal-C. It can be seen that GFAP is present on only a few cells whatever the age of the animal used. A2B5 is present on circa. 80% of cells in very young animals declining to 20-25% at 30 days and this fall is mirrored by an increase in Gal-C positive cells. We were able to detect only very few RT97 positive cells, regardless of the age of the animals used for the cell isolation.

From the combined results of the morphological and immunological characterisation of the cells we conclude that the preparations were essentially free from large-scale contamination by neurones, astroglia, ependyma or erythrocytes. The results from the two techniques were in disagreement, however, over many of the cells present in the preparations from the older animals. Morphologically (in agreement with Snyder et al, 1980), we conclude that the preparations from animals aged ten days and above resemble highly purified oligodendroglia. Immunologically, however, only about half the cells isolated from 30 day old rats are Gal-C positive. The morphologically unidentified cells from the younger animals are clearly mainly A2B5 positive and may be glioblasts.

LIPID SYNTHESIS

The main purpose of this investigation was to study lipidsynthesis in oligodendroglia and their precursors. We examined the uptake of a number of radioactive lipid precursors into "oligodendroglial" lipids (glycerol, linolenate, inositol, glucose) before choosing glucose for most of our experiments. This decision was made for three reasons,

(1) It is a general precursor and is incorporated into all the classes of lipid of interest in this study (cerebrosides and phospholipids).

(2) Its conversion to lipid requires the participation of many enzyme systems in the mitochondria, cytosol and endoplasmic reticulum. Formation of lipids from glucose should thus require a good degree of cellular integrity.

(3) It is a physiological substrate for lipid formation in the brain (Geiger, 1958) and has some relevance to studies in culture as well as in vivo.

The incubation protocol which we used was developed by Andrew Gibson, a graduate student in this laboratory for work on ox brain oligodendroglia (Gibson and Brammer, 1984) and briefly consists of exposing cells to ^{14}C-glucose (10μCi, 5.5mM) in isotonic pH 7.4 isolation medium supplemented with 10% foetal calf serum and antibiotics (gentamicin and amphotericin B). The incubations were carried out at 37°C with oxygenation and gentle shaking and were terminated by lipid extraction by the method of Bligh and Dyer (1959).

For the purposes of the present investigation, we have considered the incorporation of radioactive glucose into four groups of lipids, namely (i) serine plus inositol glycerophospholipids (PS+PI); (ii) choline glycerophospholipids (PC); (iii) ethanolamine glycerophospholipids (PE) and (iv) cerebrosides. Entry of glucose into these four groups of lipids was found to be a linear function of cell number (up to at least 6 million cells/incubation) and incubation time (up to 4h) at 37°C. The initial rate of entry of glucose into lipids was ascertained by varying either cell number or incubation time within these limits. Each cell preparation was sufficient for 4-8 incubations and facilitated reasonably accurate estimation of the initial rate of lipid labelling by glucose. In the present context, the "total rate of glucose incorporation into lipids" refers to the combined rate of uptake of label into PS+PI+PC+PE+cerebrosides.

It was found that incorporation of glucose was linear with time (up to about 5h) and cell number (up to 6 x 106ml). The glucose label was found (using alkaline and acid digestion) to be incorporated into the glyceryl moieties of phospholipids (>90%) and the hexose moieties of cerebrosides (>90%). Glucose is, of course, a general precursor for the different moieties in lipid molecules (Burton et al, 1958; Miller et al, 1977) and it is known that several days after intracranial injection of ^{14}C-glucose, label is found mainly in the acyl chains of phospholipids (Miller et al, 1977) but in the short term, it appears predominantly in the glyceryl backbone (Brauning and Gercken, 1976). Our experiments also showed that incorporation of ^{14}C-glucose into lipids was highly correlated with uptake of ^{3}H-glycerol uptake in double label experiments thus reinforcing our contention that, in short-term incubations, glucose uptake indicates de novo synthesis of phospholipids.

Glucose incorporation into lipids was inhibited by phloretin and by metabolic poisons (cyanide, iodoactetate) and thus requires uptake into cells and active energy metabolism. The fact that cell integrity is required is reinforced by the observation that "phase-dark" cells do not incorporate glucose into lipids.

All these preliminary obserations using cells isolated from rat brain were in agreement with data obtained using highly purified bovine oligodendroglia (Gibson and Brammer, 1984).

Our object was the study of the age-dependence of lipid synthesis from glucose in our isolated cell preparations and it can be seen that there is no significant age-correlated alteration in total incorporation of glucose into the lipids studied ($r = 0.157$, $t = 0.856$, $p = 0.4$, 31 preparations). The mean total rate of incorporation was 144.7 ± 15.8 pmol./h./106 cells (mean ± SEM, 31 preparations, see Fig. 3).

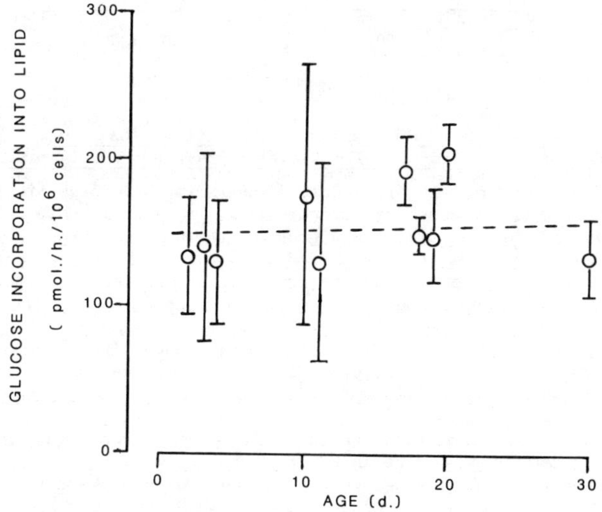

Fig. 3 Age dependence of glucose incorporation into total oligodendroglial lipids

Values are means (± SD's) of three separate experiments

However, when the percentage distribution of the label between different lipid classes is studied as a function of age, some interesting observations can be made (Fig. 4).

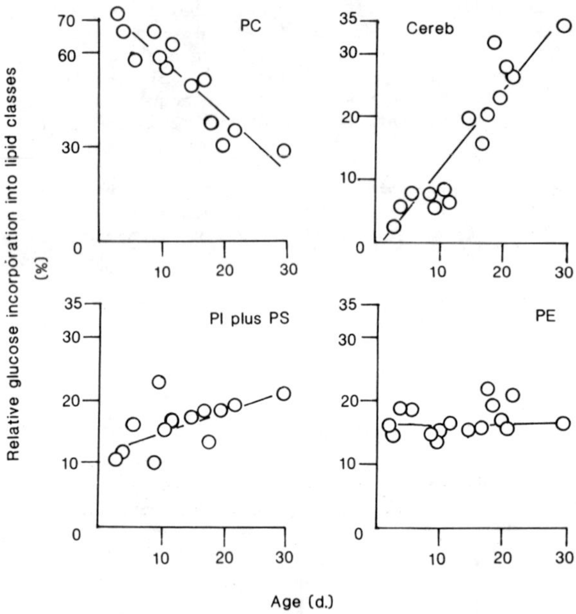

Fig. 4 The relative incorporation of glucose into different lipid classes of cells isolated from developing rat brain

The ordinate is the rate of glucose incorporation into particular group of lipids expressed as a percentage of the total rate of incorporation into all the lipid groups investigated (PS+PI+PC+PE+ cerebroside), see text for further details. Each point represents the mean of two or three separate experiments carried out with animals of a particular age.

Firstly, there was a highly significant age-related negative correlation with relative PC synthesis ($r = 0.9113$, $p < 0.001$) and a positive correlation with cerebroside synthesis ($r = 0.9229$, $p < 0.001$).

There was a barely significant change in PS+PI relative labelling ($r = 0.5964$, $p < 0.05$) and no significant change in PE labelling ($r = 0.2508$ $p = 0.3 - 0.4$). Thus the major alteration with age is a shift in glucose uptake from PC (a general membrane constituent) to cerebroside (a much more specific myelin component). When the ratio of incorporation of glucose into nfa and hfa cerebrosides was studied it also appeared that this ratio declined markedly with age. This is an interesting observation as hydroxy fatty acids are highly enriched in myelin galactocerebrosides. Analysis of the conversion of UDP-glucose to cerebrosides by oligodendroglial homogenates (Fig. 5) using methods previously described in this laboratory (Brammer, 1984) showed that only glucocerebroside was formed by cells isolated from rats younger than about 7-8 days whereas galactocerebroside formation peaked around the time of maximum myelination. Thus it is tempting to speculate that the "glial precursors" synthesise primarily glucocerebroside (and hence gangliosides such as A2B5) while the appearance of "mature oligodendroglia" is accompanied by utilisation of glucose (and UDP-glucose) to synthesise galactocerebrosides.

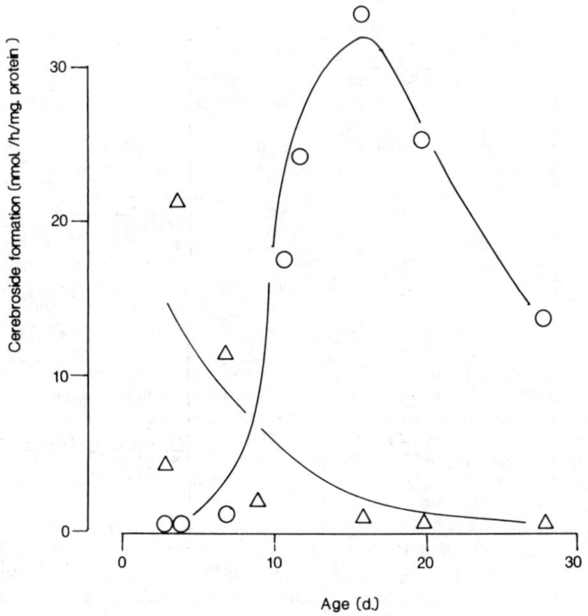

Fig. 5 Age dependence of gluco- and galactocerebroside formation in homogenates of isolated cells

The rates of formation of gluco- (triangles) and galacto- (circles) cerebrosides on incubation of cell homogenates with UDP-[^{14}C]-glucose and mixed hfa and nfa ceramides are shown.

It thus appears that the appearance of significant numbers of oligodendroglia in the preparations (as judged by immunological and morphological evidence) is accompanied by the synthesis of increasing amounts of typical myelin lipid (cerebroside).

In summary, the "oligodendroglial" isolation method of Snyder et al (1980) yields phase-bright cells which are capable of complex metabolic activities (lipid synthesis from glucose). Cells isolated immediately after birth appear to consist mainly of glial precursors (A2B5 positive) but with increasing age, an increasing number of galactocerebroside-positive cells, presumably oligodendroglia, are obtained.

ACKNOWLEDGEMENT

The author would like to express his gratitude to Mr K Weaver for help with the electron microscopy and to Miss C Hughes, Miss N Melihy, Mr A Gibson and Mrs S Mulhall for their contributions to measuring lipid synthesis.

BIBLIOGRAPHY

Bligh, E.G. and Dyer, W.J. (1959) Can.J.Biochem. 37, 911-917.

Brammer, M.J. (1984) J.Neurochem. 42, 135-141.

Brammer, M.J. and Carey, S.G. (1980) J.Neurochem. 35, 873-879.

Braunding, C. and Gercken, G. (1976) J.Neurochem. 26, 1257-1261

Bunge, R.P. (1968) Physiol.Rev. 48, 197-201.

Chao, S.W. and Rumsby, M.D. (1977) Brain Res. 124, 347-351.

Eisenbarth, G.S., Walsh, F.S. and Nirenberg, M. (1979) Proc.Natl.Acad.Sci.USA 76, 4913-4917.

Fulcrand, J., Valat, J. and Marty, R. (1975) VIIieme Congres Internationale de Neuropathologie, Budapest. Amsterdam. Excerpta Medica, pp 733-735.

Geiger, A. (1958) Physiol.Rev. 38, 1-20.

Gibson, A. and Brammer, M. J. (1984) J.Neurochem. 43, 619-626.

McCarthy, K.D. and de Vellis, J. (1980) J.Cell.Biol. 85, 890-902.

Miller, S.L., Benjamins, J.H. and Morell, P. (1977) J.Biol.Chem. 252, 4025-4037.

Pfeiffer, S.E. (1984) In "Oligodendroglia", Advances in Neurochemistry Vol. 5. (Norton, W.T., ed) pp 233-298 Plenum Press, New York and London.

Poduslo, S.E. (1981) In Research Methods in Neurochemistry Vol. 5. (Rodnight, R. and Marks, N. eds) pp 113-125. Plenum Press, New York.

Privat, A., Valat, J. and Fulcrand, J. (1981) J.Neuropath.Exp.Neurol. 40, 46-60.

Raine, C. S., Poduslo, S. E. and Norton, W. T. (1971) Brain Res. 27, 11-24.

Ranscht, B., Clapshaw, P.A., Price, J., Noble, M. and Seifert, W. (1982) Proc.Natl.Acad.Sci.USA 79, 2709-2713.

Roback, H.N. and Scherer, H.J. (1935) Virchows Arch.Path.Anat. 294, 365-413.

Schonbach, J., Hu, K.H. and Friede, R.L. (1968) J.Comp.Neurol. 134, 21-38.

Skoff, R.P., Price, D.L. and Stocks, A. (1976) J.Comp.Neurol. 169, 291-312.

Snyder, D.S., Raine, C.S., Farooq, M. and Norton, W.T. (1980) J.Neurochem. 34, 1614-1621.

Tennekoon, G.I., Cohen, S.R., Price, D.L. and McKhann, G.M. (1977) J.Cell Biol. 72, 604-616.

Phospholipid Metabolism in Glia Cell Culture

B. Witter and H. Debuch

Institut für Physiologische Chemie, Universität zu Köln, Joseph-Stelzmann-Str. 52, D-5000 Köln 41, FRG

ABSTRACT

Preparations of plasma membranes were obtained from glial cell primary cultures by hyposmotical disruption. After incubation of the cell cultures with 1-alkyl- and 1-alkenyl-glycerophosphoethanolamine, both labelled in the hydrophilic head as well as in the hydrophobic tail, the products were isolated. Conclusions on the metabolism of these etherlipids could be drawn according to the ratios of the two different labels.

KEYWORDS

Glial cells; primary culture; plasma membranes; ether phospholipids; acylation; methylation.

The myelin sheath of mammalian brain is one of the richest sources of ethanolamine-plasmalogen, amounting in the adult brain to almost 90% of the total phosphatidylethanolamine fraction (O'Brien and Sampson, 1965). During the time of myelination the biosynthesis of this phospholipid is especially increased.

In order to get more information about the phospholipid metabolism of oligodendroglia we established a glia cell culture. This culture was performed from newborn rat brain according to the method of Booher and Sensenbrenner (1972). The cultures were grown in the presence of brain extract (Sensenbrenner, et al,1972), and consisted of astrocytes as a bottom layer with oligodendroglia laying on them as a second one. The presence of both cell types was proved by histochemistry and by immunohistochemical reactions for GFAP (glial fibrillary acidic protein) specific to astrocytes (Witter and Debuch, 1982) and for antigens "O1" and "O4" specific to oligodendroglia (Witter , et al, 1983). 16 days old cell cultures were employed for biochemical investigations.

The phospholipid fractions of brain show an intramolecular asymmetry of the hydrophobic residues (Clarke and Dawson, 1981) such that each phospholipid is composed of three different molecules (Fig.1): e.g. the ethanolamine-phosphoglycerides (EPG) fraction contains 40% of the diacyl species, bearing 2 fatty acids, 56% is found as plasmalogen, bearing in 1-position an aldehyde in enoletherlinkage and 4% as alkyl-acyl species containing an alcohol in an ether-linkage in 1-position. In the choline-phosphoglycerides (CPG) 92% of the molecules are found as diacyl-species, only 3% as plasmalogen, the alkyl-acyl-species amounts to 5%.

Another intramolecular variability is expressed by the interconversion of the three bases, ethanolamine, choline and serine. The most important modulation of the polar head group occurs through base exchange, a nonenergy-requiring step. A second possibility exists in the N-methylation of ethanolamine to choline, a reaction occurring only to a lower degree. The variability in the hydrophobic moiety as well as in the hydrophilic portion of the phospholipids increases the number of permutations within these molecules as response to membrane proteins or neighbouring lipids.

The average phospholipid pattern of the cultured glial cells showed about 50% of CPG-fraction and approximately 25% of EPG-fraction (Witter and Debuch, 1982). The plasmalogen content came to 7,5%. The lipoproteins of the fetal calf serum contained only traces of EPG lipids, thus assuring that the cells were able to synthesize this lipid-fraction by themselves.

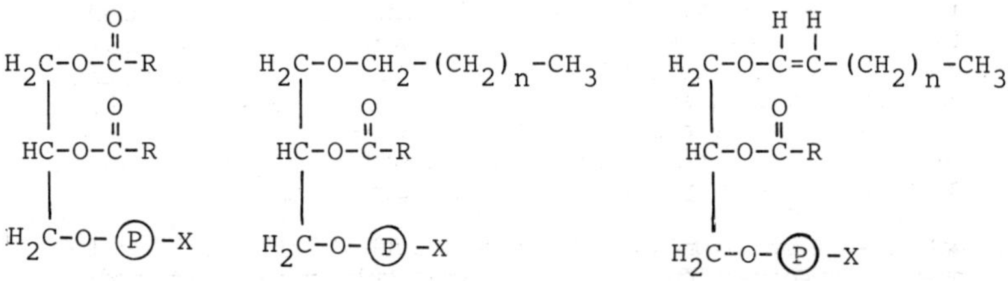

Diacyl GPE	Alkyl-acyl GPE	Alkenyl-acyl GPE
4o	4	56
Diacyl GPC	Alkyl-acyl GPC	Alkenyl-acyl GPC
92	5	3

Fig.1. Intramolecular asymmetry of phosphoglycerides. Figures in %, recalculated from P-analyses of the individual phospholipid fractions of whole rat brain (Clarke and Dawson,1981).R=long chain fatty acids, X= bases: ethanolamine, choline

Our recent investigations of plasmalogen biosynthesis concerned the hydrophobic moiety of the phospholipids. Two series of experiments were performed (Fig.2). The first (Witter and Debuch,1982) was done with 1-alkylGPE (lyso-etherphosphatidylethanolamine), on which the desaturation, the introduction of the double bond to form 1-alkenylGPE (lyso-plasmalogen), was studied. For comparison the second experiment was performed with 1-alkenylGPE, containing already the enolether linkage (Illig, et al, 1982a).

Referring only to the 3h incubations the results with varying substrate concentrations were as follow:
AlkylGPE was always internalized to a higher amount than alkenylGPE. The unaltered substrate concentration within the cells was always higher after alkylGPE application, alkenylGPE was present only in very low amounts. The acylation of both substrates reached a maximum after 3h. Apparent K_M's were 5,0 nmol/mg cell protein for alkylGPE and 1,7 nmol/mg cell protein for alkenylGPE. The desaturation of the ether-linkage to the enolether-linkage occurred only to 2-5 nmol/mg cell protein. Furthermore, small amounts of radioactivity could be found in the CPG-fraction in the 3h experiments. After 12h incubations however, almost 40-50% of the total internalized radioactivity was present in the CPG-fraction with both substrates (Illig, et al, 1982b; Witter, et al, 1983).

SUBSTRATES

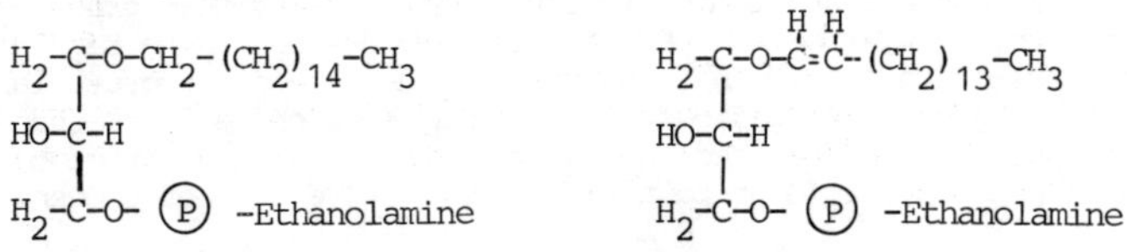

1-AlkylGPE 1-AlkenylGPE

Fig.2. Labelling of the substrates: 1-[3H]alkylGPE, 1-[3H]alkenylGPE. [3H]-labelled ethanolamine-analogues.

These observations prompted new experiments with ethanolamine-labelled substrate analogues (Witter, et al, 1983). After 12h incubations with these substrates, almost 50% of the label was present in the medium as watersoluble substances. Approximately 60-70% of the watersoluble radioactivity were found as ethanolamine and 20-30% as phosphoryl-ethanolamine. From these results it could be assumed that the substrates were hydrolyzed possibly at the plasma membranes either by an action of a lyso-phospholipase D followed by a lyso-phosphatidic acid phosphohydrolase or to a lyso-phospholipase C activity.

In order to get more information about the enzyme activities embedded in the plasma membranes, a fractionation of the glial cell culture was performed (Schimmel, et al, 1973). 5 fractions were obtained, which were characterized by marker enzyme estimation (Tab.1).

Tab.1. Enzymic Properties of Glial Cell Subfractions

Fraction	Protein (mg)	5'Nucleo-tidase[x]	Arylsul-fatase[x]	NADPH[x] Cytochr.c Oxido-re-ductase	Succin.[x] Cytochr.c Oxido-re-ductase
Fract.1	2,6±0,57	198,5±54,5o	13,0±0,45	8,1±0,74	1,7(n=1)
Fract.2	3,7±0,33	302,5±25,21	11,6±1,09	8,8±2,36	1,2±0,49 (n=3)
Fract.3	2,8±0,28	244,8±25,76	27,1±1,10	13,7±2,81	4,1±1,96
Fract.4	5,1±0,32	81,0± 5,72	2o,2±0,92	8,2±1,52	44,0±10,74
Fract.5	2,6±0,18	62,0± 7,65	21,o±2,94	7,1±1,49	33,1±8,99

The values are reported as the mean ± S.D. of 4 observations, except those in parentheses.
(X = nmol x min-1 x mg protein-1)

The first fraction shows some 5'nucleotidase activity already. The second and the third fractions are mostly composed of plasma membranes. Whereas the contamination with other cell organelles is very low in fraction 2 (F_2), fraction 3 (F_3) contains some microsomes and lysosomes.

In all our studies we observed that the best results for initial anabolizing enzyme activities could be obtained in experiments up to 3h incubation time. Therefore prior to fractionation the cells were incubated for 0,5, 1 and 3h in one set of experiment with alkylGPE and in another one with alkenylGPE.

Both substrates (Fig.2) were labelled either in the radyl-chain - in the hydrophobic tail - or in the ethanolamine moiety - in the polar head group - of the molecule. Thus the radyl-labelled radioactivity recovered can be regarded as alkyl- or alkenyl-chain containing products. In the case of products labelled in the polar head group, the radioactivity could represent either ethanolamine or a metabolic product such as choline. Whether the polar head group belongs still to an etherphospholipid or to a diacyl-derivative was not proven.

During all the experiments the substrate concentration was kept constant between 110-115 nmol/mg cell protein with approximately 30000 dpm/dish and the ratio of the radyl- to ethanolamine-labelling was approximately 1:1.

The first section will deal with data obtained with 1-alkylGPE as substrate (Fig.3).

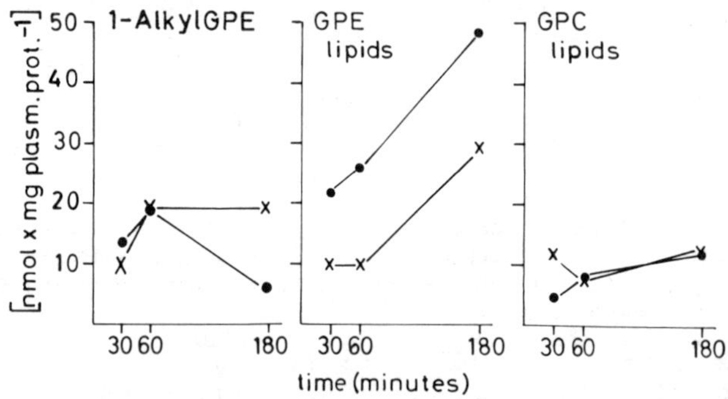

Fig.3. 1-[3H]alkylGPE and 1-alkylGP[3H]E in the plasma membranes (F_2) of cultured glial cells and the products 1-alkyl,2-acylGPE and 1-alkyl,2-acylGPC. (For experimental conditions see text). ● [3H] alkyl-, x [3H] ethanolamine-label.

Looking at the intramembrane unaltered substrate it can be seen that during the initial penetration up to 1h some ethanolamine has been lost. However, after 3h a decrease of the alkyl-chain is observed, which is in line with our previous experiments (Witter and Debuch, 1982). Since the ethanolamine-label at 3h is much higher than the hydrophobic one, the presence of an endogenously available lipid acceptor seems very likely.

As we showed in earlier experiments (Witter and Debuch, 1982), the acylation to alkyl-acylGPE is a rapid process reaching a maximum after 3h of incubation. Indeed, the recovery of both labels follows a very similar course. However, the recovery of the ethanolamine is always 40-60% less than that of the alkyl-chain, e.g. 3o nmol/mg plasma membrane protein in comparison to 50 nmol/mg plasma membrane protein after 3h. This indicates that half of the acylated substrate has lost its labelled hydrophilic head, but was recombined with endogenous ethanolamine (reverse reaction ?).

A second product of the substrate can be detected within the CPG-fraction. After 1 and 3h almost equal amounts of both labels can be found, pointing to a direct conversion by methylation of the alkyl-acylGPE to the choline-analogue. One important observation is the high methylation of ethanolamine after 0,5h of incubation, which is introduced into the GPC-lipids (Fig.3).

1-Alkyl-glycerol is a degradation product of the substrate. It comes to 4 to 11 nmol/mg plasma membrane protein during the observation time (not shown) and is probably the product of different enzyme actions. These enzymes were found by us to be most active in microsomal membranes (Vierbuchen, et al, 1979).

The analogous experiments performed with 1-alkenylGPE (lyso-plasmalogen) reveal in general similar results and are in line with previous experimental data (Fig.4):

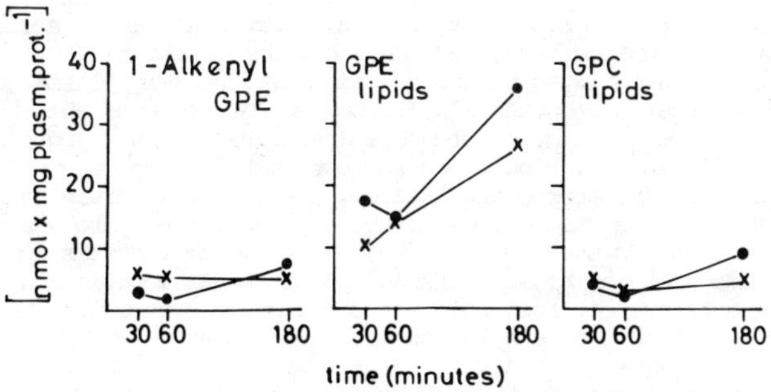

Fig.4. 1-[14C]alkenylGPE and 1-alkenylGP[3H]E in the plasma membranes of cultured glial cells and the products 1-alkenyl, 2-acylGPE and 1-alkenyl,2-acylGPC. (For experimental conditions see text).
● [14C]alkenyl-, x [3H]ethanolamine-label.

i) The incorporation of the substrate into the plasma membranes is only about 5 nmol/mg plasma membrane protein. During the whole incubation period the amount of the unaltered substrate remains almost constant.

ii) Here again the acylation to alkenyl-acylGPE, the plasmalogen, is the predominant reaction with the maximum after 3h. The sum of the total alkenyl-labelled EPG-fraction comes only to 35 nmol/mg plasma membrane protein.

iii) The labelling of the CPG-fraction, that means, the choline-plasmalogen formation, shows pretty close similarity with that of the intramembrane substrate. Since this phospholipid is bearing equal amounts of radioactivity at both parts of the molecule, it can be assumed that the ethanolamine plasmalogens are converted directly to the choline-analogues by methylation.

iiii) As a third product of the substrate free aldehydes are determinable between 4 and 7 nmol/mg plasma membrane protein during time course (not shown). In experiments on lysoplasmalogenase even a higher activity of this enzyme was estimated in cultured glial cells than in the microsomal fraction of young rat brain (Gunawan, et al, 1983).

Comparing the results of both types of experiments we see, that the amount of alkylGPE uptake is much higher than that of alkenylGPE. Therefore the content of the acylated substrate is also higher with the first mentioned substrate, although probably a reversed reaction took place. Furthermore the amount of choline phospholipids derived from the ethanolamine precursors is also increased. However, after 1h of incubation the ratio of the substrate to the corresponding acylation product is 1:1,4 in the case of the alkyl derivative, whereas it is 1:4 after incubation of alkenylGPE.

At last it is mentioned with some reservation, that there is some radioactivity determinable in the serine-phosphoglyceride fraction after incubations with the ethanolamine-labelled alkenylGPE. This, too, is in agreement with hints in the literature, that in brain there should exist serine-plasmalogens (Taki and Kanfer, 1978).

Furthermore, in order to get some information about the methylation activity of the methyltransferases, both radyl-labelled substrates were incubated for 0,5h in the presence of methionine (Tab.2). For comparison the data of the experiments with the ethanolamine-labelled substrates are added. It can be seen, that the mere methylation is unimportant, 0,3 nmol/mg plasma membrane protein. The application of alkylGPE stimulates the methylation to 2 nmol/mg plasma membrane protein. The difference between the alkyl-labelled CGP-compound and the CGP-formation derived from methylation of ethanolamine comes to 7 nmol/mg plasma membrane protein within 30 min, pointing to an unknown methyl-group donor. It is known from the literature (Diliberto and Axelrod, 1976) that in brain cells, in glia and in neurons, a stimulated carboxymethylase is able to hydrolyze methyl-groups linked to a protein. However, one has to keep in mind, that the concentration of exogenously added methionine is a crucial point in all methylation experiments. Perhaps these data are obtained with a suboptimal concentration.The alkenyl-chain has no effect on the methylation activity so far.

Tab.2. Recovery of Radioactivity in Phosphatidylcholines and Choline Plasmalogen of Glial Cell Plasma Membranes after Incubation (30 min) with Ethanolamine containing Analogues.

Substrates	CPG-Lipids^{x}	Substrates	CPG-Lipids^{x}
Methionine*	0,3	Methionine*	0,3
1-Alkyl*-GPE + Methionine*	2,1	1-Alkenyl*-GPE + Methionine*	0,2
1-Alkyl*-GPE	4,7	1-Alkenyl*-GPE	3,6
1-Alkyl-GPE*	11,3	1-Alkenyl-GPE*	4,4

Substrate concentrations: 1-alkylGPE 110-115 nmol/mg cell protein; 1-alkenylGPE 110-115 nmol/mg cell protein; methionine 35o nmol/mg cell protein; X = nmol/mg plasma membrane protein. * radioactive labelling.

CONCLUSION

In summary it can be said, that both substrates (alkylGPE and alkenylGPE,respectively) are incorporated into the plasma membranes of glial cells. During the process of penetration probably both substrates are partially hydrolyzed -alkyl-GPE more than alkenylGPE- , expressed by reduced recovery of labelled polar head groups.The different radyl-chains seem to define the amount of incorporation as well as perhaps a different deposition into the membrane. This fact seems to provoke in different ways enzyme activities residing in the plasma membrane. The acylation of the two substrates shows different profiles concerning the hydrophobic as well as the hydrophilic moiety of the molecules. Both substrates are converted into choline analogues. Here again the radyl-chain seems to determine the way and amount of the conversion: the alkyl-chain stimulates an unknown methylation step. The choline plasmalogens are formed to a smaller degree by methylation. A methylation experiment with exogenously added methionine shows a slight stimulation in the presence of the alkyl-chain containing substrate. The mere methylation is unimportant.

ACKNOWLEDGEMENTS

The authors thank Dr. J. Gunawan for providing the substrates and for the marker enzyme estimations. Supported by a grant of the Deutsche Forschungsgemeinschaft.

REFERENCES

Booher, J.,and M. Sensenbrenner (1972). Neurobiology 2, 97-105.
Clarke, N.G., and R.M.C. Dawson (1981). Biochem. J. 195, 301-306.
Diliberto, E.J., and J. Axelrod (1976). J. Neurochem. 26, 1159-1165.
Gunawan, J., H. Debuch and B. Witter (1983). J. Neurochem. 41, Supl. S86D.
Illig, H.K., B.Witter and H.Debuch (1982a). Neurochem. Res. 7, 1257-1268.
Illig, H.K., B.Witter, J.Gunawan, P.Ahrens and H.Debuch (1982b). Hoppe-Seyler's Z. Physiol. Chem. 363, 709-716.
O'Brien, J.S., and E.L. Sampson (1965). J. Lipid Res. 6, 537-544.
Schimmel, S.D., C.Kant, R. Bischoff and P.R. Vagelos (1973). Proc. Nat. Acad. Sci. USA 7o, 3195-3199.
Sensenbrenner, M., N.Springer, J.Booher and P.Mandel (1972). Neurobiology 2, 49-60.
Taki, T., and J.N.Kanfer (1978). Biochim. Biophys. Acta 528, 309-317.
Vierbuchen, M., J.Gunawan and H.Debuch (1979). Hoppe-Seyler's Z. Physiol. Chem. 360, 1091-1097.
Witter, B., and H.Debuch (1982). J. Neurochem. 38, 1029-1037.
Witter, B., J.Gunawan and H.Debuch (1983). J. Neurochem. 40, 64-69.
Witter, B., M.Fissenewert, J.Gunawan and H.Debuch (1983). Hoppe-Seyler's Z. Physiol. Chem. 364, 1284.

Lipid Protein Metabolism by Schwann Cells *in vivo*

R. M. Gould, G. Mattingly and W. D. Spivack

Institute for Basic Research in Development Disabilities, 1050 Forest Hill Road, Staten Island, New York 10314, USA

ABSTRACT

An interplay of biochemical and autoradiographic procedures was used to demonstrate two properties of myelinating Schwann cells. 1) Protein synthesis, unlike that of phospholipid, is largely restricted to perinuclear cytoplasm. 2) Phospholipids concentrated in this tissue, in the myelin sheaths, are synthesized in a manner which suggests that the terminal enzymes may be in a separate intracellular organelle than other enzymes of phospholipid synthesis.

KEYWORDS

Schwann cells; myelination; phospholipids; proteins; synthesis; autoradiography; teased fibers.

INTRODUCTION

In mature peripheral nerves, Schwann cells develop and maintain two types of stable structural relationships with axons. Each myelinating Schwann cell forms and maintains a single myelin sheath around an individual large caliber axon, whereas a non-myelinating Schwann cell surrounds bundles of small caliber (unmyelinated) axons, separately enclosing each within fingerlike processes. This presentation will focus on the myelin-forming/maintaining Schwann cells and consider two main points relevant to their metabolism of proteins and lipids. 1) Phospholipid synthesis occurs at sites where protein synthesis cannot be detected, and 2) enzymes which produce the phospholipids that are concentrated in peripheral myelin sheaths are physically separated, i.e. on distinct organelles, from those producing the phospholipids more widely distributed in Schwann cell membranes.

Before discussing evidence that demonstrates distinct sites of phospholipid synthesis in myelinating Schwann cells, some features of the regional distribution of Schwann cell cytoplasm along myelinated internodes and of the techniques which are used to localize metabolism to regional sites will be reviewed.

The Schwann cell nucleus and surrounding organelle-rich perinuclear region are located external to compact myelin layers, midway along the internode, i.e. the length (approx. one millimeter) of nerve spanned by a single myelin-forming Schwann cell (Fig. 1). In transverse section, perinuclear regions usually bulge

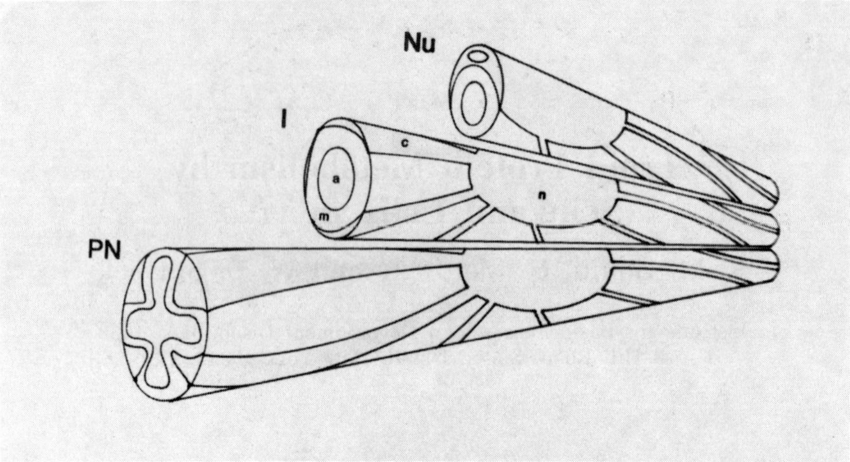

Fig. 1. Schematic diagram illustrating the relationship of teased fiber vs. cross-sectional views of myelinated nerve fiber. Section through nucleus and perinuclear region, Nu, internode, I, and paranode, PN, are illustrated. Actual dimensions of a large caliber fiber would be approx. 1 mm in length and 30 μm diameter. Symbols: c, cytoplasmic channel; n, perinuclear region; a, axon; m, myelin sheath.

out somewhat from the myelinated fiber; they may also cause an indentation in the myelin sheath and underlying axon. Extending from the perinuclear cytoplasm are long narrow strands of cytoplasm that pass uninterrupted to the ends of the internode, the location of the nodes of Ranvier. In cross-section, these so-called surface cytoplasmic channels are seen as rather evenly-spaced furrows in the underlying myelin sheaths, giving individual myelin profiles a "cog-wheel" appearance (Berthold, 1978, not emphasized in this fig.). As these channels approach the nodes, in the paranodal areas, the amount of cytoplasm in each furrow increases. This alteration compresses the underlying myelin sheath and axon into a maltese-cross shaped form. It is important to realize that paranodal profiles contain Schwann cell cytoplasm that is quite far (on the order of 0.5 mm) from the nuclear region (Fig. 1). For further description of peripheral nerve fiber morphology reviews by Landon and Hall (1976), Berthold (1978), Landon (1981) and Gould et al. (1982) should be consulted.

To locate the sites where protein biosynthesis occurs, an _in vivo_ autoradiographic approach was used (Gould and Dawson, 1976). Proline was chosen as the protein precursor because it is not extensively metabolized and it lacks a primary amine rendering it less reactive with glutaraldehyde than amino acids. Rat sciatic nerves were exposed, (care being taken to leave the blood supply intact) and tritiated proline was microinjected through the perineurium and into the nerve fascicles. Following the injection, rats were allowed to move about freely until death. Fixation/perfusion with buffered aldehydes served the purpose of immediately arresting the metabolism, cross-linking the labeled proteins, though not precursors or other labeled metabolites, and preserving the histological structure of the injected nerve. Subsequent washing steps removed unfixed water-soluble radioactivity. Following embedment in plastic, thin (transverse) sections were cut, placed on slides and coated with a nuclear emulsion (Kodak NTB-2 or Ilford K.5). During exposure in light tight boxes, beta radiations emitted from the radioactive products in the section were captured by nearby overlying silver halide crystals in the emulsion. Upon development of the slides, silver grains were formed that provided localization

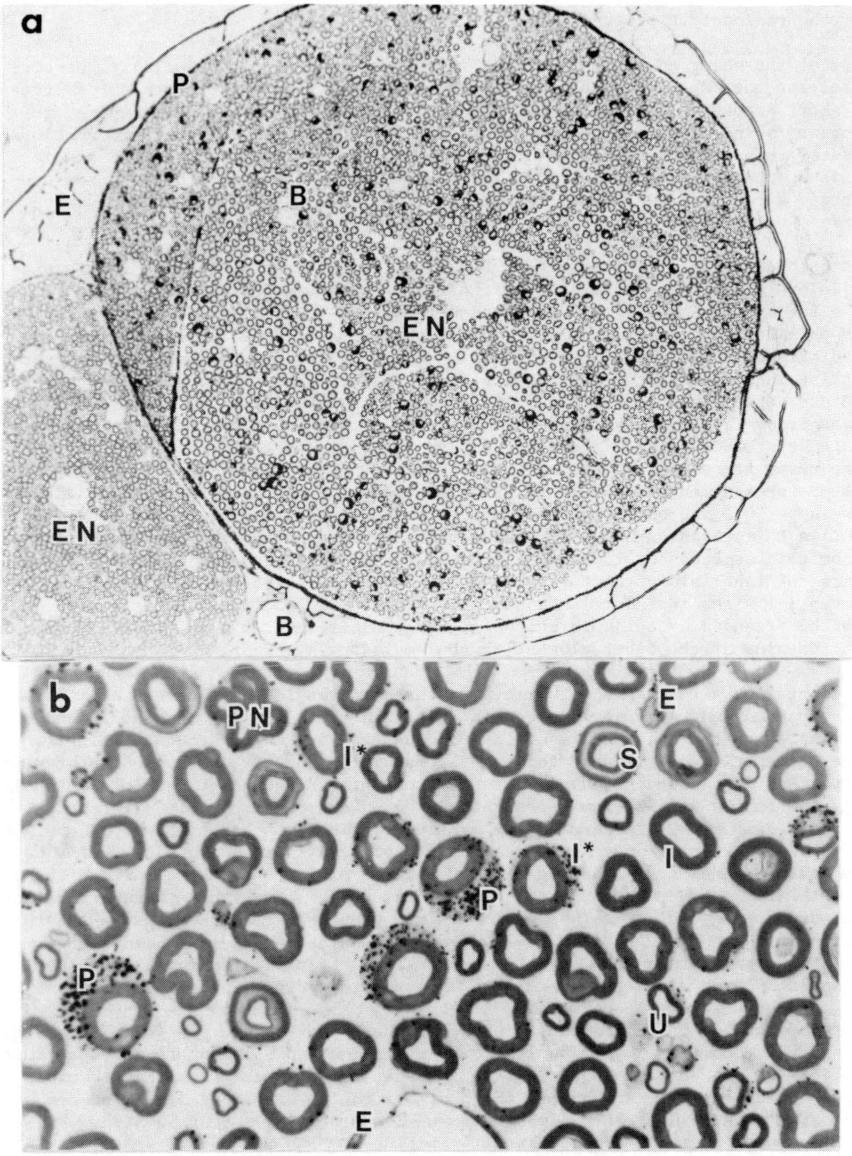

Fig. 2. LM autoradiographs of transversely sectioned rat sciatic nerve labeled with proline for 20 min. a) Low (200 X mag.) and b) higher power (1500 mag.) micrographs near the injection site. Clusters of silver grains marking the sites of protein synthesis are seen in (a), and individual silver grains are seen in (b), shorter exposure. Symbols: a) E, epineurium; P, perineurium; EN, endoneurium; B, blood vessel; b) P, perinuclear Schwann cell cytoplasm; I*, labeled internode; I, unlabeled internode; PN, paranode; S, Schmidt-Lanterman incisure; E, endothelial cell; U, unmyelinated fiber bundle.

of the radioactive proteins (Fig. 2).

One of the major advantages of this *in vivo* approach is the simplicity of controlling the exposure/incubation times of the nerve with the precursor. Indeed, a short exposure time allows the precise localization of the sites where precursor is incorporated into macromolecular product, whereas longer incubations reveal the sites where products subsequently accumulate. In previous studies this approach was used to follow the movement of lipids and proteins from synthetic sites in Schwann cell cytoplasm into myelin sheaths (Gould and Dawson, 1976; Gould, 1976; Gould, 1977).

LM autoradiographs of transversely-sectioned peripheral myelinated fibers provide one view of the relationship that exists between individual Schwann cells and the myelin sheaths they form and metabolically maintain. A different view is obtained when individually teased nerve fibers are the subject of the autoradiographs (Gould and Sinatra, 1981). In practice, this method involves separating individual large caliber myelinated fibers from the labeled nerves. Before separating the fibers by teasing, the injected labeled nerves are fixed, washed, osmicated, dehydrated and infiltrated with Spurr's embedding resin. Once separated, teased fibers are placed on slides that are heated to polymerize the resin, then emulsion is applied as in conventional LM autoradiography. Since the beta-emissions are of low energy, during the exposure only radioactivity in the upper 0.2 microns of a fiber are recorded in the overlying emulsion layer. Unlike LM autoradiography of conventional sections which display radioactivity from thin sections of individual fibers, teased fiber autoradiography displays sites of label along the entire length, i.e. several internodes of a nerve. Potential sites include: Schwann cell cytoplasm in the perinuclear region and in the extended cytoplasmic channels, and the outermost myelin lamellae, *all* on the side of the fiber closest to the emulsion layer.

A short incubation period (20 min.) was used to label sites where protein synthesis occurred. A survey micrograph (Fig. 2a) reveals silver grains around the perimeter of the injected fascicle and clustered in small patches within it. The perineurial cell layer, forming the fascicle boundary, both incorporates the precursor into protein and prevents it from entering adjacent fascicles. Labeling within the endoneurium seems restricted to crescent-shaped profiles of Schwann cell cytoplasm surrounding myelinated fibers. A higher power micrograph (Fig. 2b) better demonstrates localization of newly-formed proteins to the perinuclear regions of myelinating Schwann cells. Myelin sheaths, axons and the extracellular space around the fibers contains far less label than perinuclear Schwann cell cytoplasm. In addition, some, though not all, Schwann cell cytoplasm in internodal profiles is labeled, whereas cytoplasm in Schmidt-Lanterman incisures and paranodes is not labeled. Endothelial cells lining the blood vessels and unmyelinated fibers also incorporate proline into protein.

Autoradiographs of teased fibers show that newly formed protein (Fig. 3a and b) is highly localized to perinuclear Schwann cell cytoplasm and the more proximal regions of the surface channels. This result is different from that observed when choline incorporation into phosphatidylcholine was studied with teased fiber autoradiography (Fig. 3c, Gould and Sinatra, 1981). Not only the perinuclear and proximal surface channels incorporate choline into lipid, the *entire* internode contains choline lipid synthetic activity. The more limited protein synthesis, limited to more proximal regions of the surface channel, also explains why only some internodal Schwann cell cytoplasm is labeled with proline (Fig. 2b).

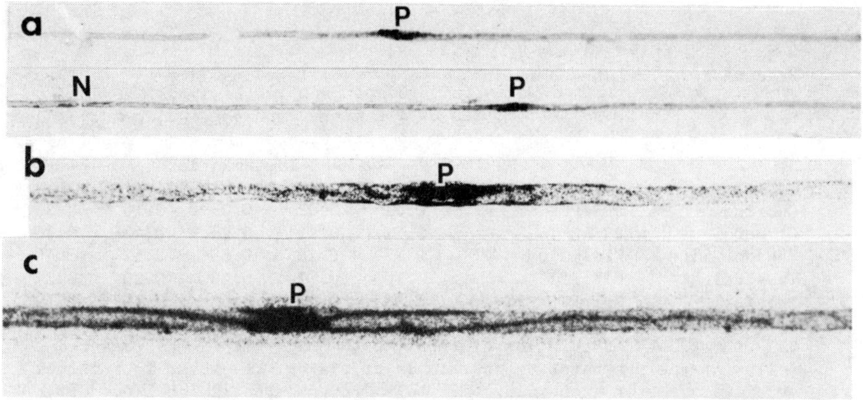

Fig. 3. LM autoradiographs of teased fibers labeled for (a) 20 min. or (b) 1 hr. with tritiated proline, or (c) 20 min. with tritiated choline. Silver grains represent newly formed protein (a and b) or phosphatidylcholine (c). With proline, silver grains are restricted to the perinuclear regions midway between the nodes, whereas choline lipid is formed both in perinuclear areas and in cytoplasmic channels extending to the nodes. Symbols: P, perinuclear area, N, nodes of Ranvier. Magnification: a) 160 X, b and c) 300 X.

It looks as if the protein label spreads into the cytoplasmic channels between 20 min. and 1 hr., compare Figs. 3a and b. This result may indicate that some labeled proteins migrate from the perinuclear regions outward along the surface channels. We have not as yet characterized this process, but plan to determine whether inhibitors known to block protein movement in fast axonal transport will block the movement of labeled proteins along the Schwann cell surface channels.

Myelin sheaths have a unique phospholipid composition. Sheltawy and Dawson (1966) identified four phospholipids that are characteristic of myelinated peripheral nerve. They showed that phospholipid compositions of peripheral nerves containing myelin had higher levels of sphingomyelin, phosphatidylserine, ethanolamine plasmalogen and phosphatidylinositol 4,5 bisphosphate than nerves that were devoid of myelin.

A hypothetical mechanism Schwann cells might employ to concentrate these lipids in their myelin sheaths has recently been proposed (Gould et al., 1986a). This proposition is based solely on the time dependency of incorporation of precursors into these 'myelin' phospholipids, and was devised to explain apparent lags in the incorporation of precursors into these phospholipids relative to other phospholipids. The observations were simply that the incorporation of tritiated precursors favored one or some phospholipids at short (30 min. to 2 hr.) labeling periods, while after longer periods (1 to 14 days), the phospholipids, characteristic of myelinated peripheral nerve, became proportionally better labeled. Therefore, choline incorporation into SPH followed behind labeling of PC, also (Gould and Dawson, 1976), glycerol labeled PE and PC earlier than PS, ethanolamine labeled PE (includes alkyl acyl PE) before ethanolamine plasmalogen, and tritiated inositol labeled PI before the polyphosphoinositides (also Goswami and Gould, 1985). A summary of these findings is presented in the Table.

Table

Precursor	Labeled Lipid	Labeling Time				
		5 m	30 m	120m	1 d	14 d
		Lipid Ratio				
Choline	SPH/PC	0.02	0.02	0.03	0.10	0.33
Glycerol	PS/PC	0.21	0.12	0.28	0.42	0.80
Ethanolamine	Plasmalogen/PE	0.27	0.28	0.63	2.52	6.10
Inositol	PIP_2/PI	0.10	0.11	0.26	0.20	0.23
Inositol	PolyPI's/PI	0.38	0.40	0.55	0.70	0.79

Values represent the ratios of label in the major lipids as measured following their separation on thin layer plates. At least four nerves were analyzed for each value. Symbols: PC, phosphatidylcholine; SPH, sphingomyelin; PS, phosphatidylserine; PE, phosphatidylethanolamine; PI, phosphatidylinositol; PIP_2, phosphatidylinositol 4,5-bisphosphate; m, minutes; d, days.

We ascribe the lags in synthesis of myelin-directed lipids to a separation of their biosynthetic organelles from those where general Schwann cell phospholipid synthesis occurs. The latter reactions, which include the de novo synthesis of PC and PE, presumably occur in rough and smooth endoplasmic reticulum of Schwann cells, as they do in other cells (e.g. review of Pelech and Vance, 1984). We postulate that PC is the phosphorylcholine donor in sphingomyelin synthesis, and that the enzyme catalyzing this transfer is located in Schwann cell plasma membrane. A number of recent studies have demonstrated both that PC is the phosphorylcholine donor in SPH synthesis and that this cholinephosphotransferase is located in the Golgi apparatus and/or the plasma membrane and not in endoplasmic reticulum (Marggraf et al., 1981, 1982; Voelker and Kennedy, 1982; Lipsky and Pagano, 1985; van den Hill et al., 1985). Neither this pathway nor the reaction site has as yet been demonstrated in brain (Ullman and Radin, 1974) or other nervous tissue preparation. The alternate (original) pathway involving CDP choline as the intermediate (Sribney and Kennedy, 1958; Stoffel and Melzner, 1980) does not so readily explain the kinetic data. With separate sites of phosphatidylcholine and sphingomyelin synthesis in myelinating Schwann cells, we explain the delay in incorporation of labeled choline into sphingomyelin as the time required to transfer labeled PC from its synthetic site to the site where sphingomyelin synthesis occurs, realizing that this membrane system contains its own endogenous (unlabeled) PC.

The delays in incorporation of tritiated glycerol into phosphatidylserine, ethanolamine into ethanolamine plasmalogen and to a lesser extent inositol into PIP_2 (Table) could all be explained by a similar mechanism, although current evidence is less compelling and outside the scope of this chapter. The two (or multi) site model is illustrated in Fig. 4. Precursor phospholipids, namely phosphatidylcholine for sphingomyelin, PC and/or PE for phosphatidylserine (via base exchange), alkyl acyl-PE for ethanolamine plasmalogen and phosphatidylinositol for PIP_2 are synthesized in rough and smooth endoplasmic reticulum within Schwann cell cytoplasm. Synthesis would probably occur both in the perinuclear area and in surface cytoplasmic channels (not shown). These precursor lipids would then have to transfer (individually via exchange lipoproteins and/or via bulk membrane flow, Holtzman and Mercurio, 1980; Ledeen, 1985) into Golgi apparatus and/or plasma membrane sites to be converted into the product lipids that concentrate in the myelin sheath. Since these latter sites are metabolically closer to where the phospholipids are incorporated into compact myelin, they would preferentially enter myelin since phospholipids formed in the endoplasmic reticulum would have a greater likelihood of being degraded by cytosolic and organelle associated phospholipases of Schwann cell cytoplasm.

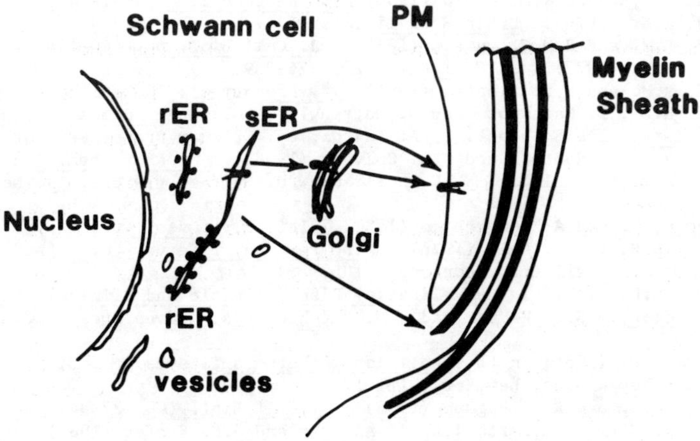

Fig. 4. Artists representation of the relationship between membranes in the Schwann cell perinuclear region. We hypothesize that PC, PE and PI synthesis occurs in the rough and smooth endoplasmic reticulum (rER and sER) and that these lipids must enter the Golgi, plasma membrane (PM) and/or compact myelin by protein catalyzed exchange or vesicular movement to become substrates for the synthesis of sphingomyelin, phosphatidylserine, ethanolamine plasmalogen and phosphatidylinositol 4,5-bisphosphate. From these sites they rapidly enter compact myelin (if they are not formed there) leaving them less likely to be catabolized than the parent (PC, PE and PI) phospholipids.

In summary, LM autoradiography of cross-sectioned nerve and teased nerve fibers has demonstrated differences in the disposition of sites where lipid and protein syntheses occur in myelinating Schwann cells. Precursor incorporation by a system primarily engaged in myelination and myelin maintenance (Gould et al., 1986b), has provided the basis for a proposal that phospholipids that are concentrated in peripheral myelin sheaths do so because they are made on organelles close to where myelin formation occurs.

ACKNOWLEDGEMENT

This work was supported in part from NIH grants NS 13980 and NS 16305 to R.M. Gould. Edna Troiano typed the manuscript, John Holshek helped prepare the figures, and Lucille Donadio prepared Fig. 4.

REFERENCES

Berthold, C.-H. (1978) in Physiology and Pathobiology of Axons. S.G. Waxman, ed. Raven Press, New York, 3-63.

Goswami, S.K. and R.M. Gould (1985). J. Neurochem. 44, 941-946.
Gould, R.M. (1976). Brain Res. 117, 169-174.
Gould, R.M. and R.M.C. Dawson (1976). J. Cell Biol. 68, 480-496.
Gould, R.M. (1977). J. Cell Biol. 75, 326-339.
Gould, R.M. and R.S. Sinatra (1981). J. Neurocytol. 10, 161-167.
Gould, R.M., D. Matsumoto and G. Mattingly (1982) in Handbook of Neurochemistry, 2nd Ed., vol. 1, A. Lajtha, ed. Plenum Press, New York 397-414.
Gould, R.M., F. Connell and W.D. Spivack (1986a). J. Neurochem. (in press).
Gould, R.M., J. Holshek, F. Connell and W.D. Spivack (1986b). J. Neurochem. (in press).
Holtzman, E. and A.M. Mercurio (1980). Int. Rev. Cytol. 67, 1-67.
Landon, D.N. and S. Hall (1976) in The Peripheral Nerve. D.N. Landon, ed. Chapman and Hall Ltd., London, 1-105.
Landon, D.N. (1981) in Demyelinating Disease: Basic and Clinical Electrophysiology. S.G. Waxman and J.M. Ritchie, eds. Raven Press, New York, 25-49.
Ledeen, R.W. (1985) in Phospholipids in Nervous Tissues. J. Eichberg, ed. John Wiley & Sons, New York, 135-172.
Lipsky, N.G. and R.E. Pagano (1985). J. Cell Biol. 100, 27-34.
Marggraf, W.-D., R. Zertani, F.A. Anderer and J.N. Kanfer (1982). Biochim. Biophys. Acta 710, 314-323.
Marggraf, W.-D., F.A. Anderer and J.N. Kanfer (1981). Biochim. Biophys. Acta 664, 61-73.
Pelech, S.L. and D.E. Vance (1984). Biochim. Biophys. Acta 779, 217-251.
Sheltawy, A. and R.M.C. Dawson (1966). Biochem. J. 100, 12-18.
Sribney, M. and E.P. Kennedy (1958). J. Biol. Chem. 233, 1315-1322.
Stoffel, W. and I. Melzner (1980). Hoppe-Seyler's Z. Physiol. Chem. 361, 755-771.
Ullman, M.D. and N.S. Radin (1974). J. Biol. Chem. 249, 1506-1512.
van den Hill, A., G.P.H. van Heusden and K.W.A. Wirtz (1985). Biochim. Biophys. Acta 833, 354-357.
Voelker, D.R. and E.P. Kennedy (1982). Biochemistry 21, 2753-2759.

Regulation of Schwann Cell Expression of Myelin-specific Glycoconjugates

J. F. Poduslo

Membrane Biochemistry Laboratory, Neurology and Biochemistry, Mayo Clinic and Mayo Foundation, Rochester, Minnesota 55905, USA

KEYWORDS: P_0 glycoprotein; galacto- and glucocerebrosides; sulfatides

I. ABSTRACT:

Schwann cell biosynthesis of myelin-specific glycoconjugates has been investigated in the permanently transected, adult rat sciatic nerve, where there is no myelin assembly, and in the adult crushed nerve, where there is myelin assembly. In this brief review, the biosynthesis of the major myelin glycoprotein (P_0) is compared with the biosynthesis of both the galactocerebrosides and sulfatides in the presence and absence of myelin assembly. In addition, this biosynthesis is placed in perspective with other known modifications of proteins, such as glycosylation, phosphorylation, sulfation and acylation, and known modifications of glycolipids, particularly oligosaccharide chain elongation.

Although the levels of expression of P_0 are substantially reduced after nerve transection, processing of the oligosaccharide chain on P_0 can be demonstrated with the accumulation of the high mannose-type oligosaccharide as the predominant species. Other processing events occur; however, the extent of processing of this oligosaccharide as well as other post-translational modifications of P_0 remain to be determined. The present evidence suggests that the controlling mechanisms in the processing of this down-regulated glycoprotein occur within the Golgi apparatus.

In contrast, the regulatory mechanisms for the biosynthesis of the cerebrosides and sulfatides appear to occur in the endoplasmic reticulum. A switch in the biosynthesis of the galactocerebrosides to form the glucocerebrosides occurs after nerve transection. This results in the biosynthesis of a series of oligohexosylceramides which likely undergo further processing in the Golgi apparatus. A working model is presented in this review which summarizes these observations and suggests independent regulatory mechanisms for the biosynthesis of these myelin-specific glycoconjugates.

II. POST-TRANSLATIONAL PROTEIN MODIFICATION

A. GLYCOSYLATION:

The targeting of proteins to their proper intra- or extracellular compartments where they fulfill specific biological functions is often uniquely determined by specific post-translational modifications. The mechanisms that regulate these post-translational protein modifications are poorly understood.

Glycosylation is one important processing event that is highly developed and specifically coordinated; however, the mechanisms by which glycosylation events are controlled are just beginning to be understood. The biosynthesis of asparagine-linked glycoproteins has been shown to follow complex multistep and multiorganelle pathways where the glycoproteins are processed and sorted (Schachter et al, 1978; Lennarz, 1980; Hanover and Lennarz, 1981; Hubbard and Ivatt, 1981; Rothman, 1981). Potential sites of regulation include 1) the biosynthesis of the lipid carrier dolichol phosphate; 2) the biosynthesis of the glycosyl donors, the sugar nucleotides; 3) the glycosyl transferases that catalyze these reactions resulting in the synthesis of the oligosaccharide-P-P-dolichol; 4) the co-translational transfer of the precursor oligosaccharide to the acceptor apoprotein via a membrane-bound oligosaccharide transferring enzyme; 5) the accessibility of the tripeptide acceptor sequence for oligosaccharide attachment; 6) the regeneration of dolichol phosphate from dolichol pyrophosphate; 7) the ordered oligosaccharide trimming by glucosidases followed by α-mannosidases; 8) and the post-translational oligosaccharide elongation by specific glycosyl transferases to form the diverse array of complex glycoproteins.

Recent investigations suggest that both oligosaccharide trimming and elongation serve to control final oligosaccharide assembly. GlcNAc-transferase I which adds GlcNAc in (1→2)- β linkage to the Man-α-(1→3)-Man-β-(1→terminus) plays an important control function in the subsequent removal of two mannose residues as well as the addition of fucose to the core asparagine-linked GlcNAc residue (Harpaz and Schachter, 1980b; Longmore and Schachter, 1982). In addition, GlcNAc-transferase III which adds a "bisecting" GlcNAc residue in β1→4 linkage to the β-linked Man plays important control functions in the synthesis of complex-type oligosaccharides (Narasimhan, 1982) including the inactivation of GlcNAc-transferase II (Harpaz and Schachter, 1980a), GlcNAc transferase I-dependent α mannosidase (Harpaz and Schachter, 1980b), α6-fucosyl-transferase (Longmore and Schachter, 1982) and GlcNAc-transferase IV (Gleeson and Schachter, 1983). The removal or addition of a single sugar residue may be sufficient to alter the three-dimensional shape of the molecule and therefore affect the action of subsequent glucosidases or transferases (Harpaz and Schachter, 1980b). Consequently, the assembly process can be directed along a particular pathway.

Ivatt (1981) has suggested that competition among rival glycosyl transferases is regulated at the supramolecular level by the formation of specific complexes of component glycosyl transferases that are spatially organized such that the product of one glycosyl transferase serves as a subsequent substrate of another. Such specific complexes could explain the mechanism by which a cell could express its diverse array of membrane surface glycoproteins. The regulation of this complex formation could provide a flexible mechanism for the correct expression of membrane surface carbohydrates as differentiation and normal development proceed. Recently, Goldberg and Kornfeld have provided evidence for a high degree of subcellular organization within the Golgi complex where compartmentalization of processing enzymes may regulate the competing reactions and hence control the processing of asparagine-linked oligosaccharides. In addition, the steric accessibility of oligosaccharide chains to processing enzymes may determine the extent of processing (Hsieh et al., 1983).

Since the major myelin glycoprotein contains an asparagine-linked oligosaccharide chain (Ishaque et al., 1980), its synthesis and assembly follow a unique pathway involving preassembly of the oligosaccharide chain on a lipid anchor followed by transfer to protein and subsequent modification to form the complex-type glycoprotein. This is supported by the observation of Uyemura et al. (1981), who have demonstrated that the glycosylation of this myelin glycoprotein occurs via lipid intermediates, as well as by tunicamycin inhibition studies for this glycoprotein (Smith and Sternberger, 1982; Rapaport et al., 1982; Poduslo, unpublished observations).

B. PHOSPHORYLATION:

The post-translational phosphorylation-dephosphorylation of proteins has been implicated in the regulation of numerous intracellular processes. These include glycogenolysis, gluconeogenesis, glycolysis, lipolysis, nucleic acid transcription, protein translation, contractile protein interactions, and cellular excitation (for reviews see Flockhart and Corbin, 1982; Shacter-Noiman et al., 1983; Levitan et al., 1983, La Porte and Koshland, 1983; Ingebritsen and Cohen, 1983; and Nestler and Greengard, 1983). About 35 enzymes and countless other proteins are now known to be regulated in this manner, and protein phosphorylation is clearly a general mechanism by which intracellular events respond to external physiological stimuli.

In purified peripheral nerve myelin, Singh and Spritz (1976) have described at least 2 protein kinase systems. When highly purified myelin from rat sciatic nerve is incubated with [γ-^{32}P]ATP, a variety of protein components of the membrane were phosphorylated indicating the presence of both the substrate and an endogenous kinase in the membrane. Polyacrylamide gel electrophoresis followed by scintillation counting of gel slices demonstrated proteins with molecular weights of 28,000, 23,000, and 19,000 that were phosphorylated at the serine residues with the [^{32}P] from [γ-^{32}P]ATP. More recently, Wiggins and Morell (1980) evaluated rat sciatic nerves incubated with [^{32}P]orthophosphate and found that after electrophoresis the [^{32}P]phosphate appeared to label a variety of myelin proteins. Specifically, the [^{32}P]phosphate was incorporated into proteins in the region of the polyacrylamide gel which coincided with [^{3}H]fucose incorporation into P_0 glycoprotein. Their data also suggest that only P_0 protein that is incorporated into myelin is modified by phosphorylation.

A novel form of post-translational processing is the mannose-6-phosphate recognition marker acquired by acid hydrolases which allows binding to the phosphomannosyl enzyme receptors and subsequent transport to lysosomes (for reviews, see Neufeld and Ashwell, 1980; Sly, 1982; and Kornfeld et al., 1982). The main function of the mannose-6-phosphate recognition marker on acid hydrolases and its receptors on mammalian cell membranes is to provide an intracellular transport system for the delivery of enzymes to lysozymes. The phosphorylation of mannose involves a two step process. First, N-acetylglucosamine-1-phosphate is transferred to an acceptor mannose at the 6 position on high mannose-type oligosaccharides. Next, an α-N-acetylglucosaminyl phosphodiesterase removes the N-acetylglucosamine, leaving the phosphate in monoester linkage with the mannose residue. This phosphodiesterase follows galactosyl transferase activity on subcellular fractionation, suggesting a Golgi localization (Varki and Kornfeld, 1980). This phosphomannosyl residue then serves as the lysosomal recognition marker. In I-cell disease, the first enzyme is absent resulting in targeting of newly synthesized acid hydrolases not to lysosomes but for secretion into the extracellular milieu. Hence, the mannose-6-phosphate receptor-mediated transport pathway is believed to be a branch of the secretory pathway for glycoproteins where the targeting in this case is to the lysosome.

The generality of the mannose-6-phosphate system is not yet known. Presumably, any glycoprotein (secretory, intracellular, plasma membrane, etc.) with this appropriate structure could be cleared by this system. The basal levels of expression of P_0 that occur in the permanent nerve transection model in the absence of myelin assembly (Poduslo, 1984), the accumulation of the high mannose-type intermediates (Poduslo et al., 1985), and the decrease in radioactive label after pulse-chase experiments (Poduslo, 1985) could indicate that this glycotein is being relocated from the Golgi apparatus to the lysosome for degradation.

C. SULFATION:

Protein sulfation on tyrosine residues appears to be a widespread covalent modification and may have important roles in cell function (for review, see

Huttner, 1982). During the past few years, sulfated asparagine-linked sugar chains of glycoproteins (as distinguished from sulfated glycosaminoglycans and sulfated mucins) have been described in developing and differentiating systems (Heifetz and Lennarz, 1979; Heifetz et al., 1980) and in viral-infected cells (Nakamura and Compans, 1978; Prehm et al., 1979). Sulfated carbohydrates have been shown to resist the action of glycosidases (Kent et al., 1978; Martin et al., 1968), impart a resistance to proteolysis to the protein core (Mikumi-Takagaki and Hotta, 1979), inhibit the action of neuraminidase (Mian et al., 1979), facilitate the concentration of secretory proteins in the Golgi condensation vacuoles (Reggio and Palade, 1978), affect the pattern of synthesis of oligosaccharide chains (Kent et al., 1978), and enhance the stability of certain hormones (Parsons and Pierce, 1980). It is clear that sulfated oligosaccharides play an important role in recognition phenomena at the cell surface or in directing intracellular movement of glycoprotein molecules.

Matthieu et al. (1975) have indicated from [^{35}S]sulfate incorporation studies in the rat that the major myelin glycoprotein, P_0, is sulfated and that this sulfate group is probably on the sugar residues. In addition, Simpson et al. (1976) described the presence of high molecular weight sulfated glycoproteins which are structurally distinct from brain glycosaminoglycans in a purified synaptic plasma membrane and myelin membrane from rat brain. The sulfated glycoproteins were shown to consist predominantly of a class of glycoproteins containing sulfate in ester linkage to oligosaccharide chains. In addition, Margolis and Margolis (1970) have isolated and characterized a sulfated glycopeptide fraction from whole rat brain following extraction of brain with chloroform-methanol. Miller and Waechter (1979) have also described the *in vitro* sulfation of a high molecular weight glycoprotein associated with pig brain membranes. Utilizing PAPS [^{35}S] in the presence of NaF, they provide evidence indicating that PAPS can also serve as a sulfate donor in the sulfation of membrane glycoproteins.

D. ACYLATION:

Recent studies have provided evidence that some membrane glycoproteins can undergo the post-translational modification of acylation by long chain fatty acids. Examples include virus spike proteins in infected cells (Schmidt et al., 1979), transferrin receptor protein in cultured human leukemic T-cells (Omary and Trowbridge, 1981a,b), and myelin proteolipid proteins in rat brain (Agrawal et al., 1982). Most experiments examined the incorporation of [^{3}H]palmitic acid into proteins of whole cells. Criteria for acylation include resistance to extraction with organic solvents, resistance to dissociation under denaturing conditions (such as SDS-PAGE), lability to alkali, and lability to hydroxylamine with the resulting formation of [^{3}H]palmitoyl hydroxamate. Acylation appears to be selective even among membrane glycoproteins since not all surface proteins are labeled. The functions of protein acylation are not clear, but its selectivity suggests possible roles in the insertion, interactions, or orientations of proteins in membranes. Three hypotheses have been presented concerning the potential function of acylation, including the anchor, transport, and fusion trigger hypotheses (for reviews, see Magee and Schlesinger, 1982, and Schmidt, 1983). The precise location of the fatty acid linkage in the amino acid sequence has not been determined for any acylated protein; however, there is substantial evidence which suggests that lipid is bound in the domain of the protein which interacts strongly with the lipid bilayer and that this linkage probably involves ester bonds to the side chains of hydroxyl-containing amino acids (Ser, Thr, Tyr).

More recently, Agrawal et al. (1983) described the incorporation of [^{3}H]palmitic acid into P_0 glycoprotein after direct injection into rat sciatic nerve. Fluorography of gels after electrophoresis revealed that P_0 was acylated *in vivo*. All the radioactivity of purified P_0 was released into the organic phase after treatment with methanolic sodium hydroxide. This suggests that P_0 glycoprotein is also acylated in addition to being glycosylated, phosphorylated, and sulfated. It is likely that all these modifications involve post-translational processing, possibly within the Golgi apparatus. By evaluating the permanently transected nerve, it should be possible to determine the extent to which acylation of P_0 glycoprotein occurs in this down regulated system. In addition, by comparing the crush and the permanently transected models it should be possible to evaluate the sequence of the acylation event relative to the glycosylation, phosphorylation, and sulfation events. Such a comparison should also permit an assessment of the potential function of these modification events. For example, it is possible that the function of protein bound fatty acids is to permit anchoring within the membrane bilayer as has been proposed by Schmidt (1983). Such anchoring may become important as a last post-translational modification step which triggers the relocation from the Golgi apparatus via a vesicle to the site of myelin assembly, or for the actual myelin assembly.

III. P_0 PROCESSING BY SCHWANN CELLS:

A comparision of the extent of post-translational processing of a membrane specific glycoprotein that is normally expressed during active membrane assembly with that expressed by quiescent cells in the absence of membrane assembly provides an approach to understanding how protein modification is regulated. An evaluation of the mechanism by which quiescent cells down-regulate this expression should permit the definition of not only important controlling steps in the biosynthetic pathway but also an identification of the rate-limiting steps.

Schwann cells from the adult animal can be forced into a quiescent state or an actively myelinating state by either permanently transecting the nerve and hence preventing axonal regeneration and remyelination in the distal nerve segment or by crushing the nerve and inducing axonal regeneration and remyelination (Poduslo, 1984). In the transected nerve, quiescent Schwann cells down regulate the production of myelin constituents, whereas in the crushed nerve Schwann cells are actively producing myelin components and assembling them into the myelin membrane. Our interests have focused on the major myelin glycoprotein, designated P_0, which comprises 50-60% of the total myelin protein. We have demonstrated that quiescent Schwann cells express basal levels of this glycoprotein as determined by radioiodinated lectin binding after electrophoresis and by precursor incorporation (Poduslo, 1984). Positive identification of this glycoprotein has been obtained by assessment of its characteristic shift in M_r in the absence of reducing agents after SDS-pore gradient electrophoresis for both the lectins and radioactive precursors (Poduslo, 1984), immunological verification using antisera to P_0 as determined by direct immune overlay after electrophoresis (Poduslo et al., 1984), and direct immunoprecipitation (Poduslo, 1985).

In addition, an investigation of the temporal sequence of events that occur during the transition of the Schwann cell from a myelin maintaining state to a quiescent state in the transected nerve revealed two M_r species of P_0 (28,500 and 27,700) at 4 and 7 days after nerve transection (Poduslo et al., 1985). By 14 days after nerve transection, the 27,700 species was the predominant component. This M_r change was characterized by a biosynthetic switch from the complex-type oligosaccharide structure as an end product to the predominantly high mannose-type oligosaccharide structure as a biosynthetic intermediate (Poduslo et al., 1985).

Figure 1 illustrates the oligosaccharide processing that occurs on this major myelin glycoprotein in the presence and absence of myelin assembly in these two experimental animal models of neuropathy. After the addition of the core oligosaccharide from the dolichol phosphate onto the asparagine residue of P_0 as it is being translated, the terminal glucose sugars are removed in the ER since

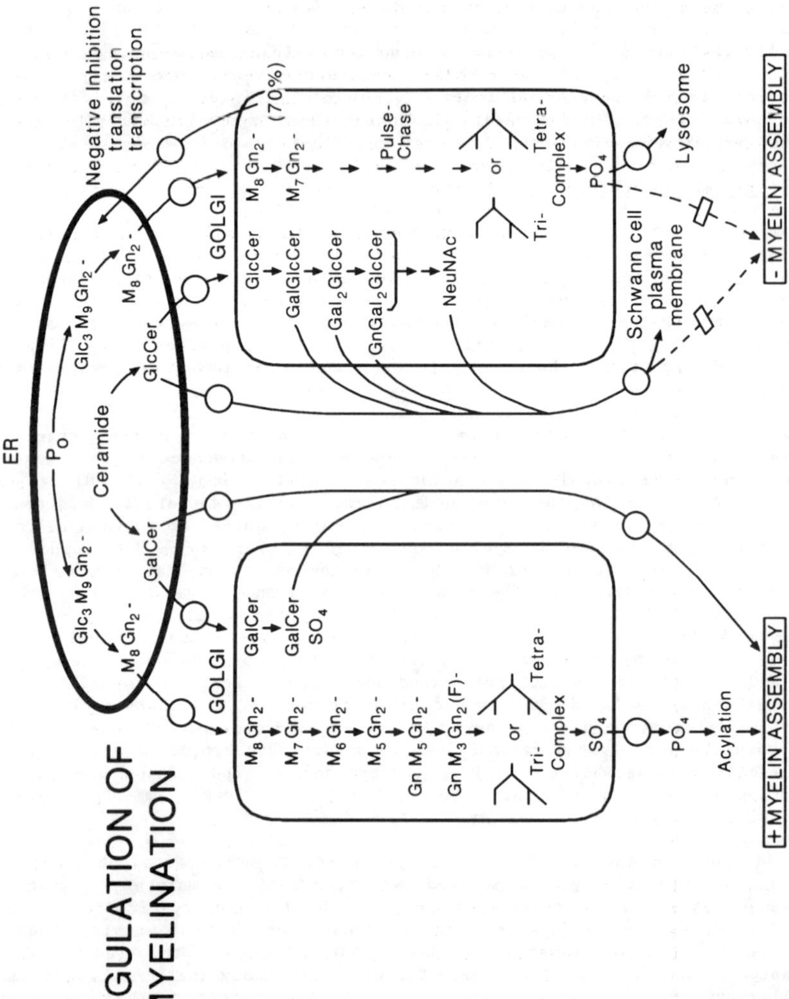
REGULATION OF MYELINATION
ER
P_0
Ceramide
$Glc_3M_9Gn_2$-
M_8Gn_2-
GalCer
GlcCer
Negative Inhibition
translation
transcription
GOLGI
GOLGI
M_8Gn_2-
M_7Gn_2-
M_6Gn_2-
M_5Gn_2-
GnM_5Gn_2-
GnM_3Gn_2(F)-
or
Tri-
Tetra-
Complex
SO_4
GalCer
GalCer
SO_4
PO_4
Acylation
+MYELIN ASSEMBLY
GlcCer
GalGlcCer
$Gal_2GlcCer$
$GnGal_2GlcCer$
NeuNAc
(70%)
Pulse-
Chase
PO_4
Lysosome
Schwann cell
plasma
membrane
- MYELIN ASSEMBLY

the three glucosidases have been shown to have an ER location (Grinna and Robbins, 1979; Elting et al., 1980). In addition, we have proposed that the first mannose residue is also cleaved by a mannosidase having an ER localization. This is based on the observation that no early processing intermediates ($Glc_3Man_9GlcNAc_2$ to $Man_9GlcNAc_2$) have been observed after separation of the high mannose-type oligosaccharides on Biogel P4 (Poduslo, 1985). In support of this observation are the studies of Godelaine et al. (1981) who provided evidence that a single mannose residue is removed in the endoplasmic reticulum prior to glycoprotein transfer to the Golgi complex.

The circles that are outside the ER and the Golgi apparatus indicated in Figure 1 are used to designate vescicular transport systems within the cytoplasm. Proteins undergo relocation from the ER to the Golgi apparatus in such a transport system. Within the Golgi apparatus during myelin assembly, the high mannose-type oligosaccharide undergoes modification to form a tri- or tetraantennary complex-type oligosaccharide chain on P_0. The formation of these complex-type oligosaccharide structures is based on <u>in vitro</u> labeled P_0 glycopeptide binding to Concanavalin A affinity columns after pronase digestion of P_0 obtained after SDS-pore gradient electrophoresis (Poduslo, 1985). In this model, we hypothesize that sulfation of the oligosaccharide chain occurs in the Golgi apparatus, possibly as a last processing step which permits the release of the glycoprotein from the Gogli. As the glycoprotein is transported from the Golgi to the site of myelin assembly, it is further hypothesized that phosphorylation of P_0 by protein kinases occurs. In this model, we propose that acylation is occuring at the site of myelin assembly either during the insertion process of P_0 into the myelin membrane or during the compaction of the multilamellar structure.

Also shown in Figure 1 is a summary of the processing that has been observed on the major myelin glycoprotein in the permanently transected sciatic nerve of adult rats at 35 days after transection where there is no myelin assembly. The regulatory mechanisms associated with the biosynthesis of this glycoprotein by these quiescent Schwann cells are characterized by the formation and accumulation of the high mannose-type oligosaccharide intermediates (Poduslo et al., 1985). These high mannose-type intermediates, consisting of $Man_8GlcNAc_2$ and $Man_7GlcNAc_2$ (Poduslo, 1985), demonstrate a lower molecular weight (27,700) after SDS-PGE which contrasts with the complex-type oligosaccharide chain on the mature glycoprotein expressed during myelin assembly having a molecular weight of 28,500 (Poduslo et al., 1985). The accumulation of the $Man_8GlcNAc_2$ and $Man_7GlcNAc_2$ oligosaccharide chain on P_0 suggests that the next processing step involving the removal of mannose residues to give $Man_6GlcNAc_2$ and $Man_5GlcNAc_2$ occurs more slowly and therefore is likely to be a rate limiting step. The conversion of $Man_9GlcNAc_2$ to $Man_5GlcNAc_2$ has been shown to involve two α-1,2-mannosidases which have been prepared and partially characterized from Golgi membranes. These include a mannosidase 1A (Tabas and Kornfeld, 1979) and a second mannosidase 1B (Tulsiani et al., 1982). Although these two α-1,2-mannosidases have very similar catalytic properties, they are distinguishable by their behavior on cellulose phosphate chromatography, their thermolability and storage sensitivity, and their activity toward $Man_5GlcNAc$ and p-nitrophenyl α-D-mannoside (Tulsiani et al., 1982). The structural studies on the high mannose-type oligosaccharides of the down-regulated P_0 suggest such a differential function for these two mannosidases in the processing of this glycoprotein (Poduslo, 1985). The accumulation of these high mannose-type intermediates likely results in an inhibition of further translation and/or transcription, resulting in only low levels of expression of this glycoprotein by the Schwann cells as indicated in Figure 1.

Additional pulse-chase experiments, however, have demonstrated that these high mannose-type intermediates can undergo further processing to form the biantennary complex-type oligosaccharides, which in turn form the tri- or tetraantennary complex-type oligosaccharide. Another alternative explanation is that the biantennary complex-type oligosaccharide might undergo other modifications (for example, sulfation, phosphorylation, etc.) such that it no longer binds to concanavalin A Sepharose.

An important question that needs to be addressed is the reason why the low levels of this glycoprotein that are being expressed undergo any post-translational translational modification in the Golgi apparatus at all. One potential hypothesis that can be experimentally addressed is that P_0 processing in the Golgi apparatus is required to provide a targeting mechanism by which P_0 is directed to the lysosome for degradation. Modification of the high mannose-type oligosaccharide by the addition of phosphate to form mannose-6-phosphate might provide a mechanism for lysosomal delivery.

Additional oligosaccharide processing of P_0 after nerve transection is further characterized by the addition of fucose to the core GlcNAc residue and also by a molecular weight shift from the high mannose-type oligosaccharide chain on P_0 at 27,000 to the predominantly complex-type oligosaccharide chain at 28,500 with subsequent pulse-chase (Poduslo, 1985). It is possible that the fucosylation itself accounts for this molecular weight shift; however, since P_0 is known to undergo further modifications to form the complex-type oligosaccharide as well as other post-translational modifications including acylation, sulfation, and phosphorylation, further studies are necessary to account for this molecular weight shift. The definition of specific controlling steps in the biosynthetic pathway for this down regulated major myelin glycoprotein will have to await further determination of the extent of oligosaccharide processing, as well as the determination of other post-translational modifications.

IV. GLYCOLIPID MODIFICATION:

Although glycolipids are enriched in plasma membrane and Golgi rich fractions, regulatory mechanisms which control their biosynthesis have not been rigorously evaluated. The intracellular compartments containing the enzyme complements required for glycolipid synthesis are also not completely resolved. Following the addition of glucose or galactose to ceramide, a broad range of processing pathways are available depending on the tissue evaluated. The sugar transferases that are involved in glycolipid synthesis are thought to be distinct from the transferases involved in glycoprotein synthesis. The glycolipid sugar transferases responsible for the synthesis of gangliosides have been found enriched in preparative Golgi fractions of rat liver, kidney, and bovine thyroid (Eppler et al., 1980; Keenan et al., 1974; Pacuszka et al., 1978). Several gangliosides have been shown in biosynthetic studies of cultured neuroblastoma cells to be transported from an intracellular site to the cell surface with a half-time of 20 minutes (Miller-Podraza and Fishman, 1982). Pulse-chase studies of rat glioma cells with tritiated galactose show a precursor-product relationship between glucosylceramide and more complex glycosphingolipids. A recent study by Miller-Podraza and Fishman (1983) on cultured neurotumor cells suggests that the initial glucosylation of ceramide might take place in the cis-Gogli cisternae or in the endoplasmic reticulum, and further glycosylations in the trans-Golgi. This suggests that glucosylceramide, rather than ceramide itself, is a precursor of gangliosides.

Studies by Saito et al. (1984) using monensin, a monovalent cationophore, on cultured human fibroblasts indicate a rapid accumulation of glucosylceramide and lactosylceramide when treated with micromolar concentrations. Simultaneously, a decreased incorporation of labeled galactose was observed in the more highly glycosylated neutral glycosphingolipids and sialoglycosphingolipids. These results indicate that monensin greatly increases the simplest glycosphingolipids that are precursors to the major plasma membrane glycosphingolipids. Since monensin is known to disrupt the Gogli apparatus and since the stepwise addition of monosaccharides to the mono- and dihexose glycosphingolipids which result in the highly glycosylated compounds occurs mainly in the Golgi apparatus, these investigators suggest that glucosylceramide and lactosylceramide synthesis probably occurs in the monensin-insensitive, anabolic endoplasmic reticulum. The higher glycosylation occurs in the monensin-sensitive Golgi apparatus. It is clear that further investigations are necessary to resolve the sites of intracellular biosynthesis of glycolipids and the mechanisms by which they are transported from one subcellular structure to another. It is likely that specific glycolipid transfer proteins may play crucial roles in the relocation of the glycolipids from one site to another (Metz and Radin, 1982; Abe et al., 1982; Yamada and Saskaki, 1982; Wong et al., 1984).

V. GLYCOLIPID PROCESSING BY SCHWANN CELLS:

Based on the known glycolipid modifications as described in the previous section, a working model for glycolipid biosynthesis by Schwann cells in the presence and absence of myelin assembly is also presented in Figure 1. As reported by Yao and Poduslo (1985), [^{3}H]galactose was primarily incorporated in the normal adult nerve into endoneurial monogalactosyl diacylglycerol (MGDG), cerebrosides (hydroxy and nonhydroxy), and phosphatidylcholine, and [^{35}S]sulfate was incorporated into sulfatides (hydroxy and nonhydroxy) as determined by fluorography after high performance-thin layer chromatography (HPTLC). During nerve remyelination, as in the crush model, there was a marked increase of both [^{3}H]galactose incorporation into MGDG, cerebrosides, and sulfatides, and [^{35}]sulfate incorporation into sulfatides and a relative decrease of [^{3}H]galactose incorporation into phosphatidylcholine. In contrast, the incorporation of [^{3}H]galactose into MGDG and cerebrosides, but not the [^{35}]sulfate incorporation into sulfatides, was barely detectable in the absence of myelin assembly. Instead a more complicated profile of ^{3}H-labeled glycosphingolipids, possibly containing ceramide oligohexosides, was observed. These de novo synthesized oligohexosylceramides probably originated from glucocerebroside based on pulse-chase experiments. The presence of [^{35}S]sulfate incorporation into endoneurial sulfatides and the lack of [^{3}H]galactose incorporation into cerebrosides in the transected nerve suggest that endogenous cerebrosides rather than the newly synthesized cerebrosides served as the substrates for the sulfotransferase reaction. The absence of [^{3}H]galactosylcerebrosides suggest that the UDP-galactose:ceramide galactosyltransferase reaction which transfers galactose to ceramide may play a regulatory role in the biosynthesis of peripheral nerve myelin glycolipids. The enhancement of the UDP-glucose:ceramide glucosyltransferase reaction in the absence of myelin assembly may be stimulated by specific genetic control, specific glycolipid transfer proteins which relocate the substrate to the specific transferase, or distinct compartmentalization of these two enzymes.

Studies by Townsend et al. (1984) on the effects of monensin and colchicine on myelin galactolipids have demonstrated a reduced appearance of [^{35}S]sulfatide in a myelin fraction after treatment with 0.1 μM monensin while the appearance of [^{3}H]cerebroside was not significantly reduced. The inhibition of sulfatide synthesis in myelin by monensin is consistent with the view that sulfation of cerebroside occurs in the Golgi and that the sulfatide is transported via Golgi derived vesicles to the forming myelin membrane. Since the synthesis of cerebroside and its appearance in myelin were not inhibited by monensin, these investigators have suggested that the cerebroside destined for myelin is not processed through the Golgi apparatus. As a result, two intracellular transport vesicles are depicted in Figure 1 (+ myelin assembly) which demonstrate that the transport of the galactocerebroside is directed to either the site of myelin assembly or to the Golgi apparatus where sulfation occurs by the sulfotransferase. This model is further supported by developmental studies of cerebrosides and sulfatides by Shimomura et al. (1984) and Burkart et al. (1982, 1983).

In the absence of myelin assembly (Fig. 1), there is a shift from the formation of the galactocerebrosides to form glucocerebroside which then is likely relocated into the Golgi apparatus for further glycosylation to form a series of oligohexosides. These may also become modified by addition of sialic acid to form a series of gangliosides. It is hypothesized that these complex glycosphingolipids and/or gangliosides are important constituents of the Schwann cell plasma membrane. It is thought that they are actively being synthesized as Schwann cell surface components during the time when the Schwann cells have undergone active proliferation as a result of the permanent nerve transection.

VI. CONCLUSION:

Although many aspects of this model are speculative, a direct comparison of the biosynthesis of myelin specific glycoconjugates by Schwann cells in the presence and absence of myelin assembly will aid in understanding the regulatory roles associated with the biosynthesis of these components. This information should aid in our understanding the mechanism associated with the regulation of myelination.

VII. ACKNOWLEDGEMENTS:

Supported by grants from the NINCDS (NS20551, NS14304) and by the Borchard Fund.

VIII. REFERENCES:

Abe, A., K. Yamada, and T. Sasaki (1982) Biochim. Biophys. Res. Commun. 104, 1386.
Agrawal, H.C., C.L. Randle, and D. Agrawal (1982) J. Biol. Chem. 257, 4588-4592.
Agrawal, H.C., R.E. Schmidt, and D. Agrawal (1982) J. Biol. Chem. 258, 6556-6560.
Burkart, T., L. Caimi, H.P. Siegrist, N.N. Herschkowitz, and U.N. Wiesmann (1982) J. Biol. Chem. 257, 3151-3156.
Burkart, T., L. Caimi, and U.N. Wiesmann (1983) Biochim. Biophys. Acta 753, 294-299.
Elting, J.J., W.W. Chen, and W.J. Lennarz (1980) J. Biol. Chem. 255, 2325-2331.
Eppler, C., D. Morre, and T. Keenan (1980) Biochim. Biophys. Acta 619, 318-331.
Flockhart, D.A. and J.D. Corbin (1982) CRC Crit. Rev. Biochem. 133-186.
Gleeson, P.A. and H. Schachter (1983) J. Biol. Chem. 258, 6162-6173.
Godelaine, D., M.J. Spiro, and R.G. Spiro (1981) J. Biol. Chem. 256, 10161-10168.
Goldberg, D.W. and S. Kornfeld (1983) J. Biol. Chem. 258, 3159-3165.
Grinna, L.S. and P.W. Robbins (1979) J. Biol. Chem. 254, 8814-8818.
Hanover, J.A. and W.J. Lennarz (1981) Arch. Biochem. Biophys. 211, 1-19.
Harpaz, N. and H. Schachter (1980a) J. Biol. Chem. 255, 4885-4893.
Harpaz, N. and H. Schachter (1980b) J. Biol. Chem. 255, 4894-4902.
Heifetz, A. and W.J. Lennarz (1979) J. Biol. Chem. 254, 6119-6127.
Heifetz, A., W.H. Kinsey, and W.J. Lennarz (1980) J. Biol. Chem. 255, 4528-4534.
Hsieh, P., M.R. Rosner, and P.W. Robbins (1983) J. Biol. Chem. 258, 2548-2554.
Hubbard, S.C. and R.J. Ivatt (1981) Annu. Rev. Biochem. 50, 555-583.
Huttner, W.B. (1982) Nature 299, 273-276.
Ingebritsen, T.A. and P. Cohen (1983) Science 221, 331-338.
Ishaque A., M.W. Roomi, I. Szymanska, S. Kowalski, and E.H. Eylar (1980) Can. J. Biochem. 58, 913-921.
Ivatt, R.J. (1981) Proc. Natl. Acad. Sci. 78, 4021-4025.
Keenan, T., D. Morre, and S. Basu (1974) J. Biol. Chem. 249, 310-315.
Kent, P.W., C.J. Coles, J.R. Cooper, and N.R. Mian (1978) In Am. Chem. Soc. Symp. Series 77, Carbohydrate Sulfates (Schweiger, R. G. ed.) pp. 29-43.
Kornfeld, S., M.L. Reitman, A. Varki, D. Goldberg, and C.A. Gabel (1982) In Membrane Recycling. Pitman Books Ltd., London (Ciba Foundation Symp. 92) p. 138-156.
LaPorte, D.C. and D.E. Koshland (1983) Nature 305, 286-290.
Lennarz, W.J. (1980) In The Biochemistry of Glycoproteins and Proteoglycans. Plenum Press, New York.
Levitan, I.B., J.R. Lemos, and I. Novak-Hofer (1983) TINS 496-499.
Longmore, G.D. and H. Schachter (1982) Carbohydr. Res. 100, 365-392.
Magee, A.I. and M.J. Schlesinger (1982) Biochim. Biophys. Acta 694, 279-289.

Margolis, R.K. and R.U. Margolis (1970) Biochem. 9, 4389-4396.
Martin, F., R. Lambert, M. Martin, and J.L. Vuez (1968) Arch. Fr. Mal. Appar. Dig. 57, 5-14.
Matthieu, J.-M., J.L. Everly, R.O. Brady, and R.H. Quarles (1975) Biochim. Biophys. Acta 167-174.
Metz, R.J. and N.S. Radin (1982) J. Biol. Chem. 257, 12901-12907.
Mian, N., C.E. Anderson, and P.W. Kent (1979) Biochem. J. 181, 377-385.
Mian, N., C.E. Anderson, and P.W. Kent (1979) Biochem. J. 181, 387-399.
Mikuni-Takagaki, Y. and K. Hotta (1979) Biochim. Biophys. Acta 584, 288-297.
Miller, R.R., and C.J. Waechter (1979) Arch. Biochem. Biophys. 198, 31-41.
Miller-Podraza, H. and P.H. Fishman (1982) Biochem. 21, 3265-3270.
Miller-Podraza, H. and P.H. Fishman (1983) J. Neurochem. 41, 860-867.
Nakamura, K. and R.W. Compans (1978) Virol. 84, 303-319.
Narasimhan, S.(1982) J. Biol. Chem. 257, 10235-10242.
Nestler, E.J. and P. Greengard (1983) Nature 305, 583-588.
Neufeld, E.F. and G. Ashwell (1980) In The Biochemistry of Glycoproteins and Proteoglycans (W. T. Lennarz, ed.), Plenum, New York, pp. 241-266.
Omary, M.B. and I.S. Trowbridge (1981a) J. Biol. Chem. 256, 4715-4718.
Omary, M.B. and I.S. Trowbridge (1981b) J. Biol. Chem. 256, 12888-12892.
Pacuszka, T., R. Duffard, R. Nishimura, R. Brady, and P. Fishman (1978) J. Biol. Chem. 253, 5839-5846.
Parsons, T.F. and J.G. Pierce (1980) Proc. Natl. Acad. Sci. U.S.A. 77, 7089-7093.
Poduslo, J.F. (1984) J. Neurochem. 42, 493-503.
Poduslo, J.F. (1985) J. Neurochem. 44, 1194-1206.
Poduslo, J.F., C.T. Berg, and P.J. Dyck (1984) Proc. Natl. Acad. Sci. U.S.A. 81, 1864-1866.
Poduslo, J.F., P.J. Dyck, and C.T. Berg (1985) J. Neurochem. 44, 388-400.
Prehm, P., A. Scheid, and P.W. Choppin (1979) J. Biol. Chem. 254, 9669-9677.
Rapaport, R.N., J.A. Benjamins, and R.P. Skoff (1982) J. Neurochem. 39, 1101-1110.
Reggio, H.A. and G.E. Palade (1978) J. Cell Biol. 77, 288-314.
Rothman, J.E. (1981) Science 213, 1212-1219.
Saito, M., M. Saito, and A. Rosenberg (1984) Biochem. 23, 1043-1046.
Schachter, H., S. Narasimhan, and J.R. Wilson (1978) In Glycoproteins and Glycolipids in Disease Processes (Walborg, E. F. Jr., ed.), pp. 21-46. American Chemical Society, Washington, D.C.
Schmidt, M.F.G., M. Bracha, and M.J. Schlesinger (1979) Proc. Natl. Acad. Sci. USA 76, 1678-1691.
Schmidt, M.F.G. (1983) In Current topics in Microbiology and Immunology, Vol. 102, Springer-Verlag Berlin Heidelberg, New York, pp. 101-129.
Shacter-Noiman, E., P.B. Chock, and E.R. Stadtman (1983) Phil. Trans. R. Soc. Lond. B. 312, 157-166.
Shimomura, K., S. Yahara, Y. Kishimoto, and J.A. Benjamins (1984) Biochim. Biophys. Acta 795, 265-270.
Simpson, D.A., D.R. Thorne, and H.H. Loh (1976) Biochem. 15, 5449-5457.
Singh, H. and N. Spritz (1976) Biochim. Biophys. Acta 448, 325-337.
Sly, W.S. (1982) In The Glycoconjugates, Vol. IV, Part B, (M. I. Horowitz, ed.), pp. 3-25, Academic Press.
Smith, M.E. and N.H. Sternberger (1982) J. Neurochem. 38, 1044-1049.
Tabas, I. and S. Kornfeld (1979) J. Biol. Chem. 254, 11655-11663.
Tulsiani, D.R.P., S.C. Hubbard, P.W. Robbins, and O. Touster, (1982) J. Biol. Chem. 257, 3660-3668.
Townsend, L.E., J.A. Benjamins, and R.P. Skoff (1984) J. Neurochem. 43, 139-145.
Uyemura, K., K. Horie, M. Suzuki, and K. Kitamura (1981) Neurochem. Res. 6, 959-968.
Varki, A. and S. Kornfeld (1980) J. Biol. Chem. 255, 8398-8401.
Wiggins, R.C. and P. Morell (1980) J. Neurochem. 34, 627-634.
Wong, M., R.H. Brown, Y. Barenholz, and T.E. Thompson (1984) Biochem. 23, 6498-6505.
Yamada, K. and T. Sasaki (1982) Biochim. Biophys. Acta 687, 195-203.
Yao, J.K. and J.F. Poduslo (1985) Trans. Am. Soc. Neurochem. 16, 259.

Cellular Development and Myelin-related Membranes in Monolayer Primary Cultures of Dissociated Brain Cells from 14-day-old Embryonic Mice

M. Fabre*, J. Hummel*, M. T. Strosser†, A. Lowenthal‡, M. Noppe‡, L. Bologa§, K. Langley‖, A. Ebel‖, V. Ferret-Sena‖ and L. L. Sarliève‖

*Service Central de Microscopie Electronique, Faculté de Médicine;
†Laboratoire de Physiologie Générale, Université Louis Pasteur, Strasbourg, France
‡Department of Neurochemistry, Universitaire Instelling, Antwerp, Belgium
§Department of Neurobiochemistry, Beckman Institute of the City of Hope, National Medical Center, Duarte, California, USA
‖U44 de l'INSERM and Centre de Neurochimie du CNRS Strasbourg, France

ABSTRACT

Correlated morphological and biochemical studies were performed on dissociated-attached cell cultures from embryonic mouse cerebral hemispheres grown for periods up to 40 days. Early in culture, cell contacts formed occasionally between neuronal processes and oligodendrocytes. Multilayered membranous material was identified only after 30 days in culture and displayed an alternation of electron-lucent and electron-dense lamellae with a periodicity of 10.2 nm.

KEYWORDS

Brain cells ; primary cultures ; scanning and transmission electron microscopy ; maturation of oligodendrocytes ; axon-glial interaction ; myelin-related structures.

Occurence of myelin-like structures in monolayer cultures of dissociated brain cells from 14-day-old embryonic mice has been demonstrated both by ultrastructural and biochemical observations after 30 to 40 days in culture (DIC) (Sarliève, et al., 1981a, 1983). At this period, oligodendrocytes were numerous, but neurones were never observed (Sarliève, et al., 1981b). Since contacts between oligodendrocyte processes and axons appear to be prerequisite for myelin elaboration by oligodendrocytes, it seemed necessary to examine in detail how the various cell populations differentiate and develop under our culture conditions.
Primary cultures from embryonic mouse cerebral hemispheres were grown for periods up to 40 days, and prepared for scanning and transmission electron microscopy and observation with Nomarski optics (Sarliève, et al., 1983). In parallel, immunohistochemical and quantitative biochemical studies were performed on the cultures at various periods.
During development of the cultures 3 periods could be defined. During the first 10 DIC cells proliferated significantly and neurones differentiated though few of them were able to establish synapses. The oligodendroblasts lay along the axonal prolongations of the neuroblasts and offered contacts with these. Typical endocytotic profiles showing coated-pits were observed From 10-20 DIC neurones progressively disappeared and differentiated oligodendrocytes as well as astrocytes were predominant. Oligodendrocytes were identified by immunocytochemistry using the staining with antibodies to galactocerebroside (Bologa, 1985) and carbonic anhydrase isoenzyme II (CAII) (Sarliève, et al., 1981b). In

addition, total CAII level and 2',3'-cyclic nucleotide 3'-phosphohydrolase (CNP) activity, two oligodendroglial markers, increased steeply in our cultured cells between the 4th and the 16th day by 7.5 and 3.5 times respectively. Furthermore an 11-fold increase of the amount of glial fibrillary acidic protein (an astroglial marker) was observed from 12 to 22 DIC (Fig. 1). In parallel choline acetyltransferase activity (Fig. 1) and the density of muscarinic receptors, used as pre- and post-synaptic markers respectively, declined thus confirming the loss of neurones, as observed at the ultrastructural level (Sarliève, et al., 1981) and by using antibodies to tyrosine hydroxylase and glutamic acid decarboxylase. After 20 DIC, three cellular types were commonly encountered : oligodendrocytes, astrocytes and macrophages (microglial cells).

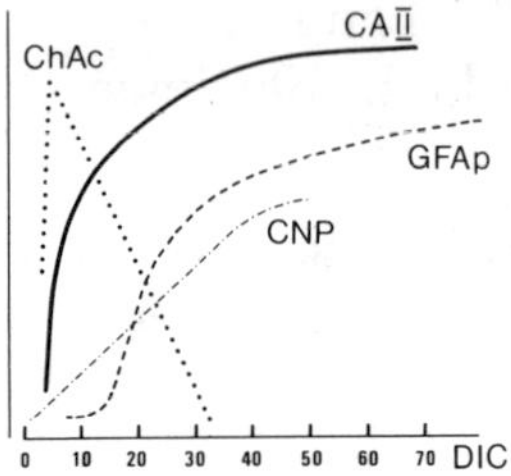

Fig. 1. Activities of choline acetyltransferase (ChAc), 2',3'-cyclic nucleotide 3'-phosphohydrolase (CNP), total carbonic anhydrase isoenzyme II (CAII) and glial fibrillary acidic protein (GFAp) in dissociated brain cells from 14-day-old embryonic mice after various periods in culture.

The ability of the cells to incorporate [^{3}H]Galactose into 4 myelin-associated lipids (Sarliève, et al., 1980) increased 6-fold on average between 20 and 41 DIC. This observation prompted us to search at an ultrastructural level to see if the oligodendrocytes were able to elaborate compact myelin (Sarliève, et al., 1983). Before 30 DIC it was never possible to identify such structures. However, from 30 to 41 DIC, membranes of 2 to about 30 concentric layers with the same periodicity (10.2 nm) as mature compact myelin, containing a distinctive intraperiod line were infrequently observed. It is interesting to note that these multilamellar structures were always associated with oligodendrocytes and free of axons. Since most of the neurones had disappeared after 20 DIC, it can be assumed that oligodendrocytes were exposed to neuronal inductive signals early in culture, for the initiation of myelin membranes. Cell contacts occasionally seen between neurones and oligodendrocytes at 10 DIC suggest such a hypothesis.

ACKNOWLEDGEMENTS

This work was supported by grants from INSERM (CRL n° 80.6.026) and from CNRS (ATP n° 3181). Thanks are due to Prof. N. Herschkowitz (Inselspital Bern, Universitäts-Kinderklinik, Freiburgstrasse 23, Bern, Switzerland) for laboratory facilities, to Dr. M. Sensenbrenner for her helpful comments, to Dr. H. Downey for linguistic assistance and to Mrs. C. Thomassin-Orphanides for typing this article.

REFERENCES

Bologa, L. (1985) J. Neurosci. Res., 14, 1-20.

Sarliève, L.L., J.P. Delaunoy, A. Dierich, A. Ebel, M. Fabre, P. Mandel, G. Rebel, G. Vincendon, M. Wintzerith and A.N.K. Yusufi (1981b) J. Neurosci. Res., 6, 659-683.

Sarliève, L.L., M. Fabre, J.P. Delaunoy, R.A. Pieringer and G. Rebel (1980) Neurological Mutations Affecting Myelination, INSERM Symposium n° 14 (N. Baumann, Ed.), pp. 489-500.

Sarliève, L.L., M. Fabre, G. Rebel, G. Vincendon and J.M. Matthieu (1981a) Proc. Int. Soc. Neurochem., 8, 154.

Sarliève, L.L., M. Fabre, J. Susz and J.M. Matthieu (1983) J. Neurosci. Res., 10, 191-210.

Characterisation of a Myelin-like Membrane Produced by Oligodendrocytes in Subcultures of Rat Brain Cells

J. Chapman and M. Rumsby

Department of Biology, University of York, York YO1 5DD, UK

ABSTRACT

A myelin-like membrane present in subcultures of rat brain cells has been isolated and characterised, in comparison with normal myelin, for polypeptides, Concanavalin A-binding, presence of CNPase and MBP, and lipid composition.

KEYWORDS

Myelin-like membrane, rat brain cell subcultures, normal myelin, polypeptides, Concanavalin A, lipids.

INTRODUCTION

We have recently reported that a myelin-like membrane is synthesised in subcultures of rat brain cells after some 3-4 weeks in culture (Walker et al., 1984, 1985). The subcultures contain neurones, Type 2 astrocytes and oligodendrocytes; myelin-like membrane is synthesised following interactions of oligodendrocytes with neurones. The morphological appearance of the myelin-like membrane, which sometimes surrounds axons, is of many loosely-associated bilayers of membrane which show compaction in some places. In regions of compaction main dense and intraperiod dense lines are apparent on electron microscopic examination. Production, and sometimes the analysis, of myelin-like membrane in cultures of brain cells has been reported by others (for example, Matthieu et al.,1977; Sheppard et al.,1978; Bradel and Prince,1983; Sarlieve et al.,1983). It was thus of interest to isolate and characterise the myelin-like membrane produced in our rat brain subcultures.

METHODS

Normal myelin was isolated from brains of about 250g rats for comparison. Myelin and myelin-like membrane were collected by density gradient centrifugation at the 0.32M/0.85M sucrose boundary of simple discontinuous gradient systems. Polypeptides were resolved by SDS-polyacrylamide gel electrophoresis on 6-30% gradient gels, subsequently stained with Coomassie Blue R250. For Concanavalin A-binding polypeptides after SDS-PAGE were transferred to nitrocellulose membrane by electroblotting (Bio-Rad Transblot cell). Biotinylated Con A (Sigma) was used with a Streptavidin-peroxidase system (Amersham Int.). Blots on nitrocellulose were probed for myelin basic protein (MBP) and CNPase using affinity-purified antibodies. Lipid analysis was undertaken on chloroform:methanol

Advances in the Biosciences Vol. 61

extracts of the membranes by thin-layer chromatography on silica gel H (solvent system, chloroform:methanol, 185/15v/v) and on silica gel H slurried in 0.1M Na_2CO_3 (solvent system, chloroform:methanol:acetic acid:water, 65/25/4/2,v/v).

RESULTS AND DISCUSSION

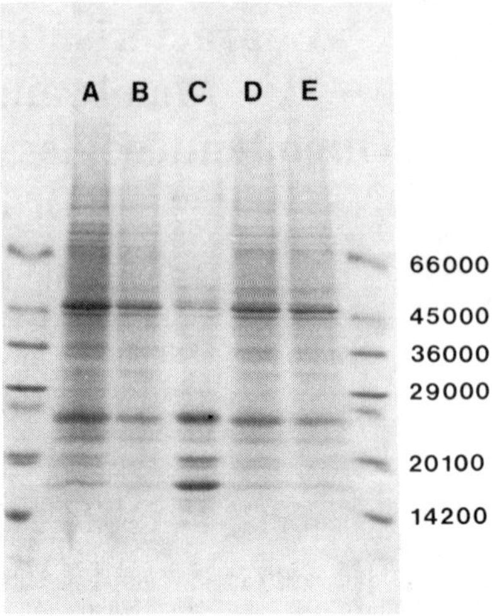

FIG.1. SDS-PAGE analysis of proteins in four different preparations of myelin-like membrane(A,B,D,E) and control myelin (C) with standards, stained with Coomassie Blue. Protein applied: A,8µg, B,3.5µg, D,4µg,E,3µg for myelin-like membrane and C,4µg for control myelin.

SDS-PAGE analysis (Fig.1) revealed that the myelin-like membrane differed from mature myelin in its content of polypeptides. Overall, the myelin-like membrane has many more polypeptides. While bands which may correspond to MBP, PLP and the Wolfgram components are present in the myelin-like membrane clearly other bands are present, especially of a high molecular weight. Yet the myelin-like membrane floats on 0.85M sucrose, like mature myelin. Immunoblotting showed that MBP is present in the myelin-like material in agreement with results on subcultured cells by immunofluorescence (Walker et al.,1985). Immunoblotting also showed that CNPase (Wolfgram components) was present in the myelin-like membrane. A notable difference between the two membranes was that the myelin-associated glycoprotein (MAG) was absent from the myelin-like membrane as revealed by Coomassie staining of gels (Fig.1) and by Con A-binding. This finding is of interest in that a role for MAG in oligodendrocyte-axolemma interactions has been postulated. Lipid differences were also detected between the two membrane systems. The myelin-like membrane had a lower glycolipid content when compared with mature myelin. Further, the HFA component of the cerebroside doublet on tlc was reduced in the myelin-like membrane compared with myelin. Mature myelin has ethanolamine phosphoglycerides as the major phospholipid class; choline phosphoglycerides exceeded ethanolamine phosphoglycerides in the myelin-like material. The analyses reveal that the myelin-like membrane produced by oligodendrocytes in the subcultures is not comparable with normal myelin. Certain signals to stimulate synthesis of normal myelin are still missing even though oligodendrocytes are in contact with neurones in the subculture system.

REFERENCES

Bradel, E.J. and F.P. Prince (1983). J.Neurosci.Res. 9, 381-392.

Matthieu, J-M., P. Honegger, B.D. Trapp, S.R. Cohen and H. deF. Webster (1978). Neuroscience 3, 565-572.

Sarlieve, L.L., M. Fabre, J. Susz and J.M. Matthieu (1983). J.Neurosci.Res.10, 191-210.

Sheppard, J.R., D. Brus and J.M. Wehner (1978). J.Neurobiol.9, 309-315.

Walker, A.G., J.A. Chapman, C.B. Bruce and M.G. Rumsby (1984). J.Neuroimmunol.7, 1-20.

Walker, A.G., J.A. Chapman and M.G. Rumsby (1985). J.Neuroimmunol.9, 159-177.

ACKNOWLEDGEMENTS: We thank Dr. P. Brophy, University of Stirling, for MBP and CNPase blots. This work is supported by the Multiple Sclerosis Society of G.B. & N.I.

Session IV

THE BIOCHEMICAL LESIONS AND PHARMACOLOGY OF GLIA CELLS

Co-chair : E. SCHOFFENIELS (*Liège, Belgium*)
D. B. TOWER (*Maryland, USA*)
D. M. WOODBURY (*Salt Lake City, USA*)

Role of Glial Carbonic Anhydrase in Focal and Generalized Tonic-clonic seizures

D. M. Woodbury[1], F. L. Engstrom, J. K. McQueen, R. E. Anderson, H. S. White and Y. C. Yen-Chow

University of Utah School of Medicine, Department of Physiology, Division of Neuropharmacology and Epileptology, 410 Chipeta Way, Room 167, Salt Lake City, Utah 84108, USA

ABSTRACT

Carbonic anhydrase (CA) in the central nervous system is found only in glial cells and inhibition of its activity results in profound changes in neuronal excitability. Thus, it is evident that glial cell CA is involved in regulation of neuronal excitability. CA is distributed in cytoplasm and in membranes in astrocytes and in cytoplasm, mitochondria and myelin of oligodendrocytes. DBA mice that are susceptible to audiogenic seizures have a higher activity than do C57 control mice. Also, oligodendrocytes have higher CA activity than do astrocytes. Changes in the pattern of maximal electroshock seizures induced by acetazolamide, were shown to correlate directly with inhibition of CA in the brainstem cytosolic fraction, whereas changes in the electroshock seizure threshold correlated inversely as a power function with inhibition of CA in the cerebral cortical myelin fraction. Thus, glial cells are directly involved in regulation of excitability. With respect to focal epileptogenic seizures induced by local cobalt implantation, the gliosis produced by this procedure as a result of increased neuronal activity enhances CA activity in glial cells and increases the ability of these cells to handle the increased metabolic production of CO_2 and, as a result, decreases the severity of the induced seizures and polyspike activity in the electrocorticogram. Inhibition of glial CA activity by acetazolamide under these conditions enhances excitability. Further evidence for a role of CA in seizures was obtained in a comparison of DBA and C57 mice. The DBA strain has a lower EST and a higher activity of CA than does the C57 strain. However, acetazolamide prevents audiogenic seizures in DBA mice, an indication that CA is etiological in such seizures. Also in DBA mice acetazolamide decreases excitability whereas in C57 mice it increases excitability by blocking the glial protective effect of CA. Explanation of these findings in terms of the amount and activity of CA in the two strains and of the effects of acetazolamide are discussed.

KEY WORDS

Carbonic anhydrase; glial cells; epilepsy; focal seizures; generalized seizures; acetazolamide; tolerance

INTRODUCTION

The pathological changes that occur in epileptogenic foci in experimental animals and man are characterized mainly by glial proliferation and loss of nerve cells. Because of the observation demonstrated by many workers that hyperactivity of neurons is a stimulus for glial proliferation, and since this is a major event in the development of a seizure focus, it is of importance to elucidate the role of glia in the seizure process. This glial response appears to be a mechanism to increase the capacity of these cells to modulate the enhanced release of neuronal depolarization products (K^+, H^+, CO_2, NH_3, and various neurotransmitters) that result from seizure activity. Glial cells have active cation and anion transport systems that make it possible to regulate the concentrations of these products in the narrow interstitial spaces of the brain and thereby facilitate the maintenance of ionic and neurotransmitter homeostasis in the brain.

The role of carbonic anhydrase (CA) in the regulation of transport by glia becomes apparent by the facts that this enzyme is found only in these cells and that inhibition of its activity results in profound changes in neuronal excitability, suppression of seizures induced in animals by electrical and chemical stimulation of the brain, and protection from epileptic seizures in man. That glial proliferation is present in a seizure focus is demonstrated by anatomical studies showing increased numbers of astrocytes surrounding neurons in the area of the focus and also by biochemical studies that demonstrate an increase in DNA and in carbonic anhydrase activity in the same areas. The increase in CA activity is evidence of glial differentiation since its expression is a marker of this process and a deficiency of CA in quaking mice during development is associated with abnormal oligodendroglial development. The expression of CA in the CNS is confined to glia with the activity in oligodendroglia much greater than in astroglia, as will be discussed later.

Thus, it is apparent that CA has a central role in the important regulatory role of glial cells in central nervous system function. This role is concerned with the cooperative metabolism of neurons and their closely contacting glia during regulation of anion and acid-base balance of brain cells, interstitial fluid and cerebrospinal fluid. Thus, CA in glia depends for its substrate on metabolites that are mostly generated and released during neuronal discharge, and it is directly involved in the maintenance of the internal Cl^- in glia and indirectly regulates internal K^+, both of which are involved in regulation of neuronal excitability.

This chapter will, thus, be concerned with elucidation of the role of carbonic anhydrase in epilepsy as derived mostly from research in my laboratory carried out over the last 30 years since publication of our first paper on acetazolamide (Millichap *et al.*, 1955).

CA promotes the hydration of CO_2 to carbonic acid (H_2CO_3), which spontaneously dissociates into H^+ and HCO_3^-, or at high pH:

$$\begin{array}{l} CO_2 + OH^- = HCO_3^- \\ \qquad\ \ \downarrow\uparrow \\ \qquad\ OH^- + H^+ = HOH \end{array}$$

with H^+ left from the dissociation of H_2O. This enzyme thus regulates the amount of CO_2 formed and the concentrations of H^+ and HCO_3^-. Since CO_2 generation is primarily the function of neuronal activity (glucose = H_2O + CO_2) and CA is only present in adjacent glial structures, it follows that both H^+ and HCO_3^- concentrations within astroglia reflect neuronal activity. To handle this load of CO_2 from neurons, glial membranes have an anion counterport system that exchanges external Cl^- for internally accumulated HCO_3^- that is derived from the CA-catalyzed hydration of CO_2.

The first question that arises with respect to the role of CA in the CNS and in epilepsy which is key to its role is, where is it distributed in brain?

Previous studies (Anderson, *et al.*, 1984) from this laboratory have examined electron micrographs of cerebral and cerebellar cortices from DBA and C57 mice stained by the PAP method of Sternberger with anti-human CAII antiserum made in goats. The mice were 25-30 days old, a time when the DBA mice are still susceptible to audiogenic seizures. In astrocytes from C57 and DBA strains, CA was found to be present in the cytoplasm, on the outer surface of intracellular vesicles, and on the inner surface of the plasma membrane; mitochondria were not stained. More staining was seen in the astrocytes from DBA than in these cells from C57 mice. In oligodendrocytes from C57 and DBA mice, CA was present in the cytoplasm and in the myelin; other membranes were not stained. The mitochondria of the oligodendrocytes were also stained. The DBA had more staining in these cells than did the C57 mice. Also, as has been shown by others, the oligodendrocytes had more staining than did the astrocytes.

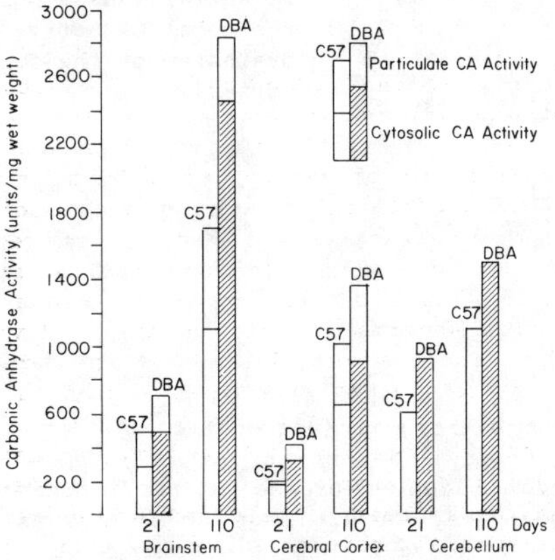

Fig. 1. Activity of carbonic anhydrase (CA) in cytosolic and total particulate fractions of cerebral cortex and brainstem and in the total cerebellum homogenate of C57 and DBA mice 21 and 110 days of age. Activity of CA is expressed in Units per mg protein.

In Fig. 1 is shown the activity of this enzyme in cytosolic and total particulate fractions of cerebral cortex, and brainstem and the total activity in the cerebellum of C57 vs DBA mice at 21 days of age when the DBA are maximally sensitive to audiogenic seizures and at 110 days when they are resistant to such seizures (Engstrom, et al., 1984b). The cytosolic CA is higher in the DBA as compared with the C57 mice at both 21 and 110 days, but the difference is greater at 21 than at 110 days. The activity of the total particulate enzyme is also greater in the DBA than in the C57 but the increase with age is greater in the C57 mice. CA activity is highest in brainstem, next highest in cerebellum and lowest in cerebral cortex, but on a DNA (per cell) basis the cerebellum is lower than the cortex. In all three tissues total CA activity is higher in DBA than in C57 mice.

In Fig. 2 is shown the distribution of CA in the particulate fraction of the cerebral cortex and brainstem (Engstrom, et al., 1984b). The enzyme is present in the microsomes (membranes), myelin and the mitochondria. The total activity is generally higher in the DBA than in the C57 mice but the % of total activity in each fraction is lower in the DBA as compared with the C57 mice. In the cerebral cortex, the specific enzyme activity is highest in the cytosol, next highest in the myelin and lowest in the mitochondria. In the brainstem, the specific activity is highest in the cytosol, next highest in the microsomes followed by the myelin and lowest in the mitochondria. Since, as described above, CA in oligodendrocytes is found only in the cytosolic, myelin and mitochondrial fractions and that in astrocytes is only in the cytosolic and microsomal fractions, the data shown in this figure allow us to distinguish the differences in the cellular localization of CA and the effects of aging on the CA activity of these two cell types in the two strains.

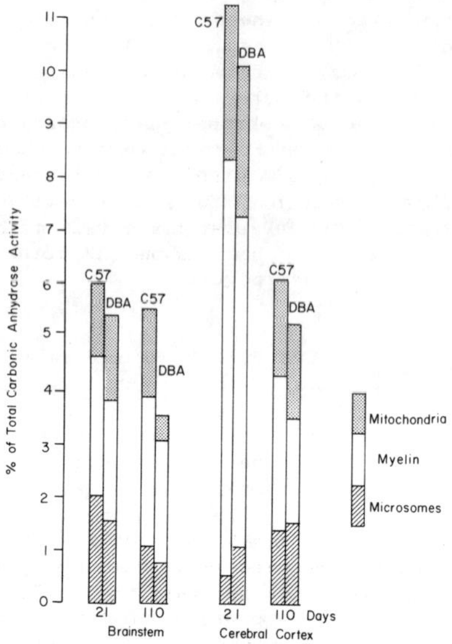

Fig. 2. Distribution of CA in the particulate fractions (microsomes, myelin and mitochondria) of cerebral cortex and brainstem of C57 and DBA mice at 21 and 110 days of age. The ordinate is activity as a percent of the total CA activity in the brainstem or the cerebral cortex.

In the brainstem, the percentage of microsomal (astrocytic) CA decreases with age in both strains; the % activity was also less in DBA as compared with C57 mice at both ages. There was only a slight decrease or little change in the myelin and mitochondrial (oligodendroglia) CA with age or in DBA as compared with C57 mice, except for the marked decrease in DBA mitochondria at 110 days of age. The decrease in the microsomal CA activity in the DBA as compared with the C57 strain is greater in older than that in younger mice. This correlates with a lower EST in DBA as compared with C57 mice at 110 days of age as discussed below. In the cerebral cortex, the microsomal (astrocyte) CA activity increased slightly with age in both strains and the DBA had a higher percentage than did the C57 at both ages. In the case of the oligodendroglial myelin CA the percentage of total activity decreased markedly with age in both strains and the DBA myelin CA was less than that in the C57 at both ages. There was little change in mitochondrial CA. Thus, as judged from CA activity measurements, in the cerebral cortex there appears to be an increase in astrocytes and a decrease in oligodendroglia with age that is more marked in the DBA strain. These data suggest that the membrane (microsomal) CA, which is involved in active anion transport in astrocytes, as described later, and the myelin CA, which appears to be involved in both CO_2 fixation, a process that requires HCO_3^- ion and results in synthesis of fatty acid and myelin, as well as CO_2 (HCO_3^-) transport across the myelin membrane surrounding neurons are deficient in DBA mice. This results in an increased sensitivity to seizures. Further evidence for this will be discussed later.

A summary of the distribution of CA in glial cells and its relation to neuronal metabolism is shown in Fig. 3. In this figure are depicted the formation of HCO_3^- from CO_2 in oligodendroglial mitochondria catalyzed by mitochondrial CA and the subsequent fixation of this HCO_3^- to form malonyl CoA by reacting with acetyl CoA and its further synthesis to fatty acids and myelin. This process probably defines the role of CA in these cells. The myelin enzyme is probably located in or between the myelin membranes and may act as postulated by Wistrand (1984) as a channel to facilitate CO_2 and H^+ movement out of the neurons, about which the myelin wraps. Inhibition of this enzyme would cause CO_2 to accumulate and raise the threshold for excitation of the neurons. This is described later. In the astrocytes the cytoplasmic CA is involved in the production of

HCO_3^- from neuronally derived CO_2. The H^+ left over from the splitting of H_2O

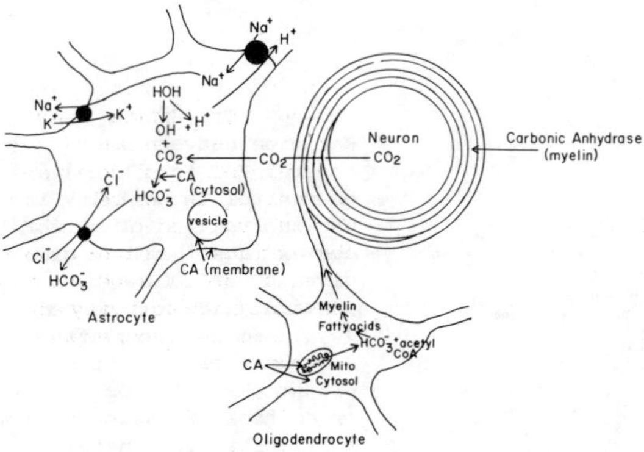

Fig. 3. Schematic drawing of the distribution of CA in astrocytes, oligodendroglia and myelin of mice and its relation to CO_2 metabolism in neurons and glia.

to H^+ and OH^-, the latter of which with CO_2 is catalyzed by CA to form HCO_3^-, exchanges with Na^+ to regulate H^+ concentration in cells. The HCO_3^- exchanges for Cl^-, a process which is probably catalyzed by membrane CA. Inhibition of the enzymes at these sites by acetazolamide causes marked changes in excitability, the nature of which will now be discussed.

The effects of CA inhibition by sulfanilamide and acetazolamide on maximal electroshock seizure (MES) pattern in mice was determined by Millichap et al. (1956). As with phenytoin, both these agents abolished the tonic extensor component of the MES, an evidence of an anticonvulsant effect. The ED_{50} for acetazolamide (74 mg/kg) was one-half that for sulfanilamide (140 mg/kg). However, acetazolamide is 100 x more potent as an inhibitor of CA. Examination of the concentration of the two drugs in the brain, showed that the concentrations of sulfanilamide was 50 times higher in the brain than that of acetazolamide. Thus, the relative potency of the two drugs against MES could be accounted for exactly by their inhibition of CA. The other point about these data was that it required about 98% inhibition of the enzyme to reach the ED_{50} of either drug for blocking the tonic-extensor phase of the MES. For an ED_{99} + over 99.5% inhibition of the enzyme was required. The maximum degree of enzyme inhibition of these two drugs corresponded with their time of peak anticonvulsant effect.

Previous experiments in our laboratory (Koch and Woodbury, 1958; see also Woodbury and Esplin, 1959; Woodbury and Karler, 1960) have shown that acetazolamide and carbon dioxide act in the same manner, because acetazolamide causes accumulation of CO_2 in the brain, as revealed by conventional laboratory tests. The effects of these agents on spinal cord synaptic transmission was carried out by Esplin and Rosenstein, (1963) who showed that acetazolamide had a different spectrum of action than did phenytoin and trimethadione, both in the laboratory tests and on the spinal cord parameters. Only acetazolamide and carbon dioxide of all the drugs tested produced a reduction in the monosynaptic spike without affecting polysynaptic activity and PTP. Thus these two drugs (acetazolamide and CO_2) are unique. Also, tolerance to CO_2 develops pari-passu with development of tolerance to acetazolamide (Koch and Woodbury, 1958).

To define which region of the brain and which glial pool of CA is concerned with its anticonvulsant (both threshold raising and anti MES) effects in mice, the time course of inhibition by 40 and 200 mg/kg doses of acetazolamide of the CA in the cytosolic, microsomal, myelin and mitochondrial fractions of cerebral cortex, cerebellum, and brainstem were measured and correlated with changes in electroshock seizure threshold (EST) and MES pattern determined at the same times (Anderson, R.E., and Woodbury, unpublished observations). The results are shown in Fig. 4.

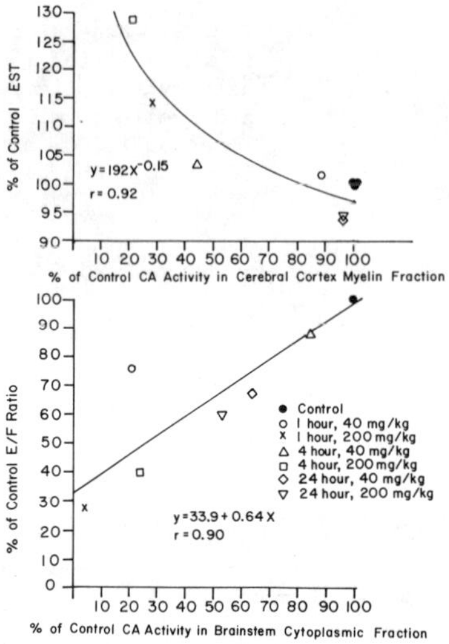

Fig. 4. Top Graph: Relation between % of control EST (ordinate) and % of control CA activity in myelin fraction of cerebral cortex (abscissa) in mice given 40 or 200 mg/kg acetazolamide and tested at 1, 4, and 24 hours after administration. Bottom Graph: Relation between % of control Extension/Flexion (E/F) ratio (ordinate) and % of control CA activity in brainstem cytoplasmic fraction (abscissa) in mice given 40 or 200 mg/kg acetazolamide and tested at 1, 4, and 24 hours after administration.

The results show that changes in the MES pattern correlate directly with inhibition of brainstem cytosolic CA whereas the EST changes correlate inversely as a power function with inhibition of cerebral cortical myelin CA.

Thus, the accumulation of CO_2 in neurons that causes the anti-MES effect, as just described, is due to the inhibition of brainstem cytosolic CA which decreases the hydration of CO_2 to HCO_3^- in glial cells and causes it to back up and accumulate in neurons. The myelin fraction was inhibited at a slower rate than was the cytosol but the inhibition lasted for a longer period of time. It is postulated that the increase in EST is due to inhibition of transport of neuronally derived CO_2 across the myelin surrounding the neurons, an effect that would raise the threshold for firing of the neurons. Wistrand (1984) has presented evidence that CA is incorporated in the membrane to form a channel that is concerned with HCO_3^- (or other anion) movement across the membrane. Inhibition of CA could disrupt the function of this channel. Increased CO_2 would also decrease pH and increase ionized Ca^{++} concentration which can alter conductances across the membrane.

The evidence from the CA inhibition data supports the hypothesis that glial CA is directly involved in regulation of excitability. CA inhibition by acetazolamide results in decreased excitability. However, this is not the complete picture. Evidence now to be discussed from our studies with cobalt-induced focal epileptogenic lesions in the frontal cortex of rats, suggests that the glial CA response to increased neuronal activity caused by seizures enhances the ability of these cells to handle the increased metabolic production of CO_2, and that inhibition of this activity by acetazolamide under these conditions enhances excitability.

Cobalt implantation in the right frontal cerebral cortex in rats results in the development of a semi-chronic epileptic state with persistent epileptogenic foci, but only rare spontaneous seizures. The presence of both primary and secondary foci makes this a particularly useful model for studying the pathogenesis of focal epilepsy. Since the primary cobalt lesion is characterized by gliosis and neuronal loss, the onset of gliosis coincides with that of epileptic activity in the electrocorticogram (ECoG), and since CA is found exclusively in glial cells, the role of CA in the development of the focal seizure state was assessed. A role of CA (and glia) in seizures is evident, as mentioned above, from the fact that sulfonamide inhibitors of this enzyme are anticonvulsants and from data on audiogenic seizure mice, that there is a direct correlation between seizure events and CA activity in different parts of the brain (see Woodbury et al., 1984).

Single epileptiform spikes first appeared in the ECoG 4-5 days after cobalt implantation. At 12 days, the time of peak effect there were frequent bursts of bilateral polyspikes in both frontal regions with some spread to parietal areas. These discharges were associated with clonic movements of the jaw and of the forelimb on the side opposite the implant. Seizure activity was markedly attenuated at 19 days and the electrical activity of the epileptogenic focus had disappeared by 30 days. Examinations of different areas of the brain showed that there was a significant increase in CA activity of the secondary focus on the contralateral cortex that was first noted 5 days after cobalt implantation (Fig. 5). At 12 days these changes were greater and more widespread, involving frontal and parietal areas, but not the occipital cortex, which also did not exhibit polyspike activity. Similar, but smaller, increases in the activity of CA were observed at 19 days but by 30 days the activity in all areas was not different from control values (Fig. 5). The increases in CA activity were clearly associated with the presence of abnormally discharging neurons.

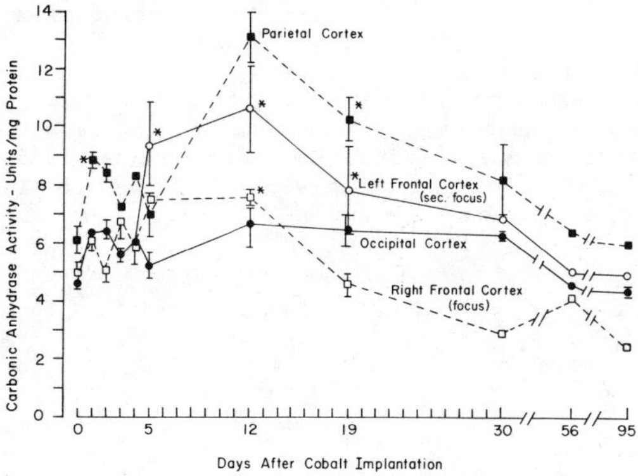

Fig. 5. Changes in CA activity (Units/mg protein) (ordinate) in 3 different areas of the left and right cerebral cortex with time in days (abscissa) after cobalt implantation into the right frontal cortex of rats.

The CA inhibitor acetazolamide was used to assess whether the changes in CA were the cause or the result of seizure discharges. Acutely, 100 mg/kg acetazolamide increased polyspike activity of 12-day implanted cobalt rats and caused spontaneous seizures lasting 15 seconds with continuous bilateral spiking at 3-4 spikes/second. CA activity was decreased in all areas by 25-30% (actual brain enzyme inhibition of 99.0-99.8%) (Fig. 6). Since, the seizure discharges were enhanced not supressed, CA is not involved in maintenance of the foci. Chronic administration of 20 mg/kg acetazolamide b.i.d. for 14 days starting 2 days before cobalt implantation showed that the drug did not prevent development of spike discharges. In fact, foci tended to appear earlier (2-4 days) and both the ECoG and behavioral manifestations of cobalt implantation were more severe than in non-treated cobalt-implanted rats. CA was completely inhibited in all areas of the brain (Fig. 7). Following cessation of acetazolamide therapy at 12 days spiking and behavioral manifestations of seizures rapidly disappeared. CA activities at 19 days of age in these rats were slightly higher in all areas of the brain (Fig. 7). In the secondary focus they were 40% higher than in the 19-day post-operative cobalt rats. These data suggest that the increase in CA activity is a protective response to abnormal firing in the seizure focus and that chronic treatment with acetazolamide induces an adaptive increase in CA in brain which counteracts the inhibition of the enzyme and leads to drug tolerance.

To determine whether CO_2 *per se* might also induce the increase in CA activity, the effects of elevating tissue CO_2 levels were measured. Rats were placed in

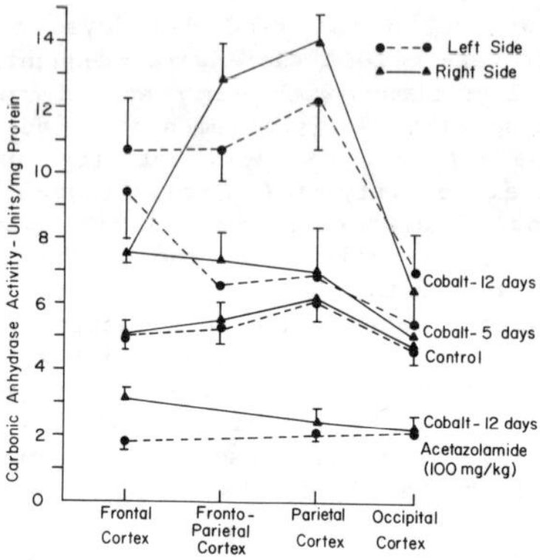

Fig. 6. CA activity (Units/mg protein) (ordinate) in 4 different areas of the left and right cerebral cortex (abscissa) of control rats and of rats implanted with cobalt in the right frontal cortex for 5 and 12 days (shown on the right side of the lines connecting the CA activity for the 4 areas studied). Also shown is the effect of acetazolamide (100 mg/kg in one dose) on the CA activity in the various areas of the brain of rats implanted for 12 days with cobalt and given the CA inhibitor on the 12th day

an airtight, plastic box with food and water and 9-10% CO_2 was flowed through at 100 ml/min. After 2-5 days exposure the animals were killed and the same areas of the brain were assayed for CA activity. Arterial pCO_2 was increased to 50-55 mm Hg but the enzyme levels in both blood and brain were not changed. Thus, local stimulation of glia would appear to be required for initiating new CA synthesis or activation.

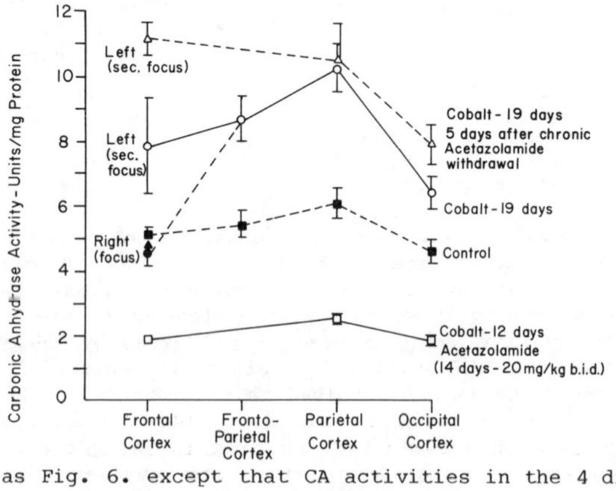

Fig. 7. Same as Fig. 6. except that CA activities in the 4 different cerebral cortical areas are shown for the control rats and for those in which cobalt had been implanted for 19 days. Also shown are the CA activities of rats implanted for 12 days with cobalt but given acetazolamide for 14 days at a dose of 20 mg/kg b./i.d., starting 2 days before implantation (bottom line), and for animals 5 days after withdrawal of acetazolamide given for 14 days prior to its withdrawal on the 14th day after cobalt implantation. They were thus killed on the 19th day after cobalt implantation (top line).

In view of previous reports that cyclic AMP increases CA amounts in cultured glial cells (Narumi et al., 1979) the concentration of cyclic AMP in the various areas of the brain were examined after cobalt implantation. Cyclic AMP concentration in the primary lesion (focus) was significantly reduced after 4 days, probably reflecting the tissue destruction in that area. The only other change was a significant elevation in the secondary focus at a time (19 days) when seizure discharges were becoming less frequent and CA activity was decreasing. An explanation for this effect is not apparent at present.

Cobalt implantation for 12 days decreased total free $^{14}CO_2$ space (23%) and increased $^{14}CO_2$-fixation (44%) in the projected (secondary) focus of the left frontal cortex. Acetazolamide had the opposite effect i.e., increased total free $^{14}CO_2$ and decreased CO_2 fixation. Thus, glial cell CA modulates brain excitability by regulating the CO_2-$H_2CO_3^-$$HCO_3^-$ system and appears to regulate CO_2 entrance into Krebs cycle via pyruvate carboxylase, pyrimidine synthesis, etc (one carbon metabolism).

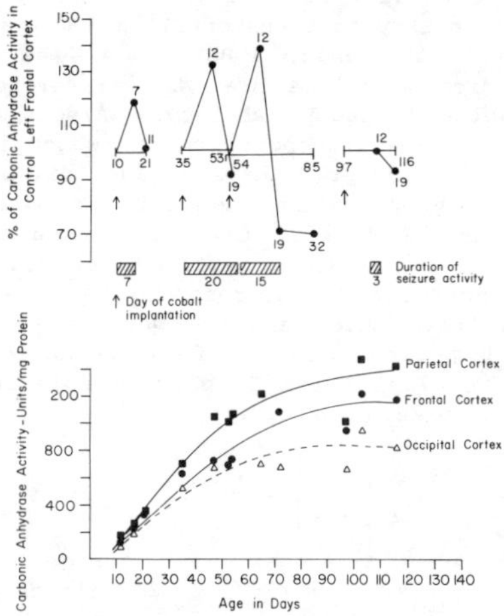

Fig. 8. Top Graph: changes in CA activity in the left frontal cortex (projected focus) as a percent of the control CA activity in this area at various days after cobalt implantation in the right frontal cortex of rats of various ages. The numbers next to the symbol (o) are the days after cobalt implantation. The numbers at each end of the horizontal lines are the age of the animals. The vertical arrows are the days of age when the cobalt was implanted. The horizontal hatched bars and the numbers below them depict the duration of seizure activity as measured by the polyspike activity in the electrocorticogram. Bottom Graph: Changes in the activity of CA (Units/g wet weight) (ordinate) in parietal, frontal, and occipital cortex of rats at different days of age (abscissa).

Aging factors that involve glial CA are also important in determining susceptibility to cobalt-induced focal seizures in rats as depicted in Fig. 8. The polyspike activity in the ECoG in response to cobalt was highest in 35-day-old rats and decreased to almost zero responsiveness in 97-day-old rats. Seizure discharges after cobalt implantation started earlier and were of longer duration (20 days) in 35-day- as compared with 53- (14 days) and 97- (2 days) day-old rats. Similarly, the CA activity increase in response to cobalt implantation was highest in the 35- and 53-day-old rats, whereas the 97-day-old rats showed no increase in CA activity in response to cobalt. The number of glial cells and the activity of CA increases with age (Fig. 8). However, both have plateaued by the 97th day and these cells have reached their maximal ability to respond to enhanced neuronal activity (protective effect); CA activity can not, therefore, further increase. Consequently, the seizure response to cobalt is maximally attenuated as was observed. These data demonstrate that glial cell CA plays a role in protecting against focal seizures induced by cobalt. The increased neuronal activity induces this enzyme by an as yet unknown mechanism that appears not to involve cyclic AMP nor CO_2. It is postulated that K^+ or other depolarization products probably are involved in this mechanism.

Further evidence for a role of CA in epilepsy is derived from our studies on genetic strains of audiogenic seizure mice, namely, the DBA as compared with C57 strains. In DBA/2J mice susceptibility to audiogenic seizures is highest between 18-22 days of age and has disappeared by the 60th day.

As discussed above, the CA activity of glial cells is increased markedly in the DBA audiogenically susceptible as compared with the resistant C57 mice. It has been demonstrated that the glial cells and the choroid plexus of the DBA mice however, have a decreased ability to transport I^- and Cl^- as compared with those in C57 mice (Engstrom et al., 1984a, 1986b). As the animals age, seizure susceptibility to sound disappears, as does the difference in CA activity in brain and the reduced anion transport ability. Also, the difference in the ability of the thyroid gland to transport iodide, which is much higher in the DBA as compared with the C57 mice at 21 days of age, disappears pari-passu with decreased anion transport in the brain and with susceptibility to audiogenic seizures, thus demonstrating a role of thyroid hormone in audiogenic seizure susceptibility (Engstrom et al., 1984a, Seyfried, et al., 1979). It appears that the increased CA activity is a response to the reduction in anion transport, an adaptation that would tend to overcome the defect by providing more enzyme (or increasing its activity) for the active transport process. These data suggest that in glial cells CA plays a role in maintaining anionic homeostasis and regulating seizure susceptibility.

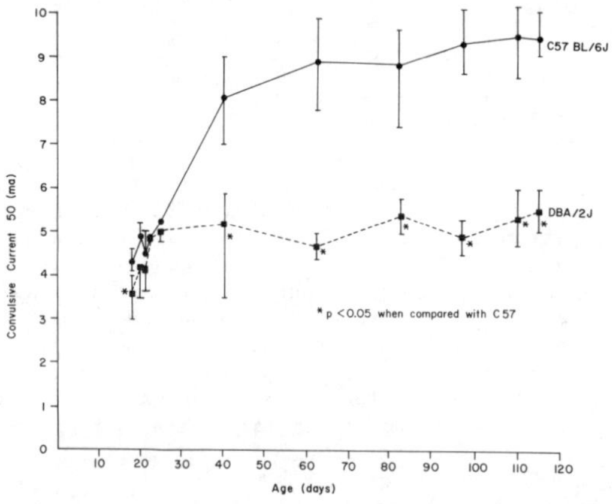

Fig. 9. Minimal electroshock seizure threshold for DBA and C57 mice between 18 and 115 days of age. Each point represents the current (ma) necessary to produce a minimal seizure in 50% of the mice tested at each age. Ninety-five percent confidence limits are shown. (From Engstrom et al., 1986 with permission)

In order to define further the role of CA in this strain of mice we studied their susceptibility to another kind of seizure, namely that produced by electrical stimulation, rather than sound. In Fig. 9 is shown the electroshock seizure threshold (EST) in DBA and C57 mice with age (Engstrom et al., 1986a). The threshold of the DBA IAS strain, a one gene mutation variant of the DBA strain that does not exhibit audiogenic seizures was also tested. Except at 21 days of age, when the DBA strain is maximally sensitive to audiogenic seizures, the EST was lower in them than in the C57 mice at all ages. The EST of the IAS strain is in between, but closer to that of the DBA mice. It is evident that the DBA mice have lower seizure thresholds as compared to the C57 mice throughout life, despite the fact that audiogenic seizure susceptibility disappears by the 60th day. They are thus a genetic model of hyperexcitability. The data already presented suggest that the hyperexcitability is related to reduction of membrane and myelin bound CA in the cerebral cortical and brainstem areas of the brain with consequent compromise of anion transport mechanisms.

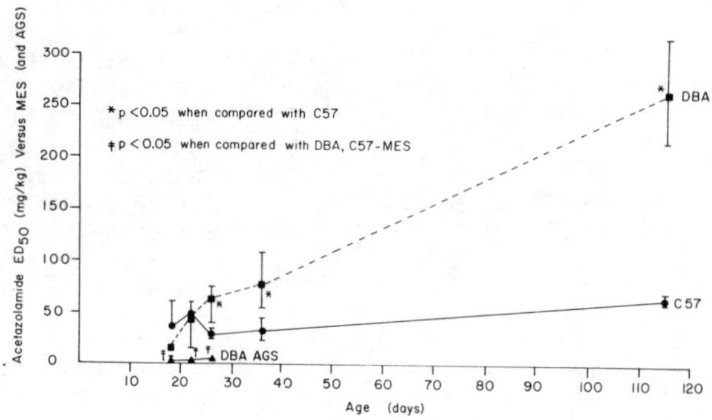

Fig. 10. Acetazolamide (ACTZ) effective dose 50 values (ED_{50}s) versus maximal electroshock seizures (MES) in DBA and C57 mice at various ages between 18 and 115 days of age; and ACTZ ED_{50}s against audiogenic seizures (AGSs) in DBA mice at 18, 22, and 26 days of age. Each point represents the dose of ACTZ (mg/kg) at which maximal seizures in response to MES or sound were inhibited in 50% of the mice tested. Ninety-five percent confidence limits are shown. (From Engstrom et al., 1986a with permission)

The next question concerns the effects of acetazolamide on audiogenic seizures in DBA mice and on maximal electroshock seizures in DBA as compared with C57 mice. These data are shown in Fig. 10 Engstrom et al. (1986a). The ordinate is the ED_{50} for acetazolamide against MES or audiogenic seizures. The abscissa is age of the mice in days. The acetazolamide ED_{50} for abolition of the tonic extensor phase of the MES in C57 mice increased only slightly with age and was the same as that for DBA mice at 21 days of age, the time of maximal audiogenic seizure susceptibility. The ED_{50} for DBA mice increased markedly with age in parallel with the increase in cytosolic CA activity. This is to be expected since acetazolamide has a very low inhibitory constant (K_i) (high affinity) for this enzyme. The K_i for glia is 1×10^{-10}M, (H.S White, unpublished observations) hence the drug present in the brain is almost completely bound to the enzyme as the EI complex. The acetazolamide ED_{50} for protection against audiogenic seizures in DBA mice is considerably lower than that for MES (around 2.5 mg/kg) and increases only slightly with age. Acetazolamide not only abolished the tonic extensor component of the maximal audiogenic seizure, as was the case with MES, but completely blocked the clonic and running phases as well. Thus, the audiogenic seizure is much more sensitive to the effects of CA inhibition than is the MES. Furthermore, the protection from these seizures is complete, an indication that CA is etiological. This is also borne out by the fact that acetazolamide lowers EST of C57 mice whereas it raises the EST of DBA mice. It would appear that an increase in glial cell CA activity with age in C57 mice is an adaptive response to maintain ionic homeostasis in the brain, but that in the DBA mice the markedly increased CA activity that is present appears originally to be an adaptive response to the genetically based increased excitability but it becomes maladaptive because of some defect related to the activity of CA in the membrane and myelin fractions. Thus, in the C57 mice acetazolamide, although still abolishing the tonic extensor component of the MES, increases excitability by blocking the glial protective effect of CA but in DBA mice it decreases excitability by inhibiting the maladaptive effect of CA that increases susceptibility to seizures.

The data shown in Fig. 11 support this conclusion (Engstrom et al., 1986a). They demonstrate that C57 mice given acetazolamide for 4 days and then tested for the ED_{50} against MES have a higher ED_{50} than do animals given a single

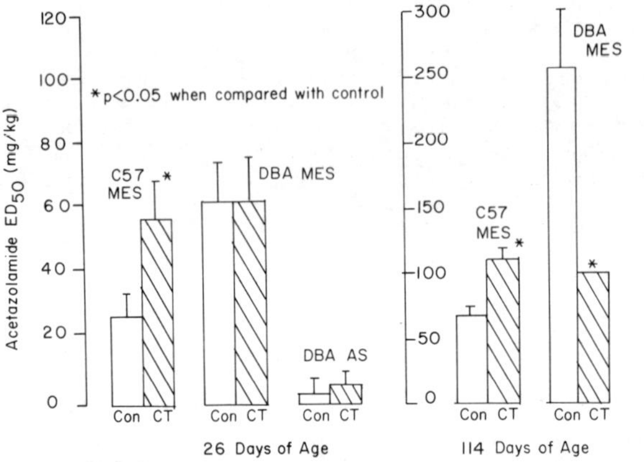

Fig. 11. Acetazolamide (ACTZ) ED_{50}s versus MES in control (Con) and chronically treated (CT) DBA and C57 mice at 26 and 114 days of age; and ACTZ ED_{50}s versus AGSs in control and CT DBA mice at 26 days of age. (From Engstrom et al., 1986a with permission)

dose. Thus, tolerance develops to the anticonvulsant effects of this drug as measured by MES, as we have previously demonstrated (see Woodbury and Karler, 1960) and as has been shown in patients with epilepsy (see Woodbury, 1980). In contrast, however, in DBA mice given acetazolamide chronically, their ED_{50}'s against MES as well as against audiogenic seizures are the same as those receiving one dose of the drug. Thus, tolerance to acetazolamide does not develop in this strain. Further experiments (Engstrom et al., 1986a) have shown that the tolerance in the C57 mice is due to an increase in the activity of CA in the brain, but in the DBA mice this does not occur, perhaps because CA activity is already maximally increased in these mice and further enhanced synthesis and activation of this enzyme cannot occur. Studies in this laboratory with fluorescent antibodies of mouse CA II have shown that DBA mice have increased amounts as well as activities of CA in their glia but that DBA mice increase the amount but not the activity of CA on multiple treatment with acetazolamide to a greater extent than do C57 mice (Banks, Anderson and Woodbury, unpublished observations). Thus, since acetazolamide does not increase activity on multiple treatment of DBA mice, the results demonstrate that DBA mice cannot readily activate their CA. Since there is evidence that glial cell CA is activated by epinephrine, dibutyryl cyclic AMP, and possibly other neurotransmitters and inhibited by cholinergic agents it is possible that the DBA mice may have deficiencies in transmitters, growth factors or second messengers that have a role in regulating CA synthesis and activity.

ACKNOWLEDGEMENTS

This work was supported by grants 1-R01-NS-21255 from the National Institutes of Neurological and Communicative Disorders and Stroke from the National Institutes of Health, Bethesda, MD, USA and the Epilepsy Foundation of America

[1]Dr. Dixon M. Woodbury is the recipient of a Research Career Award (5-K06-NS-13838) from the National Institute of Neurological and Communicative Disorders and Stroke, NIH, Bethesda MD.

REFERENCES

Anderson, R.E., F.L. Engstrom and D.M. Woodbury (1984). Ann. N.Y. Acad. Sci., 429, 502-504.

Engstrom, F.L., S.Y. Chow, J.W. Kemp and D.M. Woodbury (1984a). Epilepsia, 25, 518-525.

Engstrom, F.L., J.W. Kemp and D.M. Woodbury (1984b). Epilepsia, 25, 759-764.

Engstrom, F.L., H.S. White, J.W. Kemp and D.M. Woodbury (1986a). Epilepsia, 27, In press.

Engstrom, F.L., J.W. Kemp, S.Y. Chow and D.M. Woodbury (1986b). Epilepsia, 27, In press.

Esplin, D.W. and R. Rosenstein (1963). Arch. Int. Pharmacodyn., 143, 498-513.

Koch, A. and D.M. Woodbury (1958). J. Pharmacol. Exp. Ther., 122, 335-342.

Millichap, J.G., D.M. Woodbury and L.S. Goodman (1955). J. Pharmacol. Exp. Ther., 115, 251-258.

Narumi, S., H.K. Kimelberg and R.S. Bourke (1978). J. Neurochem., 31, 1479-1490.

Seyfried, T.N., G.H. Glaser and R.K. Yu (1979). Science, 205, 598-600.

Wistrand, P.J. (1984). Ann. N.Y. Acad. Sci., 429, 195-206.

Woodbury, D.M. (1980). Adv. Neurol., 27, 617-633.

Woodbury, D.M., F.L. Engstrom, H.S. White, C.F. Chen, J.W. Kemp and S.Y. Chow (1984). Ann Neurol. 16 (suppl.), S135-S144.

Woodbury, D.M. and D.W. Esplin (1959). Proc. Assoc. Nerv. Ment. Dis., 37, 24-56.

Woodbury, D.M. and R. Karler (1960). Anesthesiology, 21, 686-702.

Alterations of Glial (Na^+, K^+)-ATPases in Experimental and Human Epilepsy

T. Grisar*,†, D. Guillaume*,† and
A. V. Delgado-Escueta†

*Laboratory of Neurochemistry, University of Liège, Belgium
†Neuroscience Molecular Laboratory, Comprehensive Epilepsy Program, Department of Neurology, U.C.L.A.–School of Medicine, Los Angeles, USA

Abstract

Because high extracellular K^+ concentrations increased glial Na^+ - and K^+ - dependent adenosine triphosphatase activities, while this increase was not observed in neuronal preparations, it is hypothetized that the K^+ released in the extracellular space during neuronal firing is actively taken up by glial cells. In acute and chronic epileptogenic lesions of cats, glial (Na^+ + K^+)-ATPase dramatically decreased when compared to both control animals and the perifocal area, while its activation by extracellular K^+ in concentrations between 3 and 18 mM was absent 3, 6 and up to 45 days after production of freezing lesions. Similar results were observed in 18 specimens of anterolateral temporal neocortex obtained during temporal lobectomies in patients with intractable temporal lobe epilepsy, compared with post-mortem human specimen (n = 8) or control brain tissues (n = 5). Hence, a glial (Na^+, K^+)-ATPase abnormality exists in epileptogenic tissue. The hypothesis that this glial abnormality may favor the transition from interictal episodes to ictal phenomena is proposed.

Key words

Glial cells and Epilepsy; (Na^+, K^+)-ATPase; Human Epilepsy.

Introduction

In the area of cerebral cortex where an epileptogenic spike is recorded, a striking astrogliosis is very often present both in experimental models in animals and in humans.
Because during the last decade several authors have emphasized the importance of glial cells in the control of neuronal excitability, it was reasonable to suspect that astrogliosis present in human epilepsy might have an important role in the generation or spread of partial epilepsy.

Numerous investigators give glial cells an important role in controlling extracellular K^+ concentrations in the nervous system (see Franck et al., 1978; Sykova, 1983 for review). Glia can limit the rise of potassium released during neuronal firing either by passive diffusion through the glial "syncytium" ("spatial buffering") or by an active transport process. Glial (Na^+, K^+)-ATPase of adult brain has been shown, for instance, to be specifically activated between 3-5 and 18-20 m MK^+, this phenomenon was not seen in neuronal preparations (Henn et al., 1972, Grisar et al., 1978;1980). In 1970, Trachtenberg and Pollen and Franck independently proposed that a significant factor in the development of focal epilepsy could be a degradation of this spatial buffering function of the glial cells.

Since this time, electrophysiological results showed conflicting datas, glial cells K^+ spatial buffering function being either intact (Pedley et al., 1976) or diminished (Lewis et al., 1977) or even improved (Ward, 1978) in epileptic foci.

Hence, we decided to analyze the K^+ activation mechanisms of the glial vs neuronal (Na^+, K^+)-ATPase from gliotic and epileptogenic tissue both in animals and humans.

1. THE EPILEPTOGENIC FOCI OF EXPERIMENTAL FREEZING LESION

Freezing a focal area of the cerebral cortex has proven to be an effective way of producing actively discharging acute and chronic epileptogenic lesions in the cat (Escueta et al., 1974).

An abnormal decrease in the pump flux of the K^+ and Na^+ and an abnormal increase in the downhill movement of ions in synaptosomes isolated by discontinuous ficoll gradient from epileptogenic foci of cold lesions in cat, was demonstrated by Escueta et al. (1974). These results prompted several laboratories to investigate the variations of (Na^+, K^+)-ATPase specific activities in the epileptic foci.

Changes in (Na^+, K^+)-ATPase activities within the whole brain fraction have been reported to be either increased (Lewin and Mc Crimon 1967) or decreased (Rapport et al., 1975) and various interpretations have been given to explain these observations. In whole brain fractions of acute freezing lesions, total (Na^+, K^+)-ATPase activity increased with total brain sodium. Total brain potassium decreased (Escueta and Horan, 1980). The increase of (Na^+, K^+)-ATPase activity was considered to be a compensatory change resulting from increased total brain sodium. Consequently, the Na^+, K^+ pump was considered to change as an effect of and not as a cause of epileptogenic discharges. A second explanation proposed for these observations was that all the observed findings were the result of cerebral oedema which is quite prominent in an 8 hour old lesion (Escueta and Horan, 1980). In chronic lesions, reduced enzyme activities had been suspected to be the result mainly of lesion production but similar decreases in the mirror focus argued in favor of reduced enzyme activity being linked to epileptogenesis. Thus, epileptogenic discharges, cerebral oedema, and tissue destruction have all been considered important variables that determine (Na^+, K^+)-ATPase activity. However, the relative importance of these processes, as well as the specific contribution of neurons, synapses and glia to the observed (Na^+, K^+)-ATPase changes, remained uncertain.

To approach this problem more critically, we carried out some experimental modifications (Grisar et al., 1982). Larger freezing lesions and thus more astrogliosis were produced which allowed the successful harvest of glia during cell separation.The rate of ATP

hydrolysis was measured by the rate of liberation of H^+ ions and the method of Grisar et al. (1978) was used for the kinetic analysis of (Na^+, K^+)-ATPase. In particular, $(K^+)_o$ activation of the enzyme was analyzed, the focal gliotic lesion site was studied separately from the surrounding region or the perifocal site and mirror focus. Biochemical analysis was carried out during continuous and sustained 3/sec focal epileptogenic discharges in the primary and mirror focus (Grisar et al., 1982).
Using the hysteretic model for analyzing (Na^+, K^+)-ATPase, the increase in synaptosomal enzyme activities in the acute cold lesion and their decrease in the subacute and chronic epileptogenic states were verified. Several differences were noticeable, however; the enzyme shifts were not as marked and 45 days after lesion production activities increased towards normal levels.
No significant changes were observed in the secondary focus during the acute and chronic epileptogenic states, whereas a slight decrease in enzyme activities was seen in the perifocal area.
Striking changes in the glial (Na^+, K^+)-ATPase activities occured; enzyme activities dramatically decreased in both the primary and secondary foci of acute and chronic lesions. $(K^+)_o$ activation of the glial enzyme in the primary and secondary foci was absent 3, 6 and up to 45 days after lesion production.
Reaction velocities of the glial enzyme between 3 and 18 mM K^+ decreased. Both Vmax and Kmapp were significantly decreased compared to both control animals and the perifocal area (figure 1).
In the perifocal area, glial (Na^+, K^+)-ATPase was also decreased 1-13 days after lesion production, but in subsequent weeks these activities returned to normal levels. K^+ activation of enzyme activities even increased in the perifocal areas after 45 days of seizure activity.
Studies on K^+ activation mechanisms in synaptosomes from the actively firing chronic epileptogenic foci revealed further important observations. K^+ activation constant of synaptosomal (Na^+, K^+) -ATPase was significantly decreased in the actively firing chronic focus. The apparent affinity of the synaptosomal enzyme for K^+, therefore, appeared to be increased within epileptic tissue that was actively firing. Interestingly, while sustained focal paro-

Interestingly, while sustained focal paroxysms could raise synaptosomal (Na^+, K^+)-ATPase, glial (Na^+, K^+)-ATPase and its activation by $(K^+)_o$ remained decreased and this poor response of glial enzyme to $(K^+)_o$ never reversed to "normal" values.

Hence, these experiments provided new information that glial (Na^+, K^+)-ATPase responds to K^+ in a different manner when compared to synaptic enzyme. Glial ATPase and its activation by $(K^+)_o$ remain decreased in either actively discharging acute lesions or in the indolent chronic foci. This might indicate a reduction of the protective ability of glial membranes in $(K^+)_o$ homeostasis.

As already suggested by Dichter et al. (1972), the impairment in glial control of elevated $(K^+)_o$ could be mainly responsible for the transition of discharges to total episodes, with the primary and the secondary foci. Interestingly enough, the inverse interpretation could be proposed for the normal or perifocal area (PF) where the physiological efficiency of glial enzyme remained optimal even at high K^+ concentrations, thus indicating an increase rather than a reduction of the protective ability of glial membranes.

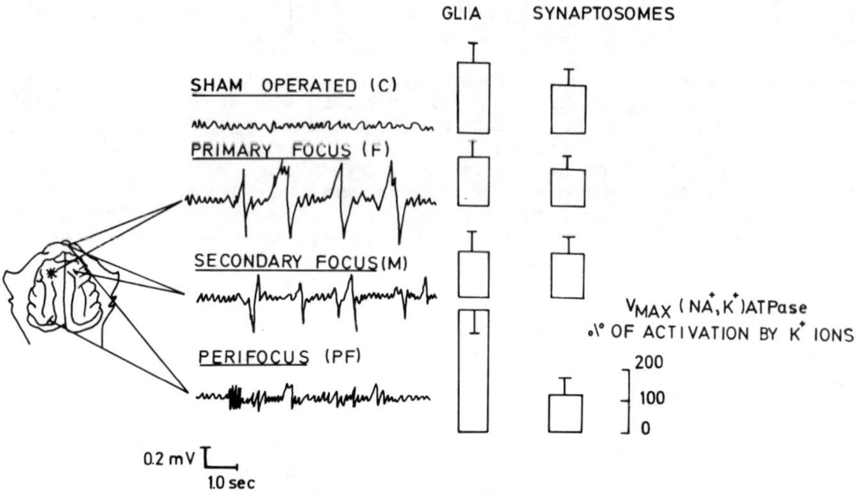

Figure 1.

Left : Electrocorticographic (ECOG) recording of epileptogenic discharges within the primary (F) and secondary or "mirror" (M) focus of the cat with freeze lesion. Note the absence of discharges in control (C) sham operated animals and in PF, i.e., the posterior suprasylvian gyrus homolateral to the lesion.

Right : Histogramm of the maximum velocities (V_{max}) expressed in % of activation of (Na^+, K^+)-ATPase activities measured in presence of 3mM K^+ and considered equal to 100. Values are obtained from Michaelis-Menten analysis (Linear regression from inverse Lineweaver-Bürke plots) obtained from each experiments when enzyme activities are determined in presence of various K^+ concentrations (see Grisar et al., 1982 for details).
Vertical bars represent ± SD.
Note the decrease of K^+ activation of glial enzyme in F and PF and its increase in PF. (See text for discussion).

2. (NA^+, K^+)-ATPASE IN HUMAN TEMPORAL LOBE EPILEPSY.

Since abnormal ATPase activities were found in experimental models of epilepsies, we attempted to determine the K^+-activation mechanisms of (Na^+, K^+)-ATPase in glial and synaptosomes enriched fractions isolated from human temporal lobes which contained a well identified epileptogenic focus.

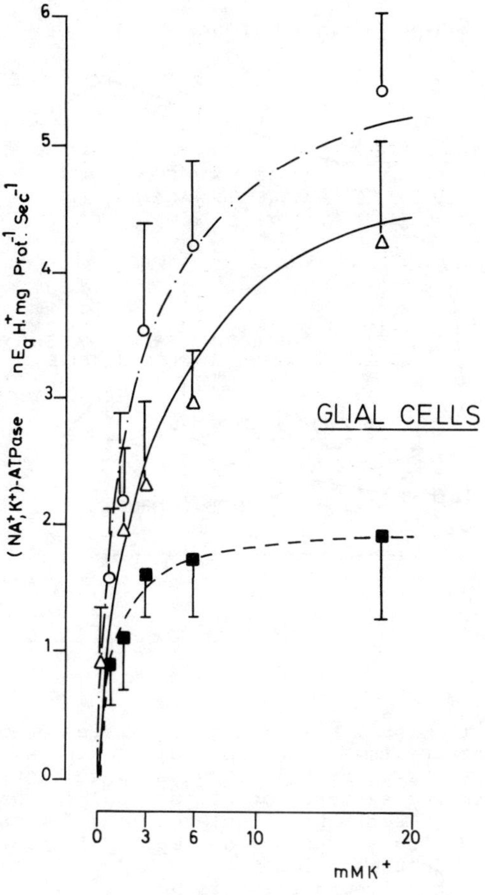

Figure 2.

K^+ influence on the (Na^+, K^+)-ATPase activities of glial cells enriched fractions isolated from frontal (○—·—○) and temporal (△———△) neocortices of post-mortem human brain (n = 8) compared with glial cells isolated from 18 neocortices of temporal lobe origin of patients with intractable partial complex seizures. (●———●)

Note the lack of K^+ sensitivity of the glial enzyme from epileptic patients.

Results are expressed in nEq H^+.mg $prot^{-1}$.sec^{-1}.

Vertical bars represent ± SD.

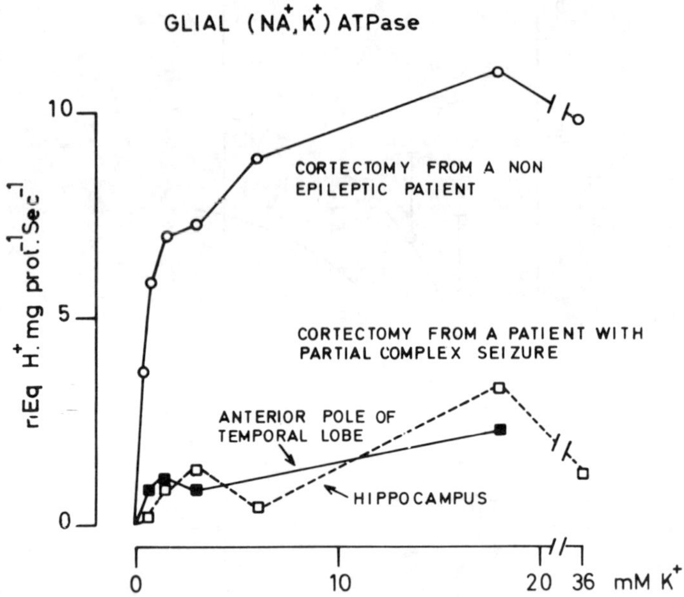

Figure 3.

K^+ influence on the (Na^+, K^+)-ATPase activities of glial cells isolated from hippocampi (□-------□) and anterior poles (■———■) of 2 patients with EM demonstrated hippocampal sclerosis and intractable partial complex seizures of hippocampal-amygdalar origin, compared with enzyme activities isolated from "normal" neocortices. (n = 3) excised during surgery on non-epileptic patients (○———○). Once again, a net decrease of both enzyme activities and sensitivity is observed in epileptogenic tissue.

Our experiments, using highly enriched glial cell fractions isolated from the temporal neocortices of 18 patients suffering of partial complex epilepsy and who have well identified hippocampal-amygdalar or medial temporal and hippocampal sclerosis, successfully demonstrate that glial (Na^+, K^+)-ATPase decreased its total activities and lost or decreased its response to a range of K^+ concentrations previously shown to activate (i.e. between 3 and 18 mM K^+) (Figure 2).

Figure 3 also illustrates that similar results were obtained when comparing glial enzyme activities from hippocampi excised during surgery in epileptic patients with "normal" neocortices excised during surgery in non epileptic patients.
Although synaptosomal (Na^+,K^+)-ATPase also decreased total activities, no significant changes in its reactivity to external K^+ were observed (Grisar et al., 1986, a, b). Using a different technique for preparing glial cells, a similar loss of reactivity by glial (Na^+, K^+)-ATPase was recently reported in human cerebral

epileptogenic foci produced by brain tumors (Kawano et al., 1983). These authors also reported a high degree of correlation between the degree of epileptogenesis of cortical areas with the decreased sensitivity of glial (Na^+, K^+)-ATPase to K^+.

Several possibilities could explain the abnormal (Na^+, K^+)-ATPase observed in these cellular fractions.-

(1) The antiepileptic drug patients were receiving or halothane anaesthesia used during surgery could inhibit the enzyme or alter its regulation by K^+ ions. Although further experimental data are needed before definite and final conclusions can be made, our observations from two additional cases suggest the high probability that these factors are not responsible for the altered glial enzyme property.
Both cases had normal neuropathological findings, were receiving antiepileptic drugs and underwent halothane, endotracheal intubation and the same surgical procedure used in the 18 patients who have abnormal rate reactions for glial (Na^+, K^+)-ATPase. These cases in spite of the antiepileptic drugs, halothane anaesthesia and anterior temporal lobectomy, exhibited K^+ activation mechanisms of glial (Na^+, K^+)-ATPase similar to those reported in normal cats and rabbits brain and in autopsy brains of neurologically normal patients. Interestingly enough these two cases have been reclassified as patients with partial complex seizure of extratemporal origin.

(2) Ischemic hypoxia induced by surgical interruption of pial arterial blood supply prior to en bloc resection could also impair (Na^+, K^+)-ATPase activity. This possibility is also unlikely because glial enzyme had already decreased the total level of glial enzyme activities and its sensitivity to K^+ ions in the temporal neocortex excised before ligation of arterioles and anterior temporal lobectomy (Grisar, 1984).

(3) The third and more likely explanation is that glial and synaptosomal (Na^+, K^+)-ATPase activities are both decreased within the temporal neocortices of human anterior temporal lobes that contained both hippocampal amygdalar epileptogenic foci and hippocampal sclerosis.

Most important perhaps, glial (Na^+, K^+)-ATPase lost its sensitivity to extracellular K^+. These observed abnormalities in enzyme activities are probably part of the biochemical lesions of the underlying disease process.
According to our previous model (Grisar et al., 1982), these findings suggest that glial membranes in human temporal neocortex from patients with partial complex seizures of temporal origin have reduced their capacities to actively control elevated concentrations of external potassium during epileptiform paroxysms.
This may be of great importance in the spread of seizures and accordingly in clinical manifestations of epilepsy. An unresolved issue is whether these biochemical lesions cause epilepsy or are the result of the epileptogenic paroxysms. Our previous experiments on the epileptogenic foci of freezing lesions in cats showing the close relations of ictal epileptogenic activities and impaired glial (Na^+, K^+)-ATPase favor the concept that impaired glial functions play a role in the genesis of epilepsy .

CONCLUSIONS, RESPECTIVE ROLES OF GLIA AND NEURON IN EPILEPTOGENESIS

At the neuronal level, the generation of focal cortical epilepsy is generally attributed either to enhanced physiological responses or to paroxysmal depolarization shifts that lead to the formation of icterictal electroencephalographic spikes. The molecular events that underlie these phenomena remain unknown.

This fundamental biochemical lesion in epilepsy could be of genetic origin. Experimental models of genetic epilepsy and studies of human genetic epidemiology show that single genes can exert major effects on cellular excitability within discrete central neural pathways. In addition, single-locus mutations are now known to produce abnormal gene products in neurons that subsequently generate interictal EEG spikes.
However, neither neuronal paroxysmal depolarization shifts nor interictal EEG spikes are sufficient to produce epilepsy. The clinical expression of epilepsy necessitates recruitment of the surrounding normal neurons, utilization of pathways prepared for the propagation of seizures, and breakdown of mechanisms that limit the spread of seizures. From our experiments, we feel that abnormal glial cells, which contain a defect in their active spatial buffering function, could trigger the transition from the interictal to the ictal state.

Acknowledgements

We are grateful to our patients who allowed us the opportunity to study their epilepsy by consent for removal and study of surgical specimen. We thank A. MINET and M. VERGNIOLLES-BURETTE for their excellent technical assistance and we are grateful to V. LOUIS for word processing technology.
This project has been funded at least in part with Federal funds from the Department of Health and Human Services under contract N° N01-NS-0-2332.

References

1. Escueta AV Delgado, Davidson D. and Reilly EL. (1974) : Potassium transport within synaptosomes isolated from epileptogenic foci. Brain Res. 78 : 223-237.
2. Escueta AV Delgado and Horan MP (1980) : Brain synaptosomes in epilepsy : organization of ionic channels and the Na^+-K^+ pump. In : Antiepileptic Drugs : Mechanisms of action, (eds.) Glaser GH., Penry JK. and Woodbury DM., Raven Press, New York, pp. 85-126.
3. Franck G. (1970) : Echanges cationiques au niveau des neurones et des cellules gliales du cerveau. Arch. Internat. Physiol. Biochim. 78 : 613-720.
4. Franck G., Grisar T., Moonen G. and Schoffeniels E. (1978) : Potassium transport in mammalian astroglia. In : Dynamic Properties of Glia Cells, (eds.) Schoffeniels E., Franck G., Hertz L. and Tower D.B., Pergamon Press, Oxford, pp. 315-325.
5. Grisar T. (1984) : Glial and Neuronal Na^+-K^+ pump in epilepsy. Ann. Neurol. (165) : S128-S134.
6. Grisar T. (1986) : Neuron-glia relationships in human and experimental epilepsy. A biochemical point of view in basic mecha-

nisms of the epilepsies (Escueta et al. eds.), Raven Press, in press.
7. Grisar T. and A.V. Delgado-Escueta (1986) : Glial contribution in human temporal lobe epilepsy : K^+ activation of (Na^+, K^+)-ATPase in bulk isolated glial cells and synaptosomes. Brain Research, in press.
8. Grisar T., Frère JM., Charlier-Grisar J., Franck G., Schoffeniels E. (1978) : Synaptosomal Na^+K^+ATPase is a hysteretic enzyme. FEBS Letters 89 : 173-176.
9. Grisar T., Franck G. and Schoffeniels E. (1978) : K^+ activation mechanisms of the Na^+K^+ATPase of bulk isolated glia cells and neurons. In : Dynamic Properties of glia cells, (eds.) Schoffeniels E., Franck G., Hertz L. and Tower D.B., Pergamon Press, Oxford, pp. 353-363.
10. Grisar T., Franck G. and Schoffeniels e. (1980) : Glial control of neuronal excitability in mammals. II. Enzymatic evidence : two molecular forms of the Na^+K^+ATPase in brain. Neurochem. Internat. 2 : 311-320.
11. Grisar T., Franck G., Escueta A.V. Delgado (1982) : Glial contribution to seizures : K^+ activation of Na^+K^+ATPase in bulk isolated glial cells and synaptosomes of epileptogenic cortex. Brain Research.
12. Henn F.A., Haljamme M. and Hamberger A. (1972) : Glial cell function : active control of extracellular K^+ concentration. Brain Research 43, 437-443.
13. Lewin E. and Mc Crimon A. (1967) : ATPase activity in discharging cortical lesions induced by freezing. Arch. Neurol. 16 : 321-325.
14. Lewis D.V., Mutsuga N., Schuette W.H. and Van Buren J. (1977) : Potassium clearance and reactive gliosis in the alumina gel lesion. Epilepsia 18 : 132-136.
15. Pedley T.A., Fisher R.S., Futamachi K.J. and Prince D.A. (1976) Regulation of extracellular potassium concentration in epileptogenesis. Fed. Proc. 35 : 1254-1259.
16. Rapport R.L., Harris A.B., Friel P.N. and Ojemann G.A. (1975) : Human epileptic brain NaKATPase activity and phenytoin. Arch. Neurol. 32 : 549-554.
17. Sykova E. (1983) : Extracellular K^+ accumulation in the central nervous system. Prog. Biophys. Mol. Biol. 42 : 135-189.
18. Trachtenberg M.C. and Pollen D.A. (1970). Neuroglia : Biophysical properties and physiologic function. Science 167 : 1248-1252.
18. Ward A. (1978). Glia and epilepsy. In : Dynamic Properties of glia cells, (eds.) Schoffeniels E., Franck G., Hertz L. and Tower D.B., Pergamon Press, Oxford, pp. 413-427.

Effects of Ammonia in Primary Astrocyte Cultures Morphological and Biochemical Considerations

M. D. Norenberg, L. W. Mozes,
L.-O. B. Norenberg and J. B. Gregorios

Veterans Administration Medical Center and Departments of Pathology and Neurology, University of Miami, Jackson Memorial Hospital, USA

ABSTRACT

Ammonia is believed to have a significant role in the pathogenesis of a number of neurological disorders. The exact mechanism of ammonia neurotoxicity, however, is still poorly understood. In view of current evidence implicating astrocytes in ammonia metabolism as well as morphological observations suggesting the astrocyte as the target of ammonia toxicity, we have investigated the effects of ammonia on astrocytes in primary culture. Especially with higher concentrations and longer exposure, the results indicate that ammonia induces astrocytic injury associated with functional failure. These astrocytic alterations may contribute to the pathogenesis of ammonia encephalopathy.

KEY WORDS

Ammonia, astrocytes, tissue culture, neurotransmitters, potassium, encephalopathy.

INTRODUCTION

Ammonia is a normal product of cerebral metabolism resulting from the degradation of amines, purines, pyrimidines and amino acids (Kvamme, 1983). The levels of cerebral ammonia often reflect the degree of metabolic activity in the CNS (Tsukada, 1971). Additionally, ammonia is involved in key metabolic pathways, especially those concerned with amino acid neurotransmitters-glutamate and GABA (Benjamin, 1982). Through glutamine, it is important in many biosynthetic reactions (Kvamme, 1983). Ammonia may also play a role in the regulation of pH (Kelley and Kazemi, 1974; Van Gelder, 1982) and osmolarity (Chan and Fishman, 1982) in the CNS.

The continuing interest in ammonia has been chiefly due to its great toxic potential. Abnormally high concentrations of ammonia have been documented to result in various neurologic derangements including seizures and coma (Hindfelt, 1975; Conn and Lieberthal, 1979). Elevated levels in serum (and presumably also in brain) are likewise observed in a variety of clinical conditions including hepatic encephalopathy (Conn and Lieberthal, 1979; Duffy and Plum, 1982), Reye's syndrome (Huttenlocher, et al, 1969), urea cycle enzyme deficiencies (Hsia, 1974), organic acidurias (Stumpf, et al, 1985), ischemia (Ljunggren, et al, 1974) hypoglycemia (Norberg and Siesjo, 1976), hypercapnia (Weyne, et al, 1978) and convulsive disorders (Tsukada, 1971) as well as in a number of iatrogenic conditions.

To prevent its adverse effects on the CNS, ammonia must therefore be converted into a non-toxic compound. In contrast to liver where it is eliminated through the synthesis of urea, in brain it is chiefly detoxified by the amidation of glutamate to form glutamine through the action of glutamine synthetase (GS) (Cooper, et al, 1979). To a lesser degree, the "backward" reaction of glutamate dehydrogenase may be operative resulting in the reductive amination of α-ketoglutarate to form glutamate. These reactions are, however, carried out at considerable energy costs as they consume ATP, NADH and potentially deplete α-ketoglutarate from the tricarboxylic acid cycle.

Such detoxification of ammonia has been shown to be compartmentalized in brain (Berl and Clarke, 1969). This concept, which originated from the finding of anomalous precursor-product relationships, has been significantly strengthened by the immunohistochemical localization of GS in astrocytes (Norenberg, 1983) thus identifying this compartment within the astrocyte. Studies using tissue culture (Patel, et al, 1982; Drejer, et al, 1985) and differential cell injury (Nicklas, et al, 1979) have all been supportive of this concept and the role of the astrocyte in this process. While the majority of experimental data has been in keeping with these views, some synaptosomal activity has been identified (Dennis, et al, 1980). The latter findings, however, must take into consideration the contamination known to occur with these isolation techniques (Henn, et al, 1976).

The mechanism(s) by which ammonia exerts its toxicity remains to be resolved. Bioenergetic failure was an early view originally proposed by Bessman and Bessman (1955). Such failure was postulated to result from the process of ammonia detoxification itself. While some support is available for such a view, presently, there seems to be a consensus that no such energy failure is present in brain (Hawkins, et al, 1973) except perhaps in the brain stem (McCandless and Schenker, 1981) and in certain chronic conditions where failure of the malate-aspartate shuttle has been postulated (Hindfelt, et al, 1977). Inability to identify energy failure, however, has to contend with the issue of compartmentation. Accordingly, a small but critical compartment may be affected but may not be detectable when whole brain preparations are assayed. Thus, while most contemporary studies do not support the concept of bioenergetic failure, such a possibility still cannot be excluded.

Neurotransmitter failure may occur as a result of increased levels of ammonia. Production of "false" neurotransmitters due to defective amino acid transport into the brain has been proposed (James, et al, 1979). Additionally, a decrease in glutamate, an excitatory neurotransmitter, has been found in several models of hyperammonemia (Hindfelt, et al, 1977; Hamberger, et al, 1979). This fall has been explained by its consumption during glutamine synthesis. The inhibition of glutaminase by ammonia has been offered as an alternative explanation (Bradford and Ward, 1976).

Ammonia can result in electrophysiologic derangements. In high doses it can induce depolarization (Binstock and Lecar, 1969) as well as suppression of electrical activity (Benjamin, et al, 1978). At pathophysiologic doses it impairs post-synaptic inhibitory mechanisms (Raabe and Gumnit, 1975). While the latter might explain seizure activity, it does not readily explain coma. It is also unclear if the effect on inhibition is mediated by ammonia on the postsynaptic membrane directly or through an ammonia-induced metabolic effect (Benjamin, et al, 1978; Alger and Nicoll, 1983).

A number of other effects have been attributed to ammonia including: increased spontaneous release of glutamate and decreased evoked release of glutamate (i.e. after depolarization with high doses of K^+) (Hamberger, et al, 1979), changes in tissue electrolytes (Benjamin, et al, 1978), alterations in membrane fluidity (O'Connor, et al, 1984) vasoparalysis (Kindt, et al, 1977), altered blood-brain barrier (Mans, et al, 1983), defects in protein synthesis (Bessman and Pal, 1982) and changes in glutamate receptors (Ferenci, et al, 1984). Additionally, effects have been described on various enzymes such as phosphofructokinase (Lowry and Passoneau, 1966), glutamic dehydrogenase (Kisfaludi, 1975), monoamine oxidase (Sadasivudu and Murthy, 1978), Na^+, K^+-ATPase (Skou, 1960; Schenker, et al, 1967; Subbalakshmi and Murthy, 1981), and glutaminase (Bradford and Ward, 1976).

It is clear, therefore, that ammonia has many and complex actions on the nervous system. However, what may be crucial for the development of toxicity is still unknown.

Our approach in studying the effect of ammonia on the nervous system has been to focus on the astrocyte. There is substantial morphologic data implicating astrocytes in this process (Norenberg, 1981). Additionally, the generally accepted involvement of astrocytes in ammonia metabolism suggests that astrocytes might be the target of ammonia toxicity.

Hepatic encephalopathy, a condition associated with elevated serum and CSF levels of ammonia, has as its principal morphologic feature an astrocyte alteration referred to as the Alzheimer type II change (Adams and Foley, 1953; Norenberg, 1981). Neuronal involvement has been either absent or at best inconsistent. Patients with inborn defects in urea cycle enzymes associated with hyperammonemia and with otherwise intact liver function also display similar histopathological changes (Solitare, et al, 1969; Bruton, et al, 1970). Various animal models of hyperammonemia have also shown the presence of Alzheimer type II astrocytes (Cavanagh and Kyu, 1971; Cole, et al, 1972; Gibson, et al, 1974; Gutierrez and Norenberg, 1975). The significance of this alteration, however, remains an enigma.

In a rat model of hepatic encephalopathy produced through the creation of a portocaval shunt followed by gavage feedings of an ammoniated cation exchange resin, coma and death occurred in about 5-8 days (Norenberg, 1977). The chief pathologic change in this model was the Alzheimer type II change. The electron microscopic features were biphasic. Astrocytic hypertrophy with organelle proliferation was initially seen followed by degenerative changes; the latter corresponded to the Alzheimer type II alteration seen by light microscopy. Similar observations have been noted by a number of investigators (Zamora, et al, 1973; Diemer, 1978; Pilbeam, et al, 1983).

Attempts to study the direct effects of ammonia on astrocytes todate have been meager. The use of bulk-fractionation methods after short-term ammonia treatment have shown decreased glutamic dehydrogenase and glutaminase activities as well as increased GS and aspartate aminotransferase when expressed per cell (Subbalakshmi and Murthy, 1983). However, when expressed per protein no change in GS or in aspartate aminotransferase was found. Using organotypic cultures, Mossakowski, et al (1970) found astrocytic alterations following ammonia treatment. It is unclear from these studies whether the results reflect a direct effect on astrocytes or whether the effects are secondary to a more fundamental change in some other component of the nervous system.

To clarify the significance of the morphologic findings and to more directly explore the effects of ammonia on astrocytes, it was necessary to use a model by which astrocytes could be studied independently of other cellular influences. We believed that the use of primary astrocyte cultures provided the best means to currently explore this problem. Initial studies on the fate of glutamate in primary astrocyte cultures treated with ammonia have been presented (Yu, et al, 1984).

TISSUE CULTURE STUDIES

Primary astrocyte cultures were derived from neonatal rat cortices and grown using a method similar to that described by Booher and Sensenbrenner (1972). The cultures were prepared by mechanical dissociation of rat cerebral cortical tissue, maintained in Dulbecco's modified Eagle's medium supplemented with 5-15% fetal calf serum and antibiotics, and incubated at 37°C in a humidified chamber provided with 5% CO_2 and 95% air. After two weeks of incubation, 0.5 mM dibutyryl cyclic AMP (dBcAMP) was added to the growth medium.

Three week old cultures were then maintained in culture medium containing 0.1 mM dBcAMP and treated daily with ammonium chloride (2, 5 and 10 mM) for 1, 2, 4, 7 and 10 days. Control (without ammonia) and treated cultures were then fixed with neutral formalin, stained with May-Grunwald/Giemsa and examined by light microscopy. Comparable cultures were processed for electron microscopic studies.

Light microscopic findings of control cultures consisted of a monolayered sheet of polyhedral cells with fine threadlike processes, amphophilic cytoplasm and oval or round nuclei containing fine evenly distributed chromatin and one or two small nucleoli (Fig. 1A). A few stellate cells with prominent processes were also noted. Following initial treatment with ammonia, the cultures showed greater basophilic staining and granularity, prominent cytoplasmic processes and cytoplasmic vacuolization (Fig. 1B). Increased nucleolar size and higher nucleolar/nuclear ratio were also noted. Later, the cells demonstrated loss of cytoplasmic processes, formation of many dense bodies and cytoplasmic vacuoles, followed by fragmentation and frank cellular disintegration (Fig. 1C). Alzheimer type II-like cells were sometimes observed within necrotic zones (Fig. 1D). These alterations were proportional to the concentration and duration of ammonia treatment and were reversible following cessation of ammonia treatment.

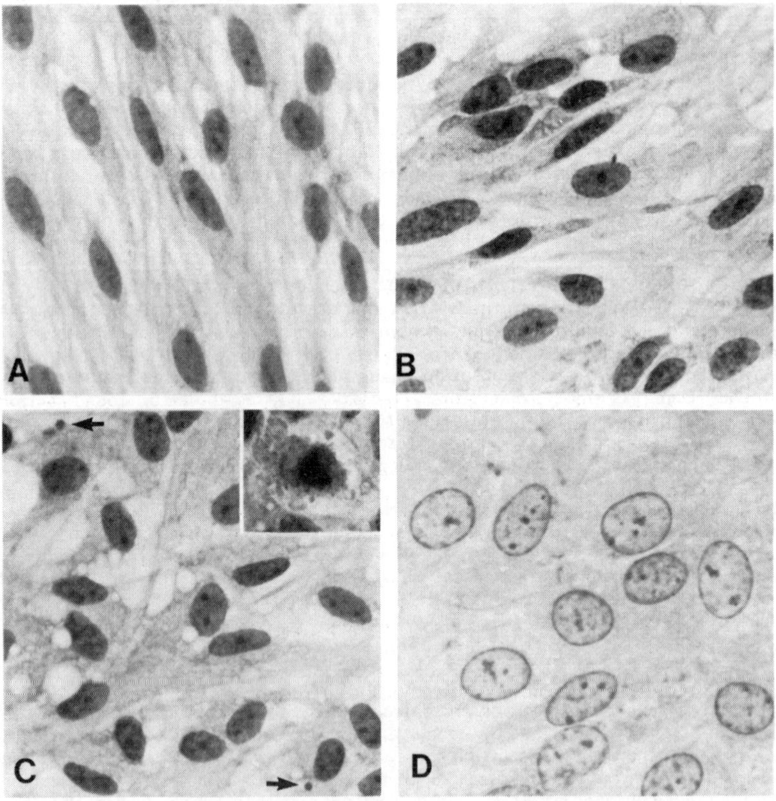

Figure 1. (A) Control culture grown in the presence of dBcAMP consists of a monolayered sheet of cells with round to oval nuclei, inconspicuous nucleoli, amphophilic cytoplasm and fine cytoplasmic processes. (B) Following treatment with 2 mM ammonia for 1 day, the cells show more basophilic cytoplasm, increased cytoplasmic granularity and prominent cytoplasmic processes. (C) With 10 mM ammonia for 1 day, intracytoplasmic vacuoles and dense bodies (arrows) are noted. Frankly disintegrating cells are also seen (inset). (D) Alzheimer type II-like cells with peripheral margination of chromatin and central clearing of the nuclei are noted within necrotic zones. All x 900.

Ultrastructural studies of control cells showed a moderately electron dense cytoplasm and round to oval regular nuclei with fine evenly dispersed chromatin (Fig. 1A). The cytoplasm contained numerous bundles of intermediate filaments and abundant polyribosomal clusters. Mitochondria were few to moderate in number. Cisterns of rough endoplasmic reticulum were inconspicuous. A few randomly scattered microtubules and some aggregates of small fibrillary or homogeneously electron-dense bodies were also noted.

Ammonia-treated cultures showed increased cytoplasmic lucency with disintegration and loss of polyribosomal clusters. Increased numbers and pleomorphism of mitochondria as well as prominent degranulated rough endoplasmic reticulum were noted. There was a significant loss of glial filaments (Fig. 1B). Cytoplasmic vacuoles, both membrane and nonmembrane-bound, were present with 2 mM ammonia and became more numerous with longer exposure. With 10 mM ammonia, the vacuoles were initially more prominent, but later diminished with time. Numerous large heterogeneous cytoplasmic electron-dense bodies were noted even with low ammonia levels, and became much more prominent with higher concentration and longer treatment (Figs. 2C,D). These structures may correspond to the lipofuscin granules present in Alzheimer type II astrocytes in both man (Adams and Foley, 1953) and experimental animals (Voorhies, et al, 1983; M. Norenberg, unpublished observations).

Thus, both light and electron microscopic findings in ammonia-treated cultures showed initial reactive morphologic alterations followed by frank degenerative changes which were directly proportional to the concentration of ammonia. In general, the changes observed in culture were very similar to those seen in vivo thus lending validity to the tissue culture observations and supporting the view of a direct effect of ammonia on astrocytes.

Based on these results, we hypothesized that the initial alteration was a reactive response, probably an adaptive reaction and perhaps reflective of the ammonia detoxification process. This interpretation is in keeping with the presumptive involvement of astrocytes in ammonia metabolism. The ensuing degenerative changes could have resulted from metabolic exhaustion during the process of ammonia detoxification itself, due to energy failure as originally suggested by Bessman and Bessman (1955) although other mechanisms obviously cannot be excluded. These degenerative changes, reflective of cellular injury, may result in an impaired ability to carry out a number of putative functions including uptake of neurotransmitters, regulation of extracellular K^+ concentration, pH regulation and other activities critical for normal activity of the nervous system (Kimelberg, 1983; Hertz, 1982). Preliminary data indeed seems to indicate the occurrence of energy failure in ammonia-treated astrocyte cultures (M. Norenberg, R. Busto and M. Ginsberg, unpublished observations).

Effect of Cyclic AMP

During the conduct of our studies, we observed that primary astrocyte cultures exposed to ammonia showed greater morphologic damage when dBcAMP was omitted from the culture medium. Thus, the findings after one day of exposure to 2 mM ammonia in non-dBcAMP-treated cells very closely resembled those in dBcAMP-treated cells given 10 mM ammonia for 4 days (Fig. 3). Similar "protective" effects were observed when the cultures were given a cyclic AMP analogue (8-bromo-cyclic AMP) as well as agents which raise intracellular levels of cyclic AMP (e.g. isoproterenol, theophylline, IBMX). The addition of butyrate had no effect.

The mechanism involved in this apparent protective action of cyclic AMP on ammonia-treated cultures is not known. Cyclic AMP has a number of well known effects including elevation of GS (Nicklas and Browning, 1983; M. Norenberg and L. Mozes, unpublished observations), carbonic anhydrase and Na^+, K^+-ATPase (Kimelberg, et al, 1978), all of which might protect the cell against ammonia toxicity or enhance the detoxification of ammonia. The possibility that such cAMP-induced effect might have relevance in vivo and its potential therapeutic implication awaits further investigation.

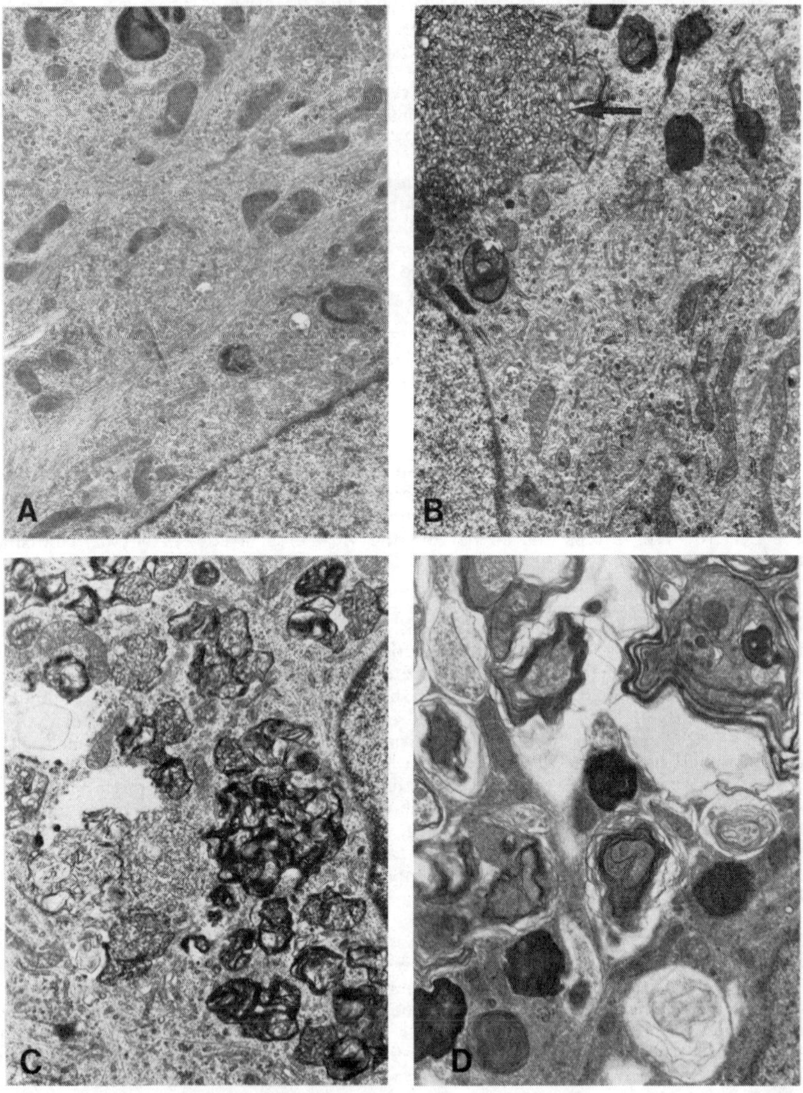

Figure 2. (A) Control cell (day 1) shows prominent bundles of intermediate filaments, polyribosomal clusters, inconspicuous RER cisterns, moderate number of mitochondria and occasional dense body. (B) Cell treated with 2 mM ammonia for 1 day shows slight mitochondrial swelling, loss of glial filaments, many degranulated RER and increased dense bodies. A large curvilinear body is also present (arrow). (C) With 5 mM ammonia for 1 day, dense bodies are more numerous and heterogeneous. Slight mitochondrial swelling, degranulated RER and occasional vacuoles are also seen. (D) After 10 day treatment with 2 mM ammonia, the cytoplasm is filled with myelinoid and dense bodies as well as some membrane-bound spaces. (All x 12,000).

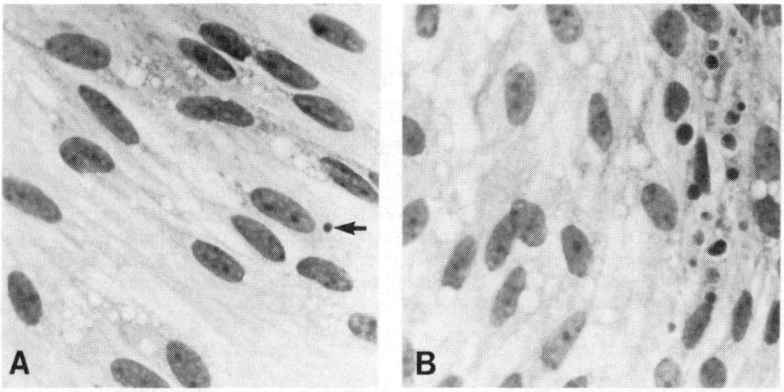

Figure 3. (A) Exposure of cultures to 2 mM ammonia for 1 day in the absence of dBcAMP causes formation of prominent intracytoplasmic vacuoles. Dense body is also present (arrow). (B) Following exposure to 10 mM ammonia for 1 day, numerous intracytoplasmic vacuoles and large dense bodies are noted. These changes are more severe when compared to cultures treated in the presence of dBcAMP (see Figures 1B and C, respectively). Both x 900.

Enzyme Studies

The effect of ammonia on several enzymes was also investigated. The enzymes initially chosen were those that were either astrocyte-specific or believed in some fashion to be involved in the neurotoxicity of ammonia.

Very little effect in GS activity was observed after one day of ammonia treatment even at 10 mM concentration, a finding which is in keeping with studies which show no effect on GS activity by ammonia (Colombo, et al, 1977; Sadasivudu, et al, 1977; Cooper, et al, 1985). However, by day 4 a significant fall was seen with 2 mM ammonia (17.5%) and with 5 mM, a 52.7% fall was noted. This goes along with the view of the impaired capacity of the hyperammonemic brain to further detoxify ammonia (Cooper, et al, 1985). The gradual accumulation of ammonia may serve to further damage other elements of the nervous system.

In contrast to GS, ATPase activity on day 1 showed a progressive rise with greater concentrations of ammonia (at 2 mM a 31.3% rise while at 10 mM, 81.3%). Since ammonia is known to specifically activate ATPase (Skou, 1960) it is unclear if these results represent increased enzyme activity or increased enzyme content. Ammonia is known to induce alterations in electrolytes (i.e., decreased K^+ and increased Na^+) so that the initial enhanced activity in Na^+, K^+-ATPase may actually be a reaction to these electrolyte changes.

Carbonic anhydrase is involved in pH, electrolyte and water regulation (Woodbury, 1980). Considerable activity is found in primary astrocyte cultures (Hertz and Sapirstein, 1978; Kimelberg, et al, 1982). In vivo, it appears to be distributed chiefly in oligodendrocytes (Ghandour, et al, 1979) although recent immunohistochemical studies indicate substantial astroglial content (Anderson, et al, 1984). Preliminary studies using 10 mM ammonia showed increased activity (29.4% at 1 day and 21.2% at 4 days). The rise in carbonic anhydrase activity may be for correcting the electrolyte derangements created by ammonia (vide supra). Additionally, as suggested by Woodbury (1980), it may be involved in anapleurotic reactions by providing HCO_3^- for the synthesis of oxaloacetate via the action of pyruvate carboxylase. This would thus replenish potential losses of tricarboxylic acid cycle intermediates.

Uptake Studies

A key function of astrocytes is believed to be the uptake of the amino acids, glutamate and GABA (Hertz, 1979; Schousboe, 1981). Accordingly, the ability of glial cells to accumulate these putative neurotransmitters was determined. Our studies showed that with 1 day ammonia treatment, the high affinity uptake of glutamate (50 μM) and GABA (5 μM) was slightly increased especially with higher levels of ammonia. However, by day 4, the uptakes were significantly decreased. At 10 mM ammonia GABA uptake decreased by about 35% while glutamate fell by 22%. Thus, as in the morphologic study, the pattern was that of initial activation followed by functional deterioration. ^{86}Rubidium was used for the determination of potassium uptake (1 mM). At 5 mM ammonia, ^{86}Rubidium uptake fell by about 35% while at 10 mM, it decreased by almost 100%. It thus appears that an early and dramatic effect of ammonia is an impaired capacity of these cells to clear K^+ from the extracellular space.

Potassium Content

The intracellular content of this ion was determined since Benjamin, et al (1978) had previously shown its fall in brain slices after ammonia treatment. Comparable results were obtained in primary astrocyte cultures. Ammonia induced a 10% fall in dBcAMP-treated cells, and an even greater loss (40%) in non-dBcAMP-treated cells. These effects were similarly noted after 1 and 4 days of ammonia exposure.

SUMMARY

Our preliminary studies indicate that the tissue culture model may provide a useful means for resolving the role of the astrocyte in ammonia encephalopathy. The morphologic changes, which are similar to those observed in vivo in hyperammonemic animals, indicate an initial reactive adaptive change followed by degenerative changes. The latter progressed with increasing dosages and prolonged exposure. Cyclic AMP seemed to exert a beneficial effect but the mechanism of this protection is unclear.

The enzymatic and uptake studies, in general, correlated with the morphologic findings showing either increased or no change in activity early on followed by depressed activity in the various parameters evaluated. These studies support the view that to a large extent, ammonia toxicity may result from astroglial functional failure. Such findings, of course, do not rule out other important effects of ammonia on the nervous system. Nevertheless, the correlation between the in vivo and in vitro findings along with enzymatic and functional observations do support the view that astrocytes may have a critical role in the pathogenesis of ammonia neurotoxicity. Future studies using neuronal and mixed glial-neuronal cultures should clarify the contribution of various cell types to this process.

REFERENCES

Adams, R.D. and J.M. Foley (1953). Res. Publ. Assoc. Nerv. Ment. Dis. 32, 198-237.
Alger, B.E. and R.A. Nicoll (1983). J. Neurophys. 49, 1381-1391.
Anderson, R.E., F.L. Engstrom and D.M. Woodbury (1984). Ann. N.Y. Acad. Sci. 429, 502-504.
Benjamin, A.M. (1982). Handbook of Neurochemistry, 2nd ed., A. Lajtha (Ed.), Plenum Press, New York, Vol 1, 117-137.
Benjamin, A.M., K. Okamoto and J.H. Quastel (1978). J. Neurochem. 30, 131-143.
Berl, S. and D.D. Clarke (1969). Handbook of Neurochemistry, 1st ed., A. Lajtha (Ed.), Plenum Press, New York, Vol 2, 447-472.
Bessman, S.P. and A.N. Bessman (1955). J. Clin. Invest. 34, 622-628.
Bessman, S.P. and N. Pal (1982). Isr. J. Med. Sci. 18, 171-175.
Binstock, L. and H. Lecar (1969). J. Gen. Physiol. 53, 342-361.
Booher, J. and M. Sensenbrenner (1972). Neurobiol. 2, 97-105.
Bradford, H.F. and H.K. Ward (1976). Brain Res. 110, 115-125.
Bruton, C.J., J.A.N. Corsellis and A. Russell (1970). Brain 93, 423-434.
Cavanagh, J.B. and M.H. Kyu (1971). J. Neurol. Sci. 12, 63-75.

Chan, P.H. and R.A. Fishman (1979). Brain Res. 161, 293-301.
Cole, M., R.B. Rutherford and F.O. Smith (1972). Arch. Neurol. 26, 130-136.
Colombo, J.P., C. Bachman, E. Peheim and J. Beruter (1977). Enzyme 22, 399-406.
Conn, H.O. and M.L. Lieberthal (1979). The Hepatic Coma Syndromes and Lactulose, Williams and Wilkins, Baltimore.
Cooper, A.J.L., J.M. McDonald, A.S. Gelbard, R.F. Gledhill and T.E. Duffy (1979). J. Biol. Chem. 254, 4982-4992.
Cooper, A.J.L., S.N. Mora, N.F. Cruz and A.S. Gelbard (1985). J. Neurochem. 44, 1716-1723.
Dennis, S.C., J.C.K. Lai and J.B. Clark (1980). Brain Res. 197, 469-475.
Diemer, N.H. (1978). Acta. Neurol. Scand. Suppl. 58, 1-144.
Drejer, J., O.M. Larsson, E. Kvamme, G. Svenneby, L. Hertz and A. Schousboe (1985). Neurochem. Res. 10, 49-62.
Duffy, T.E. and F. Plum (1982). The Liver: Biology and Pathobiology, I. Arias, et al, (Eds.), Raven Press, New York, 693-715.
Ferenci, P., S.C. Pappas, P.J. Manson and E.A. Jones (1984). Hepatology 4, 25-29.
Ghandour, M.S., O.K. Langley, G. Vincendon and G. Gombos (1979). J. Histo. Cyto. 27, 1634-1637.
Gibson, E.G., A. Zimber, L. Krook, E.P. Richardson Jr. and W.J. Visek (1974). J. Neuropathol. Exp. Neurol. 33, 201-211.
Gutierrez, J.A. and M.D. Norenberg (1975). Arch. Neurol. 32, 123-126.
Hamberger, A., B. Hedquist and B. Nystrom (1979). J. Neurochem. 33, 1295-1302.
Hawkins, R.A., A.L. Miller, R.C. Nielsen and R.L. Veech (1973). Biochem. J. 134, 1001-1008.
Henn, F.A., D.J. Anderson and D.G. Rustad (1976). Brain Res. 101, 341-344.
Hertz, L. (1979). Prog. Neurobiol. 13, 277-323.
Hertz, L. (1982). Handbook of Neurochemistry, 2nd. ed., A. Lajtha (Ed.),Plenum Press, New York, Vol 1, 319-355.
Hertz, L., V. Sapirstein (1978). In Vitro 14, 376.
Hindfelt, B. (1975). Ann. New York Acad. Sci. 252, 116-123.
Hindfelt, B., F. Plum. and T.E. Duffy (1977). J. Clin. Invest. 59, 386-396.
Hsia, Y.E. (1974). Gastroenterology 67, 347-374.
Huttenlocher, P.R., A.D. Schwartz and G. Klatskin (1969). Pediatrics 43, 443-454.
James, J.H., V. Ziparo, B. Jeppsson and J.E. Fischer (1979). Lancet 2, 772-775.
Kelley, M.A. and H. Kazemi (1974). Resp. Physiol. 22, 345-359.
Kimelberg, H.K. (1983). Cell Molec. Neurobiol. 3, 1-16.
Kimelberg, H.K., S. Narumi and R.S. Bourke (1978). Brain Res. 153, 55-77.
Kimelberg, H.K., P.E. Stieg and J.E. Mazurkiewicz (1982). J. Neurochem. 39, 734-742.
Kindt, F.W., M. Brock, L.L. Altenau and W. Poll (1977). Lancet 1, 201.
Kisfaludy, S. (1975). Acta Med. Acad. Scient. Hungar. 32, 279-287.
Kvamme, E. (1983). Progr. Neurobiol. 20, 109-132.
Ljunggren, B., H. Schultz and B.K. Siesjo (1974). Brain Res. 73, 277-289.
Lowry, O.H. and J.V. Passoneau (1966). J. Biol. Chem. 241, 2268-2279.
Mans, A.M., J.F. Biebuyck and R.A. Hawkins (L983). Am. J. Physiol. 245, C74-C77.
McCandless, D.W. and S. Schenker (1981). Exp. Brain Res. 44, 325-330.
Mossakowski, M.J., K. Renkawek, Z. Krasnicka, M. Smialek, A. Pronaszko (1970). Acta Neuropathol. 16, 1-16.
Nicklas, W.J. and E.T. Browning (1983). J. Neurochem. 41, 179-187.
Nicklas, W.J., R. Nunez and S. Berl (1979). J. Neurochem. 33, 839-844.
Norberg, K. and B.K. Siesjo (1976). J. Neurochem. 26, 345-352.
Norenberg, M.D. (1977). Lab. Invest. 36, 618-627.
Norenberg, M.D. (1981). Advances in Cellular Neurobiology, S. Fedoroff and L. Hertz (Eds.), Academic Press, New York, Vol 2, 303-352.
Norenberg, M.D. (1983). Glutamine, Glutamate and GABA in the Central Nervous System, L. Hertz, et al,(Eds.), Alan R. Liss Inc., New York, 95-111.
O'Connor, J.E., C. Guerri and S. Grisolia (1984). Biochem. Biophys. Res. Commun. 119, 516-523.
Patel, A.J., A. Hunt, R.D. Gordon and R. Balazs (1982). Develop. Brain Res. 4, 3-11.
Pilbeam, C.M., R. McD. Anderson and P.S. Bhathal (1983). J. Pathol. 140, 347-355.
Raabe, W. and R.J. Gumnit (1975). J. Neurophys. 38, 374-355.
Sadasivudu, B. and C.R.K. Murthy (1978). Arch. Int. Physiol. Biochim. 86, 67-82.

Sadasivudu, B., T.I. Ras, C. Radhakrishna and C.R. Murthy (1977). Neurochem. Res. 2, 639-655.

Schenker, S., D.W. McCandless, E. Brophy and M.S. Lewis (1967). J. Clin. Invest. 46, 838-848.

Schousboe, A. (1981). Int. Rev. Neurobiol. 22, 1-45.

Skou, J.C. (1960). Biochem. Biophys. Acta 42, 6-23.

Solitare, G.B., V.E. Shih, D.J. Nelligan and T.F. Dolan (1969). J. Ment. Defic. Res. 13, 153-170.

Stumpf, D.A., W.D. Parker Jr. and C. Angelini (1985). Neurology 35, 1041-1045.

Subbalakshmi, G.Y.C.V. and C.R.K. Murthy (1981). Biochem. Pharmacol. 30, 2127-2130.

Subbalakshmi, G.Y.C.V. and C.R.K. Murthy (1983). Neurochem. Int. 5, 593-597.

Tsukada, Y. (1971). Handbook of Neurochemistry, 1st ed., A. Lajtha (Ed.), Plenum Press, New York, Vol 5, 215-233.

Van Gelder, N.M. (1982). Adv. Exp. Med. Biol. 153, 501-508.

Voorhies, T.M., M.E. Ehrlich, T.E. Duffy, C.K. Petito and F. Plum (1983). Ped. Res. 17, 970-975.

Weyne, J., F. Van Leuven, H. Kazemi and I. Leusen (1978). J. Appl. Physiol. 44, 333-339.

Woodbury, D.M. (1980). Antiepileptic Drugs: Mechanism of Action, G.H. Glaser, et al, (Eds.), Raven Press, New York, 617-633.

Yu, A.C.H., A. Schousboe and L. Hertz (1984). J. Neurochem. 42, 594-594.

Zamora, A.J., J.B. Cavanagh and M.H. Kyu (1973). J. Neurol. Sci. 18, 25-45.

Early Metabolic Changes in Astrocytes in Experimental Hepatogenic Encephalopathy

J. Albrecht

Department of Neuropathology, Medical Research Centre, Polish Academy of Sciences, Warsaw, Poland

KEYWORDS

Rat; hepatogenic encephalopathy; thioacetamide; astrocytes; synaptosomes; RNA synthesis; protein synthesis; Na^+/K^+ ATPase; GABA transport; calcium transport.

INTRODUCTION

The term "hepatogenic encephalopathy" /HE/ describes a wide spectrum of disturbances of the central nervous system related to liver dysfunction. It is noteworthy that, depending upon the criteria applied, HE may be defined in two different ways. By clinical criteria, it is a disease of neural transmission, which in acute or terminal stages assumes the form of hepatic coma. On the other hand, by morphological criteria, it may be called a disease of astrocytes. As a rule, in the course of HE astrocytes proliferate and/or transform into different types of cells, whereas neurons remain relatively intact /for reviews see Diemer, 1977; Mossakowski, 1978; Norenberg, 1981/. In most of the reports, these two facets of the disease have been considered separately, as if they were independent phenomena. However, in light of the growing evidence of the role of astrocytes in neural control, it was tempting to speculate that, the "neuronal" and the "astrocytic" aspect of the disease are interrelated. The present studies were undertaken to verify this speculation.

BASIC ASSUMPTIONS OF THE EXPERIMENTAL APPROACH

The selection of the experimental model to study HE was based on the prediction that, if the morphological and functional changes have common causes, these are most likely to be detected in the very early stage, preceding sound morphological and clinical manifestations of the disease. In this study, early HE was produced in rats by two administrations of a strong hepatotoxin - thioacetamide. The procedure and the characteristics of HE in this model are outlined in the next section.

Since astrocytes are poorly accessible to direct in situ testing, an *in vitro* approach was adopted based on investigating cell fractions enriched in astrocytes. The fractions were obtained with a modification of the method of Farooq and Norton /1978/, yielding preparations with less than 20% contamination by neuronal elements /Albrecht, *et al*, 1982/. To evaluate the cell specificity of the early changes in HE, parallel tests were performed with highly purified preparations of synaptosomes /Troeger, *et al*, 1984/, representing the neuronal partner of astrocytes across the synaptic cleft.

CHARACTERISTICS OF EARLY HE INDUCED BY THIOACETAMIDE /TAA/

Female Wistar rats weighing 160-200 g received two injections of TAA in physiological saline /250 mg/kg body weight/ at 24 h intervals and were examined 24 h after the last injection. At that time, the clinical symptoms in the animals were confined to moderately reduced motor activity and slightly decreased reaction to audial and visual stimuli. ECoG recordings showed only slightly increased frequency and decreased amplitude, which does not correspond with advanced HE /Pluta and Albrecht, 1984/. There were scattered areas of centrilobular necrosis in the liver, but no sound morphological or ultrastructural changes in the brain, especially of degenerative type /Hilgier, *et al*, 1983; Mossakowski and Borowicz, 1985/.

The changes in blood and brain characterizing the advancement of HE in this model are listed in Table 1.

Table 1. Changes in blood and brain in early HE induced by thioacetamide.

BLOOD		
	increased:	- ammonia />200%/[1]
		- phenols />50%/[1]
BRAIN		
	increased:	- glutamine />100%/[2]
		- glutamine synthetase />25%/[1]
		- carbonic anhydrase />20%/[3]
		- octopamine />30%/[4]
		- phenylethanolamine />20%/[4]
	decreased:	- CBF /$\pm$50%/[5]
	unchanged:	- ammonia[3]
		- α-oxoglutarate[2]
		- $CMRO_2$[5]

[1]Hilgier, *et al*, 1983; [2]Hilgier, 1983; [3]Albrecht and Hilgier, 1984; [4]Hilgier, *et al*, 1985; [5]Pluta and Albrecht, submitted.

The here observed hyperammonemia is considered to be the earliest and most reproducible marker of the onset of HE, and increased glutamine synthesis in the brain is a typical reaction to increased ammonia load. In turn, the increase of blood phenols is indicative of amino acid imbalance in favor of aromatic amino acids, whereas elevation of octopamine and phenylethanolamine in the brain reflects disturbed metabolism of these amino acids /Zieve, 1981/. Interestingly, hyperammonemia was not accompanied by changes in brain ammonia, indicating efficient ammonia detoxication, which is consistent with the increase of brain glutamine. Furthermore, there was no change in the brain α-oxoglutarate content, suggesting adequate supply of tricarboxylic acid intermediates. The activation of carbonic anhydrase has been suggested to counteract the outflow of α-oxoglutarate by increasing the rate of CO_2 fixation /Albrecht and Hilgier, 1984/. The absence of severe energetic failure in the brain was also manifested by unchanged oxygen consumption, which was apparently not affected by reduction of the cerebral blood flow, already manifested at this stage. The results taken together may be interpreted to reflect the early,"metabolic" stage of HE, with relatively insignificant advancement of morphological changes, Hence, in accordance with the criteria outlined in the previous section, this experimental model may be considered suitable for examining the early pathogenesis of the disease.

RNA AND PROTEIN SYNTHESIS IN ASTROCYTES

The progressive character of morphological changes in astrocytes at advanced stages of HE gave rise to the hypothesis that, they may be preceded by changes in the potential of these cells for macromolecular syntheses. An earlier study revealed increased RNA synthesis in astroglial cell nuclei in 2 models of chronic HE /Albrecht, 1981/. Studies with the present model have confirmed this more directly with respect to early HE: astrocytes from the "diseased" rats showed an increased above control ability to synthesize RNA and proteins <u>in vitro</u>: the increase was 20% for RNA synthesis and more than 50% for protein synthesis /Albrecht, 1984/. It will be of interest to examine, whether the enhanced macromolecular synthesis in astrocytes is instrumental in increasing the activity of glutamine synthetase /Table 1/, which is an astroglia-specific protein /Norenberg and Martinez-Hernandez, 1979/.

Na^+/K^+ ATPase ACTIVITY

In the investigations described below, attention was focused on the processes involved in what may be defined as the "buffering" or "homeostatic" function of astrocytes, that is the transport of ions and neurotransmitters. At this point, a comparison with synaptosomes was carried out to evaluate the tissue-specificity of the astroglial reaction. In the first instance, the Na^+/K^+ ATPase was studied, since a/ - the sodium pump is the driving force in the majority of transport processes and b/ - the enzyme in astrocytes is particularly active and responsive to changes in the extracellular ions concentrations /this matter was reviewed at this Symposium by Dr T. Grisar/.

Induction of early HE in the present model resulted in the about 40% increase of the astroglial enzyme activity, while the activity in synaptosomes was unaffected, indicating that the activation was astroglia-specific /Table 2/.

Table 2. Na^+/K^+ ATPase activity, ouabaine-sensitive ^{86}Rb uptake and GABA uptake in astrocytes and synaptosomes from control and HE rats.

Fraction	Group	Na^+/K^+ ATPase[1]	^{86}Rb uptake[2]	GABA uptake[3]
Astrocytes	Control	41±5 /6/	1.1±0.2 /3/	330±22 /4/
	HE	56±6 /7/	4.9±0.7 /4/	600±130 /4/
Synaptosomes	Control	28±5 /9/	-	1230±9 /4/
	HE	25±2 /9/	-	1220±12 /4/

Results are means ± SD with number of experiments in parentheses. [1]µmoles Pi/mg protein/h /Albrecht, et al, 1985/; [2]nmoles/mg protein/5 min. Ouabain-sensitivity tested at 1 mM ouabain /Dąbrowiecki and Albrecht, in preparation/; [3]pmoles/mg protein/5 min. More details in Albrecht, et al /1985/.

The effect of HE appeared much more pronounced when the enzyme activity was determined more directly, by measuring the ouabain--sensitive ^{86}Rb uptake to astrocytes: an almost 5-fold increase was observed /Table 2/. The above results indicate that, astrocytes in early HE are capable of more active clearance of potassium ions from the extracellular space of the central nervous system.

GABA UPTAKE

Astrocytes have been shown to be capable of inactivating neurotransmitters - in particular the amino acid neurotransmitters GABA and glutamate - by active uptake /see the article by Dr A. Schousboe in this volume/. It was decided to study the GABA uptake in the first instance, because GABA, as an inhibitory neurotransmitter, has been implicated in the pathogenesis of HE /Jones, et al, 1984/. The effects of HE on this process were strikingly similar to those observed with Na^+/K^+ ATPase: the GABA uptake by astroglia increased by about 60%, while that by synaptosomes remained unaffected /Table 2, more details in Albrecht, et al, 1985/. It may thus be inferred that, astrocytes in early HE are stimulated towards more active removal of GABA from the extracellular space. In terms of the control of neuronal function, this result again points to the increased buffering capacity of astrocytes.

GABA RELEASE

GABA release by glial cells has been a matter of dispute. Depending upon the type of preparation used, astrocytes have been either claimed /Sellström, et al, 1977/ or denied /Pearce, et al, 1981/, to be capable of releasing GABA upon depolarization. This controversial issue was taken up by Dr Rafałowska from the Department of Neurochemistry and myself with the assumption that, if the process does operate indeed and undergoes changes due to HE, it might interfere with the GABAergic transmission.

Control astrocytes preloaded with radiolabelled GABA showed a slow basal GABA efflux at low potassium concentration, and no increase was observed at high potassium. In contrast, the astrocytes from rats with HE responded to depolarization by high K^+ with a significant release of GABA /Fig. 1A/. On the other hand, the release from synaptosomes was typically K^+-dependent and was not influenced by the pathological condition /Fig. 1B/. The results indicate that astrocytes in early HE acquire the ability to react to cell membrane depolarization with enhanced GABA release into the extracellular space and that this is an astroglia-specific response.

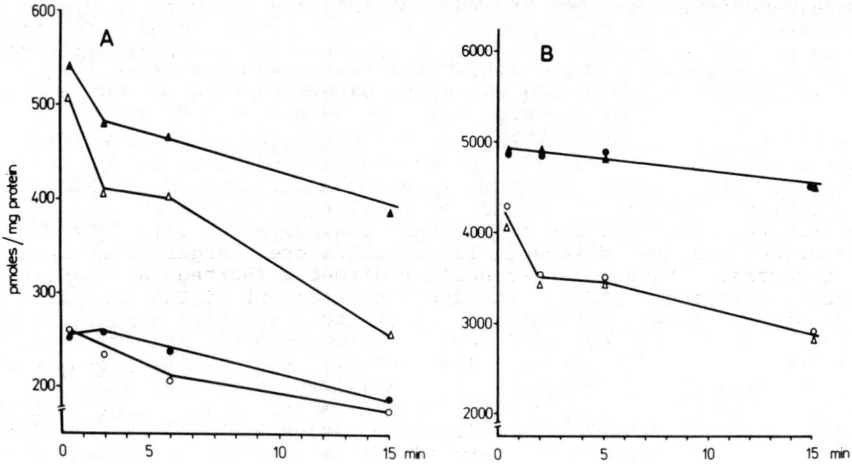

Fig. 1. GABA release from astrocytes /A/ and synaptosomes /B/ of control and TAA--treated rats at low and high concentrations of potassium. ● - control, 5 mM K^+; o - control, 75 mM K^+; ▲ - TAA, 5 mM K^+; Δ - TAA, 75 mM K^+. Results representative for 4 determinations. Astrocytes or synaptosomes were preincubated at 5 mg protein/ /ml with U-^{14}C GABA /226 mCi/mmol, Amersham/ plus 2 mM aminooxyacetic acid at 25°C for 15 min. At times indicated, 400 µl samples were withdrawn, diluted ten--fold and centrifuged through silicon oil.

As to the functional meaning of the GABA release by astrocytes, it has been hypothesized that this process may counteract generation of excessively high interglial GABA concentrations /Bowery, et al, 1979/. If it is so, the activation of GABA release during HE would serve to prevent excess accumulation of the amino acid in astrocytes, which is likely to result from inhibition of GABA transamination accompanying this disease /Ferenci, et al, 1984/. On the other hand, since the HE-induced release of GABA is several times less active than the synaptosomal release, the net effect of the former on the GABAergic transmission may not be too significant.

CALCIUM UPTAKE

By analogy to the nerve endings, it has been postulated that the potassium-dependent GABA release by astrocytes is triggered by inward calcium transport /reviewed by Sarthy, 1983/. The present study indicates that there is a good correlation of the dependences of the two processes on HE. The calcium uptake into control astrocytes was virtually the same at low and high K^+, while in astrocytes from HE rats it was stimulated 1.5 fold by high K^+. In contrast, the K^+-dependence of the synaptosomal release was not potentiated by HE; in fact there was a net reduction of the K^+ effect, which might be due to the considerable "leakage" of HE synaptosomes to calcium at low K^+ /Table 3/.

Table 3. Calcium uptake into astroglia-enriched fraction and synaptosomes of control and TAA-treated rats at low and high K^+ concentrations.

Fraction	Group	$^{45}Ca^{2+}$ uptake /nmoles/mg protein/		Uptake ratio
		5 mM K^+	75 mM K^+	75 mM K^+/5 mM K^+
Astroglia	Control	0.80±0.05[1]	0.74±0.05	0.92±0.12
	TAA	1.02±0.08**	1.57±0.20	1.54±0.32*
Synaptosomes	Control	1.96±0.17	3.39±0.05	1.72±0.17
	TAA	2.59±0.10**	3.77±0.20	1.45±0.13***

Astrocytes or synaptosomes /0.5 mg protein/ml/ were incubated with ^{45}Ca /0.19 mCi/mmol, OPiDi, Poland/ in Krebs-Henseleit medium including 1.2 mM Ca^{2+} and low /5 mM/ or high /75 mM/ K^+ for 1 min at 37°C. The incubations were terminated by EGTA-containing stopping solution and the suspension was rapidly filtered through Whatman GF/F glass microfiber filters. [1]Mean ± SD of 3 experiments; *$p < 0.05$; **$p < 0.01$; ***$p < 0.001$, evaluated with the Student's t-test /Dąbrowiecki and Albrecht, in preparation/.

Since this study has not accounted for the calcium-independent GABA release /Sarthy, 1983/, it would be premature to conclude that the increased calcium uptake is actually the trigger of GABA release in HE astrocytes. Experiments are in progress to clarify this point.

The increased calcium transport in HE astrocytes deserves detailed examination considering the presumed role of calcium in the control of membrane potential /Sihra, et al, 1984/. In this context, it is noteworthy that HE in this model leads to hyperpolarization of the astroglial cell membranes /Dąbrowiecki and Albrecht, 1985/.

CONCLUSIONS AND PERSPECTIVES

As for now, the results allow to conclude that:
- The early metabolic changes in HE appear to be astroglia-specific, which supports morphological evidence that HE primarily effects the astrocytes.
- The enhancement of macromolecular syntheses in astrocytes in early HE is consistent with the progressive character of morphological changes at more advanced stages of the disease.
- Astrocytes in early HE appear to respond more actively /in terms of their buffering function - more effectively/ to intra- and extracellular changes in the ions' and neurotransmitters' content, with possible implications for neural transmission.

As to the latter conclusion it will be of interest to see whether the changed conditions of astroglial transport concern other neurotransmitters as well. This holds in particular for the excitatory counterpart of GABA - glutamate, since imbalance between GABAergic and glutaminergic transmission has been considered a crucial factor in the pathogenesis of HE /Ferenci, et al, 1983/.

A number of basic questions remain to be answered, and this will require implementation of different experimental approaches, including in vivo studies. For instance, it is not possible to conclude from the present study, what is the quantitative impact of the activation of astroglia during HE on neural transmission. Assuming that it is of significance, it is not clear whether the changes reflect meaningful adaptation of the cells which is beneficial for the neuronal function, or uncontrolled hyperexcitation of astroglia, which would rather contribute to the further advancement of the pathological process. Detailed examination of the properties of astrocytes at more advanced stages of the disease will help to answer these questions.

ACKNOWLEDGEMENT

The following colleagues have participated in the project: Zbigniew Dąbrowiecki, Wojciech Hilgier, Zuzanna Kraśnicka, Ryszard Pluta, Urszula Rafałowska, Halina Weinrauder and Urszula Wysmyk--Cybula. Special thanks are due to Professor Mirosław Mossakowski for encouragement and stimulating discussions. Excellent technical assistance of Sławomir Januszewski, Inez Kolsicka, Aleksandra Lenkiewicz, Teresa Pańkowska and Mirosława Poławska is gratefully acknowledged. Supported by grants 10.4.05.1.1 and 10.4.05.5.1 from the Polish Academy of Sciences.

REFERENCES

Albrecht, J. /1981/. J. Neurosci. Res., 6, 553-558.
Albrecht, J. /1984/. Acta Neurol. Scand., 70, 314-316.
Albrecht, J. and W. Hilgier /1984/. Neuroscience Letters, 45, 7-10.
Albrecht, J., W. Hilgier, J. Ułas and U. Wysmyk-Cybula /1982/. Neurochem. Res., 7, 513-517.
Albrecht, J., U. Wysmyk-Cybula and U. Rafałowska /1985/. Acta Neurol. Scand., in the press.
Bowery, N.G., D.A. Brown and S. Marsh /1979/. J. Physiol. /London/, 293, 75-101.
Dąbrowiecki, Z. and J. Albrecht /1985/. Neurochem. Res., 10, 315-318.
Diemer, N.H. /1978/. Acta Neurol. Scand., 58, suppl. 71, 1-144.
Farooq, M. and W.T. Norton /1978/. J. Neurochem., 31, 887-894.

Ferenci, P., D.F. Schafer, S.C. Pappas and E.A. Jones /1983/. in New aspects of clinical nutrition /eds. Kleinberger G. and E. Deutsch/, Karger, Basel, 485-504.
Ferenci, P., R. Jacobs, S.C. Pappas, D.F. Schafer and E.A. Jones /1984/. J. Neurochem., 42, 1487-1490.
Hilgier, W. /1983/. Neuropat. Pol., 21, 511-518.
Hilgier, W., J. Albrecht and Z. Kraśnicka /1983/. Neuropat. Pol., 21, 487-494.
Jones, E.A., D.F. Schafer, P. Ferenci and S.C. Pappas /1984/. The Yale J. Biol. Med., 57, 301-316.
Mossakowski, M.J. /1978/. Riv. Patol. Nerv. Ment., 37, 25-36.
Mossakowski, M.J. and J.W. Borowicz /1985/. Neuropat. Pol., 23, in the press.
Norenberg, M.D. and A. Martinez-Hernandez /1979/. Brain Res., 161, 303-310.
Norenberg, M.D. /1981/. in Adv. Cell. Neurobiol., 2, 303-352.
Pearce, B.R., D.N. Currie, R. Beale and G.R. Dutton /1981/. Brain Res., 206, 485-489.
Pluta, R. and J. Albrecht /1984/. Neuropat. Pol., 22, 379-385.
Sarthy, P.V. /1983/. J. Neuroscience, 3, 2494-2503.
Sellström, A. and A. Hamberger /1977/. Brain Res., 119, 189-198.
Sihra, T.S., I. Scott and D.G. Nicholls /1984/. J. Neurochem., 43, 1624-1630.
Troeger, M.B., U. Rafałowska and M. Erecińska /1984/. J. Neurochem., 42, 1735-1742.
Zieve, L. /1981/. Hepatology, 1, 360-165.

Cellular and Molecular Aspects of Glial Cells with Aging

A. Vernadakis, N. Sakellaridis, D. Mangoura, D. Davies and K. Parker

Departments of Psychiatry and Pharmacology, University of Colorado, School of Medicine, Denver, Colorado, USA

ABSTRACT

Studies using a glioma cell line and primary glial cell cultures derived from newborn or aged mouse brain indicate that there is a predominance of the astrocytic glial population with cell passage in culture. This phenomenon is interpreted to represent glial aging in culture.

KEY WORDS

Glioma cells, primary glial culture, astrocytes, oligodendrocytes, glial enzymes, cell passage, aging.

INTRODUCTION

For many years the consensus has been that memory and cognitive decline of the aged was simply the result of attrition of nerve cells which are estimated to be lost at the rate of 50 - 100,000 cells per day. However, although neuronal loss must play a role in the memory and cognitive decline in aging, it is now clear that this is not the complete explanation. The neuron is only one component of the complex circuitry of the brain. Other significant cellular components are the neuroglial cells and the connective tissue cells such as endothelial cells, fibroblasts and mesenchymal cells. All of these cells, neurons, neuroglia, and connective tissue cells intercommunicate through their microenvironment and function as a unit. Age-related changes in any component of this cellular unit will shift the balance, interrupt intercellular relationships and ultimately affect neuronal function.

It is only in the last decade that consideration has been given to glial cell in aging, and the literature contains still very few specific studies. In this chapter, we discuss some of our findings on glial cell changes using a glioma cell line (C-6 glial cells 2B clone) and primary cultures derived from newborn mouse brain and 18-month old mouse brain. These studies have provided some evidence that morphological and biochemical changes occur in glial cells in culture which may represent cellular senescence.

C-6 GLIOMA CELLS IN CULTURE AND CELL PASSAGE AS A MODEL

Several studies using cell culture as a model have explored cell changes with aging. More specifically, using somatic cells as a model, changes with cell passage in culture have been interpreted as changes with aging (Strehler, 1977). Thus we have been using early passage (20-25) and late passage (50-90) C-6 glial cells as a model to study changes in glial cells with aging in culture.

C-6 glial cells, 2B clone, have been obtained through the courtesy of Dr. Jean de Vellis, Department of Anatomy, UCLA. Cells of various passages from 20 to 100 were kept frozen in liquid nitrogen. For the experimental studies cells were plated on Falcon plastic flasks ($75cm^2$) or plastic dishes containing Dulbecco's Modified Eagle's Medium (DMEM) with 10% fetal calf serum (FCS) and were incubated at 36-37° C in a humidified atmosphere of 7.5% CO_2 in air. Cells were harvested using 0.1% trypsin and cell counts were made with a hemocytometer. Protein was measured by the Lowry method (Lowry et al, 1951). 2',3' cyclic nucleotide 3' phosphohydrolase (CNP) activity was measured with the method of Prohaska et al (1973). Glutamine synthetase (GS) activity was measured with the method of Berl (1966).

Studies Using Light Microscopy. As shown in Fig. 1 A and B (Parker et al, 1980), the early (P22) and late (P88) passaged cells are morphologically different. Almost all cells of passage 22 possess small, dark nuclei, scanty perikaryon, and relatively short, thin cytoplasmic processes. In contrast, cells of passage 88 have significantly larger and paler nuclei, frequently abundant perikaryon, and considerably longer cytoplasmic processes. This cell type has never been seen in early passage cultures.

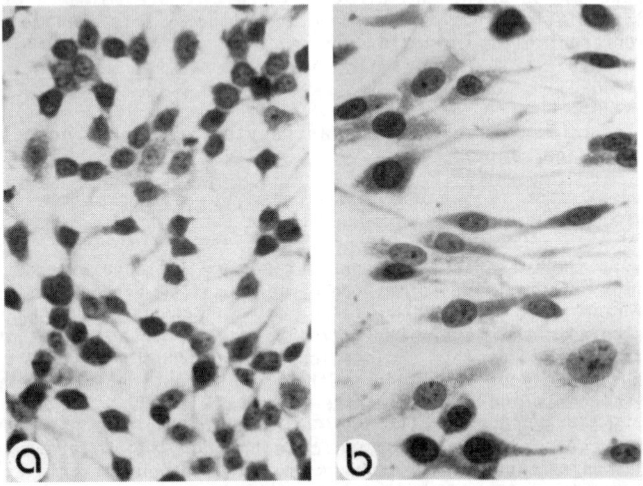

Fig. 1. Photomicrographs of Giemsa-stained C-6 glial cells grown on glass cover slips. Scale bar, 20 micrometers. A) Cells of passage 22, and B) Cells of passage 88.

Cell Growth. We have compared the doubling time in young (25 passage) and old (54 passage) C-6 glial cells plated at low density (1-1.5 x 10^5 cells/flask, 25 cm^2) or high density (1-1.5 x 10^6 cells/flask) (Vernadakis et al, 1982). Young passage cells plated at low cell density appear to proliferate at a similar rate to that of old passage cells (Figs. 2 and 3). However, when cells are plated in high cell density, old passage cells have a doubling time of approximately 12 to

16 hrs, whereas young passage cells show a doubling time of over 24 hrs. The rapid proliferation of the old passage cell was reflected in the higher protein and RNA content per cell in general (Vernadakis et al, 1982). Thus, this biochemical difference between young and old passage cells was also reflected in the cell size, the old passage cells being larger (Fig. 1).

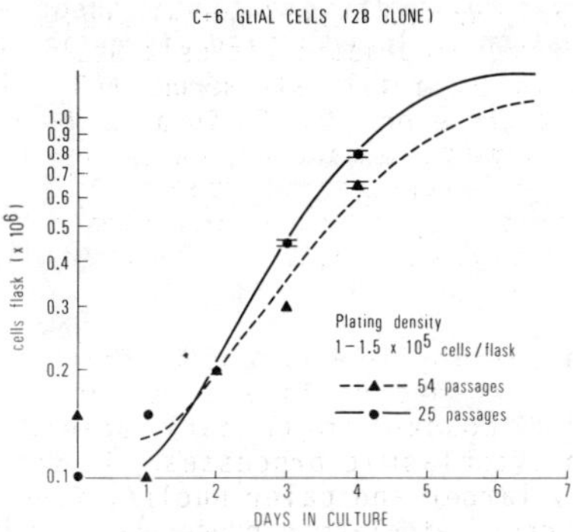

Fig. 2

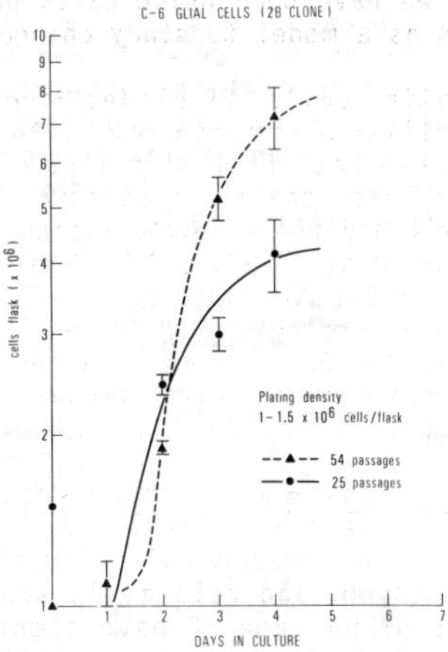

Fig. 3

Fig. 2. Cell cycle of C-6 glial cells, cell passage 25 (solid line) and 54 (broken lines plated at 1-1.5 x 10^6 cells/flask). Points represent means $\pm$ S.E. of 10-12 cultures (From Vernadakis et al, 1982).

Fig. 3. As in Fig. 2 except that cells were plated at 1-1.5 x 10^6 cells/flask.

Specific Marker Enzymes for Glial Cells. Using the enzyme markers 2',3' cyclic nucleotide, 3'phosphohydrolase (CNP) for oligodendrocytes (Poduslo and Norton, 1972) and glutamine synthetase (GS) for astrocytes (Norenberg and Martinez-Hernandez, 1979), we examined changes in the activity of these enzymes in C-6 glial cells with cell passage (Parker et al, 1980). For the CNP studies, cells of either 21 or 82 passages were used. The activity of CNP is markedly higher in cells of young passage as compared to old passage at confluency (10 days in culture); enzyme activity is similar in both cell passages at logarithmic growth (Fig. 4). Moreover, CNP activity increases in the young passage cells with days in culture perhaps reflecting the increase in synthesis of this enzyme. In contrast to the CNP activity, GS activity is markedly high in the old passage cells and low in the young passage cells (Fig. 5). Moreover, the change in GS activity with days in culture is more marked in the old passage cells. Further examination using GS immunoreactivity (Fig. 6 A,B from Estin and Vernadakis, 1984) as an astrocytic marker has revealed that late passage (P84) cells have more GS than early passage (P24) cells. Thus young passage cells express predominantly oligodendrocytic properties and old passage cells are predominantly astrocytic.

We have proposed that with aging in culture, glial cells transdifferentiate. The factors responsible for the proposed transdifferentiation from oligodendrocyte-like to astrocyte-like properties can at present only remain speculative. The fact that the cells at early and late passages are distinctly different morphologically suggests a phenotypic alteration rather than a cell selection phenomenon. It is possible that the transdifferentiation phenomenon may result from in vitro factors such as trypsinization for serial subcultivation (Waters and Walford, 1971; Maizel et al, 1975). Membrane changes that are later reflected in changes in intracellular activity may be brought about by trypsinization. The recent findings of Raff et al (1983) that cells isolated from 7-day-old rat optic nerve differentiate into a fibrous astrocyte if cultured in the presence of fetal calf serum and into an oligodendrocyte if cultured in the absence of serum supports the view that the microenvironment influences differentiation of CNS glial cells in vitro. Pritchard (1981, 1983) has reported transdifferentiation between cell types in eye tissues and has looked into similarities between light-induced retinal degeneration in vivo and the transdifferentiation of retinal cells maintained in culture. The term "metaplasia" has been used to describe the conversion of a cell from one fully-differentiated phenotype to another, a process involving dedifferentiation followed by redifferentiation (De Pomerai and Gali, 1981).

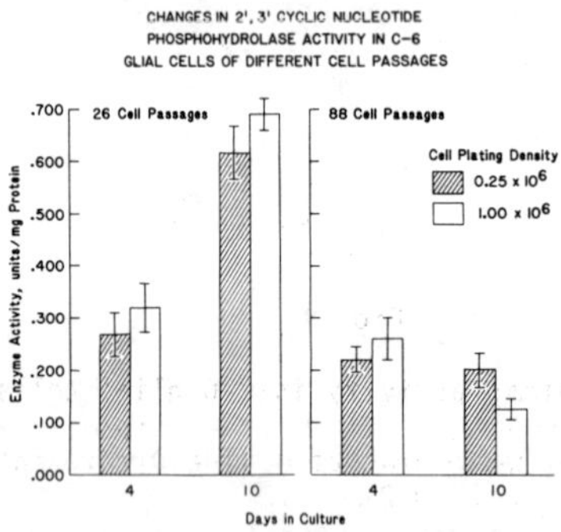

Fig. 4

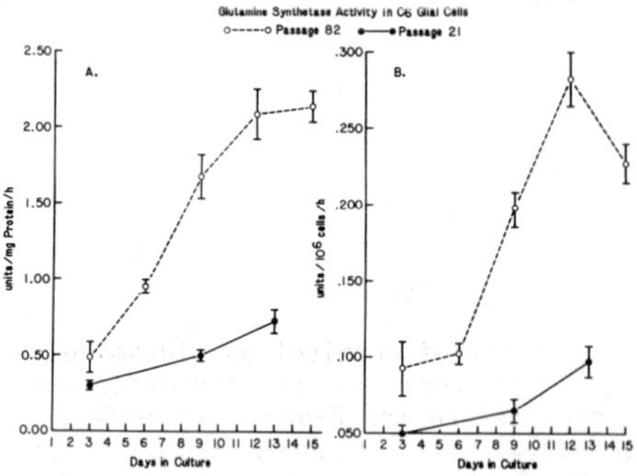

Fig. 5

Fig. 4. Changes in 2',3' cyclic nucleotide, 3' phosphohydrolase activity in C-6 glial cells of different cell passage. Activity is expressed per mg protein. Units of activity represent umoles phosphate produced per minute. Points with bracketed lines represent mean ± S.E. (From Parker et al, 1980).

Fig. 5. Changes in glutamine synthetase activity in C-6 glial cells of different cell passages. A) Activity expressed per mg protein. B) Activity expressed per 10^6 cells. Units of activity represent umoles of γ-glutamylhydroxamic acid formed. Points with bracketed lines represent means ± S.E. (From Parker et al, 1980).

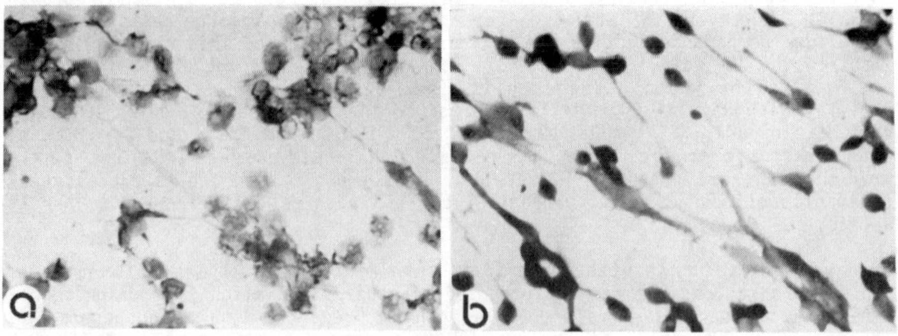

Fig. 6 Cultures of C-6 (2B clone) glioma cells, stained for glutamine synthetase using the PAP immunocytochemical method of Sternberger et al (1971). A) Cells from passage 24; B) Cells from passage 84.(From Estin and Vernadakis, 1984).

Responsiveness to Hormones. We and others have shown that hormones markedly influence glial cells in culture (Vernadakis, 1971; Bau and Vernadakis, 1981; Gremo et al, 1984; Vernadakis and Nidess, 1976; Vernadakis et al, 1978; de Vellis et al, 1971; de Vellis and Kukes, 1973; McGinnis and de Vellis, 1981; Juurlink et al, 1981). In a preliminary study we examined the response of young and old passage C-6 glial cells to cortisol or DBcAMP (Figs. 7-11).

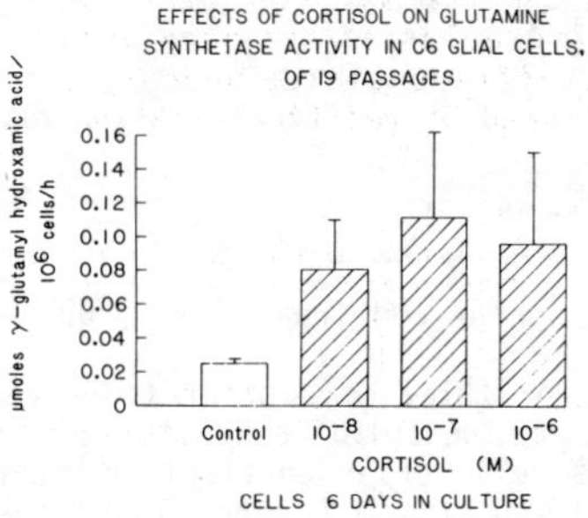

Fig. 7

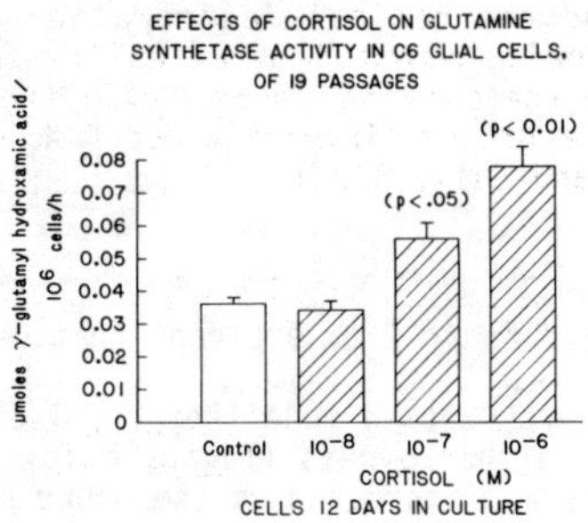

Fig. 8

Fig. 7. Effects of cortisol on glutamine synthetase activity in C-6 glial cells of 19 cell passage. Cells were exposed to cortisol in the medium for three days before the day of harvesting. Points with bracketed lines represent means ± S.E. of 4-6 cultures (From Vernadakis et al, 1982).

Fig. 8. As in Fig. 7 except for days in culture before harvesting.

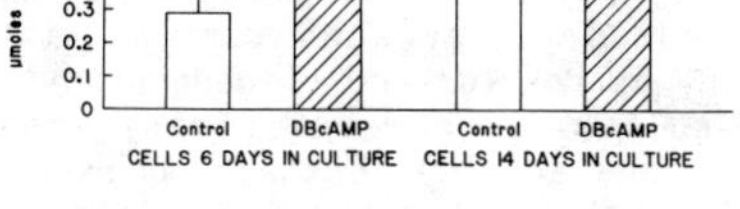

Fig. 9

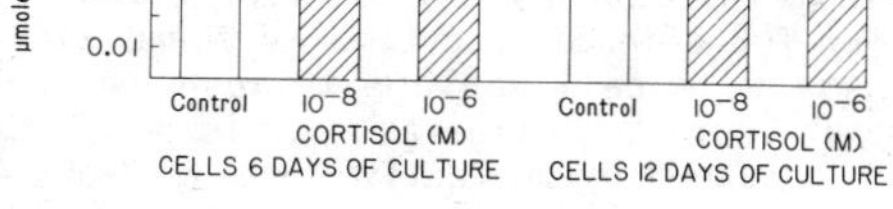

Fig. 10

Fig. 9. As in Figs. 7 and 8 except that cultures were treated with 10^{-3}M DBcAMP.

Fig. 10. As in Figs. 7 and 8 except that cells were of 79 cell passage.

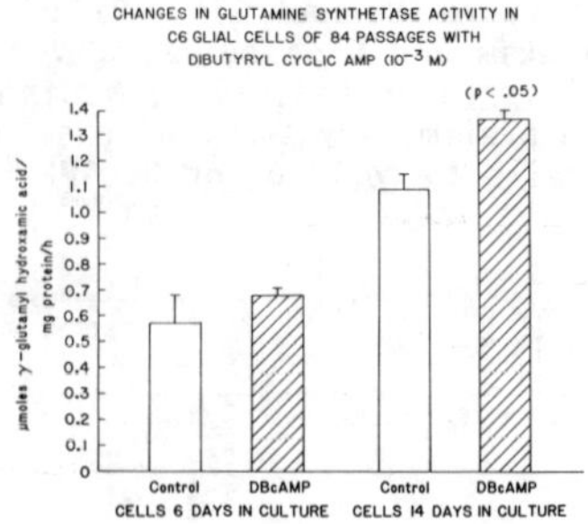

Fig. 11. As in Fig. 9 except that cells are passage 84 and treated with DBcAMP.

Cells were exposed to either cortisol or DBcAMP for three days prior to harvesting at either 6 days (logarithmic) or 12 days (confluency) of subculturing. GS activity was increased in the young passage glial cells after cortisol or DBcAMP treatment relative to controls; in contrast, GS activity in the old passage cells was increased only after DBcAMP treatment. The lack of response to cortisol of the old cells we interpret to reflect either changes in uptake of cortisol due to cell surface changes with age and/or a decrease in hormone receptors. Cortisol receptors in C-6 glial cells have been reported (Bennet et al, 1977; Holbrook et al, 1981; Vernadakis and Culver, 1980). Moreover, cortisol has been shown to produce changes in the morphology of glial cells reflecting cell surface changes (Berliner et al, 1978; Berliner and Gerschenson, 1975). In addition, changes in cell surface have been associated with glucocorticoid receptors. For example, McGinnis and de Vellis (1981) found that exposure of C-6 glial cells, normal astrocytes, and oligodendrocytes to Con A results in the loss of more than 90% of the glucocorticoid receptor activity. A decrease in glucocorticoid receptors with aging has been reported in many rat tissues and cells (Roth, 1979a,b).

PRIMARY GLIAL CELL CULTURES DERIVED FROM NEWBORN AND AGED MOUSE BRAIN

In an earlier study we compared changes occurring with days in culture and with cell passage in glial cells derived from newborn vs aged (18 mo-old) mouse cerebral hemispheres (Vernadakis et al, 1984). Using glutamine synthetase (GS) and glial fibrillary acidic protein (GFAP) as biochemical and immunocytochemical markers for astrocytes, and 2',3' cyclic nucleotide, 3' phosphohydrolase as a marker for oligodendrocytes, we found that both astrocytes and oligodendrocytes survive with cell passage in both types of cultures. However, there was a general increase in the astrocytic population and a decline in the oligodendrocytes as shown by the activities of GS and CNP in the cultures derived from the aged mouse brain. These early findings have led us to consider the changes observed with cell passage as possibly reflecting aging in vitro. In a subsequent study, we examined cell growth patterns in cultures from passage 5 to 11 (Vernadakis et al, 1986). We observed no change in CNP activity (Fig. 12), a decline in GFA positive cells, a several-fold increase in GS activity (Fig. 13) and the appearance of multinucleated cells (Fig. 14).

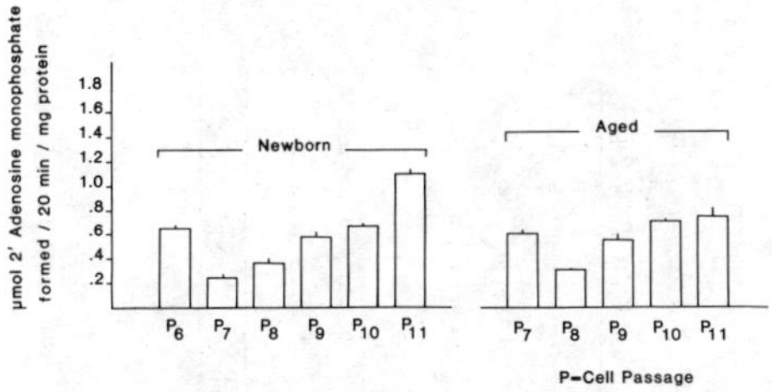

Fig. 12. Cyclic nucleotide phosphohydrolase activity in glial cell cultures prepared from newborn and aged mouse cerebral hemispheres. Cells from passage 6 to 11 (P6 to P11). Points with lines represent means ± S.E. for 3-5 cultures.

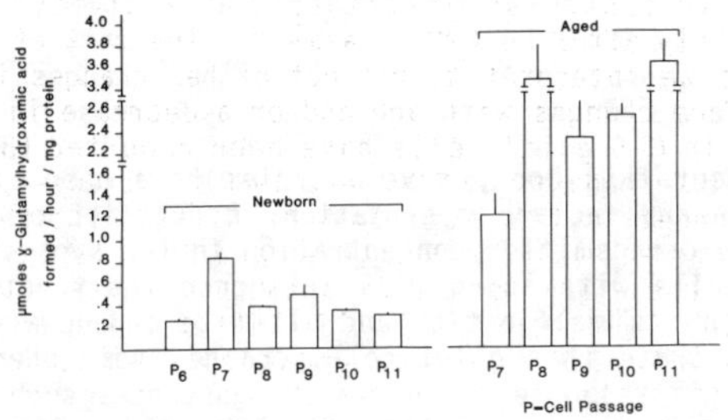

Fig. 13. Glutamine synthetase activity. Details as in Fig. 12.

Whether the increase in GS activity represents increase in activity induced by the experimental conditions or increase in the number of cells containing GS cannot be ascertained from this study. However, this marked GS activity coincides temporally with the appearance of large polyhedral multinucleated cells (Fig. 14). Lapham (1962) has reported that in reactive protoplasmic astrocytosis there occurs an increase in the number of nuclei without increased proliferation and proposes that nuclear division in reactive protoplasmic astrocytes may have a special adaptive significance with greater functional capacity. The possibility that in our study multinucleated cells are GS-containing glial cells remains to be evaluated.

Recently Hallermayer and Hamprecht (1984) reported co-localization of GS and GFA in primary cultures of dissociated cells from newborn mice. The findings of these authors demonstrate not only the heterogeneity of primary glial cells but also their plasticity _in vitro_. Extrapolating from these observations to ours, it is conceivable that the decline in GFA-positive glial cells and the increase in GS activity represents a shift towards GS continaing glial and other cells with cell passage. As discussed earlier with the findings obtained using the C-6 glial cell model, this shift in cell types may partially be due to several factors including trypsinization during subculturing, continuous exposure to serum factors, cell-stratum interactions. However, the possibility of cellular aging expressed with cell passage warrants consideration and further investigation.

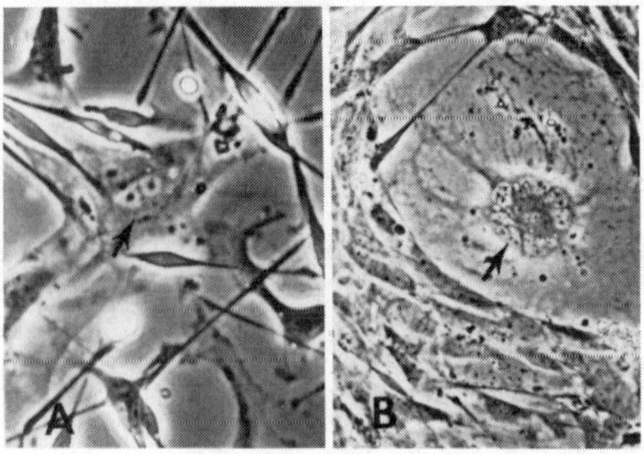

Fig. 14. Cultures prepared from mouse cerebral hemispheres. Cells were from late passage (P10 or P11). A) Cultures from newborn mouse, cell passage 10, 24 days in culture, and B) Cultures from aged mouse (18 mo), cell passage 11, 24 days in culture. Arrow indicates multinucleated cell (From Vernadakis et al, 1986).

CONCLUSIONS

As discussed throughout this book, several roles have been attributed to glial cells in CNS function and include myelination, modulation of the extraneuronal environment, i.e, neurotransmitter concentration in the synaptic cleft. Changes therefore in glial cells with aging will influence the function of the neuron and may contribute to its malfunction and ultimate degeneration. The findings discussed here demonstrate that glial cells change their phenotypic expression with age and that their response to intrinsic substances such as hormones, nucleotides and putative neurotransmitter substances is altered with age. We propose that the shifts in the glial populations, i.e., oligodendrocytes vs astrocytes, observed in these in culture studies may partially parallel the astrogliosis observed with aging. This phenomenon may be one of the intrinsic adaptive mechanisms of the brain to the aging process.

ACKNOWLEDGEMENT

The authors would like to express their gratitude to Jocelyn Bailey for her editorial assistance.

REFERENCES

Bau, D. and A. Vernadakis (1981). The Pharmacologist 23, 218.
Bennet, K., J. McGinnis and J. de Vellis (1977). J. Cell Physiol. 93, 247-260.
Berl, S. (1966). Biochemistry 5, 916-922.
Berliner, J.A. and L.E. Gerschenson (1975). J. Cell Physiol. 86, 523-532.
Berliner, J.A., K. Bennett and J. de Vellis (1978). J. Cell Physiol. 94, 321-334.
De Pomerai, D.I. and M.A.H. Gali (1981). J. Embryol. Exp. Morph. 62, 291-308.
de Vellis, J. and G. Kukes (1973). Texas Reports on Biology and Medicine 31, 271-273.
de Vellis, J., D. Inglish, R. Cole and J. Molson (1971). In: Influence of Hormones on the Nervous System, S. Karger, Basel.
Estin, C.H. and Vernadakis (1984). In, Developmental and Circulatory Aspects of Brain Metabolism, Serbian Academy of Science and Art, Siro, Belgrade.

Gremo, F., S. Porru and A. Vernadakis (1984). Devel. Brain Res. 16, 27-35.
Hallermayer, K. and B. Hamprecht (1984). Brain Res. 295, 1-11.
Holbrook, N., R.J. Grasso and J.F. Hackney (1981). J. Neurosci. 6, 75-88.
Juurlink, B.H.J., A. Schousboe, O.S. Jorgenson and L. Hertz (1981). J. Neurochem 36, 136-142.
Lapham, L.W. (1962). Pathol. 21, 1-21.
Lowry, D.H., N.J. Rosenbrough, A.L. Farr and R.J. Randall (1951) J. Biol. Chem. 93, 265-275.
Maizel, A., C. Nicoline and R. Beserga (1975). J. Cell Physiol. 86, 71.
McGinnis, J.F. and J. de Vellis (1981). Proc. Natl. Acad. Sci. USA 78, 1288-1292.
Norenberg, M.D. and A. Martinez-Hernandez (1979). Brain Res. 161, 303-310.
Parker, K., M.D. Norenberg and A. Vernadakis (1980). Science 208, 179-181.
Poduslo, S.E. and W.T. Norton (1972). J. Neurochem. 19, 727-736.
Pritchard, D.J. (1981). J. Embryol. Exp. Morph. 62, 47-62.
Pritchard, D.J. (1983). Exp. Eye Res. 37, 315-326.
Prohaska, J.R., P.A. Clark and W.W. wells (1973). Anat. Biochem. 56, 275-282.
Raff, M.C., R.H. Miller and M. Noble (1983). Nature 303(2), 390-396.
Roth, G.S. (1979a) Mech. Aging Devel. 9, 497-514.
Roth, G.S. (1979b) Fed. Proc. 38, 1910-1914.
Sternberger, L.A., P.H. Hardy, Jr., J.J. Cuculis and H.G. Meyer (1970). J. Histochem. Cytochem. 18, 315-333.
Strehler, B.L. (1977)Time, Cells and Aging, Academic Press.
Vernadakis, A. (1971). In, Enzymes in the Nervous System, Plenum Press, New York.
Vernadakis, A. and B. Culver (1980) In, Neural Tissue Culture: A Biochemical Tool, Pergamon Press, Oxford.
Vernadakis, A. and R. Nidess (1976). Neurochem. Res. 1, 385-402.
Vernadakis, A. B. Culver and R. Nidess (1978) Psychoneuroendocrinol. 3, 47-64.
Vernadakis, A., D. Mangoura, N. Sakellaridis and S. Linderholm (1984). J. Neurosci. Res. 11, 253-262.
Vernadakis, A., D. Davies, N. Sakellaridis and D. Mangoura (1986 in press). J. Neurosci. Res.
Vernadakis, A. R. Nidess, B. Culver and E.B. Arnold (1979). Mech. Aging Devel. 9, 553-566.
Vernadakis, A., K. Parker, E.B. Arnold and M.D. Norenberg (1982). In The Aging Brain: Cellular and Molecular Mechanisms of Aging in the Nervous System, Raven Press, New York.
Waters, H. and R.L. Walford (1971) Proc. Soc. Exp. Biol. Med. 137, 523.

Glial Contribution to Experimental Allergic Encephalomyelitis

L. F. Eng*, M. E. Smith[†] and B. Gerstl*

Departments of Pathology * and Neurology[†], Stanford University School of Medicine, and Veterans Administration Medical Center, Palo Alto, California 94304, USA

ABSTRACT

Experimental allergic encephalomyelitis (EAE) (the acute and chronic relapsing forms) is a cell-mediated autoimmune disease which has been the major experimental model for studying demyelinating diseases such as multiple sclerosis (MS). The pathologic features of MS and EAE reside with the glial cells -- loss of myelin elaborated by the oligodendrocytes and fibrous gliosis in the astrocytes. Our studies have focused on the activation and metabolism of astrocytes in EAE. Some of the activation factors (or mediators) which have been suggested are cell-derived (T-cells, macrophages, and endothelial cells), growth factors, mitogenic factors, matrix factors, polyamines, free radicals, proteolytic enzymes, and factors in edema fluid. In the acute and chronic relapsing model of EAE, cellular activation was characterized by increased immunostaining for GFAP in the astrocytes and by a generalized increase of radioactive amino acid uptake in all of the cytoskeletal proteins. There was no correlation between GFAP staining and the perivascular lesions in acute EAE, however, in chronic-relapsing EAE, the lesioned areas were heavily gliosed and stained for GFAP. While no change in GFAP content was detected in the spinal cord of the rat with acute EAE, an increased GFAP content was found in the spinal cord of the chronic-relapsing EAE mouse.

KEY WORDS

Experimental allergic encephalomyelitis (EAE), acute EAE, chronic-relapsing EAE, cell-mediated autoimmunity, glial fibrillary acidic protein (GFAP), astrocyte, fibrous gliosis, demyelination.

INTRODUCTION

Our interest in glial cells stems from longstanding studies of central nervous system (CNS) diseases and aging, especially multiple sclerosis (MS), a debilitating disease which has enormous social and economic consequences. The most prominent events which occur in CNS disease or injury are astrocyte activation and resulting fibrous gliosis -- the production of glial fibrillary acidic protein (GFAP), the major protein constituent of glial filaments in differentiated astrocytes (Eng, 1985). Experimental allergic encephalomyelitis (EAE) (the acute and chronic relapsing forms) is a cell-mediated autoimmune disease which has been implicated in MS and has been the major experimental model for studying the etiology, pathogenesis, and therapy of chronic MS. The pathologic features of MS and EAE reside with the glial cells -- loss of myelin elaborated by the oligodendrocytes and fibrous gliosis in the astrocytes. Extensive studies have investigated the nature and specificity of the neural antigen which induces disease, the genetic abnormalities in the immunoregulation which permits the evolution of the chronic relapsing process, the relative roles of T-cell-mediated immunity and various antibody isotypes in producing the essential lesion, and recently the intervention of the disease with antibodies to T-helper cells. (For reviews see Lassmann, 1983; Waxman and Reynolds, 1984; Raine, 1982). Our studies have focused on the intermediate filaments of the reactive astrocyte.

<u>Astrocytes of the CNS</u>: Each of the two major types of astrocytes has a specific morphology , and in addition, fibrous astrocytes are characterized as having more intermediate filaments (GFAP) than protoplasmic astrocytes (DeArmond et al., 1980). A third type of astrocyte, the reactive astrocyte, is induced by a variety of CNS injuries and by disease. These cells are larger than normal astrocytes; have a larger nuclear diameter (Hortega and Penfield, 1927; Cavanagh, 1970), an increased content of DNA (Lapham and Johnstone, 1964), intermediate filaments (Nathaniel and Nathaniel, 1977), GFAP (Bignami and Dahl, 1976; Amaducci et al., 1981), vimentin (Dahl et al., 1981, 1982), glutamine synthetase (Norenberg, 1983), glycogen (Nathaniel and Nathaniel, 1981); as well as increased oxidoreductive enzyme activity (Oehmichen, 1980.) They participate in scar tissue formation and probably to some extent in phagocytosis, especially the removal of degenerating myelin (Nathaniel and Nathaniel, 1977). Reactive astrocyte formation <u>in vivo</u> probably originate by hyperplasia and hypertrophy of normal astrocytes (Hortega and Penfield, 1927; Latov et al., 1979; Oehmichen, 1980; Barrett et al., 1981; Nathaniel and Nathaniel, 1981; Polak et al., 1982).

<u>Mechanisms for Astrocytes Activation</u>: The current view is that the acute EAE lesion is a pure T-cell-mediated reaction (helper subset of T cells that carries the L3T4 surface marker (Pettinelli and McFarlin, 1981; Dialynas et al., 1983; Lando and Ben-Nun, 1984; Wekerle, 1984; Brostoff and Mason, 1984; Waldor et al., 1985), and is supported by the report that EAE can be induced with small MBP peptides without trace of antibody formation (Hashim et al., 1973).

In acute EAE, vascular leakage has been demonstrated by staining tissue sections for fibrin (Paterson, 1976) by measuring extravasation of radiolabeled macromolecules (Linthicum, 1982; Juhler et al., 1984) or immunocytochemical staining for serum proteins (Eng, Gibbs and Smith, unpublished data). Edema commonly accompanies intense T-cell-mediated reactions and may be due to an immune complex lesion of the vessel wall, with triggering of the complement cascade. The inhibition of acute EAE lesion formation by protease inhibitors (Smith, 1980; Brosnan et al., 1980) or alpha adrenergic antagonists (Goldmuntz et al., 1985) have been reported. Activated T cells can recruit various inflammatory cells by release of lymphokines (Parker, 1980). Activated monocytes (macrophages) appear to be the predominant population of secondary cells and other inflammatory cells may also be recruited and activated, both by lymphokines and by antigen-antibody complexes (Prineas and

Wright, 1978). These activated cells (T cells, monocytes, mast cells, basophils, polymorphonuclear cells, platelets, and endothelial cells) produce a broad range of mediators (lymophokines, lysosomal enzymes, monokines, and prostaglandins) which affect vascular permeability as well as behavior of the other cells in the total infiltrate. Prostaglandin production is also increased in the stimulated monocytes which enter circulation. In addition to producing local effects, the activated cells may influence both the phenotype and the behavior of the other circulating T cells and natural killer cells. Thromboxanes, prostacyclin, leukotrienes, histamines, serotonin, and peroxides are other potential mediators.

Binding of serotonin (5-HT) to astrocytes has been demonstrated by Tardy et al. (1982) and its correlation with adenylate cyclase activity by Hertz et al. (1979). Thus, effect of 5-HT on astrocytes deserves attention. Cyclic AMP synthesis by 5-HT-stimulated astrocytes induces ornithine decarboxylase activity (Bachrach, 1977). Polyamines are increased in brain during development (Bachrach, 1980) and proliferation of glioma cells is closely related to spermidine synthesis (Sobue and Nakajima, 1977). Therefore, binding of 5-HT to astrocytes and the sequentia mentioned could be factors leading to astrocytic proliferation. Perturbation of metabolic equilibrium due to activation of adenylate cyclase, however, also has been reported. Cyclic-AMP is involved in the regulation of lymphokine production by glial cells (Fontana et al., 1982a) leading to glial cell division and implicated in the phosphorylation of GFAP and vimentin in glioma (Groppi and Browning, 1980; Browning and Ruina, 1984) and astrocyte cultures (McCarthy et al., 1985). Lycke et al. (1973) observed reduced levels of 5-HT in the CNS of guinea pigs with EAE. Scott (1982), basing his experiments on Lycke's findings, could demonstrate prolonged survival due to drugs increasing 5-HT levels in the CNS, i.e. imipramine. These results, however, are at variance with those obtained by Brosnan and Tansey, (1984) who noticed delay of onset of EAN and reduction of mononuclear cells infiltrating the peripheral nervous system in rats treated with reserpine which is known for its 5-HT-reducing effect. The variance of results obtained with 5-HT increasing or decreasing levels may be related to the presence of two types of 5-HT-receptors, one linked to adenylcyclase concentration, the other independent of this (D.L. Nelson et al., 1980).

Immunologic Role of Astrocytes in EAE: An immunological role for astrocytes was first suggested by the report that they release interleukin-1 after stimulation by various antigens (Fontana et al., 1982b), and was further implicated by the demonstration, with cultured newborn rat GFAP positive cells (Fontana et al., 1984), that astrocytes act as antigen-specific, major-histocompatibility-complex-restricted, antigen-presenting cells. In these experiments, GFAP-positive cells adhered to syngeneic myelin-basic-protein-sensitized T lymphocyte lines cultured in the presence of myelin basic protein, and adherent T cells subsequently underwent blast-cell transformation; however, allogeneic T cells did not associate with astrocytes or undergo blastogenesis. The density of surface Ia antigen increased on astrocytes exposed to sensitized syngeneic but not allogeneic T cells and myelin basic protein. These observations on the role of astrocytes may be relevant to EAE and to MS, which has been considered a primary demyelinating disease. The myelin damage in MS is associated with CNS infiltration by IgG secreting cells (Esiri, 1980) and T lymphocytes predominantly of the cytotoxic/suppressor subset; the latter are concentrated at the edge of lesions and extend into unaffected tissue (Traugott et al., 1983).

Other Astrocyte Reactions: Astrocytes have been observed in both EAE and MS to phagocytize myelin and oligodendrocytes (Raine, 1982). The proliferation of astrocytes is also a prominent feature in EAE and MS. The astrocytic reaction which is of prime interest to us and which remains to be explained is the rapid increase in GFAP staining following the breakdown of the blood barrier. Is this due to disassembly of the intact glial filament, the proteolytic degradation of the GFAP, the increase in synthesis of GFAP, increased metabolic turnover, or change in the state of phosphorylation (kinase or phosphatase activity). The experiments reported in the following have raised these questions which remain to be answered.

Reactive Astrocytes in Acute EAE: Acute EAE was induced in Lewis rats by injection of 0.5 ml of adjuvant containing 1 mg lyophilized guinea pig CNS myelin and 1.5 mg Mycobacterium tuberculosis H37Ra (Smith et al., 1983). Spinal cord sections from Lewis rats with acute EAE stained immunocytochemically for GFAP showed greatly increased staining in astrocytes. Fibrous processes in white matter were heavily stained early in the course of the disease when paralysis was first evident (10-12 days after injection of guinea pig spinal cord myelin), then protoplasmic astrocytes were stained in the gray matter and became more heavily stained at 20 days post-injection. The stained astrocytes were evenly distributed throughout the tissue, and did not correspond to the sites of the lesions. Spinal cord slices of control and EAE rats were incubated with [^{3}H] amino acids, then cytoskeletal proteins were prepared in an enriched fraction, separated by sodium dodecyl sulfate-polyacrylamide gel electrophoresis (SDS-PAGE), and the protein bands counted for radioactivity. In the EAE rat all cytoskeletal proteins, including the neurofilaments, vimentin, microtubules, GFAP and actin, showed increased uptake of radioactive amino acids. The activation of GFAP protein staining and metabolism in acute EAE might serve as a model for studying the mechanism of activated growth of astrocyte processes which lead to the severe gliosis seen in multiple sclerosis.

Reactive Astrocytes in Chronic Relapsing EAE in the SJL/J Mouse: The Brown and McFarlin (1981) method for inducing chronic relapsing EAE was modified and used in SJL/J mice to study the increase in immunocytochemical staining of astrocytes with GFAP antiserum and increase metabolism of GFAP (Smith et al., 1984a, 1984b). After three injections with mouse spinal cord, the SJL mice had initial attacks of EAE, and one or two relapses followed within several months. When the mice were reimmunized, relapses followed as early as seven days. GFAP in the CNS of fifty mice with chronic relapsing EAE was studied over a period of 11 months. Immunocytochemical staining of CNS for GFAP showed areas of intense staining, especially in the spinal cord. These areas were usually associated with lesions, in contrast to the generalized staining of astrocytes and glial processes observed in the Lewis rat with acute EAE. The cytoskeletal proteins were separated by SDS-PAGE and the stained cytoskeletal proteins were measured by densitometry. The 70,000 neurofilament protein (NF) which was quantitatively constant was used as an internal standard. GFAP/NF ratios in animals with chronic EAE ranged from 1.14 to 3.5, while those of the controls were 0.66-0.91. The ratio in mice with no distinct relapses were also higher than the control six months after the initial immunization. Increased uptake of radioactive amino acids into cytoskeletal proteins including GFAP could be demonstrated in incubated CNS slices of the chronic relapsing EAE mice. This increase was more evident during acute relapses. The increase in GFAP and fibrous gliosis in EAE lesions of the SJL mouse is very similar to that seen in gliotic multiple sclerosis plaques. The chronic relapsing mouse may serve as a good model for the study of GFAP metabolism and accumulation leading to the gliotic scar.

Recently we immunostained paraffin-embedded freeze-substituted sections of spinal cord of Lewis rats at the initial stage of EAE. Antibodies to vimentin stained the monocytes and T cells in the perivascular lesion and endothelial cells in the blood vessels. Few if any of the astrocytes within the lesion or other areas of the spinal cord stained for vimentin (Fig. 1) (Eng, Gibbs and Smith, unpublished results). Immunofluorescent studies have reported that reactive astrocytes stain for vimentin (Dahl et al., 1981; Fedoroff et al., 1984). The astrocytes throughout the tissue section stained heavily for GFAP and glutamine synthetase.

DISCUSSION

A metabolic study of chronic relapsing EAE in strain 13 guinea pigs has yielded results comparable to our studies reported here (Linington et al., 1984). A recent report by Goldmuntz et al., (1984) on acute EAE in the Lewis rat has confirmed our findings of increased immunostaining for GFAP in astrocytes and a general increase in synthesis of cytoskeletal proteins. By autoradiography of two-dimensional electrophoresis gels, they found that vimentin was the most heavily labled protein. This is consistent with our finding that the infiltrating hematogenous cells in the perivascular lesions were heavily stained for vimentin. While they were able to demonstrate an increase in synthesis of both total and cytoskeletal proteins by *in vitro* radiolabeling of spinal cord, the degree of radiolabeling of the GFAP was insufficient to be detected by radioautography of two-dimensional electrophoresis gels.

Our recent studies indicate that the normal astrocyte, which is distant from the lesion and which is activated, does not contain immunostainable vimentin in contrast to the heavy immunostain for GFAP. These astrocytes may be activated by some components in the edema fluid. At the site of the early lesion, the monocytes and T cells stain intensely for vimentin, therefore it was difficult to demonstrate astrocytes which contain both vimentin and GFAP. It is possible that only the dividing astrocytes pass through the normal developmental stages where vimentin is expressed before GFAP. It seems there was no increase in the number of astrocytes and no demonstrable proliferation of astrocytes in the early lesion. Astrocytes in culture respond to beta-adrenergic agonists such as isoperterenol or forskolin by accumulation of cyclic AMP and phosphorylation of GFAP and vimentin (McCarthy et al., 1985). The time course of phosphorylation is rapid and comparable to that observed for increased GFAP staining in astrocytes following CNS trauma. One might speculate that the initial increase in GFAP staining might reflect the increase in phosphorylated GFAP epitopes. This mechanism would not require an increase in antigenic sites resulting from disassembly or degradation of intact or bundles of GFAP filaments or new synthesis of GFAP. The existence of monoclonal antibodies which recognize phosphorylated or non-phosphorylated epitopes of neurofilaments has been well-documented (Sternberger et al., 1983).

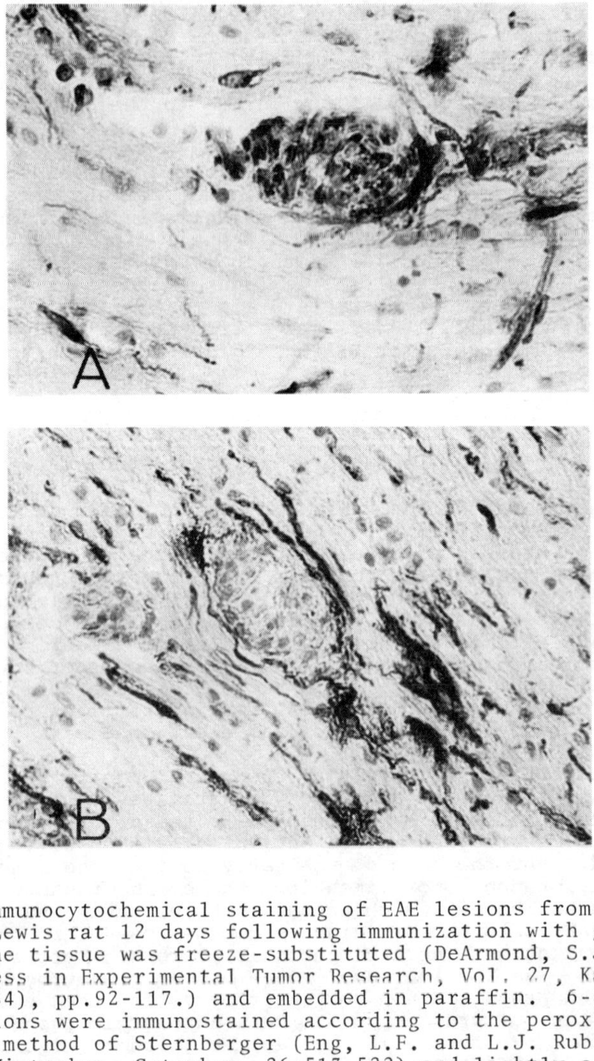

FIG. 1. Immunocytochemical staining of EAE lesions from a spinal cord of a Lewis rat 12 days following immunization with guinea pig myelin. The tissue was freeze-substituted (DeArmond, S.J. and L.F. Eng, Progress in Experimental Tumor Research, Vol. 27, Karger, Basel, (1984), pp.92-117.) and embedded in paraffin. 6-8 micron thick sections were immunostained according to the peroxidase-anti-peroxidase method of Sternberger (Eng, L.F. and L.J. Rubinstein (1978) J. Histochem. Cytochem. 26:513-522) and lightly counter-stained with hematoxyln. A) Section stained with monoclonal anti-body to vimentin provided by S. Blose. Note the intense staining of the macrophages and lymphocytes in the vascular lesion. B) Section stained with polyclonal antibody to human MS plaque GFAP. Note the intense staining of the astrocytes and astrocytic processes around the vascular lesion.

ACKNOWLEDGEMENT

The authors would like to express their gratitude to Donna Buckley for preparation of the manuscript and Michael Gibbs for technical assistance. Portions of the work presented were supported by the Veterans Administration, NINCDS, and Multiple Sclerosis Society.

REFERENCES

Amaducci, L.A., K.I. Forno and L.F. Eng (1981). Neurosci. Lett., 21, 27-32.
Bachrach, U. (1977). FEBS Lett., 75, 201-204.
Bachrach, U. (1980). Biochem. J., 188, 387-392.
Barrett, C.P., L. Guth, E.J. Donati and J.G. Krikorian (1981). Exp. Neurol., 73, 365-377.
Bignami, A, and D. Dahl (1976). Neuropathol. Appl. Neurobiol., 2, 99-111.
Brosnan, C.F., W. Cammer, W.T. Norton and B.R. Bloom (1980). Nature (London), 285, 235.
Brosnan, C.F., and F.A. Tansey (1984). J. Neuropathol. Exp. Neurol., 43, 84-93.
Brostoff, S.W., and D.W. Mason (1984). Trans. Am. Soc. Neurochem., 15, 198.
Brown, A.M., and D.E. McFarlin (1981). Lab. Invest., 45, 278.
Browning, E.T., and M. Ruina (1984). J. Neurochem., 42, 718-726.
Cavanagh, J.B. (1970). J. Anat., 106, 471-487.
Dahl, D., C.J. Cosby and A. Bignami (1981). Exp. Neurol., 71, 421-430.
Dahl, D., P. Strocchi and A. Bignami (1982). Differentiation, 22, 185-190.
DeArmond, S.J., L.F. Eng and L.J. Rubinstein (1980). Pathol. Res. Pract., 168, 374-394.
Dialynas, D.P., et al. (1983). Immunol. Rev., 74, 29.
Eng, L.F. (1985). J. Neuroimmunol., 8, 203-214.
Esiri, M.M. (1980). Neuropathol. Appl. Neurobiol., 6, 9-21.
Fontana, A., R. Jost, S. Balsiger, P. Grob and A. Grieder (1982a) Dev. Brain Res., 2, 505-511.
Fontana, A., F. Kristensen, R. Dubs, D. Gemsa and E. Weber (1982b) J. Immunol., 129, 2413-2419.
Fontana, A., W. Fierz and H. Wekerle (1984). Nature, 307, 273-276.
Goldmuntz, E.A., C.F. Brosnan, F-C. Chiu, W. Cammer and W.T. Norton (1984). Trans. Am. Soc. Neurochem., 15, 230.
Goldmuntz, E.A., C.F. Brosnan, W. Cammer and W.T. Norton (1985). Am. Soc. Neurochem., 16, 216.
Groppi, V.E., Jr., and E.T. Browning (1980). Mol. Pharmacol., 18, 427-437.
Hashim, G.A., F.H. Wang and F.J. Schilling (1973). Arch. Biochem. Biophys., 156, 298-309.
Hertz, L., F. Baldwin, and A. Schousboe (1979). Can. J. Physiol. Pharmacol., 57, 223-226.
Hortega, P. del Rio, and W. Penfield (1927). Bull. Johns Hopkins Hosp., 31, 278-303.
Juhler, M., D.I. Barry, H. Offner, G. Konat, L. Klinken and O.B. Paulson (1984). Brain Res., 302, 347-355.
Lando, Z., and A, Ben-Nun (1984). Clin. Immunopathol., 30, 290.
Lapham, L.W., and M.A. Johnstone (1964). J. Neuropathol. Exp. Neurol., 23, 419-430.
Lassmann, H. (1983). J. Neurol., 229, 207-220.
Latov, N., G. Nilaver and A. Zimmerman (1979). Dev. Biol., 72, 381-384.

Linington, C., A.J. Suckling, M.D. Weir and M.L. Cuzner (1984). Neurochem. Int., 6, 393-401.
Linthicum, D.S. (1982). Immunobiology, 162, 211-220.
Lycke, E., and B. Roos (1973). Int. Arch. Allergy, 45, 341.
McCarthy, K.D., J. Prime, T. Harmon and R. Pollenz (1985). J. Neurochem., 44, 723-730.
Nathaniel, E.J.H., and D.R. Nathaniel (1977). Exp. Neurol., 54, 60-76.
Nathaniel, E.J.H., and D.R. Nathaniel (1981). Advances in Cellular Neurobiology, Vol. 2, Academic Press, New York, 249-301.
Nelson, D.L., A. Herbet, A. Enjalbert, J. Bockaert and M. Hamon (1980). Biochem. Pharmacol., 29, 2445-2453.
Norenberg, M.D. (1983). Glutamine, Glutamate, and GABA in the Central Nervous System, Alan R. Liss, New York, 95-111.
Ochmichen, M. (1980). Pathol. Res. Pract., 168, 344-373.
Parker, C.W. (1980). Clinical Immunology, Vols. 1 and 2, Saunders, Philadelphia.
Paterson, P.Y. (1976). Textbook of Immunopathology, Grune and Stratton, New York, 179-213.
Pettinelli, C.B., and D.E. McFarlin (1981). J. Immunol., 127, 1420.
Polak, M., W. Haymaker, J.E. Johnson and F. D'Amelio (1982). Histology and Histopathology of the Nervous System, Vol. 1, Charles C. Thomas Pub., Springfield, Illinois, 363-480.
Prineas, J.W., and R.G. Wright (1978). Lab. Invest., 38, 409-421.
Raine, C.S. (1982). Lab. Invest., 50, 608-635.
Scott, C.F., Jr., N. Cashman and L.E. Spitler (1982-1983). J. Immunopharmacol., 4, 153-162.
Smith, M.E. (1980). The Suppression of Experimental Allergic Encephalomyelitis and Multiple Sclerosis, Academic Press, London, 211-226.
Smith, M.E., F.P. Somera and L.F. Eng (1983). Brain Res., 264, 241-253.
Smith, M.E., F.P. Somera, K. Swanson and L.F. Eng (1984a). Experimental Allergic Encephalomyelitis: A Useful Model for Multiple Sclerosis, Alan R. Liss, New York, 139-144.
Smith, M.E., F.P. Somera and L.F. Eng (1984b). Trans. Am. Soc. Neurochem., 15, 154.
Sobue, K., and T. Nakajima (1977). J. Biochem. (Tokyo), 82, 1121-1126.
Sternberger, L.A., and N.H. Sternberger (1983). Proc. Natl. Acad. Sci. USA, 80, 6126-6130.
Tardy, M., M.F.D. Costa, C. Fages, J. Bardakdjian, and P. Gonnard (1982). Dev. Neurosci., 5, 19-26.
Traugott, U., E.L. Reinherz and C.S. Raine (1983). Science, 219, 308-310.
Waksman, B.H., and W.E. Reynolds (1984). Proc. Soc. Exp. Biol. Med., 175, 282-294.
Waldor, M.K., S. Sriram, R. Hardy, L.A. Herzenberg, L.A. Herzenberg, L. Lanier, M. Lim and L. Steinman (1985). Science, 227, 415-417.
Wekerle, H. (1984). Lab. Invest., 51, 199-205.

Astrocyte Response to Cerebral Ischemia: Ultrastructural and Autoradiographic Studies

C. K. Petito*, W. A. Pulsinelli† and F. Plum†

Departments of Pathology* and Neurology†, The New York Hospital, Cornell Medical Center, New York, USA

ABSTRACT

Proliferative changes in post-ischemic astrocytes were evaluated by electron microscopy and autoradiography in the rat model of 4-vessel occlusion. Within 3 hours following ischemia, astrocytes in areas of ischemic neuronal necrosis progressively enlarged, and their mitochondria significantly increased in numbers and total area. Total number of astrocytes also significantly increased by 3 hours. The astrocytic proliferation was not associated with ^{3}H-thymidine incorporation but was accompanied by a decrease in perineuronal medium-light oligodendrocytes which fell from 50% of all perineuronal glia to 6.5% between 2 and 3 hours after ischemia. Astrocytes in areas of brain with only reversible neuronal damage showed similar changes that were mild and transient. We hypothesize that the reactive changes in astrocytes and the transformation in the perineuronal glia that occurs within several hours after cerebral ischemia may be stimulated by ionic changes in the extracellular fluid or by alterations in or loss of neuronal trophic factors.

KEYWORDS

Cerebral ischemia; astrocytes; perineuronal glia; electron microscopy; autoradiography.

INTRODUCTION

Astrocytes are important metabolic regulators of the central nervous system. *In vitro* studies of cultured astrocytes have demonstrated their ability to actively take up extracellular ions (Herz, 1978; Kimelberg, *et al*, 1979) and neurotransmittors (for review, see Duffy, 1983). Astrocytes are one of the first cells to undergo morphologic changes following cerebral ischemia and their early cytoplasmic swelling may relate in part to alterations in the extracellular fluid composition (Petito, *et al*, 1982). Astrocytes may protect brain with ischemic neuronal necrosis from developing into infarcts (Plum, 1983) if they retain their capacity to buffer intraneuronal and interstitial hydrogen ions (Kraig, *et al*, 1985).

Advances in the Biosciences Vol. 61

We have studied the astrocytic response to cerebral ischemia by sequential ultrastructural studies of post-ischemic brain. Cerebral ischemia was produced in fasted adult male rats by permanent occlusion of the vertebral arteries and 30 minute occlusion of the common carotid arteries (Pulsinelli, et al, 1979). Following release of the carotid clasps, the animals were allowed to live from 3 minutes to 3 days. They were sacrificed by perfusion-fixation and pieces of dorsolateral striatum and of paramedian cortex were embedded in plastic and prepared for light and electron microscopy. A parallel series of animals were injected with [^{3}H]-thymidine prior to or immediately following ischemia and were allowed to survive for 3 hours or for 3 days. This model of 4-vessel occlusion produces severe irreversible injury to the small and medium sized neurons of the dorsolateral striatum and Ca1 zone of the hippocampus, leaving neurons in the paramedian cerebral cortex and Ca3 hippocampus undamaged (Pulsinelli, et al, 1982). Although cortical neurons appear normal by light microscopy, ultrastructural studies show initial mitochondrial swelling followed by loss of Golgi cisterns, disaggregation of polyribosomes, and reduction in rough endoplasmic reticulum; ultrastructural morphology returns to normal by 24 hours (Petito, et al, 1984). Irreversibly injured neurons show similar ultrastructural changes, but subsequent dilatation of Golgi vesicles and plasma membrane defects preceded shrinkage and eventual cell death. Both neuronal changes occur within the first 30 minutes following the ischemic insult.

RESULTS

Proliferative changes were confined to interstitial astrocytes and perineuronal glia in the dorsolateral striatum (Petito, et al, 1982; Petito, 1985). Progressive swelling of astrocyte cytoplasm was documented by significantly increased cytoplasmic area in the astrocytic perikarya by 40 minutes after ischemia and in the perivascular foot processes by 3 hours. Mitochondria became pleomorphic and their internal structure changed from an orthodox to a condensed configurational pattern. The number of mitochondrial profiles and the total mitochondrial area were significantly increased in the perikarya and perivascular foot processes at 3 hours after ischemia (Table 1). Rough endoplasmic reticulum (RER) appeared increased.

TABLE ONE

MITOCHONDRIAL CHANGES IN ASTROCYTES

	PERIKARYA		FOOT PROCESSES	
	Number Mito Profiles	Total Mito. Area (u2)	Number Mito. Profiles	Total Mito. Area (u2)
Control	18.86±1.1	2.96±0.98	0.78±0.55	0.12±0.08
40 Minutes Post Ischemia	20.80±7.6	4.63±0.77	7.39±3.26	1.94±0.89
3 Hours Post Ischemia	31.38±8.4**	7.71±2.67*	9.73±6.84*	2.58±2.08*

Values are expressed as mean±S.D.; Mito: mitochondria
* $p \leq 0.05$; ** $p \leq 0.01$

The number of astrocytes (expressed as the ratio between astrocytes to the total number of astrocytes and neurons) significantly increased from 10.8 $\pm$ 1.5% in controls to 17.8 $\pm$ 1.6% at 40 minutes after ischemia ($p \leq 0.01$) and 19.9 $\pm$ 2.9% at 3 hours ($p \leq 0.01$). Astrocytic mitoses were extremely rare and [^{3}H]-thymidine autoradiography failed to show mitotic activity in brain parenchyma either in the striatum or cortex at 3 hours following ischemia. Mitotic activity in glial and endothelial cells was present in the striatum at 3 days.

The contribution of the perineuronal (PN) glia to the proliferative astrocyte response in ischemia was evaluated in the same series of animals by photographing all PN glia per ultrathin section; the number of each cell type (Peters, et al, 1976) was expressed as percent of all glial cells. In the dorsolateral striatum (irreversible neuronal injury), alterations were prominent in the medium to light (M-L) oligodendrocytes and in the astrocytes, but dark oligodendrocytes showed only mild and transient swelling of Golgi and RER cisterns. Perineuronal astrocytes were dilated and showed reactive changes similar to those described above. The M-L oligodendrocytes also were dilated with significant increases in cell size, an accumulation of tubovesicular profiles, and the eventual identification of a few intermediate filaments in their dilated processes. Nuclear changes in these cells included pleomorphism, mild enlargement, and loss of the heterochromatin clumps characteristic of oligodendrocyte nuclei.

Intermediate glia include cells unable to be classified because of a paucity of microtubules or intermediate filaments in their perikarya or because of combined features of oligodendrocytes as well as astrocytes. These were first seen at 15 minutes following ischemia. Their cytoplasm was pale and contained small collections of tubovesicular profiles as well as numerous microtubules. Nuclei had dispersed chromatin and appeared similar to astrocytic nuclei. Phagocytic cells were noted first at 3 hours after ischemia and contained myelin figures and lipid inclusion droplets.

Perineuronal glia in the paramedian cortex (reversible ischemic neuronal injury) were also examined. Dark oligodendroglia were indistinguishable from controls in the post ischemic periods. Perineuronal astrocytes showed a mild but statistically insignificant increase in cell size and had a mild increase in intermediate filaments at 2 hours following ischemia, but by 24 hours after ischemia, astrocytes were similar to controls. The medium-light oligodendrocytes showed a mild and significant cell size at 30 and at 120 minutes accompanied by a increased lucency to the cytoplasm and the accumulation of a few tubovesicular profiles. These changes were transient and M-L oligodendrocytes were indistinguishable from controls by 24 hours.

A dramatic change in the glial cell distribution occurred between 2 and 3 hours after ischemia (Table 2). The number of perineuronal M-L oligodendrocytes in the dorsolateral striatum fell from 54.5% in controls and during the first 2 hours after ischemia to 6.5% at 3 hours. At the same time, the number of astrocytes increased to 23% from a mean of 8% in controls and during the first 2 hours of reperfusion (range 0 to 15%). The number of perineuronal M-L oligodendrocytes in the paramedian cortex transiently fell from 21% and 34% in controls and at 30 minutes to 9% at 3 hours and was accompanied by an increased number of astrocytes. These changes were transient and at 24 hours, the distribution of PN glia in the cerebral cortex was similar to control values.

TABLE TWO

DISTRIBUTION OF PERINEURONAL GLIA

CELL TYPE	CONTROL	POST-ISCHEMIC INTERVAL 3 MIN	15 MIN	30 MIN	2 HOUR	3 HOUR	24 HOUR
DORSOLATERAL STRIATUM - IRREVERSIBLE NEURONAL INJURY							
	(45)	(34)	(42)	(33)	(24)	(60)	
Dark Oligo	35.5%	26%	29%	33%	33.5%	55%	
M-L Oligo	53.5%	59%	54.5%	51.5%	54%	6.5%	
Astro	11%	15%	-	3%	12.5%	23%	
Int. Glia			17%	12%		6.5%	
Phago. Glia						8.5%	
PARAMEDIAN CORTEX - REVERSIBLE NEURONAL INJURY							
	(62)			(71)		(79)	(38)
Dark Oligo	61%			44%		56%	47%
M-L Oligo	21%			34%		9%	40%
Astro	16%			14%		26%	11%
Int. Glia	2%			8%		9%	2%

Numbers in parentheses refer to total number of glial cells per group. Abbreviations: Dark oligo: Dark oligodendrocyte; M-L oligo: Medium-light oligodendrocyte; Astro: Astrocyte; Int. glia: Intermediate glia; phago. glia: Phagocytic glia.

DISCUSSION

These studies show rapid and profound structural alterations in cerebral astrocytes in response to ischemic neuronal injury. The glial response parallels the severity of the neuronal injury and only transient alterations in the PN glia occur in brain regions in which neurons are reversibly injured.

The stimuli for the proliferative glial reaction are unknown. We previously suggested that elevations of extracellular ions normally regulated by astrocytes might induce these reactive changes (Petito, et al, 1982). Precedent for such ionic stimuli inducing proliferative changes in astrocytes and apparent transformation of oligodendrocytes into astrocytes is seen in conditions in which serum ammonium is elevated (Norenberg, 1977; Diemer, et al, 1977).

Alternative mechanisms responsible for the early post-ischemic glial reaction may lie in the loss of the normal neuron-glia interaction. *In vitro* studies indicate an important trophic influence by neurons on glial cells. The co-cultivation of neurons or their plasma membrane fragments with astrocytic cultures inhibits astrocyte proliferation and alters astrocyte morphology; removal of the neuronal elements induces astrocytic proliferation and morphologic changes (Sobue, et al, 1984). A similar loss of neuronal trophism on glia may occur in post-ischemic brain.

Although the results of our present ultrastructural studies show a loss of M-L perineuronal oligodendrocytes with an accompanying increase in PN astrocytes, they do not necessarily indicate that the PN glia that transform into reactive astrocytes are, in fact, mature oligodendrocytes. More likely, the transforming PN glia may be either astrocytes whose identification as such was not possible due to an absence of intermediate filaments and the presence of microtubules in the perikarya. Alternatively, the transforming cells may have been what has been termed "multipotential" glia or "intermediate" glia. Immunohistochemical or molecular biological studies are needed for further, more precise identification of these glia in control and ischemic animals.

ACKNOWLEDGEMENTS

This work was supported by NINCDS grant NS-03346, Cerebrovascular Disease Research Center.

REFERENCES

1. Diemer, N.H., J. Klee, H. Schroder, and L. Klinken (1977). Acta Neuropath., 39, 59-68.

2. Duffy, P.E. (1983). Astrocytes: Normal, Reactive and Neoplastic, Raven Press, New York, 49-54.

3. Herz, L. (1978). Brain Res., 145, 202-208.

4. Kimelberg, H.K., S. Biddlecomb, and R.S. Bourke (1979). Brain Res., 173, 111-124.

5. Kraig, R.P., W.A. Pulsinelli, and F. Plum (1985). Am. J. Physiol., in press.

6. Norenberg, M.D. (1977). Lab. Invest., 618-627.

7. Peters, A., S.L. Palay, and H.deF. Webster (1976). The Fine Structure of the Nervous System, Saunders, Philadelphia.

8. Petito, C.K. (1985). Soc. Neurosci., 11, 396.

9. Petito, C.K., and T. Babiak (1982). Ann. Neurol., 11, 510-518.

10. Petito, C.K., and W.A. Pulsinell (1984). J. Neuropath. Exp. Neurol., 43, 141-153.

11. Plum, F. (1983). Neurol., 33, 222-233.

12. Pulsinelli, W.A., and J.B. Brierley (1979). Stroke, 10, 267-272.

13. Pulsinelli, W.A., J.B. Brierley, and F. Plum (1982). Ann. Neurol., 11, 491-499.

14. Sobue, G., and D. Pleasure (1984). Brain Res., 324, 175-179

Astroglia: Receptors and their Influence on Intracellular Processes

K. D. McCarthy, J. Prime, R. Pollenz, S. Burgess, P. Trimmer, L. Lerea and C. Ingraham

Department of Pharmacology, School of Medicine, University of North Carolina at Chapel Hill, Chapel Hill, North Carolina 27514, USA

INTRODUCTION

Many of the proposed functions of astrocytes in developing and mature brain are dependent on the ability of these cells to recognize and respond to specific neuronal signals. Neuronal signals which have been proposed to mediate neuronal-glial interactions include ions (Nichlas and Browning, 1983; Pentreath and Kai-Kai, 1982; Latzkovits et al., 1982), neurotransmitters/neuromodulators (neuromediators) McCarthy and de Vellis, 1978; van Calker et al., 1983; Evans et al., 1984; Harden and McCarthy, 1982), and macromolecules (Unsicker et al., 1984; Banker, 1980; Hatten et al., 1984). Much of the information available concerning the "responsiveness" of atrocytes to neuronal signals has been obtained using either primary cultures of glia or cell lines presumed to be of glial origin. It is clear that, using these cell culture models, astroglia exhibit a large number of different receptor systems which may be important in sensing neuronal signals. The studies described below focus on the following series of questions. First what neuromediator receptors are expressed by cultured astroglia? Second, what biochemical events are influenced in astroglia following receptor stimulation? Third, is there heterogeneity in the expression of receptors among cultured astroglia? Fourth, does the expression of receptors by cultured astroglia reflect 1) a culture phenomenon, 2) a developmental phenomenon, or 3) a characteristic of mature astrocytes in brain?

METHODS

Primary cultures of astroglia were prepared from neonatal (unless otherwise stated) rat cerebral cortical tissue as described by McCarthy and de Vellis, 1980). Cultures used to assess population parameters (i.e. receptors via "grind and bind" assays, second messengers, or regulation of protein phosphorylation) were routinely stained for glial fibrillary acidic protein (GFAP) and fibronectin to determine the percentage of astroglial cells and fibroblasts, respectively. Only cultures which were determined to be in excess of 95% astroglia (GFAP+) were used.

Receptor binding studies were completed as described previously (McCarthy and Harden, 1981; Harden and McCarthy, 1982; Trimmer et al., 1984). The influence of receptor stimulation on cyclic AMP metabolism was monitored as described previously (McCarthy and de Vellis, 1978; Evans et al., 1984). The influence of receptor stimulation on phosphatidylinositide metabolism was monitored as described by Masters et al., 1984. The expression of receptors on individual immunocytochemically defined cells via receptor autoradiography was examined previously described (McCarthy, 1983; Burgess and McCarthy, 1985).

Advances in the Biosciences Vol. 61

RESULTS

During the past ten years a large number of studies have been published which indicate that glia exhibit membrane receptors. The evidence in support of the expression of receptors by glia comes from a number of laboratories using a variety of different experimental approaches. Research from our laboratory which relates to the expression of receptors by cultured astroglia is summarized in Table 1. It should be emphasized that this probably represents a partial list of receptors expressed by cultured glia. All of the receptor systems presented in Table 1 have been shown to influence the metabolism of either cyclic AMP or phosphatidylinositide. Hence, these receptor systems represent more than binding sites. It is worth noting that astroglia exhibit the beta-1 subtype of the beta-adrenergic receptor (B-AR) and that this receptor subtype appears to be regulated by neuronal activity in the CNS (Minneman et al., 1979). This observation suggests that astrocytes may represent target cells for norepinephrine released in the CNS. Interestingly, we have never observed B-AR associated with neurons cultured from the cerebral cortical tissue of immature rats by either radioligand population analyses (McCarthy and Harden, 1982) or radioligand single-cell analyses (Burgess et al., 1985). In summary, cultured astroglia exhibit a large number of different receptor systems which regulate intracellular second messenger systems. Whether or not these receptor systems are expressed by astrocytes _in vivo_ remains an important question yet to be resolved.

Table 1

Receptors Expressed by Astroglia Cultured From Neonatal Rat Cerebral Cortical Tissue.

Receptor	Response	Reference
Beta-1 Adrenergic	Increase Cyclic AMP Metabolism	McCarthy and de Vellis, 1978 Harden and McCarthy, 1982
Alpha-2 Adrenergic	Decrease Cyclic AMP Metabolism	McCarthy and de Vellis, 1978
Alpha-1 Adrenergic	Increase PI Metabolism	De George et al., 1986 Pollenz and McCarthy, 1986
5-Hydroxytryptamine	Increase PI Metabolism	Pollenz and McCarthy, 1986
Prostaglandin E1	Increase Cyclic AMP Metabolism	McCarthy and de Vellis, 1978
Adenosine	Increase Cyclic AMP Metabolism	McCarthy and de Vellis, 1978
Muscarinicach	Increase PI Metabolism	De George et al., 1986
Histamine	Increase PI Metabolism	De George et al., 1986
Glutamate	Increase PI Metabolism	De George et al., 1986
VIP	Increase Cyclic AMP Metabolism	Evans et al., 1985
Alpha-MSH	Increase Cyclic AMP Metabolism	Evans et al., 1985
ACTH	Increase Cyclic AMP Metabolism	Evans et al., 1985
Somatostatin	Decrease Cyclic AMP Metabolism	Evans et al., 1985

The fact that cultured astroglia exhibit such a diverse set of receptor systems raises several questions. For example, is there heterogeneity among cultured astroglia with respect to receptor expression? Likewise, do astroglia isolated from more mature animals continue to exhibit a similar array of neuromediator receptors? For a number of reasons these questions have been difficult to address using standard techniques. This is not surprising when one realizes that most of the information concerning the heterogeneity of neurons comes from single cell analyses such as electrophysiological and histochemical studies. The problem in establishing whether or not astrocytes from older animals exhibit receptors is complicated by the fact that it remains very difficult to prepare astroglial cultures from animals older that 7 days postnatal that are not heavily contaminated with fibroblasts. To circumvent these difficulties, we have developed a method for assessing receptors on single, immunocytochemically-defined cell types using quantitative receptor autoradiography (McCarthy, 1983). Using this technique we have been able to show that there are two morphologically-distinct populations of GFAP+ atroglia which differ in their expression of B-AR (McCarthy 1983; Burgess and McCarthy, 1985). The predominant type of GFAP+ cell in purified astroglial cell cultures is a polygonally shaped, mitotic cell which exhibits a large number of B-AR. This cell type, refered to as type 1 astrocytes by Raff and coworkers (Raff et al., 1983), expresses far more B-AR (either per cell or per square micron) than either fibroblasts, neurons, or oligodendrocytes prepared from cerebral cortical tissue (Table 2) (Burgess et al., 1985).

Table 2

Comparison of the number of *ICYP autoradiographic grains over various types of immunocytochemically identified neural cells

	Cell type			
	Astroglia (GFAP)	Fibroblast (fibronectin)	Oligodendroglia (galacto-cerebroside)	Neurons (A2B5)
Total	94.7 ± 9.9	18.5 ± 2.1	3.4 ± 1.0	1.3 ± 0.7
Non-specific	5.2 ± 1.4	7.1 ± 2.0	2.1 ± 1.1	0.7 ± 0.2
Specific	84.8	11.4	1.3	0.6

Polygonal astroglia, fibroblasts, oligodendroglia and neurons were obtained and immunocytochemically identified. To measure total binding, the cells were exposed to 32pM ^{125}I-CYP (+ICYP); non-specific binding was measured by addition of 1.0 μM O,L-propranolol during *ICYP binding. The silver grains associated with each type of cell were quantified using a microcomputer-based video digitizing system. Values for background (silver grains on cell-free areas of the slide) were subtracted. Data are presented as density of silver grains (number of autoradiographic grains per 1000 μM^2 of cell membrane) for at least 25 cells of each type. Data from polygonal astroglia and fibroblasts were obtained from the same experiments (these cells were grown on the same slides) while the oligodendroglia and neurons were grown on separate slides.

We have also found that polygonal astroglia cultured from cerebellum and hippocampal tissue exhibit a large number of B-AR. The second type of GFAP$^+$ cell present in these cultures is a process-bearing cell which rarely divides and fails to exhibit B-AR (Table 3) (Burgess et al., 1985). This second type of GFAP$^+$ cell has been referred to as Type 2 astrocytes by Raff and coworkers (Raff et al., 1983) and can give rise to cells exhibiting oligodendroglial markers such as galatocerbroside. The relationship between these two cell types (polygonal and process-bearing GFAP$^+$ cells) has not been firmly defined. The results of these studies indicate that there are pharmacologically-distinct types of GFAP$^+$ astroglia in culture which can be distinguished on the basis of their expression of the B-AR.

Table 3

Comparisons of *ICYP autoradiographic grains over morphological variants of GFAP positie astroglia

	Type and morphology		
	Type I, polygonal	Type II, intermediate	Type III, process bearing
Total grains	133.7 ± 19.3	10.5 ± 1.7	5.5 ± 1.4
Non-specific	5.2 ± 1.4	1.4 ± 0.6	0.9 ± 0.5
Specific	128.5	9.1	4.6
% of Type I	100%	7.1%	3.6%

The cells were obtained and analyzed as in Table 2 using 64 pM *ICYP, and quantified using a video digitizer. The data presented were obtained from 9-24 cells in a single experiment, and are representative of five similar experiments.

The difficulty in preparing homogeneous cultures of astroglia from older animals (3-4 days postnatal and older) has hampered efforts to determine if astroglia isolated from older animals exhibit receptors (Trimmer et al., 1984). Using the method of combined immunocytochemistry, receptor autoradiography described above, experiments were completed to examine the expression of B-AR on astroglia isolated from rat cerebral cortical tissue ranging in age from 16-days gestation to 28-days postnatal (Trimmer and McCarthy, 1986). The results presented in figure 1 indicate that astroglia isolated from all ages of animals examined exhibit B-AR. It is important to note that astroglia exhibited B-AR even when they had been in culture for only a few days. Given the slow turnover of B-AR (Hughes et al., 1982), these results strongly suggest that the majority of B-AR were not synthesized in culture but were present on the astroglia when isolated from brain.

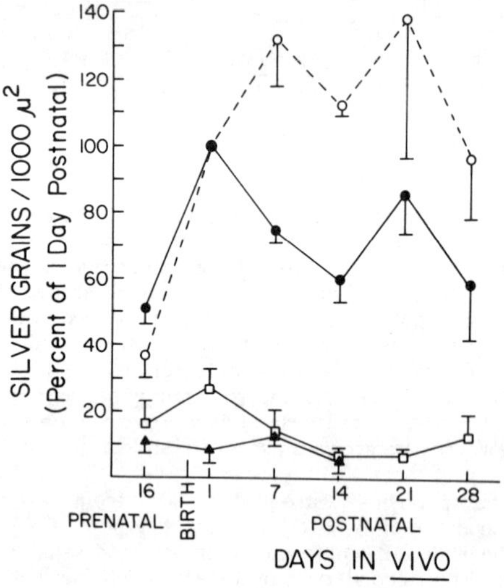

Figure 1. The density of silver grains associated with immunocytochemically-defined polygonal astroglia (long term culture-8-22 days in vitro), polygonal astroglia (short term culture-1-5 days in vitro), process-bearing astroglia and fibroblasts. Using epifluorescence optics, single cells were selected according to morphological and staining criteria. Scan boundries for the microcomputer based video digitizing system were placed so that the grain density over 70 to 90% of the cell surface was quantitated. The grain counts were expessed per 1000 square microns of cell surface. Up to 25 cells per well were counted. Autoradiographic background was determined per 1000 square microns and subtracted from the grain count for each cell. Since there were differences in ^{125}I-CYP concentration and autoradiographic conditions between culture preparations, the mean number of silver grains for each age or cell type in a given culture was expressed as a percent of the mean number of silver grains on polygonal cells from animals 1 DPN.

Polygonal Astroglia - 8-22 days in vitro (solid circles)
Each point represents the mean silver grain density from 4 (21 or 28 DPN) or 5 (16 days in gestation - 14 DPN) separate culture preparations.

Polygonal Astroglia - 1-5 days in vitro (open circles, broken line)
Each point represents the mean silver grain density from 3 separate culture preparations except at 21 DPN where ony two were counted.

Process-Bearing Astroglia (triangles)
Since these cells were not consistently present at all ages in all cultures only two cultures preparations are shown.

Fibroblasts (squares)
Each point graphed represents the mean silver grain density from 2(1, 21 DPN) or 3 (16 days in gestation, 7, 14, 28 DPN) separate culture preparations.

More recently, we have begun to examine the expression of alpha-1 adrenergic receptors (A1-AR) by cultured neural cells. While preliminary, the results of these studies suggest that cultured astroglia exhibit a high number of A1-AR relative to neurons. In contrast to the expression of B1-AR which appear on all polygonal astroglia, the A1-AR appears restricted to a subpopulation of polygonal astroglia. If these preliminary results are correct, this will be the first indication of receptor heterogeneity among the polygonal astroglia cultured from cerebral cortical tissue.

Given that cultured astroglia exhibit receptors for a variety of neuromediators, the question arises to what astroglial events are regulated by these receptors. To address this question we have examined the influence of receptor stimulation on the phosphorylation of astroglial proteins. The rationale behind this approach was that most of the receptors known to be present on cultured astroglia appear to either increase or decrease cyclic AMP levels (see Table 1). As the only known function of cyclic AMP in eucaryotic cells is the activation of cyclic AMP-dependent protein kinase, it seemed likely that the phosphorylation of specific proteins might be altered by receptor stimulation. More recent studies indicate that certain receptors present on astroglia increase phosphatidylinositide (PI) metabolism (Pearce et al., 1985; De George et al., 1986) and as a result might be expected to stimulate protein kinase C via the elevation of diacylglycerol. Studies from our laboratory (McCarthy et al., 1985) and that of Browning and coworkers (Browning and Ruina, 1984) indicate that the predominant effect of receptor agonists which increase cyclic AMP levels in astroglia is an increase in the phosphorylation of the IF-proteins GFAP and vimentin (Figure 2). The identity of the proteins phosphorylated by B-AR stimulation was shown to be GFAP and vimentin through immunoblot studies (McCarthy, et al., 1985).

More recent studies have verified that the increase in phosphorylation of both GFAP and vimentin was due to an increase in the net number of phosphate groups associated with the proteins rather than an increase in the turnover of phosphates (Pollenz and McCarthy, 1986). Similar results have been obtained using 5-hydroxytryptamine (5-HT) which stimulates PI metabolism (rather than cyclic AMP metabolism) in these cells. Together, these studies indicate that the primary influence of receptor agonists on protein phosphorylation in cultured astroglia involves an increase in the phosphorylation of the intermediate filament proteins GFAP and vimentin.

A number of investigators have shown that increasing astroglial cyclic AMP levels results in the conversion of these cells from a polygonal to a process-bearing morphology. Studies were completed to examine the possibility that the phosphorylation of IF-proteins mediated the change in astroglial morphology (Pollenz and McCarthy, 1986). We were able to show that treatment with brain extract converted polygonal astroglia to process-bearing cells without an increase in the phosphorylation of IF-proteins (Pollenz and McCarthy, 1986). In addition, treatment with forskolin at room temperature maximally increased the phosphorylation of the IF-proteins without any apparent influence on the morphology of the astroglia. The key results of this study are presented in figures 3 and 4. The most likely explanation of these results is that the two processes are not related, that is, IF-protein phosphorylation is not necessary for the conversion of polygonal astroglia to process bearing astroglia. However, it remains possible that the mechanisms by which brain extract and forskolin induce shape changes are distinct and that the phosphorylation of IF-proteins may be required for the shape change observed with forskolin treatment at 37°C.

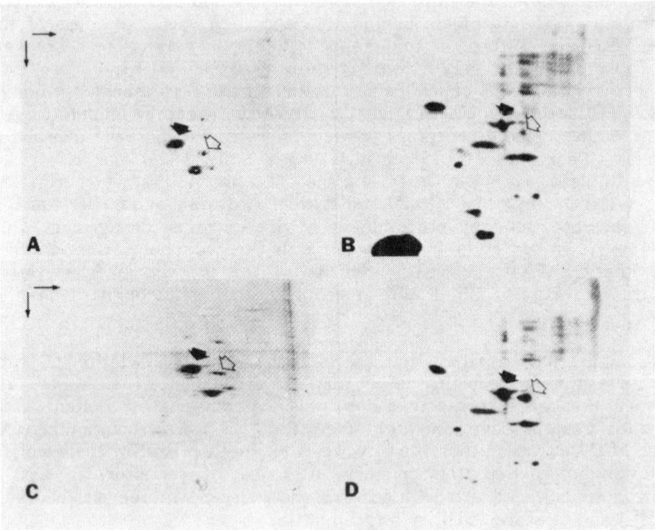

Figure 2. Effect of forskolin on the phosphorylation of astroglial proteins. Purified cultures of polygonal astroglia in 35-mm culture dishes were incubated in low-phosphate MEM supplemented with 20 mM HEPES and 0.4 mCi ^{32}P/ml medium at 37°C. Either forskolin (10^{-5} M) or vehicle was included in the incubation medium during the final 20 min of incubation. The incubation was terminated and proteins separated. Gels were stained with Coomassie Blue, dried, and autoradiograms were prepared by overlaying the dried gels with film for 12 h. The arrows on the gel and autoradiogam indicate the proteins that were phosphorylated in response to treatment with forskolin. The open arrow indicates GFAP. The solid arrow indicates vimentin. **A:** Coomassie Blue-stained gel from control astroglia. **B:** Autoradiogram prepared from control astroglia. **C:** Coomassie Blue-stained gel from forskolin-treated astrolia. **D:** Autoradiogram prepared from forskolin-treated astroglia.

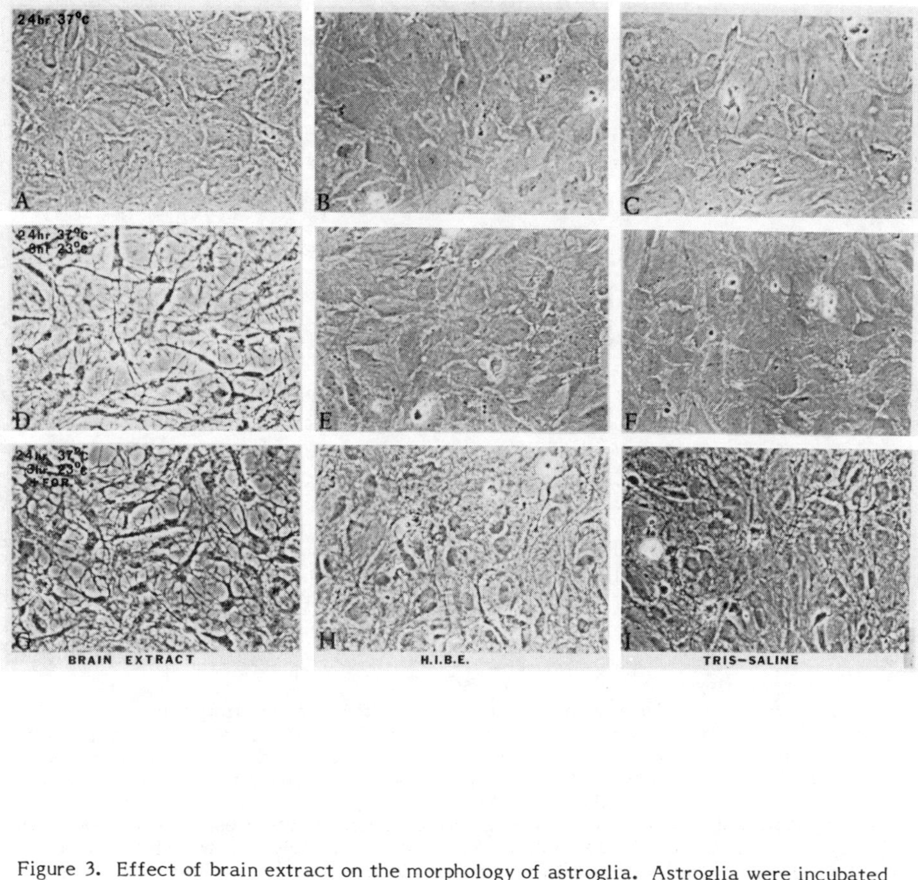

Figure 3. Effect of brain extract on the morphology of astroglia. Astroglia were incubated in media supplemented with brain extact (BE) (30% vol/vol), heat inactivated brain extract (HIBE) (30% vol/vol), or Tris-saline (30% vol/vol) at 37°C for 24 h. The cells were then incubated at 23°C for an additional 3 h. Forskolin (10 μM) was present in selected samples during the 1st hour of incubation. **A–C:** Phase-contrast micrographs of BE-, HIBE-, and Tris/saline-treated cells after a 24-h incubation at 37°C (A, BE-treated; B, HIBE-treated; C, Tris/saline-treated). **D–F:** Phase-contrast micrographs of BE-, HIBE-, and Tris/saline). **G–I:** Phase-contrast micrographs of BE-, HIBE- and Tris/saline-treated cells after a 24-h incubation at 37°C and an additional incubation at 23°C during which forskolin was present for the final hour (G, BE-treated; H, HIBE-treated; I, Tris/saline-treated).

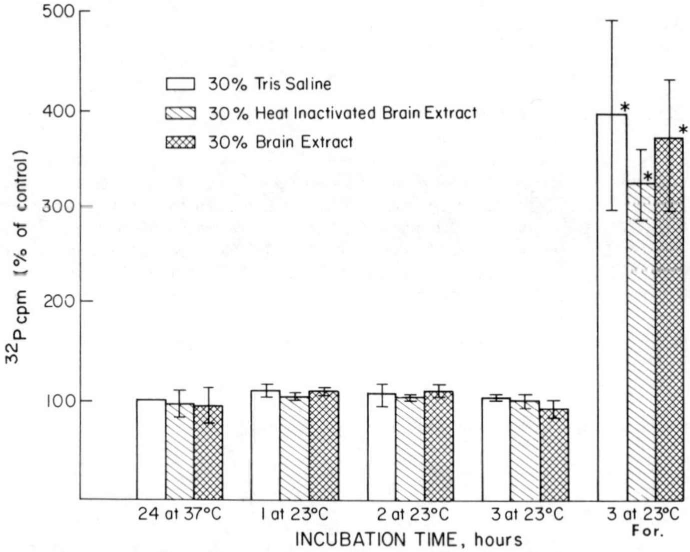

Figure 4. Effect of brain extract on the phosphorylation of GFAP. Astroglia were incubated in ^{32}P-labeling media supplemented with BE (30% vol/vol), HIBE (30vol/vol), or Tris/saline (30% vol/vol) at 37°C for 24 h. The cells were then incubated at 23°C for an additional 3 h with forskolin (10 μM) present in the indicated samples (For.) during the final hour. Protein was extracted, separated, and normalized. The histogram represents results from three separate experiments and is expressed as the mean ± SE as compared to the Tris-saline 24-h time point from each experiment (100%). *Statistically different from control ($p \leq 0.01$).

DISCUSSION

Research from this and many other laboratories indicate that cultured astroglia exhibit many different receptor systems for neuromediators. While it has been more difficult to demonstrate heterogeneity among atroglia with respect to receptor expression, such heterogeneity appears to exist. For example, there are two distinct classes of astroglia in culture which can be distinguished on the basis of their morphology and expression of B-AR. In addition, it appears that polygonal astroglia may be heterogeneous with respect to their expression of A1-AR. From these results it seems likely that astroglia may be far more complex in receptor expression than previously thought.

Receptor agonists which increase either cyclic AMP or PI metabolism result in an increase in the number of phosphate groups associated with GFAP and vimentin. While it is clear that other astroglial proteins may also be phosphorylated following receptor stimulation, their effect on IF-proteins is by far the most prominent. The effect of increasing the phosphorylation of astroglial IF-proteins remains unknown. Experiments completed suggest, but do not rule out the possibility, that the phosphorylation of IF-proteins does not mediate the morphological conversion of these cells which occurs following rises in astroglial cyclic AMP levels.

It is important to note that most of what we know concerning the pharmacology of astrocytes has been obtained using cultured cells derived from immature brain. This raises the question of whether or not the information obtained using this model reflects 1) a culture phenomenon, 2) a developmental phenomenon, or 3) a set of properties exhibited by mature astrocytes in brain. The possibility that the properties displayed by cultured astroglia reflect a culture phenomenon seems unlikely. The best argument against this possibility is the many examples of fidelity between the expression of cell-specific characteristics of cultured cells and the same cells studied *in vivo* (Pollack, 1981). However, one goal of investigators using this model should be the demonstration that the parameters characterized for astroglia in culture are exhibited *in vivo.*

The question of whether or not the pharmacological properties of cultured astroglia reflect the characteristics immature cells rather than mature astrocytes has been difficult to address. To examine this question we have studied the expression of B-AR on individual GFAP+ cells isolated from 28-day old rat cerebral cortical tissue and maintained in culture for short periods. While such astroglia exhibit B-AR, it remains possible that tissue dissociation results in the selection of immature astroglia and that these cells, like astroglia from neonates, exhibit B-AR. Much of the effort of this laboratory is directed at the question of whether or not the pharmacological properties displayed by cultured astroglia reflect the characteristics of astroglia in immature and/or mature brain.

ACKNOWLEDGEMENT
Supported by PHS Grants NS 1696 and NS 20212.

REFERENCES

Banker, G.A. (1980). *Science 209*, 809-810.

Browning, E.T. and M. Ruina (1984). *J. Neurochem. 42*, 718-726.

Burgess, S.K., P.A. Trimmer and K.D. McCarthy (1985). *Brain Res. 335* 11-19.

Burgess, S.K. and K.D. McCarthy (1985). *Brain Res. 335*, 1-9.

De George, J.J., P. Morell, K.D. McCarthy and E. Lapetina (1986). Submitted.

Evans,T., K.D. McCarthy, and T.K. Harden (1984). *J. Neurochem. 43*, 131-138.

Harden, T.K. and K.D. McCarthy (1982). *J. Pharm. Exp. Therap. 222*, 600-605.

Hatten, M.E., R.K.H. Liem and C.A. Mason (1984). *J. Cell Biol 98*, 193-204.

Hughes, R.J., M.R. Boyle, R.D. Brown, P. Taylor and P.A. Insel (1982). *Mol. Pharm. 22*, 255-266.

Latzkovitz, L., A. Rimanoczy, A. Juhasz, C. Torday, M. Sensebrenner (1982). *Dev. Neurosci. 5*, 92-100.

Masters, S.B. T.K. Harden, J.H. Brown (1984). *Mol. Pharm. 26*, 149-155.

McCarthy, K.D. and J. de Vellis (1978). *J. Cyclic Nucl. Res. 4*, 15-26.

McCarthy, K.D., and J. de Vellis (1980). *J. Cell Biol. 85*, 890-902.

McCarthy, K.D. and T.K. Harden (1982). *J. Pharm. Exp. Therap. 216*, 183-191.

McCarthy, K.D. (1983). *J. Pharm. Exp. Therap. 226*, 282-290.

McCarthy, K.D. J. Prime, T. Harmon and R. Pollenz (1985). *J. Neurochem. 44*, 723-730.

Minneman, K.P., M.D. Dibner, B.B. Wolfe and P.B. Molinoff (1979). Science 204, 866-868.

Nicklas, W.J., and E.T. Browning (1983). J. Neurochem. 41, 179-187.

Pearce, B. M. Cambray-Deakin, C. Morrow, J. Grimble and S. Murphy (1985). J. Neurochem. 45, 1534-1540.

Pentreath, V.W., and M.M. Kai-Kai (1982). Nature 295, 59-61.

Pollack, R. (1981). Readings in Mammalian Cell Culture. Cold Spring Harbor Laboratory.

Pollenz, R.P. and K.D. McCarthy (1986). J. Neurochem., in press.

Raff, M.C., E.R. Abney, J. Cohen, R. Lindsay and M. Nobel (1983). J. Neurosci. 3, 1289-1300.

Trimmer, P.A., T. Evans, M.M. Smith, T.K. Harden and K.D. McCarthy (1984). J. Neurosci. 4, 1598-1606.

Trimmer, P.A. and K.D. McCarthy (1986). Brain Res., in press.

Unsicker, K., J. Vey, H-D. Hofmann, T.H. Muller and A.J. Wilson (1984). Proc. Nat. Acad. Sci. 81, 2242-2246.

van Calker, D., F. Loffler, and B. Hamprecht (1983). J. Neurochem. 40, 418-427.

Benzodiazepine Receptors in Astrocytes

A. S. Bender and L. Hertz

Department of Pharmacology, University of Saskatchewan,
Saskatoon, Saskatchewan, S7N 0W0, Canada

ABSTRACT

Astrocytes have benzodiazepine receptors which show a high affinity binding not only for RO 5-4864, a ligand for a so-called "peripheral type" benzodiazepine receptor, but also for clinically relevant drugs like flunitrazepam and diazepam. Kinetic and pharmacological characteristics of this receptor are discussed and it is suggested that it may be associated with a calcium channel.

KEY WORDS

Astrocytes; benzodiazepine receptor; calcium channel; diazepam; flunitrazepam; RO 5-4864.

INTRODUCTION

It is now becoming accepted that two different benzodiazepine receptors are found in the central nervous system (CNS), i.e., the so called "central type" benzodiazepine receptor from which benzodiazepines are potently displaced by clonazepam but very little by RO 5-4864 (Braestrup and Nielsen, 1983; Tallman and Gallager, 1985), and the so-called "peripheral type" benzodiazepine receptor from which the benzodiazepines are potently displaced by RO 5-4864 but less potently by clonazepam (Gallager et al, 1981; Marangos et al, 1982; Schoemaker et al, 1983). The studies of the latter receptor have been much facilitated by the use of labelled RO 5-4864, a benzodiazepine which binds specifically to the "peripheral type" benzodiazepine receptor in the periphery (e.g., heart, kidney) and in the CNS (Marangos et al, 1982; Schoemaker et al, 1983; Weissman et al, 1984a; Anholt et al, 1985). It was long thought that this receptor had little functional importance in the CNS since RO 5-4864 was believed to be behaviorally inert. However, recent experiments have shown that this benzodiazepine does indeed exert sedative, convulsant and anxiogenic effects in rodents (Weissman et al, 1983; Pellow and File, 1984). Autoradiographic studies have shown that the binding of RO 5-4864 in the CNS is highest in the glial-rich olfactory nerve and that these binding sites are diffusely distributed throughout the brain (Richards et al, 1982; Gehlert et al, 1983; Mohler and Richards, 1983; Snyder, 1983; Anholt et al, 1984). Moreover, Schoemaker et al, 1982 observed that in brains of patients with Huntington's disease and after kainic acid destruction of neurons in rat brains there is a significant increase in RO 5-4864 binding sites in the brain, associated primarily with glial cells. In contrast, it has often been assumed that pharmacologically relevant benzodiazepines e.g., flunitrazepam and diazepam show little or no binding to "peripheral type" benzodiazepine receptors in the CNS (McCarthy and Harden, 1981; Richards et al, 1982; Tallman, 1983). Evidence to the contrary was first obtained using cells prepared by gradient centrifugation (Henn and Henke, 1978). More recently, detailed studies have been performed with primary cultures of astrocytes (Dudai et al, 1979; Hertz and Mukerji, 1980; Tardy et al, 1981,1985; Bender and Hertz, 1984, 1985a, 1986; Sher and Machen, 1984) . These will be discussed in the present review. Since both the "central" and "peripheral type" benzodiazepine receptors are localized in the CNS although on different cell types, i.e., neurons and astrocytes, respectively, it may be more appropriate to name these receptors neuronal and astrocytic. This terminology will be used in the present paper.

KINETICS OF BENZODIAZEPINE BINDING

The time course for binding of [^{3}H]-RO 5-4864 (1.8 nM) to intact astrocytes remaining attached to the culture dishes, is shown in Fig. 1. The total binding is 50% completed in 10 min and it reaches an equilibrium of 3,000-4,000 fmol/mg protein at about 100 min. The non-specific binding, measured in the presence of excess (10 μM) non-labelled RO 5-4864 is considerably faster

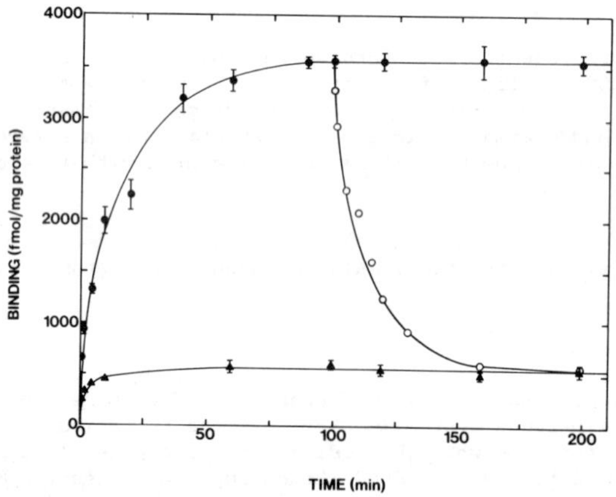

Fig. 1. Time-course for binding and dissociation of [^{3}H]RO 5-4864 in intact primary cultures of astrocytes. Total (●) and non-specific (▲) binding. Dissociation of binding of [^{3}H]RO 5-4864 (O). From Bender and Hertz, 1985a.

and reaches a maximum after 5-10 min. At this time it comprises ≃ 500 fmol/mg protein and amounts to ≃ 15% of the total binding (Bender and Hertz, 1985a). Thus, the specific binding constitutes ≃ 85% of the total binding. After equilibrium had been obtained the dissociation of labelled RO 5-4864 from its astrocytic binding sites was studied by adding excess (10 μM) non-labelled RO 5-4864 (Fig. 1). The dissociation followed first order kinetics (Fig. 2) with a rate constant of 0.06 min^{-1} which is quite close to values that have been reported for dissociation of RO 5-4864 from brain membranes (Bender and Hertz, 1985a). This similarity and the pharmacological characteristics for displacement of RO 5-4864 binding in brain membranes and in cultured astrocytes (Bender and Hertz, 1985a) are indications that the present cultures constitute a good model for their *in vivo* counterparts. Fig. 2 also shows a considerably faster dissociation for flunitrazepam (rate constant 0.20 min^{-1}). This dissociation is about four times faster than that of flunitrazepam in neurons (rate constant 0.05 min^{-1}) (Bender and Hertz, 1984).

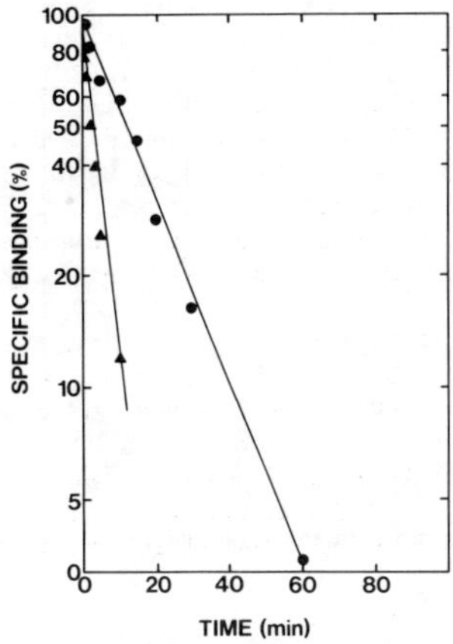

Fig. 2. Dissociation of specific [^{3}H]-flunitrazepam (▲) and [^{3}H]-Ro 5-4864 (●) binding in intact primary cultures of astrocytes.

A comparison between binding to intact cultures and to a homogenate of the cultured cells is shown in Table 1. After homogenization the [^{3}H]-RO 5-4864 binding in astrocytes is reduced by 65% when determined by a centrifugation assay and by 90% when a filtration assay is used (Bender and Hertz, 1985a). The situation for flunitrazepam is even worse since a 93% reduction was found using a centrifugation method and a 97% reduction was observed when a filtration assay was employed (Bender and Hertz, 1984; Hertz and Richardson, 1984). This difference is probably at least partly due to the very fast dissociation of flunitrazepam from its astrocytic binding sites. Thus, intact cultures are simply rapidly washed after the binding has taken place, whereas in a centrifugation assay the pellet is resuspended and washed two times and the total procedure lasts at least 10 min., i.e., a time which is long enough to cause very pronounced dissociation (Fig. 2). This does not explain the loss of binding when a filtration assay is used but a probable explanation would be that the binding of flunitrazepam to astrocytes occurs to subcellular structures which are able to pass the filters used or that the suction pressure applied may enhance dissociation. This effect of homogenization may explain why very little binding of flunitrazepam was found to an astrocytic membrane

Table 1. Binding of benzodiazepines (1.8 nM) to various preparations of astrocytes in primary cultures.

	Specific binding (fmol/mg protein)		
	RO 5-4864[a]	Flunitrazepam[b]	Diazepam[c]
Intact astrocytes	3,519±225	1,160±102	1,270±134
Homogenized astrocytes			
centrifugation assay	1,226±137	85±22.2	
filtration assay	328±78	29±1.9	

a) Bender and Hertz, 1985a;b) Bender and Hertz, 1984;
c) Bender and Hertz, 1986

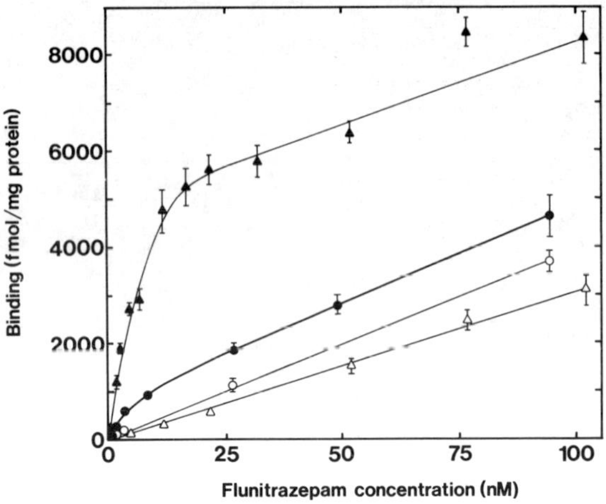

Fig. 3. Saturation of [^{3}H]-flunitrazepam binding to intact astrocytes and neurons in primary cultures; (▲) represents total binding and (△) nonspecific binding to intact astrocytes; (●) represents total binding and (0) nonspecific binding to intact neurons. Modified from Bender and Hertz, 1984.

preparation by McCarthy and Harden (1981). The somewhat slower dissociation of flunitrazepam from neuronal binding sites (see above) may explain why a larger fraction of neuronal flunitrazepam binding is recovered after homogenization and filtration (McCarthy and Harden, 1981; Bender and Hertz, 1984). This is probably the explanation why such drugs as flunitrazepam have been claimed to interact only with the "central-type" (i.e., neuronal) benzodiazepine receptor in brain membrane preparations.

Table 2. Kinetic parameters (K_D and B_{max}) for binding of benzodiazepines to astrocytes obtained by gradient centrifugation* or grown in primary cultures.

Preparation	Ligand	K_D (nM)	B_{max} (fmol/mg protein)	Reference
intact*	Diazepam	5.0	4390	Henn and Henke, 1978
intact	Diazepam	30.0	19500	Hertz and Mukerji, 1980
membranes	Diazepam	5.0	100	Tardy et al, 1981
intact	Flunitrazepam	6.6	6033	Bender and Hertz, 1984
homogenate	Flunitrazepam	13.4	1027	Bender and Hertz, 1984
intact	Diazepam	13.8	13000	Sher and Machen, 1984
intact	RO 5-4864	6.7	12228	Bender and Hertz, 1985a
membranes	Flunitrazepam	5.0	1500	Tardy et al, 1985
intact	Diazepam	12.7	8770	Bender and Hertz, 1986

As illustrated in Fig. 3 the binding of flunitrazepam to astrocytes in primary cultures is saturable within the nanomolar range and the density of the binding sites is higher than in corresponding cultures of neurons (Bender and Hertz, 1984). The Scatchard plot is linear (Bender and Hertz, 1984), indicating binding to a homogenous population of binding sites. Table 2 shows that both flunitrazepam, diazepam and RO 5-4864 bind to intact and homogenized astrocytes with a high affinity (K_D values 5-30 nM) and that the effect of homogenization on K_d is marginal. This table also shows that the decreased binding after homogenization demonstrated in Table 1 is mimicked by a substantial decrease in B_{max} values (100-1,500 fmol/mg protein after homogenization; 4,390 fmol/mg protein in astrocytes obtained by gradient centrifugation; and 6,000-19,500 fmol/mg protein in intact astrocytes in primary cultures). The binding density for diazepam is also higher than in intact neurons (Sher and Machen, 1984; Bender and Hertz, 1986). An intense binding to cocultures of neurons and astrocytes has also been demonstrated (Walker and Peacock, 1981; Sher et al, 1982). The pharmacological characteristics of this binding (see below) suggests that it mainly occurs to the astrocytes.

PHARMACOLOGICAL SPECIFICITY OF ASTROCYTIC BENZODIAZEPINE BINDING

Pharmacological differences between the astrocytic and neuronal benzodiazepine binding site have long been recognized since clonazepam is more potent in displacing benzodiazepines from neuronal than from astrocytic binding sites whereas the reverse is true for RO 5-4864 (e.g., Braestrup et al, 1978; Gallager et al, 1981; McCarthy and Harden, 1981). We have recently chosen diazepam for a detailed pharmacological study in astrocytes and neurons in primary cultures since this ligand binds well and with a high affinity to both astrocytes and neurons (Bender and Hertz, 1986). Fourteen different benzodiazepines, two benzodiazepine antagonists, and three anti-epileptics were studied. The pharmacological specificity was assessed on the basis of the IC_{50} values (concentration required to inhibit 50% of specific binding) of these agents. The [^{3}H]-diazepam binding was potently inhibited by benzodiazepines in both astrocytes and in neurons (Table 3). The IC_{50} values were in the nanomolar range (i.e., below the therapeutic plasma level of $\simeq 1\ \mu M$ (Booker and Cellesia, 1973; Garrattini et al, 1973; Greenblatt et al, 1978; Hillestad et al, 1974; R.J. Herman, personal communication) with the exception of clonazepam, flurazepam, RO 11-3128, RO 11-3624, RO 5-2181 and chlordiazepoxide in astrocytes and RO 5-4864 and RO 5-2181 in neurons where they were in the low micromolar range. For many of the drugs there was a tendency for the IC_{50} values to be slightly (e.g., flunitrazepam, midazolam) or considerably (e.g., clonazepam, flurazepam) lower in neurons than in astrocytes but exceptions to this are the two drugs RO 5-4864 and RO 7-3351 which are much more potent in astrocytes than in neurons (Table 3). On the whole, the correlation between inhibition of benzodiazepine binding in neurons and in astrocytes is poor (correlation coefficient 0.24) (Bender and Hertz, 1986). Another striking difference between astrocytic and neuronal benzodiazepine binding is the stereospecificity of the neuronal binding site and the lack of stereospecificity at the astrocytic binding site. Thus, when the two pairs of stereoisomers RO 11-6896(+) and RO 11-6893(-) and RO 11-3128(+) and RO 11-3624(-) were tested, the (+) isomers (RO 11-6896 and RO 11-3128) were significantly more potent than the (-) isomers (RO 11-6893 and RO 11-3624) in displacing neuronal diazepam binding whereas the two stereoisomers were equally potent in displacing diazepam from astrocytic binding sites. Such a lack of stereospecificty has also been observed when [^{3}H]-RO 5-4864 is used as a ligand in brain membrane preparations (Marangos et al, 1982; Schoemaker et al, 1983; Weissman et al, 1984a; Wang et al, 1984) whereas stereospecificity was observed when a "central type" (neuronal) benzodiazepine receptor is labelled in brain membranes by such drugs as flunitrazepam and diazepam (Costa et al, 1978; Braestrup and Nielsen, 1983; Wang et al, 1984). Another striking difference is that the two benzodiazepine antagonists CGS-8216 and RO 15-1788 (flumazepil) which inhibit the "central type" benzodiazepine receptor (Hunkeler, et al 1981; Czernik, 1982) are more than 1,000 times more potent in inhibiting neuronal benzodiazepine binding than in inhibiting astrocytic benzodiazepine binding (Table 3).

Table 3. Inhibition of [^{3}H]-diazepam binding in astrocytes and neurons in primary cultures.

Benzodiazepines		Astrocytes IC_{50}	Neurons IC_{50}
1.	RO 5-4864	4.98 nM	2.38 μM
2.	Flunitrazepam	10.03 nM	5.72 nM
3.	Diazepam	22.26 nM	15.22 nM
4.	RO 7-3351	27.24 nM	215.29 nM
5.	Midazolam	106.39 nM	40.19 nM
6.	RO 11-6896 (+)	170.47 nM	20.51 nM
7.	RO 11-6893 (-)	171.75 nM	79.92 nM
8.	Alprazolam	0.51 μM	21.83 nM
9.	Clonazepam	1.46 μM	4.80 nM
10.	Flurazepam	2.26 μM	40.03 nM
11.	RO 11-3128 (Meclonazepam)	3.55 μM	15.43 nM
12.	RO 11-3624	4.39 μM	695.91 nM
13.	RO 5-2181	11.53 μM	3.25 μM
14.	Chlordiazepoxide	28.33 μM	960.00 nM
Benzodiazepine Antagonists:			
1.	CGS-8216	15.07 μM	6.12 nM
2.	RO 15-1788 (Flumazepil)	73.04 μM	11.01 nM
Antiepileptic Agents			
1.	Carbamazepine	9.95 μM	97.86 μM
2.	Phenytoin	91.59 μM	93.18 μM
3.	Phenobarbital	119.37 μM	1021.81 μM

The anticonvulsants carbamazepine, phenytoin and phenobarbital all inhibited diazepam binding in both neurons and astrocytes (Table 3). Phenytoin is equipotent in the two cell types whereas carbamazepine and phenobarbital are almost ten times more potent in astrocytes than in neurons. This is in agreement with the ability of carbamazepine potently to displace RO 5-4864 binding from brain membranes (Skeritt _et al_, 1983; Marangos _et al_, 1983) and from astrocytes in primary cultures (Bender and Hertz, 1985a). An astrocytic target for carbamazepine is also in agreement with the finding of Weiss _et al_, (1985) that the anticonvulsant effect of this drug is inhibited by RO 5-4864. Another indication that the glial benzodiazepine binding site may be involved in convulsions and in the action of anticonvulsants is that the mouse strain BALB/cBy which is hyporesponsive to the seizure inducing action of RO 5-4864 has a lower density of binding sites for this ligand in the brain than do control animals (Seale _et al_, 1985). A further indication of the pharmacological relevance of the inhibition of astrocytic diazepam binding by the anticonvulsants carbamazepine, phenytoin, and phenobarbital is that the rank order of potency correlates with the ED_{50} values of these compounds against maximum electroshock seizures (Swinyard and Woodhead, 1982) and that the IC_{50} values are below or at the upper level of therapeutic plasma concentrations (carbamazepine: 50 μM (Johannessen and Strandjord, 1973; Morselli _et al_, 1977); phenytoin: 80 μM (Kutt, 1982); and phenobarbital: 170 μM (Booker, 1982).

GABA does not enhance benzodiazepine binding in astrocytes (Hertz _et al_, 1980; Tardy _et al_, 1981; Walker and Peacock, 1981; Mallorga _et al_, 1983; Bender and Hertz, 1985a) which is consistent with the fact that it does also not have any significant effect on [^{3}H]-RO 5-4864 binding in brain membrane preparations (Marangos _et al_, 1982; Patel and Marangos, 1982; Schoemaker _et al_, 1983).

FUNCTIONAL ROLE OF THE ASTROCYTIC BENZODIAZEPINE RECEPTOR

Evidence that the peripheral benzodiazepine receptor is associated with a voltage-dependent calcium channel has been obtained in experiments using guinea pig papillary muscle (Mestre et al, 1985) and ileal longitudinal muscle (Hullihan et al, 1983). Thus, RO 5-4864 and diazepam decrease the tension of the papillary muscle, and diazepam has a similar effect on the ileal muscle, and in both cases this effect could be reversed by increasing the extracellular concentration of calcium (Mestre et al, 1985; Hullihan et al, 1983). Conversely the calcium channel blockers nitrendepine, verapamil and diltiazem as well as RO 5-4864 decrease the action potential duration and the force of contraction and these effects are antagonized by PK 11195, a peripheral benzodiazepine receptor antagonist (Mestre et al, 1984, 1985). Finally, diazepam is able to displace binding of the calcium channel ligand [^{3}H]-nitrendipine to ileal longitudinal muscle (Bolger et al, 1983) and to smooth and cardiac muscle (Luchowski et al, 1984). More direct evidence that the astrocytic benzodiazepine binding site is associated with a calcium channel has come from studies showing that the calcium channel blockers nifedipine and nitrendipine are able to displace binding of [^{3}H]-RO 5-4864 from brain membranes (Cantor et al, 1984). Astrocytes in primary cultures also possess high-affinity binding sites for [^{3}H]-nitrendipine (Bender and Hertz, 1985b; Litzinger and Brenneman, 1985). The [^{3}H]-nitrendipine binding in cultured astrocytes can be displaced by benzodiazepines (Bender and Hertz, 1985b) in the following rank order (with IC_{50} values (in μM) shown in parentheses): RO 5-4864 (9.73) > diazepam (30.9) ≥ midazolam (33.5) > RO 11-3128 (38.5) > clonazepam (46.1) > RO 11-3624 (53.0) > chlordiazepoxide (202.9). Although the IC_{50} values are higher, this rank order of potency is almost identical to the rank order of potency for the same drugs in displacing diazepam binding (Table 3). The correlation between displacement of diazepam binding and of nitrendipine binding is shown in Fig. 4. The correlation coefficient is 0.88

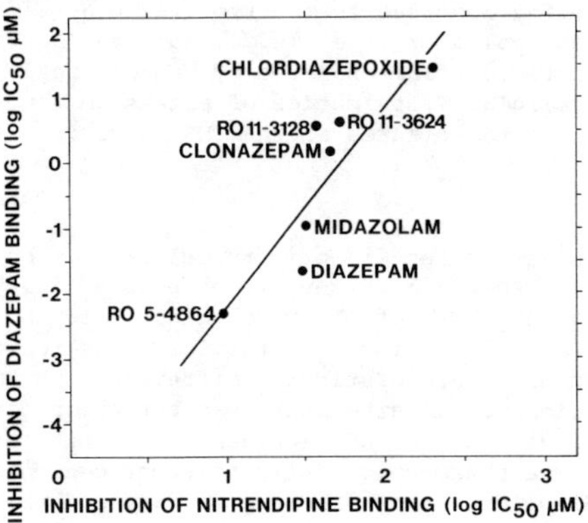

Fig. 4. Correlation between benzodiazepine induced inhibition of nitrendipine binding and of diazepam binding in astrocytes in primary cultures. Modifed from Bender and Hertz, 1985b. Benzodiazepines were kindly provided by Dr. P.F. Sorter, Hoffmann-La Roche, Nutley, N.J.

and the correlation is statistically significant ($P < 0.01$). Like displacement of the diazepam binding (Table 3) the difference between the ability of the two stereoisomers RO 11-3128 and RO 11-3624 to displace nitrendipine binding was not statistically significant. Analogously, calcium channel blockers displace both nitrendipine binding and diazepam binding in astrocytes. The rank order of potency for the displacement of nitrendipine binding (with IC_{50} values (in μM)) is: nitrendipine (1.7) > nimodipine (3.4) > felodipine (8.8) > nifedipine (21) > verapamil (46.9). The corresponding rank order for displacement of diazepam binding is: nitrendipine (0.86) > nimodipine (0.89) > nifedipine (4.7) > felodipine (5.1) > verapamil (68.4). Again, the correlation between the displacement of nitrendipine and of diazepam (Fig. 5) is good (correlation coefficient 0.92; $P < 0.05$). There is thus good evidence for an association between the "peripheral type" benzodiazepine

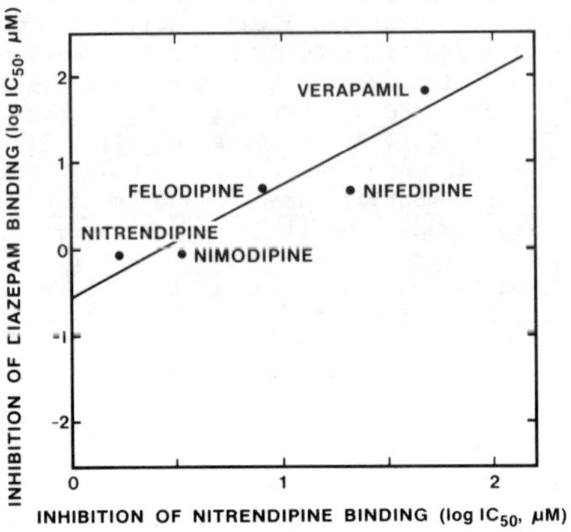

Fig. 5. Correlation between calcium channel blocker induced inhibition of nitrendipine binding and of diazepam binding in astrocytes in primary cultures. Nitrendipine and nimodipine were kindly supplied by Dr. A. Scriabine, Miles Lab., New Haven, CT.

receptor, including that in astrocytes, and a voltage-dependent calcium channel. This may be of considerable functional importance since MacVicar (1984) recently has demonstrated the presence of voltage dependent calcium channels in astrocytes. The calcium channel may, in turn be of importance for cellular distribution of other ions, e.g., by affecting calcium dependent potassium channels. It is now becoming accepted that astrocytes play a major role for potassium homeostasis at the cellular level (e.g., Walz and Hertz, 1983). An interference with astrocytic calcium channels by benzodiazepines might therefore profoundly affect cellular distribution of potassium, and thus excitability, including susceptibility to seizures in the CNS in vivo.

CONCLUDING REMARKS

There seems to be little doubt that astrocytes (in primary cultures or obtained by gradient centrifugation) possess a high density of high-affinity receptors for benzodiazepines including those of pharmacological relevance (e.g., flunitrazepam, diazepam). The pharmacological profiles of astrocytic and neuronal benzodiazepine binding sites are distinctly different with RO 5-4864 being much more potent and clonazepam being much less potent in displacing astrocytic than neuronal binding of benzodiazepines. In spite of the relatively low potency of clonazepam a pharmacologically relevant concentration ($\simeq 1.0$ μM) causes about 50% displacement of diazepam binding and the potency of flunitrazepam and diazepam is almost similar in the two cell types (Table 3). The lack of effect by GABA indicates that the mechanisms by which benzodiazepines act must be different in neurons and in astrocytes. A likely target is that the astrocytic benzodiazepine receptor may be associated with a calcium channel. Interactions of benzodiazepines with calcium channels may in turn be of major importance for regulation of excitability, including seizure susceptibility within the CNS. It is a further indication of such a mechanism of action that anticonvulsant drugs seem to interact specifically with these binding sites and that RO 5-4864 is able to modify the sensitivity of DBA/2J mice to audiogenic seizures (Benavides _et al_, 1984). Recently, it has been established that the "peripheral type" binding sites for benzodiazepines in the brain are under neural control (Bolger _et al_, 1984; Weissman _et al_, 1984b) and that they appear at a late evolutionary stage (Bolger _et al_, 1985). These findings strengthen the concept that they may represent physiologically very important structures with highly specialized functions in the CNS.

REFERENCES

Anholt, R.R.H., K.M.M. Murphy, G.E. Mack and S.H. Snyder (1984). J. Neurosci., 4, 593-603.

Anholt, R.R.H., E.B. DeSouza, L.M. Oster-Granite and S.H. Snyder (1985). J. Pharmacol. Exp. Ther., 233, 517-526.

Benavides, J., F. Guilloux, D.E. Allam, A. Uzan, J. Mizoule, C. Renault, M.C. Dubroeucq, C. Gueremy and G. LeFur (1984). Life Sci., 34, 2613-2620.

Bender, A.S. and L. Hertz (1984). J. Neurochem., 43, 1319-1327.

Bender, A.S. and L. Hertz (1985a). Brain Res., 341, 41-49.

Bender, A.S. and L. Hertz (1985b). Europ. J. Pharmacol., 110, 287-288.

Bender, A.S. and L. Hertz (1986). Submitted for publication.

Bolger, G.T., P. Gengo, R. Klockowski, E. Luchowski, H. Siegel, R.A. Janis, A.M. Triggle and D.J. Triggle (1983). J. Pharmacol. Exp. Ther., 225, 291-309.

Bolger, G.T., E. Mezey, J. Cott, B.A. Weissman, S.M. Paul and P. Skolnick (1984). Europ. J. Pharmacol., 105, 143-148.

Bolger, G.T., B.A. Weissman, H. Lueddens, A.S. Basile, C.R. Mantione, J.E. Barrett, J.M. Witkin, S.M. Paul and P. Skolnick (1985). Brain Res., 338, 366-370.

Booker, H.E. (1982). Anti-Epileptic Drugs, Raven Press, New York, pp. 341-350.

Booker, N.E. and G.G. Cellesia (1973). Arch. Neurol., 29, 191-194.

Braestrup, C. and M. Nielsen (1983). Handbook of Psychopharmacology, Vol. 17, Plenum Press, New York, pp. 285-384.

Braestrup, C., C. Nissen, R.F. Squires and A. Schousboe (1978). Neurosci. Lett., 9, 45-49.

Cantor, E.H., A. Kenessey, G. Semenuk and S. Spector (1984). Proc. Natl. Acad. Sci. U.S.A., 81, 1549-1552.

Costa, E., Guidotti, A. and G. Toffano (1978) Brit. J. Psychiat., 133, 239-249.

Czernik, A.J., B. Petrack, H.J. Kalinsky, S. Psychoyos, W.D. Cash, C. Tsai, R.K. Rinehart, F.R. Granat, R.A. Lovell, D.E. Brundish and R. Wade (1982). Life Sci., 30, 363-372.

Dudai, Y., Z. Yavin and E. Yavin (1979). Brain Res., 177, 418-422.

Gallager, D.W., P. Mallorga, W. Oertel, R. Henneberry and J. Tallman (1981). J. Neurosci., 1, 218-225.

Garattini, S., E. Mussini, F. Marcucci and A. Guaitani (1973). The Benzodiazepines, Raven Press, New York, pp. 75-98.

Gehlert, D.R., H.I. Yamamura and J.K. Wamsley (1983). Europ. J. Pharmacol., 95, 329-330.

Greenblatt, D.J., E. Woo, M.D. Allen, P.J. Orsulak and R.I. Shader (1978). JAMA, 240, 1872-1875.

Henn, F.A. and D.J. Henke (1978). Neuropharmacol., 17, 985-988.

Hertz, L. and S. Mukerji (1980). Canad. J. Physiol. Pharmacol., 58, 217-220.

Hertz, L. and J.S. Richardson (1984). Trends Pharmacol. Sci., 5, 272-276.

Hertz, L., S. Mukerji and A. Schousboe (1980). Trans. Amer. Soc. Neurochem., 11, 24.

Hillestad, L., T. Hansen, H. Melsom and A. Drivenes (1974). Clin. Pharmacol. Ther., 16, 479-484.

Hullihan, J.P., S. Spector, T. Taniguchi and J.K.T. Wang (1983). Brit. J. Pharmacol., 78, 321-327.

Hunkeler, W., H. Moehler, L. Pieri, P. Polc, E.P. Bonetti, R. Cumin, R. Schaffner and W. Haefely (1981). Nature, 209, 514-516.

Johannessen, S.I. and R.E. Strandjord (1973). Epilepsia, 14, 373-379.

Kutt, H. (1982). Anti-Epileptic Drugs, Raven Press, New York, pp. 241-246.

Litzinger, M.J. and D.E. Brenneman (1985). Abstracts, Soc. Neurosci. 15th Annual Meeting, p. 86.

Luchowski, E.M., F. Yousif, D.J. Triggle, S.C. Maurer, J.G. Sarmiento and R.A. Janis (1984). J. Pharmacol. Exp. Ther., 230, 607-613.

MacVicar, B.A. (1984). Science, 226, 1345-1347.

Mallorga, P., C. Ebel and B.F. Roth-Schechter (1983). Neurosci. Lett., 39, 45-50.

Marangos, P.J., J. Patel, J.-P. Boulenger and R. Clark-Rosenberg (1982). Mol. Pharmacol., 22, 26-32.

Marangos, P.J., R.M. Post, J. Patel, K. Zander, A. Parma and S. Weiss (1983). Europ. J. Pharmacol., 93, 175-182.

McCarthy, K.D. and T.K. Harden (1981). J. Pharmacol. Exp. Ther., 216, 183-191.

Mestre, M., T. Carriot, C. Belin, A. Uzan, C. Renault, M.C. Dubroeucq, C. Gueremy and G. LeFur (1984). Life Sci., 35, 953-962.

Mestre, M., T. Carriot, C. Belin, A. Uzan, C. Renault, M.C. Dubroeucq, C. Gueremy, A. Doble and G. LeFur (1985). Life Sci., 36, 391-400.

Mohler, H. and J.G. Richards (1983). Brit. J. Pharmacol., 79, Suppl 280P.

Morselli, P.L., A. Baruzzi, M. Gerna, L. Bossi and M. Porta (1977). Brit. J. Clin. Pharmacol., 4, 535-540.

Patel, J. and P.J. Marangos (1982). Neurosci. Lett., 30, 157-160.

Pellow, S. and S.E. File (1984). Life Sci., 35, 229-240.

Richards, J.G., H. Mohler and W. Haefely (1982). Trends Pharmacol. Sci., 3, 233-235.

Schoemaker, H., M. Morelli, P. Deshmukh and H.I. Yamamura (1982). Brain Res., 248, 396-401.

Schoemaker, H., R.G. Boles, W.D. Horst and H.I. Yamamura (1983). J. Pharmacol. Exp. Ther., 225, 61-69.

Seale, T.W., G.T. Bolger and P. Skolnick (1985). Abstracts, Soc. Neurosci. 15th Annual Meeting, p. 275.

Sher, P.K. and V.L. Machen (1984). Dev. Brain Res., 14, 1-6.

Sher, P.K., B.K. Schrier and D. Van Putten (1982). Dev. Neurosci., 5, 271-277.

Skerrit, J.H., G.A.R. Johnston and S. Chen Chow (1983). J. Pharm. Pharmacol., 35, 464-465.

Snyder, S.H. (1983). J. Med. Chem., 26, 1667-1672.

Swinyard, E.A. and J.M. Woodhead (1982). Anti-Epileptic Drugs, Raven Press, New York, pp. 111-126.

Tallman, J.F. (1983). Advances in Cellular Neurobiology, Vol. 4, Academic Press, New York, pp. 121-131.

Tallman, J.F. and D.W. Gallager (1985). Ann. Rev. Neurosci., 8, 21-44.

Tardy, M., M.F. Costa, B. Rolland, C. Fages and P. Gonnard (1981). J. Neurochem., 36, 1587-1589.

Tardy, M., C. Gages, G. Dupre, M.F. Costa, J. Bardakdjian and B. Rolland (1985). Neurochem. Res., 10, 809-817.

Walker, C.R. and J.H. Peacock (1981). Dev. Brain Res., 1, 565-578.

Walz, W. and L. Hertz (1983). Progr. Neurobiol., 20, 133-183.

Wang, J.K.T., T. Taniguchi and S. Spector (1984). Mol. Pharmacol., 25, 349-351.

Weiss, S.R.B., R.M. Post, J. Patel and P.J. Marangos (1985). Life Sci., 36, 2413-2419.

Weissman, B.A., J. Cott, S.M. Paul and P. Skolnick (1983). Eur. J. Pharmacol., 90, 149-150.

Weissman, B.A., G.T. Bolger, L. Isaac, S.M. Paul and P. Skolnick (1984a). J. Neurochem., 42, 969-976.

Weissman, B.A., P. Skolnick, and D.C. Klein (1984b). Pharmacol. Biochem. Behav., 21, 821-824.

Biochemical and Ultrastructural Study of Fructose-1, 6-Biphosphatase and Glycogen in Methionine Sulfoximine Epileptogenic Rodent Brain

T. K. Hevor, P. Delorme* and J. C. Beauvillain†

*Laboratoire de Neurobiologie Fonctionnelle, Universite de Lille 1, 59655 Villeneuve d'Ascq-CEDEX, France
†I.N.S.E.R.M., U 156, Place de Verdun, 59000 Lille, France

ABSTRACT

Methionine sulfoximine induced tonic and clonic seizures in rodents. After these "grand mal" type convulsions, glycogen content increased a great deal in central nervous system. This increase was exclusively confined to astroglial cells. The gluconeogenic enzyme fructose-1,6-biphosphatase activity also increased after the seizures. The immunostaining of this enzyme by antifructose-1,6-biphosphatase antibodies was located only in astrocytes. Gluconeogenesis may account, at least in part, for glycogen accumulation in astrocytes in methione sulfoximine epileptogenic brain.

KEY WORDS

Fructose-1,6-biphosphatase, gluconeogenesis, glycogen, glial cells, methionine sulfoximine, epilepsy.

INTRODUCTION

It is well known that the convulsive effect of methionine sulfoximine (MSO) in rodents is followed by an important increase in glycogen content of central nervous system (Folbergrova, 1973 ; Phelps, 1975 ; Bérel et al , 1977). The actual mechanism of a such increase is not well known. It would originate from an enhanced gluconeogenesis through an activation and a de novo biosynthesis of fructose-1,6-biphosphatase (FBPase, EC 3.1.3.11) (Hevor and Gayet, 1981). In order to study the cytochemical basis of glycogenesis in this kind of experimental epilepsy, we tried in this paper to correlate the localization of glycogen changes to that of FBPase.

MATERIALS AND METHODS

Male rats of Sprague Dawley strain weighing 300g and male Swiss mice of OF1 strain weighing 30g were purchased from IFFA-Credo (Arbresle). L-methionine-D,L-sulfoximime (Sigma, St-Louis; 100mg/kg) was administered intraperitoneally. The control rodents were treated with NaCl 0.9%.
Brain glycogen level and its ultrastructural localization were determined as previously described (Hevor et al , 1985; Delorme and Hevor,1985). For the FBPase activity study, the enzyme was prepared and measured as described by Hevor and Gayet (1978). FBPase was purified from rat brain and antiFBPase

raised in rabbit (Hevor and Gayet, 1981). Rats were fixed by intracardiac perfusion of phosphate-buffered picric acid paraformaldehyde mixture. For the immunocytochemical procedure, the brains were cut on a vibratome and the sections were succesively incubated in FBPase antiserum, sheep IgG raised against rabbit IgG, 3,3-diaminobenzidine and osmium. The sections were embedded in Araldite and treated for an electron microscope study.

RESULTS

In the cortex of rats and mice, the ultrastructural figures showed few beta glycogen particles in astroglial cells but not in other cell type. These particles were localized mainly in perivascular end feet but were also seen in processes and perikarya. Twenty four hours after the administration of MSO, the glycogen particles were very abundant in astrocyte perikarya and processes and more in perivascular end feet. The alpha particles were more frequent than beta one. Even in this experimental conditions, glycogen particles were always confined to astrocytes. These ultrastructural results have been correlated to the biochemical one : the glycogen content was 2.54 + 0.22 micromoles/g frozen tissue in controls versus 11.97 + 2.78 micromoles/g in MSO-treated animals. With the immunocytochemical study at the ultrastructural level, the staining obtained with an antiFBPase antibody was always very light in control rat cortex. When the same observations were made 24h after the administration of the convulsant, an obvious immunostaining was detected.This staining was located only in astroglial cells, in their end feet, processes and perikarya. FBPase activity also increased a great deal in cerebral cortex 24h after dosing : 104 + 40 nanomoles/ g tissue wet weight/ min. at 37°C in saline controls versus 480 + 56 nanomoles/ g/ min. in MSO-treated animals.

DISCUSSION

The preceding results showed that in MSO epileptogenic rodent brain, the glycogen accumulation as well as the FBPase increase occured in the same cells, the astroglial cells. Presumably, astrocytes are the cellular site of glucose pathway impairment after these convulsions. Glycogen accumulation may originate from an enhanced gluconeogenesis in these cells : the new synthesized molecules of glucose may be incorporated in glycogen. Previously, we had found that the amount of the purified FBPase increased only 2-fold after MSO administration (Hevor and Gayet, 1978). Since this moderate increase was not reliable to the higher intensity of the immunocytochemical staining after dosing, it was assumed that there was another factor which induced this marked staining. We carry on experiments to elucidate the factors which induce the FBPase reinforcement.

ACKNOWLEGMENT

This work was supported by grants from the Ministère de l'Education Nationale. We thank Mrs J. Bernard for her technical assistance.

REFERENCES

Bérel, A., P. Lehr and J. Gayet (1977). Brain Res. , 128 , 193-196.
Delorme, P. and T.K. Hevor (1985). Neuropathol.Appl.Neurobiol. , 11 , 117-128.
Folbergrova, J. (1973). J. Neurochem. , 20 , 547-557.
Hevor, T.K. and J. Gayet (1978). Brain Res. , 150 , 210-215.
Hevor, T.K. and J. Gayet (1978). Biochem. Soc. Trans. , 6 , 1029-1031.
Hevor, T.K. and J. Gayet (1981). J. Neurochem. , 36 , 949-958.
Hevor,T.K., P. Delorme and J. Gayet (1985). Neuropathol. Appl. Neurobiol. , 11 , 129-139.
Phelps, G.C. (1975). J. Neurocytol. , 4 , 479-490.

Differential Infectivity of Primary Rat Brain Cell Cultures by a Virulent Semliki Forest Virus

C. B. Bruce, J. Chapman and M. G. Rumsby

Department of Biology, University of York, York YO1 5DD, UK

ABSTRACT

The avirulent strains of semliki forest virus (SFV) A7 and M9 have been found to infect primary rat brain cell cultures although cells of the A2B5 lineage are more susceptible to infection than other cell types present.

KEYWORDS

Semliki forest virus, differential infectivity, astrocytes, oligodendrocytes, A2B5.

INTRODUCTION

The overall aim of this work is to investigate the way in which defined viruses gain entry into cells of the central nervous system, in particular the oligodendrocytes and astrocytes, and how the biochemistry of these two major glial cell classes is affected by viral infection. Viral infection of nerve cells may be the initiating event which leads, through subsequent autoimmune mechanisms, to demyelination observed in the human condition of multiple sclerosis. Using an *in vitro* system (Walker *et al.*, 1984) we have examined the effect of the avirulent strains of semliki forest virus A7 and M9 in an attempt to answer some of these questions. SFV is a neurotropic virus which has been shown to cause demyelination in some strains of mice but the two strains of virus are thought to cause this demyelination by two different mechanisms; M9 by direct oligodendrocyte death (Sheahan *et al.*, 1983) and A7 by induced immune reactions after infection (Suckling *et al.*, 1978; Kelly *et al.*, 1981).

SUMMARY OF RESULTS

Cell lysis is observed in our primary glial cultures by 24h post infection (pi). Autoradiograph of ^{35}S methionine labelled proteins during infection and separated by SDS polyacrylamide gel electrophoresis has shown that host protein synthesis is reduced to approximately 30% by 8h pi. and proteins of molecular weights corresponding to those of the virus coat proteins E1 and E2, the capsid protein as well as precursor proteins have been identified. Some of these bands also correspond to those selected by immunoblotting which suggests further that they represent virus structural proteins and their precursors. Several bands persist throughout the time period followed especially in the 50-60k molecular weight region and these are likely to be indicative of proteins synthesised by the major

Advances in the Biosciences Vol. 61

cell type present i.e. astrocytes.

Immunocytochemical methods were employed to identify which cell types were infected. Cells of the A2B5 cell marker lineage (i.e. precursor cells, oligodendrocytes and type 2 astrocytes; Raff et al., 1983) are more susceptible to infection than other cell types present. Type 1 astrocytes do become infected but express viral antigens much later in the infection period, at about 24h (Fig.1 g-i) than the A2B5 cell marker lineage cells where viral antigens can be detected at 6-8h pi (Fig.1 a-f).

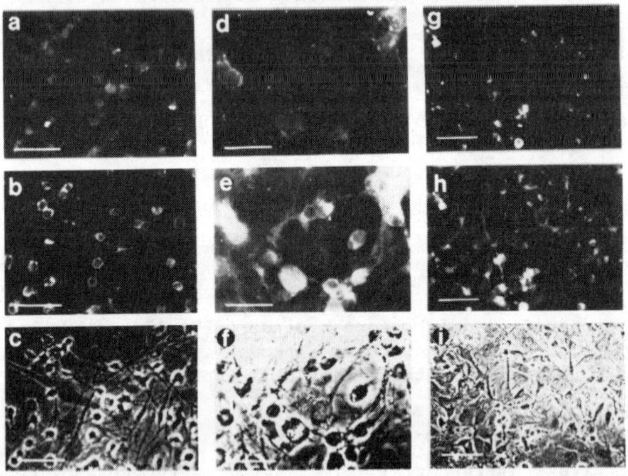

Fig.1. Expression of SFV antigen by brain cells in culture. 10-day old cultures were infected with SFV at a MOI of 0.1 pfu/cell. Cultures were processed at 8h (a-f) and 24h (g-i) for immunocytochemical identification. Bar = 50u
a. SFV^+, b. $A2B5^+$ c. phase, d. SFV^+
e. GC^+, f. phase, g. SFV^+, h. $GFAP^+$, i. phase

There was no significant difference shown between the A7 and M9 strains in their ability to infect the various cell types present, despite their reported difference in proposed mechanism in vivo (Suckling et al., 1978; Kelly et al.,1981; Sheahan et al., 1983).

ACKNOWLEDGEMENT

This work is supported by the Multiple Sclerosis Society of G.B. & N.I.

REFERENCES

Kelly, W.R., W.F. Blakemore, S. Jagelman and H.E. Webb (1981). Neuropath.Appl. Neurobiol. 8, 43-53.
Raff, M.C., R.H. Miller and M. Noble (1983). Nature 303, 390-396.
Sheahan, B.J., M.C. Gates, J.F. Caffrey and G.J. Atkins (1983). Acta Neuropath. (Berl.), 60, 257-265.
Suckling, A.J., S. Pathak, S. Jagelman and H.E. Webb (1978). J.Neurol.Sci., 39, 147-154.
Walker, A.G., J. Chapman, C.B. Bruce and M.G. Rumsby (1984/5). J.Neuroimmunol., 7, 1-20.

Characterization of Adult Human Glial Cells *in vitro*

J. Newcombe and M. L. Cuzner

The Multiple Sclerosis Society Laboratory, Department of Neurochemistry, Institute of Neurology, 33 John's Mews, London WC1N 2NS, UK

ABSTRACT

Cell types in primary cultures prepared from biopsy or autopsy adult normal human brain were identified by immunocytochemical staining.

KEYWORDS

Adult human brain; tissue culture; oligodendrocytes; astrocytes; immunocytochemistry.

INTRODUCTION

Cells cultured from biopsy or autopsy adult human brain have in most studies been identified only on morphological criteria (Pontén & Macintyre, 1968; Lumsden, 1971; Rorke, et al, 1975). In one report astrocytes were identified by indirect immunofluorescent staining with rabbit antisera directed against glial fibrillary acidic protein (GFAP) in explant cultures (Gilden, et al, 1976). In this study we have used cell-specific antibodies to immunocytochemically identify oligodendrocytes, astrocytes and fibroblasts in primary cultures from explants or dissociated tissue from adult normal human brain.

METHODS

White and grey matter was obtained from apparently normal tissue removed as a necessary part of neurosurgical procedures in treatment for tumours or aneurysms at the National Hospital for Nervous Diseases, London, and from cardiac arrest cases with postmortem times of 1-13 hours. For explant cultures finely chopped tissue (0.2-1.5g) was plated at 4mg per 13mm diameter poly-L-lysine coated coverslip in Basal Medium Eagles (modified) with Earle's salts, 10% foetal calf serum and antibiotics. Tissue was dissociated through 110μ and 45μ nylon meshes with or without prior incubation with 0.25% trypsin, and plated at 1×10^5 or 2×10^5 cells/cm^2 for white or grey matter, respectively. After 1 to 7 weeks cells were stained by indirect immunofluorescence.

RESULTS

Confluent cultures of glial and mesenchymal cells were seen in all explant and dissociated cultures, but glial cells were more evenly distributed on

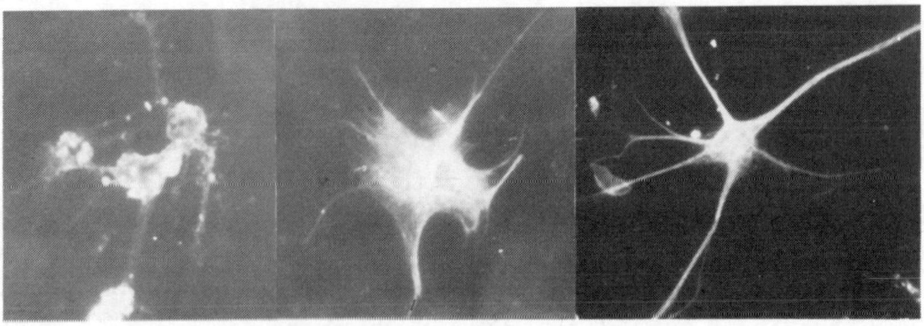

Fig. 1 A: Oligodendrocytes identified with monoclonal anti-galactocerebroside (Ranscht, *et al*, 1982). B: Protoplasmic (type 1) and C: Fibrous (type 2) astrocytes stained with rabbit anti-GFAP (raised in this laboratory); x 380.

the substratum in dissociated cultures. Macrophages were the first cell type to adhere to coverslips. They were identified by the presence of Fc-receptors and their ability to phagocytose fluorescently-labelled latex beads. Fibroblasts were labelled with anti-fibronectin and anti-Thy-1 monoclonal antibodies. From autoradiographic studies with tritiated thymidine they appeared to be the only cell type which proliferated. Oligodendrocytes with elongated irregular processes were identified in white matter cultures by antibodies directed against galactocerebroside (Fig. 1), myelin basic protein and myelin associated glycoprotein. GFAP-positive protoplasmic (type 1) and fibrous (type 2) astrocytes and cells with intermediate morphology were observed in both white and grey matter cultures, but type 2 astrocytes always predominated over type 1. In double labelling studies with the monoclonal antibody A2B5 staining was seen on only a small proportion of type 2 astrocytes.

DISCUSSION

Oligodendrocytes and astrocytes with morphologies similar to those seen in human and rodent foetal or neonatal mixed glial cells cultures were demonstrated by immunocytochemistry to be present in primary cultures of adult human brain tissue. The presence of GFAP-positive cells with intermediate morphologies in both white and grey matter cultures indicates that adult human astrocytes may undergo morphological changes in culture. Primary cultures of adult human mixed glia are potentially useful for biochemical, virological and immunopathological studies.

ACKNOWLEDGEMENT

We would like to thank neurosurgeons at the National Hospital for Nervous Diseases for providing biopsy tissue. This study was supported by The Multiple Sclerosis Society of Great Britain and Northern Ireland.

REFERENCES

Gilden, D.H., Z. Wroblewska, L.F. Eng and L.B. Rorke (1976). *J. Neurol. Sci. 29*, 177-184.

Lumsden, C.E. (1971). In: D.S. Russell and L.J. Rubinstein (Eds.), *PATHOLOGY OF TUMOURS OF THE NERVOUS SYSTEM*, 3rd edition, Arnold, London, pp.334-420.

Pontén, J. and E. Macintyre (1968). *Acta Pathol. Microbiol. Scand. 74*, 465-486.

Ranscht, B., P.A. Clapshaw, J. Price, M. Nobel and M. Seifert (1982). *Proc. Natl. Acad. Sci. USA, 79*, 2709-2713.

Rorke, L.B., D.H. Gilden, A. Wroblewska and D. Santoli (1975). *J. comp. Neurol., 161*, 329-339.

Subject Index